Magnetic Ultrathin Films, Multilayers and Surfaces

MATERIALS RESEARCH SOCIETY
SYMPOSIUM PROCEEDINGS VOLUME 384

Magnetic Ultrathin Films, Multilayers and Surfaces

Symposium held April 17-20, 1995, San Francisco, California, U.S.A.

EDITORS:

E.E. Marinero
IBM Almaden Research Center
San Jose, California, U.S.A.

B. Heinrich
Simon Fraser University
Burnaby, Canada

W.F. Egelhoff, Jr.
National Institute of Standards and Technology
Gaithersburg, Maryland, U.S.A.

A. Fert
Université de Paris-Sud
Orsay, France

H. Fujimori
Tohoku University
Sendai, Japan

G. Guntherodt
RWTH Aachen
Aachen, Germany

R.L. White
Stanford University
Stanford, California, U.S.A.

M|**R**|**S**
MATERIALS
RESEARCH
SOCIETY

PITTSBURGH, PENNSYLVANIA

This work was supported in part by the Office of Naval Research under Grant Number N00014-95-1-0569. The United Stated Government has a royalty-free license throughout the world in all copyrightable material contained herein.

Effort sponsored by the National Institute of Standards and Technology under Grant Number 43NANB423082. The views and findings of the various papers and solely those of the authors, and do not necessarily represent the policy of NIST.

CAMBRIDGE
UNIVERSITY PRESS

32 Avenue of the Americas, New York NY 10013-2473, USA

Cambridge University Press is part of the University of Cambridge.

It furthers the University's mission by disseminating knowledge in the pursuit of education, learning and research at the highest international levels of excellence.

www.cambridge.org
Information on this title: www.cambridge.org/9781558992870

CODEN: MRSPDH

A catalogue record for this publication is available from the British Library

Library of Congress Cataloguing in Publication data

Magnetic ultrathin films, multilayers and surfaces : symposium held April 17-20, 1995, San Francisco, California, U.S.A. / editors, E.E. Marinero, B. Heinrich, W.F. Egelhoff, Jr., A. Fert, H. Fujimori, G. Guntherodt, R.L. White
    p. cm.—(Materials Research Society symposium proceedings, ISSN 0272-9172 ; v. 384)
    Includes bibliographical references and index.
    ISBN: 1-55899-287-1 (alk. paper)
    1. Thin films, multilayered—Magnetic properties—Congresses. 2. Thin films—Surfaces—Magnetic properties—Congresses. 3. Magnetic films—Congreses. I. Marinero, E.E. II. Heinrich, B. III. Egelhoff, Jr., W.F. IV. Fert, A. V. Fujimori, H. VI. Guntherodt, G. VII. White, R.L. VIII. Series: Materials Research Society Symposium Proceedings ; v. 384.
QC176.9.M84M28   1995   95-31839
530.4´275—dc20   CIP

ISBN 978-1-558-99287-0 Hardback

Cambridge University Press has no responsibility for the persistence or accuracy of URLs for external or third-party internet websites referred to in this publication, and does not guarantee that any content on such websites is, or will remain, accurate or appropriate.

CONTENTS

*Invited Paper

*Invited Paper

PART IV: MAGNETIC ANISOTROPY

PART V: ULTRATHIN FILMS, MAGNETIC DOMAINS

PART VI: GIANT MAGNETORESISTANCE I

*Invited Paper

PART VII: GIANT MAGNETORESISTANCE II
AND COLOSSAL MAGNETORESISTANCE

*Invited Paper

*Invited Paper

PART IX: GRANULAR NANOSTRUCTURES

*Invited Paper

PREFACE

This volume is dedicated to the memory of Leo Falicov, whose science, creativity and magnanimous friendship touched many of us. We will miss Leo's pioneering theoretical work which is used today to interpret experimental findings not only in magnetism, but also in superconductivity, phase transitions and the electronic behavior of solids. Leo was a key participant in what has by now become a focal meeting to those of us working on thin-film magnetism. Those of us who interacted with Leo at these meetings greatly benefitted from his creative mind and his breadth of understanding of both the theory and experimental aspects of thin-film magnetism. We will certainly miss him in future gatherings. Leo Falicov passed away in his beloved Berkeley on January 24, 1995.

The papers compiled in this volume reflect the state of the art in the field of thin-film magnetism and range from insightful reviews to new results in all aspects of the field. It is the seventh meeting in the series and gave the attendees the unique opportunity to hear from participants from Europe, Japan, North America and the former Soviet Union. Original contributions were made in the following areas: novel magnetic nanostructures, growth and structure of magnetic films and interfaces, interlayer coupling, anisotropy, magnetic domains, giant magnetoresistance, spectroscopies, magneto-optical properties, and granular nanostructures.

The meeting began with a dedication to Leo which was attended by his wife, Martha, and Roger Falcone, chairman of the physics department at UC Berkeley. The technical meeting opened with key review papers on the perspectives of thin-film magnetic material devices in the storage industry. All papers published in this Festschrift were peer reviewed. The organization of the volume reflects the topical areas of the meeting that spanned over four days and included two well-attended poster sessions. Finally, we gratefully acknowledge our sponsors, ONR, NIST, Hewlett Packard, Read Rite Corporation, Komag, Sony, IBM, Seagate, and TDK Corporation, whose financial support made this exciting meeting possible.

E.E. Marinero
B. Heinrich
W.F. Egelhoff, Jr.
A. Fert
H. Fujimori
G. Guntherodt
R.L. White

June 1995

MATERIALS RESEARCH SOCIETY SYMPOSIUM PROCEEDINGS

MATERIALS RESEARCH SOCIETY SYMPOSIUM PROCEEDINGS

Prior Materials Research Society Symposium Proceedings available by contacting Materials Research Society

Part I

Novel Magnetic Nanostructures and Applications

SUPERLATTICE NANOWIRES

K. ATTENBOROUGH*, R. HART*, W. SCHWARZACHER*,
J-PH. ANSERMET**, A. BLONDEL**, B. DOUDIN** AND J.P. MEIER**
*H. H. Wills Physics Laboratory, Tyndall Avenue, Bristol, BS8 1TL. UK
**Institut de Physique Experimentale, EPFL, PHB-Ecublens, CH-1015, Lausanne, Switzerland

ABSTRACT

CoNiCu/Cu superlattice nanowires have been grown by electrodeposition in nuclear track-etched nanoporous membranes. Transmission electron microscopy (TEM) images show a good layer structure and allow an estimate of the current efficiency. Current perpendicular to plane (CPP) giant magnetoresistance of up to 22%, at ambient temperature, has been measured but appears to be limited by defects, giving rise to ferromagnetic interlayer coupling, at low non-magnetic layer thicknesses. Magnetic properties of the superlattice nanowires are influenced by in-plane anisotropy and magnetostatic coupling.

INTRODUCTION

The discovery of giant magnetoresistance (GMR) in the Fe/Cr system and others has resulted in much research into the characterisation and growth of short period metal/metal multilayers. Interest has now moved to studying GMR in the CPP (current perpendicular to plane) direction as it allows a clear separation of interface and bulk contributions to the magnetoresistance[1,2,3].

We have shown that it is possible to grow high-quality epitaxial metal/metal multilayers with individual layer thicknesses as small as ~ 6Å by the relatively simple and inexpensive technique of electrodeposition, using a single electrolyte and switching between two deposition potentials. A GMR of up to 20% was found in the CoNiCu/Cu system using this technique[4]. When an appropriate template is used, electrodeposition also provides a means of growing ultrafine wires having diameters of a few hundred Å and a length of several μm[5,6,7]. As a result of combining these techniques and using nanoporous membranes metal/metal multilayer nanowires have now been electrodeposited by various groups[8,9,10,11].

This paper will discuss the preliminary results achieved from CoNiCu/Cu nanowires

EXPERIMENTAL DETAILS

The deposition of the nanowires takes place within the pores of nuclear track-etched polycarbonate membranes (figure 1). The diameter of the pores, and thus the nanowires, was ~800Å and their length ~6μm. A thin layer of gold was evaporated onto the back of the

3

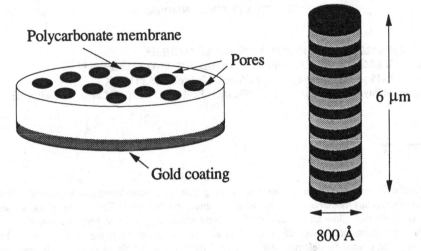

Polycarbonate membrane

Pores

Gold coating

6 μm

800 Å

Figure 1 Schematic diagram of the polycarbonate membrane and a representation of a nanowire

membrane which acted as a substrate and electrode. X-ray diffraction showed that the gold was polycrystalline with mainly (111) and (200) textured regions.

The nanowires were grown from a sulfamate electrolyte containing Co^{2+}, Ni^{2+} and Cu^{2+} ions, similar to that used for our conventional large area multilayers[4]. Experience in growing these large area multilayers has shown that the pH of the electrolyte is an important factor in the growth and influences the magnitude of the GMR in the samples[12]. The electrolyte pH was thus adjusted to 1.8 with the addition of sulfamic acid.

Each metal has a characteristic reduction potential. As copper is one of the most noble metals it requires only a small negative potential for reduction to occur, whereas Ni and Co (less noble) require a much higher negative potential. In our case we used -0.2V to deposit layers of almost pure Cu and -1.8V to deposit a magnetic alloy layer containing Ni, Co and Cu. The plating current was monitored throughout the deposition. Both potentials were measured with respect to a saturated calomel reference electrode, placed close to the membrane. The amount of Cu in the alloy layer is limited by having a low concentration of Cu in the electrolyte.

It was possible to monitor the growth of the nanowires within the pores by measuring the current[7]. As the nanowires emerge from the pores the surface area of the deposit increases and hemispherical 'caps' are formed. This 3D growth corresponds to an increase in both the Cu and the magnetic alloy currents. The current eventually starts to saturate as the hemispherical caps on the tops of the pores start to coalesce. It is at this point that plating is stopped allowing contact to small groups of wires for the magnetoresistance (MR) measurements.

After deposition the membrane is mounted on a glass support. Two gold coated spring-loaded probes are placed on the top of the membrane so that the current passes down through a group of wires, through the gold substrate and back up through another group of wires. The electrical resistance is typically 4Ω and it is estimated that approximately 30 wires are connected.

RESULTS AND DISCUSSION

Transmission electron microscopy (TEM) was used to verify the structure of the multilayer nanowires and energy dispersive x-ray (EDX) analysis was used to determine the chemical composition. The ratio of Co to Ni, in the alloy layer, was found to be 7:3, which is slightly less than that found in the planar multilayers[4]. The presence of Ni in the electrolyte appears to reduce problems associated with the dissolution of Co at the interface when the potential is switched to the less negative value appropriate to Cu deposition.

Figure 2 shows a transmission electron micrograph of a CoNiCu/Cu nanowire of diameter 800Å. The bilayer period for this wire was measured to be 30Å with the Cu layer being 10Å and the CoNiCu alloy layer being 20Å. A direct measurement of the bilayer period is needed to determine the current efficiency for metal deposition, which is less than 100% due to co-reduction of hydrogen. The current efficiency for the CoNiCu alloy was found to be 70% for this electrolyte and these growth conditions. The micrograph shows extremely good layering and shows what is possible by electrodeposition, but structurally some of the nanowires contain a high density of twin boundaries and dislocations.

Figure 2 A bright field TEM image of a 20Å CoNiCu/10Å Cu superlattice nanowire showing the contrast between the alternate layers.

A series of samples was grown with a fixed CoNiCu layer thickness of 35Å and various Cu layer thicknesses (t_{Cu}). The magnetisation and magnetoresistance properties were measured at ambient temperatures with an in-plane magnetic field. Some examples of the magnetoresistance curves are shown in figure 3. The distinct broadening of the MR curve and decrease in the magnitude of the GMR occurring for 10Å of Cu compared to 50Å of Cu may be due to pin holes in the Cu layer, causing ferromagnetic coupling between the magnetic layers. This is accompanied by a change in the easy direction of magnetisation from in the plane of the layers to along the axes of the nanowires. The decrease in GMR is expected from the increasing ratio of the Cu layer thickness to the magnetic layer thickness and the decreasing number of interfaces.

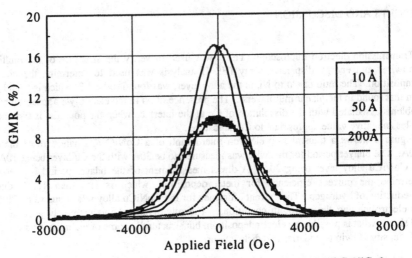

Figure 3 Magnetoresistance vs applied field for samples with a fixed CoNiCu layer
thickness of 35Å and Cu thicknesses of 10, 50,200Å respectively

Figure 4 shows the GMR as a function of t_{Cu} and it is seen that there is a maximum GMR of 17% for 50Å t_{Cu} for this set of samples. The maximum GMR achieved, to date, in CoNiCu/Cu nanowires was 22% for a sample with 24Å CoNiCu and 35Å Cu.

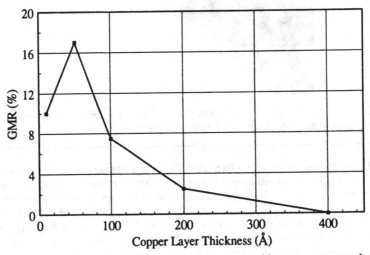

Figure 4 Giant magnetoresistance vs Cu thickness at ambient temperature and with an
in-plane applied field.

6

The shapes of the MR curves along with the high saturation fields, in figure 3, are characteristic of the nanowire samples. To explain them an understanding of the alignment and motion of the individual moments within the layers is needed.

The diameter of the wires is presumably too small to allow domain walls within the layers. The wires can therefore be considered as a column of single domain moments, which can only move by in-plane rotation. The high saturation field of the MR curves indicates that these moments cannot rotate freely. This could either be due to in-plane anisotropy, interlayer coupling or most likely to a combination of the two.

Magnetic measurements using a SQUID magnetometer showed that as the superlattice nanowires are cooled the coercivity increases dramatically, as seen from figure 5. This is attributed to anisotropy preventing the easy rotation of the individual moments or groups of moments.

Significant exchange coupling through the Cu layers will occur for thicknesses below a few tens of Å. This is commonly seen in conventional multilayers. The small diameter of these superlattice nanowires means that there should be in addition a strong magnetostatic coupling between the layers which will have a much longer range. Preliminary magnetic remanence measurements at low temperature support this conjecture as evidence is seen for an antiferromagnetic interlayer interaction even for t_{Cu} as large as 600Å [13]. This is significantly greater than the non-magnetic spin diffusion length of 400Å measured in electrodeposited Cu/Co multilayers[14].

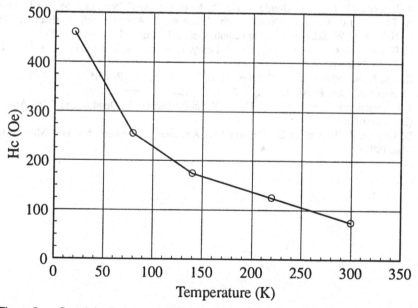

Figure 5   Coercivity from magnetisation loops vs temperature for a nanowire with a repeat of 35Å CoNiCu and 50Å Cu. The maximum in GMR occurred at a systematically larger field but followed a similar trend.

Acknowledgements

Support from the UK EPSRC and the Swiss National Science Foundation/British Council is gratefully acknowledged

References

1. Pratt Jr., S-F. Lee, J.M. Slaughter, R. Lolee, P.A. Schroeder and J. Bass, Phy. Rev. Lett. **66**, 3060 (1991).
2. M.A.M. Gijs, S.K.J. Lenczowski and J.B. Giesbers, Phy. Rev. Lett. **70**, 3343 (1993).
3. T. Valet and A. Fert, J. Magn. Magn. Mat. **121**,378 (1993).
4. M. Alper, P.S. Aplin, K. Attenborough, D.J. Dingley, R. Hart, S.J. Lane, D.S. Lashmore and W. Schwarzacher, J. Magn. Magn. Mat. **126**, 8 (1993); M Alper, K. Attenborough, R. Hart, S.J. Lane, D.S. Lashmore, C. Younes and W. Schwarzacher, Appl. Phys. Lett. **63**, 2144 (1993).
5. G.E. Possin, Rev. Sci. Inst. **41**, 772 (1990).
6. R.M. Penner and C.R. Martin, Anal. Chem. **59**, 2625 (1987).
7. T.M. Whitney, J.S. Jiang, P.C. Searson and C.L. Chien, Science **261**, 1316 (1993).
8. A. Blondel, J.P. Meier, B. Doudin And J-Ph. Ansermet, Appl. Phys. Lett. **65**, 3019 (1994).
9. A. Blondel, J.P. Meier, B. Doudin, J-Ph. Ansermet, K. Attenborough, P. Evans, R. Hart, G. Nabiyouni, W. Schwarzacher (to be published in J. Magn. Magn. Mat. 1995).
10. L. Piraux, J.M. George, J.F. Despres, C. Leroy, E. Ferain and R. Legras, Appl. Phys. Lett. **65**, 2484-2486 (1994).
11. K. Liu, K. Nagodawithana, P.C. Searson and C.L. Chien, Phy. Rev **B51**, 7381 (1995).
12. M. Alper, R. Hart, K. Attenborough, W. Schwarzacher (in preparation).
13. K Attenborough, J.P. Meier, R. Hart, W. Schwarzacher, B. Doudin and J-Ph. Ansermet (in preparation).
14. B. Voegeli, A. Blondel, B. Doudin and J-Ph. Ansermet (to be published in J. Magn. Magn. Mat. 1995).

8

TOWARDS THE SYNTHESIS OF ATOMIC SCALE WIRES

P. A. ANDERSON,* L. J. WOODALL,* A. PORCH,** A. R. ARMSTRONG,*†
I. HUSSAIN,* AND P. P. EDWARDS*
*School of Chemistry, University of Birmingham, Edgbaston, Birmingham, B15 2TT,
U. K.
**School of Electronic and Electrical Engineering, University of Birmingham, Edgbaston,
Birmingham, B15 2TT, U. K.
†Current address: School of Chemistry, University of St. Andrews, Fife, KY16 9ST,
U. K.

ABSTRACT

Recent work[1] has highlighted the possibility that through the introduction of metals into
the one-dimensional channels of zeolite L, it may be feasible to engineer charge transport along
the channels to produce a unique compound comprising a precise, assembled array of ultrafine,
atomic-scale conducting wires embedded within the aluminosilicate framework. Using electron
spin resonance (ESR), and microwave cavity perturbation measurements, we examine the
properties of these remarkable materials as a function of composition as they approach the
insulator to metal transition.

INTRODUCTION

The class of crystalline aluminosilicates known as zeolites, many of which are naturally
occurring minerals, are composed of corner-sharing SiO_4 and AlO_4 tetrahedra, arranged into
three-dimensional frameworks in such a manner that they contain regular channels and cavities
of molecular dimensions (Figure 1). Conventionally a metal is often described as a regular array
of ions embedded in a sea of itinerant electrons. Although neither exists in practice, a close
approximation to the former is a dehydrated zeolite such as zeolite L, where cations
coordinated on only one side to an anionic framework, line the inside of a series of regular
channels. The controlled and continuous doping of 'excess electrons' into these white insulating
solids is possible through their reaction with alkali metal vapour. Incoming metal atoms are
ionized by the intense electric fields within the zeolite releasing electrons to interact with the
zeolite cations.[1-6] We have noted that at some critical stage of metal loading, one expects
enhanced electron–electron interactions and the possibility of an insulator–metal transition.[6-8]

The purpose of this study is to examine the conductivity of these metal-loaded zeolites as
a function of potassium concentration and determine whether the conduction mechanism is
metallic. Since the samples are both air- and moisture-sensitive, and in powder form, a
contactless conductivity measurement is preferred. A convenient method for studying the
conductivity of such samples is the microwave cavity perturbation technique, where
dissipative eddy currents can be set up within each powder grain.[9] This dissipation is readily

9

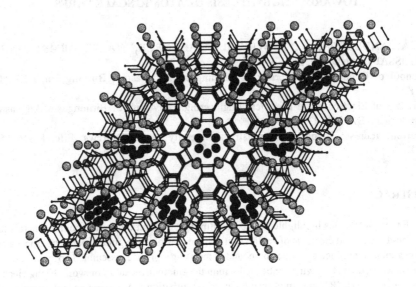

Figure 1. Representation of zeolite L with potassium metal atoms occupying the channels.

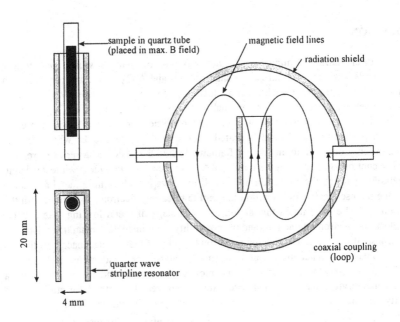

Figure 2. Schematic diagram of the 3.5 GHz copper hairpin resonator.

assessed from the resonant bandwidth of the cavity, and appropriate modelling can lead to an estimate of the powder conductivity.

EXPERIMENTAL

The compounds K_x/K_9-L (x = 1, 3, 5, 7, 9) were prepared through the reaction of dehydrated potassium zeolite LTL (K_9-L), supplied by Laporte, with an amount of potassium vapour equivalent to x potassium atoms per zeolite unit cell. The reactions were carried out at temperatures between 200 and 250 °C in sealed, evacuated quartz reaction tubes, as described previously.[1,6] Careful annealing resulted in a blue solid, with the intensity of the colour increasing with the amount of added metal. A portion of each sample was sealed in a Spectrosil side arm of the reaction tube so that ESR and microwave cavity measurements could be made without exposing the product to the atmosphere. The ESR spectra were recorded on a Bruker ESP 300 spectrometer operating at X-band frequencies (9 GHz) with 100 kHz field modulation. The microwave frequency was measured with a Hewlett-Packard 5350B frequency counter to an accuracy of ± 1 kHz, and the magnetic field with a Bruker ER 035M NMR gaussmeter to better than ± 0.1 G. Temperatures down to 4 K were attained by means of an Oxford Instruments ESR 900 continuous flow cryostat.

Microwave Cavity Perturbation Measurements

The cavity used in these measurements was a quarter-wave copper hairpin resonator, resonant at 3.5 GHz (Figure 2). This host resonator has a relatively high quality factor Q (around 1500 at room temperature, increasing to over 4000 below 20 K) resulting in a low background loss. The volume of the resonator (~ 0.8 cm^3) is very small for these high Q values, so that the effective volume filling factor of the zeolite sample was high (~ 15%), resulting in a sensitive measurement of the zeolite response. The resonator was loop coupled and all measurements were performed in transmission mode using a HP 8720A network analyser and a closed cycle cryostat in the temperature range 15–295 K. All of the zeolite samples were contained in sealed quartz tubes which were inserted into the resonator through a hole in the radiation shield (see Figure 2). One of the advantages of the hairpin resonator is that there are distinct regions of high microwave magnetic field and high microwave electric field. Since here we were looking at the eddy current losses, we placed the sample in the magnetic field antinode, close to the short circuit end of the resonator. The electric field in this position is small, and the effects of the dielectric properties of the zeolite were minimized.

RESULTS AND DISCUSSION

Figure 3 shows the effect on the resonator response of placing various metal-loaded zeolite samples (K_x/K_9–L) in the resonator at room temperature. Note that there appears to be two distinct discontinuities in this raw data. The loss increases dramatically between concentrations x = 3 and x = 5, and there seems to be a further increase between x = 7 and x = 9. A similar

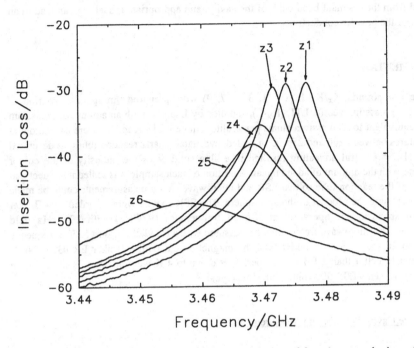

Figure 3. The change in resonant transmission as a function of frequency on the insertion of metal-loaded zeolites into the host resonator. The samples z1–6 represent the series K_x/K_9-L, with $x = 0, 1, 3, 5, 7, 9$.

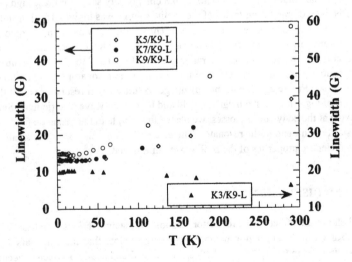

Figure 4. The ESR linewidth as a function of temperature for K_x/K_9-L ($x = 3, 5, 7, 9$).

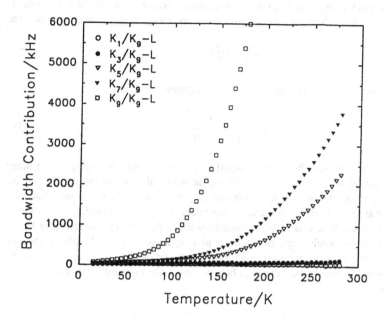

Figure 5. The subtracted bandwidth as a function of temperature for K_x/K_9-L (x = 1, 3, 5, 7, 9).

pattern is observed in Figure 4, which shows the ESR linewidth of four of the samples as a function of temperature: between $x = 3$ and $x = 5$ the behaviour changes from decreasing with increasing temperature, suggesting that the linewidth is governed by a motional averaging process, to increasing with temperature, indicative of a phonon-mediated relaxation process. This observed increasing trend strengthens markedly between $x = 7$ and $x = 9$. It is likely that these variations reflect the changing distribution of potassium within the zeolite channels as the concentration of metal increases. Figure 5 shows the resonant bandwidth as a function of temperature for each of zeolite samples. By subtracting the unloaded zeolite bandwidth from each we obtain the bandwidth contribution associated with metal-loading. Surprisingly, it appears that the low temperature bandwidth contribution for each metal concentration reaches similar values, independent of the concentration.

To relate the bandwidth to the zeolite conductivity, we use a model where we assume that the conductivity is low enough that the microwave fields penetrate each powder grain completely. This assumption is true when the microwave skin depth δ is much greater than the particle radius, and is justified by the observed ESR lineshapes. Then the dissipation within each grain is a volume effect, and when averaged over one wave cycle is given by

$$\overline{P} = \tfrac{1}{2}\sigma \int_V E^2 \, dV \tag{1}$$

13

where σ is the conductivity and E is magnitude of the induced electric field within the grain. Denoting V_{cavity} as the volume of the resonator, the stored energy averaged over one cycle can be shown to be

$$\overline{U} = \tfrac{1}{4}\mu_0 H_0^2 V_{cavity} \tag{2}$$

If Δf_B is the contribution of the sample to the resonant bandwidth, then

$$\Delta f_B = \frac{f}{Q} = \frac{1}{2\pi}\frac{\overline{P}}{\overline{U}} \tag{3}$$

As the observed bandwidth is proportional to σ, Figure 5 may be regarded as a direct reflection of changes in the conductivity of the zeolite as a result of metal loading. Despite the fact that the room temperature conductivity increases spectacularly with metal concentration, it is immediately apparent from the temperature dependence of the losses that the conductivity is not metallic. Nevertheless, it is clear that metal-loaded zeolites exhibit substantially higher conductivity than the purely ionic conductivity of dehydrated zeolites. To obtain absolute values of conductivity it is necessary to perform a full particle size analysis of the zeolite powders. This and further measurements to help determine the mechanism of conduction are already in progress.

ACKNOWLEDGEMENTS

P.A.A. is a Royal Society University Research Fellow. This work has been carried out under the auspices of the *Centre for Electronic and Magnetic Materials*, University of Birmingham.

REFERENCES

1. P. A. Anderson, A. R. Armstrong, P. P. Edwards, Angew. Chem. **106**, 669 (1994); Angew. Chem., Int. Ed. Engl. **33**, 641 (1994).
2. J. A. Rabo, C. L. Angell, P. H. Kasai, V. Schomaker, Discuss. Faraday Soc. **41**, 328 (1966).
3. P. P. Edwards, M. R. Harrison, J. Klinowski, S. Ramdas, J. M. Thomas, D. C. Johnson, C. J. Page, J. Chem. Soc., Chem. Commun., 982 (1984).
4. M. R. Harrison, P. P. Edwards, J. Klinowski, J. M. Thomas, D. C. Johnson, C. J. Page, J. Solid State Chem. **54**, 330 (1984).
5. P. A. Anderson, R. J. Singer, P. P. Edwards, J. Chem. Soc., Chem. Comm., 914 (1991); P. A. Anderson and P. P. Edwards, ibid., 915 (1991).
6. P. A. Anderson and P. P. Edwards, J. Am. Chem. Soc. **114**, 10608 (1992).
7. P. P. Edwards, L. J. Woodall, P. A. Anderson, A. R. Armstrong, M. Slaski, Chem. Soc. Rev. **22**, 305 (1993).
8. P. A. Anderson, P. P. Edwards, Phys. Rev. B **50**, 7155 (1994).
9. J. R. Waldram, A. Porch and H.-M. Cheah, Physica C **232**, 189 (1994).

NANOTESLA DETECTION USING THE PLANAR HALL EFFECT

A. SCHUHL*, F. NGUYEN VAN DAU* AND J.R. CHILDRESS**.
*Laboratoire Central de Recherches, Thomson-CSF, 91404 Orsay, France.
**Dept. of Mat. Science and Engineering, U. of Florida, Gainesville, FL 32611-2066.

ABSTRACT

A magnetic field sensor based on the planar Hall effect has been developed using epitaxial permalloy ($Ni_{80}Fe_{20}$) ultrathin films (1-10 nm). The magnetic and magnetotransport properties of these films have been studied in detail. For thicknesses above 5 nm, the resistivity of the permalloy film is below $5\mu\Omega$-cm, and its magnetoresistance ratio is 2%. By using the transverse resistivity for detection, we have reduced thermal drift effects by five orders of magnitude. We also make use of a weak uniaxial anisotropy induced in the permalloy through exchange coupling with a 6 nm-thick Fe/Pd multilayer, itself grown directly on the MgO substrate. Magnetic sensors based on these films have been used successfully to detect fields below 10 nT at 1Hz. Since the lateral dimensions of the sensing element are small ($<30\mu$m), and because of the ferromagnetic coupling with the Fe/Pd structure, it consists of a single magnetic domain. Sensitivities above 100 V/T-A have been obtained, with deviations from linearity of less than 2% over 4 decades.

INTRODUCTION

Most of the magnetic detection systems used either in magnetic recording read-heads or low-field measuring devices are based on hybrid technologies. One of the important challenges for the next few years is the production of a solid state magnetic sensor with a high level of integration. Magnetoresistive ferromagnetic thin films are one of a few candidates for such devices. In a thin film with a uniaxial magnetic anisotropy, an external magnetic field applied in the plane perpendicular to the easy axis, induces a rotation of the magnetization, which changes the resistivity of the film due to the anisotropic magnetoresistive effect (AMR)[1]. For read-head applications, high frequency rates are needed, mainly in the MHz range. The resolution of resistive measurements is then limited by the noise due to thermodynamical fluctuations, usually called Johnson noise. In the bandwidth of interest for ultra-low-field measurements (few Hz), this limitation is less severe. However, one must consider then a second type of thermal noise limitation, induced by temperature variations in the frequency range of the measurement. Through the important temperature drift of soft magnetic materials (for standard permalloy $(1/R)(dR/dT)=0.25\%$ per Kelvin), these variations lead to an important output voltage fluctuation. Indeed, this temperature drift appears to be the main factor limiting the low-field performance of magnetoresistive detectors. As an alternative solution, we have developed a sensor based on the planar Hall effect (PHE). It is actually a measurement of the AMR in a transverse configuration, so it is sensitive only to the non-isotropic part of the resistivity. The PHE voltage is given by

$$V = R_t I = \Delta R \sin(2\theta) I \qquad (1)$$

where R_t is the transverse resistivity, ΔR the total resistivity anisotropy and θ the angle between the magnetization and the current. Consequently, the temperature drift of the PHE signal is drastically reduced compared to the standard longitudinal magnetoresistance geometry, which increases the resolution in the low frequency range. The temperature dependence of R_t is small,

15

since it is originated by spin-orbit interactions. Although the PHE is a well-known phenomenon[2], it is not widely used since the transverse resistivity of metals is usually too small to produce a large enough signal. In a recent study we have shown the feasibility of growing high-quality NiFe ultrathin films (below 10 nm) by molecular beam epitaxy[3]. This material has now been used to produce nanoTesla Planar Hall sensors with very small thermal drift. Beside the suppression of the thermal drift, the PHE configuration has two other main advantages. First, the dependence of the output signal on θ (the angle between the current I and the magnetization M), is shifted by 45° compared to the longitudinal AMR geometry. The maximum sensitivity is then directly obtained at $\theta=0$, which eliminates the need for barber poles or external biasing fields. Second, the size of the sensing element may be very small (few square micrometers), which leads to a monodomain sensor.

EPITAXIAL PERMALLOY

Multilayers of metallic thin films were evaporated from separate high-temperature effusion cells onto (001)MgO substrates by Molecular Beam Epitaxy. The growth method of the multilayer is described elsewhere[4]. The multilayer structure of a typical sensor (MgO(001) / 1.5nm Fe / 0.5nm Pd / 1nm Fe / 1nm Pd / 6nm NiFe / 1nm Pd) consists of three distinct parts: a Fe/Pd multilayer as buffer layer; a magnetoresistive active layer which is a 6 nm thick layer of Permalloy; and finally, a thin layer (1.5nm) of Palladium for passivation. The initial (001)Fe/Pd multilayer is used to induce an in-plane uniaxial magnetic anisotropy. Normally, bcc (001)Fe thin films have two identical in-plane easy axes, (100) and (010), with an anisotropy field of a few hundred Oe with respect with the (110) axis. We have shown recently that when the Fe layer is deposited with a non-rotating substrate, a weak uniaxial anisotropy (anisotropy field = 20 Oe) is induced in the Fe/Pd multilayer due to the angle of incidence of the Fe atom flux with the substrate.[5] The NiFe alloy is deposited by simultaneously opening the Fe and Ni shutters, while rotating the substrate. Permalloy grows on the above Fe/Pd multilayer with a (110) texture. Its magnetotransport and magnetic properties are similar to the bulk material.[6] For thicknesses above 5 nm, the permalloy resistivity is below 25 $\mu\Omega$.cm, and its magnetoresistance is $\Delta R/R=2\%$. Since the NiFe alloy is grown on a rotating substrate, it does not show any intrinsic anisotropy. However the Fe/Pd multilayer imposes its magnetic anisotropy to the entire magnetic structure through ferromagnetic exchange coupling.[3]

PLANAR HALL EFFECT SENSORS

Devices have been fabricated from the above multilayers, by patterning the films using standard photolithography and low energy (200 eV) ion milling. Initially, a test pattern was used to simultaneously measure the transverse and the longitudinal resistivity of the device. The temperature dependence $a_T=(dR/dT)/\Delta R$ could then be compared for the two configurations on the same sample. Typically, we measured $a_T=0.5$ K^{-1} for standard longitudinal magnetoresistivity, and $a_T=0.5\times10^{-3}$ K^{-1} for the PHE. This corresponds to an equivalent magnetic noise of 10^5 nT.K^{-1} for the longitudinal measurement and 100 nT.K^{-1} for the transverse configuration.

For actual device fabrication, a different pattern was used with only two output connections corresponding to the Planar Hall signal. The resistivities of the contact legs were decreased in order to obtain both low output noise and minimal electrical dissipation. With this patterning specially designed for PHE, we obtained a better compensation, with a very low resulting thermal drift, (<10 nT.K^{-1}). Several sizes were tested for the sensing area, but we did not observe any systematic variation of the performance down to the lowest dimension (28 μm*28μm). One should note that standard magnetoresistive sensor technology always requires the use of an integrated bridge to reduce thermal drift together with the use of Barber poles or biasing fields to improve the sensitivity. The fabrication of PHE sensors is therefore much simpler, with few steps and only one lithographic mask. A second mask can be used to deposit a protection layer onto the sensing area, which avoids contamination by conducting or magnetic dust. Such a mask, however, does not need precise alignment.

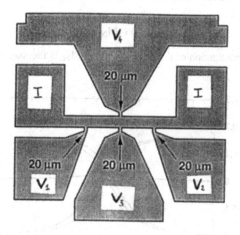

Fig 1 Pattern use for the simultaneous measurement of the longitudinal magnetoresistivity, V_1-V_2 (the voltage electrodes are sensitive to the component of the electric field parallel to the current), and the PHE signal, V_3-V_4 (the output voltage probes the component of the electric field perpendicular to the current).

The transverse magnetoresistance of a sensor is shown in figure 2, for different direction of the applied magnetic field. When a saturating field is applied along one of the two hard axes, Fe(110) and Fe(1-10), the angle between the current and the magnetization is $\theta=45°$. Therefore, the Planar Hall signal is at an extremum (minimum in one case and maximum in the other). When the field is then decreased, toward the easy direction Fe(100), the signal approaches a value in between the two extrema. When the field is applied along the easy axis direction Fe(100), the in-plane magnetization axis does not change during the entire loop, so the sensor response is constant.

One should note that the value of the transverse resistivity is not exactly equal to zero as it should be. This small difference has two different origins: First, a very small misalignment between the current and the most easy axis, for example in the patterning process, will lead to an significant signal offset. Second, a small defect in the shape of the Hall pattern would introduce a small longitudinal contribution to the magnetoresistance. This may be also the source of the small residual temperature drift observed in these PHE sensors.

SENSITIVITY AND RESOLUTION

The uniaxial anisotropy is used in the sensing process to ensure that the magnetization in zero field is along the most easy axis regardless of the magnetic history of the device. A transverse field then produces a rotation of the magnetization which leads to a transverse voltage. For a uniform magnetic layer, the PHE output signal can then be written

$$V_H = I \, \Delta R \, (H_{ext}/H_a) \tag{2},$$

where I is the current, H_{ext} is the magnetic field perpendicular to the most easy axis of the structure, ΔR is the total anisotropic resistivity and Ha is the total anisotropy field. In our structure, the anisotropy field acts directly only on the Fe/Pd structure, whereas in the field energy term (-MH), both Fe/Pd and NiFe magnetizations must be considered. Therefore, expression (2) becomes

$$V_H = I \, \Delta R \, (H_{ext}/H_a) \, (\, (M_{NiFe}+M_{Fe}) \, / \, M_{Fe} \,) \qquad (3),$$

where M_{NiFe} is the magnetization of the NiFe layer and M_{Fe} is the magnetization of the Fe/Pd structure.

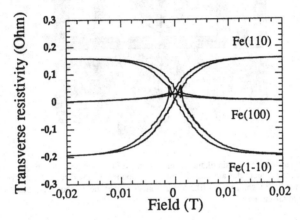

Fig 2   Planar Hall resistivity as function of external applied field in the Fe(110), Fe(1-10) and Fe(100) directions. The sensing element is a Fe/Pd/Fe/Pd/NiFe/Pd multilayer.

The saturation field H_s observed when the direction of the magnetic field is along one of the hard axes (see figure 2) is related to the value of H_a used in this expression. When the four-fold anisotropy is dominant the relation is simply $H_s=2*H_a$. The sensitivity can be then expressed as:

$$S = V_H/I \, H_{ext} = \Delta R/H_a \, (\, (M_{NiFe}+M_{Fe}) \, / \, M_{Fe} \,) \qquad (4)$$

For the structure described above, we usually measure ΔR between 0.5 and 1 Ohm. Using $\Delta R=0.5 \, \Omega$, $H_a=100$ Oe, and $(\, (M_{NiFe}+M_{Fe}) \, / \, M_{Fe} \,) = 2$, expression (4) leads to S= 100V/T-A. Figure 3 shows the Planar Hall signal versus the applied field aligned with the less easy of the two easy directions. The hysteresis is identical to values observed by SQUID magnetometry.

The linear parts correspond to rotation of the magnetization away from "easiest" (100) direction. It is the operating field range of the sensor. On Figure 3 it is seen that the linear response covers a field range on the order of 20 Oersted. The ultra-low field response was obtained using a Helmholtz coil generating a controlled field down to 10^{-6} Oe (10^{-10} Tesla). This is shown in figure 4. Since the uniaxial anisotropy is directed along the current, there is no need for a bias field to place the sensor in its optimum operating range. Moreover, the earth's magnetic field does not need to be compensated since the linear response of the sensor is spread over several Gauss. We used for the measurement a DC current of 1mA corresponding to a current density close to $5x10^5$ A/cm^2, and a time constant of 1 sec.

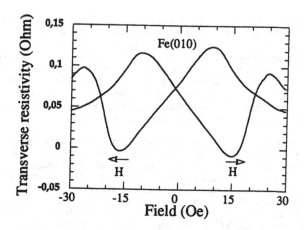

Fig. 3 Planar Hall resistivity as function of external applied field in the Fe(010) direction. The sensing element is a Fe/Pd/Fe/Pd/NiFe/Pd multilayer.

The sensitivities values deduced from Figure 3 and from Figure 4 are identical, S=70 V/T-A. Using the value of ΔR deduced from the figure 3, ΔR=0.35 Ω, and the magnetization ratio, $(M_{NiFe}+M_{Fe}) / M_{Fe} = 1.8$ deduced from the structure of the multilayer, we have calculated a value for the anisotropic field of H_a=90 Oe. This value is in relatively good agreement with the saturation field value H_S=150±30 Oe (figure 2), since a factor two is expected between Hs and Ha. Different structures have been tested with increasing NiFe thicknesses or with different Fe/Pd multilayer structure. We have always obtained a ratio H_S/H_a close to the expected value. Indeed the sensitivity does not depend on the measured current up to I=50 mA.

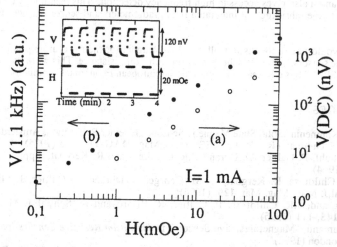

Fig 4 PHE output voltage as function of external applied field for DC (a) and AC (b) measurements (I=1mA and the applied field is perpendicular to the most easy axis). The response of the sensor to magnetic field steps of 20 mOe in amplitude is shown in the inset.

The lowest detectable field is then determined by the noise, which in our case is dominated by the electronic instrumentation noise. Using a time constant of 1 second, direct measurement of magnetic fields of the order of 10^{-3} Oe were obtained. The detection performance in the low field limit can be slightly increased using AC resistive measurements, with an excitation frequency in the 1 kHz range. This leads to an important reduction of the noise, and consequently the resolution of the sensor can be increased by an order of magnitude. The resolution (defined by a signal to noise ratio equal to unity with a time constant of one second) is observed to be below 10^{-4} Oe (10 nT). The signal is observed to be linear over more than 3 decades. Moreover, from figure 3, we can conclude that the sensor has a linear response over more than four decades.

In order to increase the sensitivity and at the same time the low field resolution, we have explored two different solutions. First, a decrease of the magnetic anisotropy would lead to an increase of the sensitivity. The NiFe magnetization is aligned with the Fe magnetization trough a strong ferromagnetic coupling. However, a decrease in the strength of this coupling will permit an independent rotation of the two magnetizations, although the coupling should stay strong enough to maintain the NiFe magnetization in the vicinity of the Fe easy axis. By increasing, up to 30 Å, the thickness of the Pd layer used as a spacer between the Fe and the NiFe layers, we have obtained sensibilities up to S=130 V/TA. A complete study of this decoupling effect will be published elsewhere[7]. Second, the adjunction of flux concentrators in the vicinity of the sensing element has lead to an important increase in the sensitivity. Preliminary results gave, S=300V/T-A, and a resolution R=3 nanoTesla. We expect further increases in the performance with this technique. Modelization of the magnetic flux lines in the system are now in progress, with the goal to optimize the design to achieve resolutions below 1 nanoTesla.

CONCLUSIONS

We have produced thin film magnetic field sensors based on the PHE, with a resolution below 10 nT and a sensitivity above 100 V/A-T. The thermal drift is observed to be very small, at least four order of magnitude below the thermal drift of longitudinal magnetoresistive sensors. The microscopic lateral dimensions of the active part of the sensor is surely the most interesting aspect of these PHE detection devices, since it allows parallel fabrication to be easily implemented on large scales, and it also gives access to high frequency detection. This technology could therefore provide lead to the fabrication of integrated field gradiometers and sensor matrices for magnetic cartography.

We would like to thanks P. Collot and A. Peugnet for their useful help in the patterning process of the sensors, and A. Wochenmayer for technical assistance. This work was supported in part by the ESPRIT Basic Research Program of the European Economic Community.

REFERENCES

1.   C. S. Roumenin "Solid State Magnetic Sensors", Elsevier Publication, Amsterdam 1994.
2.   T.R. McGuire and R.I. Plotter, IEEE trans. Mag. **MAG-11**, 1018 (1975)
3.   A. Schuhl, P. Galtier, O. Durand, J.R. Childress and R. Kergoat, Appl. Phys Lett. **65**, 913 (1994)
4.   J. R. Childress, R. Kergoat, J.-M. George, O. Durand, P. Galtier, J. Miltat and A. Schuhl,J. Mag. Mag. Mat. **129**, (1994).
5.   O. Durand, J.R. Childress, P. Galtier, R. Bisaro and A. Schuhl J. Mag. Mat. and Mag.**145**, 111 (1995).
6   P. Ciureanu, "Magnetoresistive Sensors" in *Thin-Film Resistive Sensors* page 276, IOP Ltd London (1992).
7   F. Nguyen Van Dau, J. R. Childress and A. Schuhl, to be published in the Proceeding of the European conference and sensors and actuators **"EUROSENSORS"**, Stockholm Sweden, (1995).

OBSERVATION OF MICROMAGNETIC STRUCTURE IN COMPUTER HARD DISKS BY LORENTZ TRANSMISSION ELECTRON MICROSCOPY

K. Tang, M.R. Visokay, and R. Sinclair
Dept. of Materials Science and Engineering, Stanford University, Stanford, CA 94305
C.A. Ross, R. Ranjan, and T. Yamashita
R&D, Komag Inc., Milpitas, CA 95035

ABSTRACT

We have observed micromagnetic structure in real computer hard disks with the typical structure of C/Co alloy/Cr/NiP/Al(substrate) using Lorentz transmission electron microscopy (LTEM). A chemical etching method was introduced to successfully prepare LTEM specimens directly from the computer hard disks with both smooth and mechanically textured substrates. Micromagnetic structural features, e.g., ripples and vortices, were studied in disks in bits-written, ac-demagnetized, and saturation remanent magnetic states.

INTRODUCTION

With the increase in recording density of magnetic hard disks, detailed analysis of micromagnetic structure in the media has become increasingly important. Lorentz transmission electron microscopy (LTEM) is thought to provide the highest resolution for magnetic structures.[1] However, LTEM requires specimens with large uniformly thin areas so that the deflection angle of the incident electron beam is proportional to the magnetic field within the film plane. Specimen preparation, therefore, is one of the major barriers to application of this technique to the recording media. Some authors have used model systems with specially designed substrates to facilitate preparation of LTEM specimens of Co alloy/Cr films.[2,3] Unfortunately, the microstructure of the magnetic layer can be affected by substrate character. In the present study a chemical etching method was introduced which allowed LTEM observation of micromagnetic structures in unmodified real computer hard disks with the typical C/Co alloy/Cr/NiP/Al(substrate) structures.

EXPERIMENTAL PROCEDURE

The disks studied have a smooth substrate (supersmooth) with the structure of 15nm C/29nm $Co_{84}Cr_{10}Ta_6$/50nm Cr/6μm NiP/Al(substrate). The metallic films were sputter-deposited using a dc magnetron system with a substrate bias of -200 V and a substrate temperature of 225 °C without breaking vacuum. Magnetic parameters of these films, measured using vibrating sample magnetometry (VSM), are as follows: H_c = 1590 Oe and Mrt = 1.34 memu/cm^2. The radial-to-circumferential orientation ratio of coercivity is 0.98. Magnetic bits were written in alternating direction of magnetization along tracks (in the circumferential direction of the hard disk) in a standard writing procedure for computer hard disks. A magnetoresistive head with an inductive writing element of the size of P1W/P2W = 7.5/6 μm was used at a flying height of 0.076-0.089 μm (3.0-3.5 μ"). The recording density was 585 bits/mm (15 Kfci) at the inner diameter. (The bits were written with a constant frequency at different radii, so the recording density decreased with increase of radius.) Track pitch was 10.1μm. The disk was not dc-erased before writing, so the regions between tracks were in the as-deposited magnetic state. The ac-demagnetized state was achieved by rapidly spinning the sample in the VSM machine in a magnetic field which decreased slowly from a saturation value to zero (The sample surface is parallel to this external field). The saturation remanent state was obtained by applying a saturation magnetic field to the samples along the sample surface and then removing the field.

Disks on mechanically textured substrates, having the structure of C/48nm $Co_{86}Cr_8Ta_6$/75nm Cr/NiP/Al(substrate), were also investigated. The metallic films were

21

Mat. Res. Soc. Symp. Proc. Vol. 384 ° 1995 Materials Research Society

deposited using conditions similar to those in the smooth substrate case. Magnetic parameters are as follows: H_c = 1900 Oe (in the circumferential direction of the hard disk) and Mrt = 2.5 memu/cm^2. This disk was dc-erased in the circumferential direction before bits were written.

A chemical etching method was used to produce LTEM specimens directly from the C/Co alloy/Cr/NiP/Al(substrate) computer hard disk structures. The resulting specimens typically have 2000 μm^2 or larger electron transparent areas of Co alloy/Cr films with uniform thickness. The computer hard disks were first ground from one side to remove most of the Al substrate. The thinned pieces were then mechanically cut into 3mm disks which were subsequently dimpled from the Al side into the NiP layer, which was then etched away with concentrated nitric acid at room temperature. Since both Cr and Al are insoluble in this acid, the Cr underlayer acts as a protective layer, isolating the Co alloy magnetic layer from the etchant while the Al substrate remains as mechanical support for the resulting thin metallic film. The etched samples were suitable for LTEM observation. If necessary, further removal of the Cr underlayer and C overcoat can be accomplished using low-angle ion-milling.

A Philips CM 20 FEG TEM (200 kV) was used for LTEM observation of the films. This machine is equipped with both a Twin2 Lorentz imaging lens (3nm resolution) and a SuperTwin objective lens (0.24nm resolution), which allows direct correlation between micromagnetic structural features and microstructural features of the same sample area at high spatial resolution.

RESULTS AND DISCUSSION

Figure 1a is part of a Fresnel, i.e., defocused, LTEM (FLTEM) image of the $Co_{84}Cr_{10}Ta_6$/Cr film on the smooth NiP/Al substrate in the bits-written magnetic state, which confirms that a large uniformly thin area is successfully obtained using the chemical etching method. Alternating dark and light domain walls along the track direction are observed between the bits and the regions between tracks. This is consistent with the magnetization state of the film. As described earlier, the bits were in alternating direction of magnetization along the tracks, while the regions between tracks were in the as-deposited magnetic state. Since the film was deposited using a dc magnetron system, it is possible that the magnetic field during deposition gave rise to a remanent magnetization in the radial direction of the disk, which is perpendicular to the magnetization direction within the bits. The incident electrons are deflected by the demagnetization field within the film in the direction perpendicular to the magnetization directions, generating alternating dark and light lines along the track direction in defocused images. Longitudinal magnetic ripples[4], perpendicular to local net magnetization, are found within the bits and regions between tracks. The transition regions between the bits are featured with alternating dark and light spots along the transition width (perpendicular to track direction). This observation is consistent with a previous observation by Cameron and Judy on a $Co_{86}Cr_{12}Ta_2$/Cr film deposited on a carbon-precoated Si substrate, which can be interpreted as vortices with an alternating sense of rotation.[2]

The micromagnetic structural features at the bit-transition regions can be more clearly seen at higher magnification in Figure 1b. The diameter of the magnetic vortices is estimated to be 0.1-0.2 μm based on the alternating periodicity of these spots, which is considered to be twice the dimension of the vortices across the transition width. The size of these vortices is slightly smaller than that observed by Cameron and Judy.[2] This is presumably because the films observed in this study have higher coercivity than their films. The ripples at the ends of the transition widths have two distinct structures, one being divergent from the transition region into the regions between tracks and the other being divergent from the regions between tracks into the bits. One possible interpretation can be made in terms of magnetostatic interaction between magnetic moments within the bits and those within the regions between tracks. The magnetic field generated by the regions between tracks tends to rotate adjacent magnetic moments within the bits to its own direction (perpendicular to the track direction). The rotations have two different patterns based on the relationship between the magnetization directions, as indicated in Figure 1c, which give rise to two distinct longitudinal ripple structures at the two ends of a transition width (track edge). The micromagnetic structure at the track edges can also be complicated by fringing head fields during the writing procedure.[5,6] The track edge magnetic structure is very important for future high density application and will be addressed in future studies.

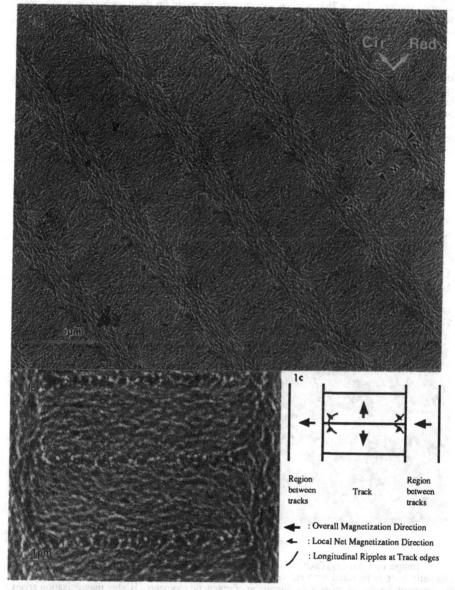

Fig 1 a) Fresnel LTEM image of $Co_{84}Cr_{10}Ta_6$/Cr film on the smooth NiP/Al substrate in the bits-written magnetic state; b) Fresnel LTEM image at a higher magnification; c) Schematic of magnetic ripples at the track edges.

Figure 2a is a FLTEM image of $Co_{84}Cr_{10}Ta_6$/Cr film (on the smooth substrate) in the ac-demagnetized state. The random pattern is consistent with randomization of the directions of magnetic moments in this state. The arrows point out "star-like" features, which we identify as magnetic vortices. Figure 2b is a schematic structure proposed for these vortices, in which small magnetic moments form clusters and these clusters then form close-fluxed vortices.[7] The size of these vortices is estimated to be around 1.0-1.5 μm, about 10 times larger than that found in the bit-transition regions in the bits-written films discussed in the previous paragraph. Because the magnetic clusters in these vortices are large, their net magnetizations are large enough to give visible walls between them. Therefore, detailed structure of the vortices can be revealed. The difference in vortex sizes in the above two cases may be rationalized as follows: in a bit-transition region the smaller vortices may be a result of constraints from adjacent bits which suppress the size of the vortices; in the ac-demagnetized state, however, there is no such constraint and consequently larger magnetic vortices may be energetically more favorable. This interesting observation has implications on the interpretations of different media noise measurement methods, such as the uniform magnetization noise and the integrated media noise, which are used to characterize transition media noise in the thin film media.[8]

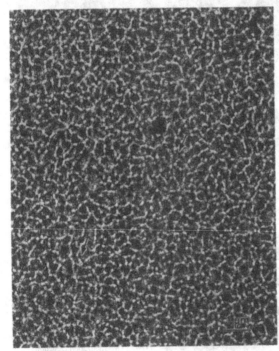

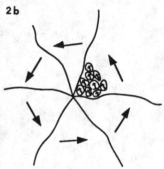

2b

Fig 2a FLTEM image of $Co_{84}Cr_{10}Ta_6$/Cr film (on the smooth substrate) in the ac-demagnetized state.

Fig 2b Schematic model for magnetic vortices observed in Fig 2a.

Image of $Co_{84}Cr_{10}Ta_6$/Cr film (on the smooth substrate) in the saturation remanent magnetic state is featured with ripples aligned largely parallel to each other (Figure 3). The alignment of the ripples reflects the alignment of magnetic moments. Higher magnetization arises from better alignment of magnetic moments and therefore is associated with better alignment of magnetic ripples. The saturation remanent state has the highest magnetization in the absence of external field, so the ripples align very well.

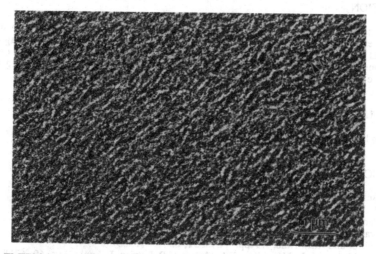

Fig 3 FLTEM image of $Co_{84}Cr_{10}Ta_6$/Cr film (on the smooth substrate) in the saturation remanent magnetic state.

Figure 4 is a FLTEM image of a $Co_{86}Cr_8Ta_6$/Cr film on the mechanically textured NiP/Al substrate in the bits-written magnetic state. Dark and light domain walls are observed between the bits and the inter-track regions for half of the bits in an alternating manner. This is because this disk was dc-erased in the circumferential direction before the bits-writing process, so half of the bits form 180° walls with the inter-track regions while the other half have the same magnetization direction as the inter-track regions and therefore no domain wall. Observation of micromagnetic structural features, i.e., ripples and vortices, in this film is somewhat complicated because contrast from the texture lines tends to obscure the more subtle magnetic contrast.

Fig 4 FLTEM image of a $Co_{86}Cr_8Ta_6$/Cr film on the mechanically textured NiP/Al substrate in the bits-written magnetic state.

CONCLUSION

A chemical etching method has been successfully introduced, which allows us to study micromagnetic structure in real computer hard disks. Some important features, such as magnetic vortices and track edge structures, are revealed. More detailed analysis and interpretation of our observations are currently underway in order to deepen understanding of the magnetic performance of the recording media.

ACKNOWLEDGMENT

Funding from Komag Inc. is greatly appreciated.

REFERENCES

1. M.R. Scheinfein, J. Unguris, D.T. Pierce, and R.J. Celotta, J. Appl. Phys. **67**, 5932 (1990).
2. G.P. Cameron and J.H. Judy, IEEE Trans. Magn. **29**, 4177 (1993).
3. T. Kawabe and J.H. Judy, IEEE Trans. Magn. **28**, 2470 (1992).
4. H.W. Fuller and M.E. Hale, J. Appl. Phys. **31**, 238 (1960).
5. T.C. Arnoldussen, L.L. Nunnelley, F.J. Martin, R.P. Ferrier, J. Appl. Phys. **69**, 4718 (1991).
6. T. Lin, J.A. Christner, T.B. Mitchell, J.-S. Gau, P.K. George, IEEE Trans. Magn. **25**, 710 (1989).
7. T. Chen, IEEE Trans. Magn. **MAG-17**, 1181 (1981).
8. R. Ranjan, W.R. Bennett, G.J. Tarnopolsky, T. Yamashita, T. Nolan, and R. Sinclair, J. Appl. Phys. **75**, 6144 (1994).

MAGNETIC PROPERTIES OF EPITAXIAL MBE-GROWN THIN Fe$_3$O$_4$ FILMS ON MgO (100)

P.A.A. VAN DER HEIJDEN[1], J.J. HAMMINK[1], PJ.H. BLOEMEN[1], R.M. WOLF[2],
M.G. VAN OPSTAL[1], P.J. VAN DER ZAAG[2], AND W.J.M. DE JONGE[1]
[1] Department of Physics, Eindhoven University of Technology (EUT), 5600 MB Eindhoven,
The Netherlands
[2] Philips Research Laboratories, Prof. Holstlaan 4, 5656 AA Eindhoven, The Netherlands

ABSTRACT

Coherent epitaxial Fe$_3$O$_4$ layers in the range of 0 to 400 Å have been grown by molecular beam epitaxy on single crystal MgO(100) substrates. The magnetic properties were studied by local magneto-optical Kerr effect experiments on a wedge shaped Fe$_3$O$_4$ layer, by ferromagnetic resonance and SQUID. The results show that the magnetic behavior of the Fe$_3$O$_4$ thin films resembles bulk Fe$_3$O$_4$ in the investigated thickness range.

INTRODUCTION

The continuing progress in thin film deposition techniques enables one nowadays to fabricate artificial layered structures with precise control over composition and thicknesses. So far the main efforts in the field of magnetic materials have been directed to metallic systems, although more complicated systems such as oxides can be deposited in a controlled fashion as well. Recently, we have shown from structural data that we were able to stabilize the correct phase of magnetite, Fe$_3$O$_4$, by MBE i.e. without unwanted phases such as Fe, FeO, and Fe$_2$O$_3$ [1]. Although Fe$_3$O$_4$ seems to be a simple oxide, its growth is not straightforward. Several different techniques have been reported. In the majority of these studies (see for instance [2]), the magnetic properties differ considerably from the bulk Fe$_3$O$_4$ behavior. Here, we focus on the characterization of thin Fe$_3$O$_4$ layers. We present ferromagnetic resonance (FMR), and SQUID data as well as the results of thickness dependent magneto-optical Kerr effect (MOKE) experiments on epitaxial Fe$_3$O$_4$ layers grown on MgO(100) single crystals by MBE and with layer thicknesses below the investigated range reported before. These studies enable a direct determination of the magnetic anisotropy constants.

EXPERIMENTAL

The Fe$_3$O$_4$ layers were grown using a UHV Balzers UMS 630 multichamber Molecular Beam Epitaxy system. The main deposition chamber consists of two differentially pumped chambers containing the evaporation sources and substrate holder, respectively. This configuration combined with the differential pumping with powerful turbo-pumps enables high oxygen pressures locally at the substrate maintaining relatively low pressures in the remaining part of the main chamber. The oxygen is supplied through a ring shaped doser located close to the substrate holder. This prevents the use of ionized oxygen generated by a Wavemet microwave/ECR plasma generator source such as in the case of Lind et al. [3]. The Fe$_3$O$_4$ layers were deposited by e-gun evaporation from Fe targets at a substrate temperature of 225° C on single crystalline MgO(100) substrates and at an oxygen pressure of 2.8×10^{-5} mbar. Before and during the deposition,

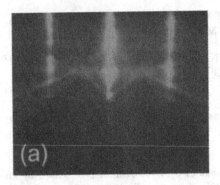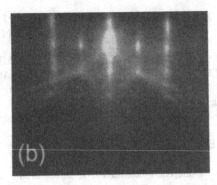

Figure 1: *RHEED patterns of MgO(100) (a) and Fe_3O_4(100) (b) at room temperature, with the e-beam parallel to the [100] direction.*

the e-gun evaporation fluxes are controlled by a cross-beam quadrupole mass-spectrometer feedback system. For further details on the preparation, see [1].

Two types of samples have been grown; samples with the Fe_3O_4 layer uniform in thickness and samples in which the Fe_3O_4 layer was deposited in the form of a wedge (from 0 to 400 Å) by slowly withdrawing a shutter located close to the substrate.

All samples were characterized in-situ with RHEED. Fig. 1 shows typical RHEED patterns of the MgO (a) and the Fe_3O_4 (b) surface at room temperature. From the photos it can be concluded that the Fe_3O_4 grows epitaxially on the MgO substrate without indications of island growth. It also appears that the MgO substrate was flat on an atomic scale, because of the clear presence of Kikuchi lines. These lines persist when depositing Fe_3O_4, indicating that the roughness does not increase significantly with Fe_3O_4 growth. Upon deposition of Fe_3O_4 and after cooling down, the observed lattice constant doubles, as follows from the appearance of additional streaks located between the MgO [0,0] and [-1,0] and [1,0] streaks. This is what one expects since the unit lattice cell length of Fe_3O_4 is about twice that of MgO; (8.396 and 4.2117 Å, respectively). A more detailed analysis of the growth will be given in [4].

The magnetic characterization of the wedge sample has been performed by means of the magneto-optic Kerr effect (MOKE) at room temperature. The MOKE studies are performed in the longitudinal geometry. The longitudinal MOKE experiments are performed with the applied field oriented along two different axes in the plane of the film, namely the [100] and [110] axes. Below thicknesses of 50 Å Fe_3O_4, the signal intensity is too low to measure hysteresis loops.

A uniform 300 Å thick Fe_3O_4 film has been investigated by means of SQUID (Quantum Design MPMS5) and FMR, employing a standard commercial FMR spectrometer with a Bruker X-band cavity (9.79 GHz) and a flow cryostat to obtain temperature dependent measurements in the range of 5 up to 300 K.

SQUID AND LONGITUDINAL MOKE EXPERIMENTS

Typical longitudinal hysteresis loops with the field applied along the in-plane [110] and [100] directions are shown in Fig. 2. The in-plane hysteresis loops reveal a low coercive field,

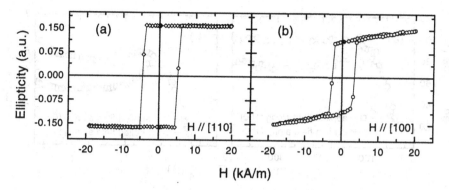

Figure 2: *Hysteresis loops for 340 Å $Fe_3O_4(100)$ measured by longitudinal MOKE with the field applied parallel to the film plane along the [110] (a) and [100] (b) axes.*

indicating the soft magnetic behavior of magnetite and show low saturation fields, indicating that the magnetization is oriented preferentially along the film plane rather than perpendicular to it. The in-plane hysteresis loop with the field applied along [110] shows a 100 % remanence, whereas with the field applied along [100] a remanence of 70 % is observed. For the latter, a field of about 20 kA/m was almost enough to obtain saturation. This indicates a cubic crystal anisotropy with [100] and [110] of the Fe_3O_4 the hard and easy in-plane directions, respectively, which is the same as for bulk Fe_3O_4. For bulk Fe_3O_4 at room temperature, the magneto-crystalline anisotropy energy is, in first approximation, given by $E = K_1(\alpha_1^2\alpha_2^2 + \alpha_2^2\alpha_3^2 + \alpha_1^2\alpha_3^2)$ with $K_1 = -1.1 \times 10^4$ J/m^3 [5]. Here α_1, α_2 and α_3 denote the direction cosines of the magnetization relative to the cubic axes. For (100) growth, this results in <110> easy and <100> hard axes in plane, as observed in the MOKE experiments.

The results of a SQUID magnetization measurement on a uniform 300 Å Fe_3O_4 single film, investigating the behavior in larger applied fields, support the above observations.

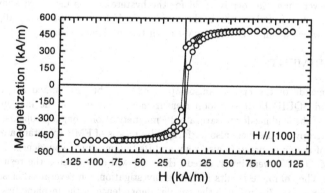

Figure 3: *The hysteresis loop of a 300 Å Fe_3O_4 single film on MgO (100) measured by a SQUID magnetometer at 300 K with the applied field along a [100] axis.*

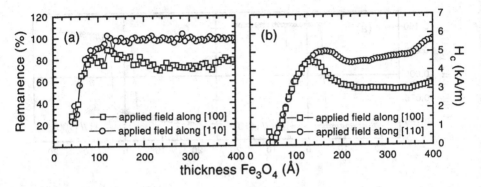

Figure 4: *The thickness dependence of the remanence (a) and the coercive field (b) obtained from the in-plane hysteresis loops with the field applied along the [100] (squares) and [110] (circles) directions.*

As can be seen from Fig. 3 again a low coercive field (of about 3 kA/m) is obtained. The field of needed to obtain saturation along the [100] axis is low and appears to be about 30 kA/m. In addition the SQUID data yield a value for the absolute saturation moment which in the present case appears to be 490 kA/m, close to the bulk value of 480 kA/m [6].

Fig. 4 shows the remaining results obtained with MOKE, i.e. the thickness dependence of the remanence and the coercive field when performing position dependent measurements along the Fe_3O_4 wedge. The [110] axis appears to be the easy axis for Fe_3O_4 thicknesses above about 120 Å as may be clear from the magnetization remanence versus the Fe_3O_4 thickness. Below 120 Å, the remanence decreases along the [110] direction. The remanence becomes equal for the [100] and [110] directions at 90 Å, suggesting a decrease of the in-plane anisotropy. For lower thicknesses the remanence of both directions decreases. Figure 4(b) shows the coercive field as a function of the magnetite thickness of the in-plane MOKE measurements for both directions of the applied field. For Fe_3O_4 thicknesses above 120 Å, the coercive field of the hysteresis loop measured along the hard axis of about 3 kA/m is, as expected, lower than the coercive field for the hysteresis loop measured along the easy axis (about 4.5 kA/m). Below 120 Å, the coercive fields for the [100] and [110] directions become equal, which again indicates a lowering of the in-plane anisotropy.

FMR EXPERIMENTS

The magnitude of the crystal anisotropy could not be determined accurately from the MOKE and SQUID hysteresis loop measurements, because of the inaccuracy in the determination of the field needed to saturate the magnetization along a [100] axis. Therefore, magnetic anisotropy studies were also performed by means of FMR to obtain a quantitative measurement of the cubic crystal anisotropy and its temperature dependence. Single layers of Fe_3O_4 on MgO (100) were investigated. Here, we will only discuss the results of a 300 Å Fe_3O_4 film. The complete results including investigations on several other samples will be published elsewhere [7]. In Fig. 5 the angular dependence of the in-plane resonance field measured at 293 K is shown. From this figure the fourfold symmetry that is expected for epitaxial growth of a cubic material in the [100] direction is clear. The minimum in the

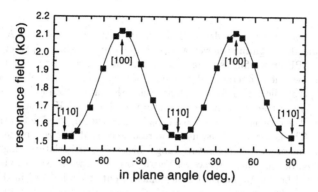

Figure 5: *Angular dependence of the in-plane resonance field measured at 9.79 GHz and 293 K of a 300 Å Fe_3O_4 film on MgO (100). The angle is defined relative to the [110] axis.*

resonance field occurs at the [110] direction, which is consistent with the easy axis direction derived from the MOKE hysteresis loop data. The solid line in the figure represents a fit of the well-known resonance condition for this geometry (see e.g., [8]). Here, the bulk g-value was assumed to be 2.12 [5]. To obtain a good fit, it appeared sufficient to take into account only the first order term of the magneto-crystalline anisotropy, K_1. A value of 28 kA/m was thus yielded for the obtained cubic anisotropy field, $-2K_1/\mu_0 M_s$. Using the value of 480 kA/m for the saturation induction $\mu_0 M_s$ of bulk Fe_3O_4 [6], an anisotropy constant K_1 of -0.9×10^4 J/m^3 is calculated, which is slightly lower than the bulk value of -1.1×10^4 J/m^3 [5]. Similar FMR experiments on a multilayer of 44 Å Fe_3O_4 / 18 Å MgO (not shown), reveal a cubic crystal anisotropy field of 6.4 kA/m. Note that this strong reduction is consistent with the MOKE hysteresis loop data, which displayed a more or less isotropic behavior for thicknesses below 120 Å (Fig. 4).

Also FMR experiments as a function of temperature have been performed.

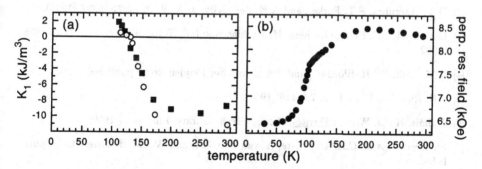

Figure 6: *FMR data on a 300 Å Fe_3O_4 film on MgO (100). (a) The temperature dependence of the cubic magneto-crystalline anisotropy constant K_1 (solid squares). Literature data for bulk [5] (open circles) are shown for comparison. (b) The temperature dependence of the resonance field with the applied field perpendicular to the film plane.*

The resulting temperature dependence of the crystal anisotropy constant K_1 is shown in Fig. 6(a). Again, the magnitude as well as the temperature at which K_1 reverses sign (T = 130 K) in this MBE-grown film is in agreement with literature data on bulk magnetite [5].

Fig. 6(b) shows the temperature dependence of the resonance field obtained for applied fields oriented perpendicular to the film plane. Here, the data extend to lower temperatures revealing the Verwey transition at a temperature of about 105 K through a drop in the resonance field. The drop is attributed to a drop in the anisotropy and not in the saturation magnetization. The latter also determines the resonance field, but is expected to be fairly constant in the investigated temperature range, because of the relatively high Curie temperature of 848K. In principle, the lowered Verwey transition with respect to the bulk value of 119 K [6] could be due to incorrect stoichiometry, but strain induced in the Fe_3O_4 film resulting from its coherent epitaxial growth on the MgO(100) single crystal may also play an important role. It is known that strained Fe_3O_4 layers could have a Verwey transition temperature, which differs from the bulk Verwey transition [9].

ACKNOWLEDGEMENTS

We would like to thank A. van Helvoort for his assistance with the FMR experiments. Part of this work was supported by the Technology Foundation (STW). The research of P.J.H. Bloemen has been made possible by a fellowship of the Royal Netherlands Academy of Arts and Sciences.

References

[1] R.M. Wolf, A.E. De Veirman, P. van der Sluis, P.J. van der Zaag, and J.B.F. aan de Stegge in Epitaxial Oxide Thin Films and Heterostructures, edited by D.K. Fork, J.M. Phillips, R. Ramesh and R.M. Wolf (Mater. Res. Soc. Symp. Proc. 341, Pittsburgh, PA, 1994) pp 23-28.

[2] D.T. Margulies, F.T. Parker, and A.E. Berkowitz, J. Appl. Phys. 75, 6097 (1994).

[3] D.M. Lind, S.D. Berry, G. Chern, H. Mathias, and L.R. Testardi, Phys. Rev. B 45 1838 (1992)

[4] R.M. Wolf, P.J.H. Bloemen, and P.A.A. van der Heijden, to be published.

[5] L.R. Bickford, Phys. Rev. 78, 449 (1950)

[6] J. Smit, H.P.J. Wijn in Ferrites, (Philips Tech. Library, Eindhoven 1959).

[7] M.G. van Opstal, C.H.W. Swüste, A. van Helvoort, and W.J.M. de Jonge, to be published.

[8] B. Henrich in Ultrathin Magnetic Structures II, edited by J.A.C. Bland and B. Heinrich (Springer-Verlag Berlin Heidelberg, 1994), pp 195.

[9] E. Lochner, K.A. Shaw, R.C. DiBari, W. Portwine, P. Stoyonov, S.D. Berry, and D.M. Lind, IEEE Trans. Mag. 30, 4912 (1994).

NANOSECOND STRUCTURAL TRANSFORMATION OF MAGNETIC THIN FILMS: PtMnSb, STRUCTURE AND MAGNETIC PROPERTIES

YUKIKO KUBOTA* ** and ERNESTO E. MARINERO*
* IBM Research Division, Almaden Research Center, 650 Harry Road, San Jose, CA 95120-6099
**Department of Materials Science and Engineering, Stanford University, Stanford, CA 94305

ABSTRACT

We report on the phase transformation of amorphous PtMnSb thin films induced by laser annealing in the nanosecond time regime. Structural and magnetic transformations are investigated by TEM, XRD, AFM and in-situ MOKE and VSM. We have established that a minimum laser fluence is required to crystallize the amorphous films and thus, to induce magnetic activity. The transformation kinetics vs number of irradiation pulses reveals that the magnetically active C1b phase is formed via an intermediate phase, namely, tetragonal-PtMn. We have also established that the thin film crystallization induced by the nanosecond laser annealing proceeds via nucleation rather than grain growth. Measurement of the lattice parameter of the C1b-PtMnSb produced by the laser quenching (LQ) indicates an essentially unstrained structure with $a = 6.17$ Å vs $a = 6.201$ Å reported for the bulk. Nevertheless, we observe the generation of large surface undulations upon laser annealing and suggest that this is the mechanism for stress relaxation concomitant with the large volumetric changes involved in the phase transformation. In addition, we observe decrements in saturation moments and Curie temperatures which are attributed to the nanocrystalline nature of the LQ specimens.

INTRODUCTION

The Heusler alloy, C1b-PtMnSb [1], is known to exhibit the largest Kerr rotation at room temperature of all known ferromagnetic materials. [2] Thus, there is considerable interest for optical storage applications and due to its unique band structure (metal for the majority spin electrons and semiconductor for minority spin electrons) [3], it is also a promising material for magneto-resistive applications.

Previous reports on the generation of the C1b crystalline phase have utilized long term, high temperature annealing employing conventional isothermal annealing. Recently Carey et al [4], reported also the usage of rapid thermal annealing to produce the C1b phase. The use of nanosecond duration laser pulses to crystallize amorphous PtMnSb was first reported by Marinero. [5] In this work we report for the first time, details pertaining the crystallization kinetics of these alloys, the nature of the structural transformation and the development of magnetic properties intimately related to the nano scale of the crystalline film produced under this non-equilibrium thermal conditions.

EXPERIMENTAL

PtMnSb thin films were deposited by DC magnetron sputtering (base pressure $< 1 \times 10^{-7}$ Torr.) onto quartz substrates from elemental targets utilizing Argon sputter pressures around 3 mTorr. The Heusler alloy was grown between SiN layers as follows: **quartz / SiN (20 nm) / PtMnSb (150 nm) / SiN (12.5 nm)**. All layers were grown in the same pumpdown and the nitride was formed by reactive sputtering. To correlate film structure to magnetic and magneto-optic properties, we utilized the following measurement techniques: FILM STOICHIOMETRY: x-ray microprobe, Rutherford backscattering (RBS), secondary ion mass spectrometry (SIMS) and Auger analysis. MICROSTRUCTURAL STUDIES: x-ray diffraction (XRD), transmission electron microscope and atomic force microscopy (AFM). MAGNETIC CHARACTERIZATION: Magneto-optic Kerr Effect (MOKE), vibrational sample magnetometer (VSM) and torque magnetometry.

Laser annealing was conducted within the pole pieces of a 20 kOe electromagnet which is part of the MOKE apparatus. No external field was applied during the sample annealing. Dielectric mirrors and beam combiners are utilized to guide both the excimer laser radiation and the probe He-Ne laser onto the same spot of the sample. The excimer laser irradiates a 6 mm diameter area (determined by the pole bore hole) whereas the He-Ne laser spot is approximately 1.5 mm in diameter and is fully confined within the irradiated area. In this fashion we are able to in-situ monitor the magnetic phase evolution of the amorphous alloy. Irradiation in our work was conducted through the SiN/air interface. The laser is a Lambda Physik (102) excimer laser running on KrF (248 nm) and it generates 16 ns pulses of around 250 mJ in energy. Control of the laser fluence is obtained by means of variable attenuator and the pulse energy is measured with a pyrolectric detector. To minimize pulse to pulse energy variations, a repetition rate of 2.5 Hz was employed. This also guaranteed that no thermal build up effects are present in our annealing experiments. In addition to LQ, we utilized also rapid thermal annealing in our experiments to compare the resulting structural and magnetic properties under these conditions and those induced by LQ. The RTA (AG Associates 610) pulse heat profile in our studies was the following: 5 s hold at room temperature, ramp-up to the target temperature at 125°C/s, 120 s hold at the target temperature and the sample was removed from the instrument after cooldown below 100° C.

EXPERIMENTAL RESULTS

A wide range of compositions around the stoichiometric 1:1:1 ratio were fabricated. Large variations in magneto-optical activity as a function of composition were observed. In this work we report only on studies of films whose elemental composition ratio was $Pt_{34.8}Mn_{33.4}Sb_{31.8}$. Oxygen and carbon traces ($\leq 1\%$) were found through the film thickness, however, the levels detected are close to the instruments' sensitivity limits. The as-deposited (AD) film structure as indicated by XRD and TEM electron diffraction (ED) is amorphous. Smooth flat surfaces and SiN/Quartz interfaces were observed with no pinholes. The amorphous films exhibit no magnetic activity.

In Figure 1, the maximum Kerr rotation or saturated rotation (θ_S) is plotted vs number of irradiation laser pulses for a fluence of 68.4 mJ/cm². This saturation value is that obtained when the external field aligns all moments out of plane of the film and was determined to be around 4.8 kOe for this hard axis orientation. The fluence threshold for observing the development of magnetic activity was determined to be 34 mJ/cm². This is correlated to the need to raise the amorphous film temperature to or close to its melting temperature. The figure shows that very little magnetic activity develops upon the first laser irradiation, whereas, the second pulse develops close to 50% of the maximum rotation obtained in the experiment. Note also the changes in reflectivity occur during the cumulative laser annealing of the sample. TEM studies utilizing plan and cross-sectional view were conducted for the as-deposited sample, after a single laser pulse exposure, two pulses and 2000 to investigate the phase transformation.

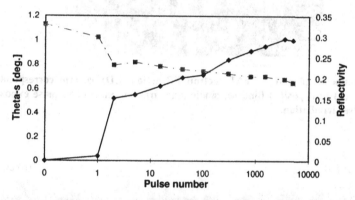

Figure 1. Typical MOKE signal evolution of PtMnSb thin films vs. number of KrF excimer laser pulses. (fluence = 68.4 mJ/cm²)

Figure 2 gives a composite electron diffraction pattern corresponding to plan view observations of: (a) as-deposited film; $\theta_S = 0°$, (b) first pulse irradiation; $\theta_S = 0.04°$, (c) two consecutive pulses; $\theta_S = 0.52°$ and (d) after 2000 pulses: $\theta_S = 1.0°$. The laser fluence utilized corresponds to Fig. 1. The typical broad halos of an amorphous structure are seen in 2(a). In contrast 2(b) shows the onset of crystallization and the index analysis indicates that this corresponds to tetragonal PtMn. Reference samples of pure PtMn treated under similar annealing conditions, exhibited no Kerr activity within experimental error. Thus, we attribute the θ_S value observed in Fig. 1 corresponding to single laser irradiation to traces of C1b phase that could be formed concurrently with the PtMn. Application of a second laser pulse 2(c) shows predominantly the Heusler C1b phase, although arcs of PtMn are still visible. Finally after 2000 pulses (2(d)), the ED pattern is all essentially C1b phase.

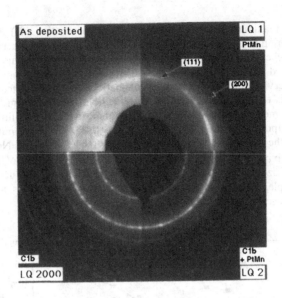

Figure 2. Composite plan view TEM electron diffraction (ED) patterns corresponding to figure 1. (a) as-deposited film, (b) single pulse irradiation, (c) two pulse exposure (d) 2000 pulses irradiation.

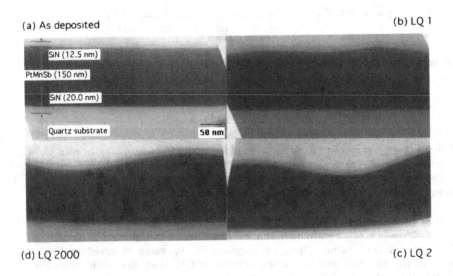

Figure 3. Cross sectional TEM bright field images corresponding to (a) - (d) of figure 2. Note the laser induced surface undulations.

Figure 3 shows cross sectional TEM (Akashi 002B, 200keV) bright field images corresponding to the samples of Fig. 2. Again, the as-deposited film, 3(a) was amorphous and exhibits smooth flat top and bottom interfaces. Following the first pulse as seen in 3(b), significant crystallization has occurred. However, note that the nanocrystallites are located in the middle of the film. Top and bottom ~100 nm thick amorphous layers remain in the vicinity of the SiN interfaces. It is clear also from 3(b) that the laser annealing leads to the generation of surface undulations. After two consecutive pulses, the film volume is filled with 10~20 nm size nanocrystals and θ_s jumped up to 50% of the maximum value. Fig. 3(d) shows the effect of 2000 pulses, the surface undulations are very apparent and it is noted that the the average grain size did not change from 2 to 2000 pulses. In contrast, θ_s gradually increased to its maximum value. The AFM study shows that the amplitude of the surface undulations is enhanced with the number of pulses, whereas the periodicity of the undulation, ~ 500 nm, is essentially unchanged.

The XRD spectrum of the film exposed to 2000 pulses is given in Fig. 4. In this, we compare its spectrum to that observed utilizing RTA annealing to 650°C for 120sec. Whereas the laser annealing produced single phase C1b-PtMnSb with a lattice parameter of 6.17 Å, the RTA led to a multi-phased crystalline film, consisting of C1b, tetragonal PtMn and cubic $PtSb_2$. A lattice parameter of a = 6.18Å is measured for the C1b phase formed by RTA. Both C1b lattice parameters in LQ and RTA film are comparable to that of the bulk value, a = 6.20Å, reported by Watanabe [1], indicative of absence of lattice strain.

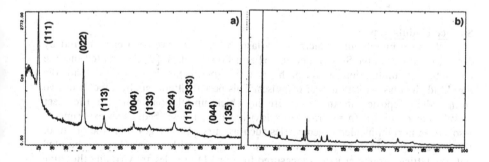

Figure 4. Comparison of XRD spectra for RTA (650°C) and laser annealed (fluence = 68.4 mJ/cm²) PtMnSb specimens. LQ samples yield single C1b PtMnSB whereas secondary phases are formed during RTA.

DISCUSSION

Intermediate PtMn Phase

The formation of the Clb PtMnSb induced by laser annealing clearly proceeds via the formation of the intermediate PtMn tetragonal phase. We observe a threshold for the onset of crystallization which we attribute to the need for liquid-like diffusion for crystallization to occur in the nanosecond time scale of our annealing. Thermodynamic driving forces are likely to facilitate the formation of the tetragonal PtMn phase more readily than that of the complex Clb structure. This intermediate phase then acts as nucleation centers for the generation of the Clb phase. The unique aspect of this transient annealing is that the grain growth is significantly arrested. Thus, as determined in our study, the grain size from 2 to 2000 pulses changes little, hence the crystallization process occurs essentially via nucleation. Such non-equilibrium conditions lead to constraints in the grain size which have a profound effect on magnetic properties. For example, we observe a significant depression in the Curie temperature for LQ samples ($T_c = 230$ °C) vs 310 °C for the RTA samples. Corresponding resulting average grain diameters are 20 nm and 50 nm for LQ and RTA respectively.

The effect of the temperature gradient during laser exposure is the reason for the results of Fig 3(b). As seen in the figure, nucleation of tetragonal PtMn occurs initially in the middle of the film with amorphous regions remaining in the vicinity of the interfaces. These interfaces provide more effective cooling for the thin film changing the quench rates significantly so that the cooldown curve near the interfaces has a larger quench rate, preventing crystallization occurring at the interfaces. In contrast, the quench rate in the film center crosses the TTT diagram so that the end result is crystallization.

Surface Undulations

Surface undulations generated by laser annealing have been encountered by others, see for example Siegman et al. [6] and Preston et al. [7] As pointed out the characteristic undulation wavelength in our experiments is ~500 nm and the amplitude increases with number of pulses. This periodicity differs by a factor of two from values reported in the literature in which undulations equal to the laser wavelength or to λ/n (n = refractive index) have been described. Such experiments also utilize normal incidence geometries such as ours. We suggest that the origin of the undulations observed in our work are associated with stress relief. As pointed out, the lattice parameter values measured for our LQ samples are virtually the same as those published for bulk values. Nevertheless, as can be expected from raising the temperature of the absorbing film from ambient to or close to its melting point in such short time scales, leads to large volume changes. In addition, crystallization under non-equilibrium conditions is realized with concomitant volume changes. In the time scales involved in our experiments, there is not sufficient time to accommodate such large volume expansions as simple tensile or compressive stress and thus, the stress is released by the formation of such surface undulations. Such an effective stress relief mechanism has been discussed by Stephen et al. [8] for silicon dioxide on silicon substrate system. Further work to quantify the periodicity and amplitude in terms of this process and as a function of laser fluence are in progress.

CONCLUSION

The phase transformation from amorphous to crystalline PtMnSb induced by laser annealing has been investigated with a variety of structural and magnetic techniques. We have established for the first time that the transformation proceeds via the intermediate phase formation of tetragonal PtMn. Temperature and time constraints lead to formation of the C1b phase predominantly via a nucleation process. Evidence for stress relaxation via the formation of surface undulations has been given and the effect of reduced dimensionality on magnetic properties has been described. We suggest that the application of such transient laser annealing techniques be utilized to synthesize unique nanostructures and the formation of metastable phases not accessible under equilibrium annealing conditions.

ACKNOWLEDGEMENTS

This work was partially supported by the NSIC/ARPA contract number MDA972-93-1-0009.

REFERENCES

1. K. Watanabe, **J. Phys. Soc. Japan, 28**, 302 (1970).
2. K. H. J. Buschow, P. G. van Engen and R. Jongerbreur, **J. Mag. Mag. Mat., 38**, 1 (1983).
3. R. A. de Groot, F. M. Mueller, P. G. van Engen and K. H. J. Buschow, **J. Appl. Phys., 55**, 2151 (1984)
4. R. Carey, H. Jenniches, D.M. Newman and B. W. J. Thomas in Proceedings of Magneto-Optical Recording International Symposium '92, (J. Magn. Soc. Japan., 17, Supple. S1, 1993) pp. 290-293
5. E. E. Marinero, **Appl. Sur. Sci., 43**, 117, (1989)
6. A. E. Siegman and P.M. Fauchet, **IEEE J. Quantam Elec., QE-22** , 1384 (1986)
7. J. S. Preston, J. E. Sipe and H. M. van Driel in Interfaces Under Laser Irradiation, edited by Launde, Bauerle and Wautelet (ASI, 1986) pp. 127-136
8. J. Stephen, B. J. Smith and N. G. Blamires in Laser and Electron Beam Processing of Materials, edited by C. W. White and P. S. Peercy (Academic, 1980) pp. 639-644

STRUCTURAL AND MAGNETIC CHARACTERIZATION OF Bi-SUBSTITUTED GARNET ON Si AND GaAs

KEN M. RING, A.L. SHAPIRO*, F. DENG, R.S. GOLDMAN, F. SPADA, F. HELLMAN*, T.L. CHEEKS, K.L. KAVANAGH
Departments of Electrical and Computer Engineering and (*) Physics
University of California at San Diego, La Jolla CA 92093-0407
TAKAO SUZUKI
IBM Almaden Research Center, 650 Harry Road, San Jose, CA 95120-6099

ABSTRACT

Novel material structures that combine magneto-optic (MO) and semiconductor devices have potential applications in monolithic microwave systems and optoelectronics. We have investigated the materials issues pertaining to the film structure, interface uniformity, and magnetic/MO properties of $(BiDy)_3(FeGa)_5O_{12}$ (Bi-DyIG) thin films sputter deposited on Si and GaAs. The rapid thermally annealed films were polycrystalline with a nominal grain size of 20 nm. The magnetic and MO properties were strongly dependent on the type of substrate such that square hysteresis loops and coercivities of 0.1 to 0.9 kOe were observed for Bi-DyIG/Si structures while Bi-DyIG/GaAs structures showed much lower coercivity values (0.03 kOe). A comparison of the magnetic properties, microstructure and substrate composition was carried out with plan-view and cross-section transmission electron microscopy, as well as electron and x-ray diffraction. The results suggest that grain orientation effects, stress, and compositional inhomogeneity due to interfacial reactions or diffusion introduced by the substrate strongly influence the magnetic and MO properties of the films.

INTRODUCTION

Monolithic integration of magnetic and magneto-optic (MO) materials with semiconductor electronics and optoelectronics has potential for a host of novel device applications that include microwave electronics [1,2], sensors [3], as well as information storage and processing [4,5]. Combining MO materials with underlying high speed III-V or Si electronic and optoelectronic device structures offers device opportunities that exploit the properties of both materials. In particular, we have investigated the possibility of monolithically combining MO and optoelectronic devices such as lasers and detectors. To realize an integrated device, compromises in growth conditions and materials compatibility may be necessary to ensure that the properties of each material are maintained.

Integrating MO materials with lasers, for example, places certain restraints on the MO material selection. These criteria include selecting an MO material with a large Faraday rotation, perpendicular magnetic anisotropy, and low absorption at the wavelength of interest. Bismuth substituted garnets satisfy these criteria [6,7]. These materials exhibit large Faraday rotation of about 1-2 °/μm at 633 nm, when sputter deposited, and more transparency than metal MO films such as TbFe [8]. However, the garnet films are typically deposited in an amorphous, non-magnetic state and require a subsequent high temperature (600-700 °C) anneal to obtain a magnetic, crystalline form [9]. During processing, these temperatures could reduce the integrity of the underlying semiconductor. Furthermore the garnet film properties are sensitive to various

41

growth and processing conditions [9], as well as substrate properties such as lattice constant and thermal expansion coefficient [10]. Hence the figures of merit of both the MO material and semiconductor can be influenced by the choice of the substrate material. The following study evaluates this issue by examining the influence of the substrate on the magnetic and MO characteristics of garnet films.

We report on the magnetic and MO properties of $(BiDy)_3(FeGa)_5O_{12}$ (Bi-DyIG) films grown on Si, GaAs, and glass. Rapid thermal annealing (RTA) has been successfully employed to crystallize Bi-DyIG resulting in small grain sizes (20 nm) and good magnetic and MO properties for films on Si and glass. Here, we discuss the relationship between microstructure, interface roughness and magnetic properties and propose possibilities for improved structures based on garnet-semiconductors.

EXPERIMENTAL

The 200-300 nm thick Bi-DyIG films were rf magnetron sputter deposited on Si (100), Si (111), GaAs (100), GaAs (111), and Corning 7059 glass substrates. The garnet target composition was $Bi_{2.0}Dy_{1.0}Fe_{4.0}Ga_{1.0}O_x$. As-deposited films were subsequently annealed using RTA at 650 °C for 3 minutes, with a ramp rate of 125 °C/s. RTA processing results in a lower grain size compared to conventionally annealed films, believed to reduce the read/write noise associated with polycrystalline garnet MO media [11]. Microstructural investigations were carried out using x-ray diffraction (XRD) with Co Kα radiation, as well as plan view and cross section transmission electron microscopy (TEM), and electron diffraction. TEM samples were prepared with mechanical polishing and Ar ion milling. Composition analysis was performed using Rutherford backscattering (RBS). The saturation magnetization and coercivity of the films were measured (in the film plane and film normal directions) with a vibrating sample magnetometer. MO characterization of the Bi-DyIG films was carried out with Kerr or Faraday rotation measurements (at 633 and 840 nm respectively). Faraday rotation measurements were carried out on films deposited on glass, while Kerr measurements were used to study films on the opaque Si and GaAs substrates.

RESULTS

Magnetic and MO Properties of Garnet films

The magnetic and MO properties of Bi-DyIG films deposited on glass, Si, and GaAs are summarized in Table I. The table shows that the saturation magnetization (M_s), coercivity (H_c) and Faraday rotation (Θ_F) of Bi-DyIG deposited on glass compare reasonably well with previously published results [12]. A typical Faraday rotation measurement from this sample shows a square loop indicating perpendicular magnetic anisotropy, shown in Figure 1(a).

Bi-DyIG films deposited on Si were found to possess somewhat smaller coercivities and Kerr rotation values (Θ_k) than films on glass. For both the Si (100) and (111) substrates the films exhibited perpendicular anisotropy and square hysteresis loops. The magnitude of Θ_k varied, however, from 0.2 to 0.7 °/μm for two separate growth runs, presumably due to a subtle difference in preparation conditions. The Bi-DyIG films grown on GaAs substrates exhibited much reduced H_c and Θ_k relative to films grown on glass and Si. Examination of the magnetic properties in and out of the film plane indicated no perpendicular anisotropy for films on GaAs.

Table I: Magnetic and MO Data of Bi-DyIG Films

Substrate	Ms (emu/cc)	Hc (kOe)	Θ (Kerr/Faraday) (°/μm)
Glass	10	1.3	1.2*
Si (100)	60	0.4	0.2-0.7
Si (111)	15	0.2	0.6
GaAs (100)	80	0.03	0.0-0.1
GaAs (111)	100	0.05	None

Note: Kerr measurements were taken at 633 nm, and (*) Faraday measurements at 840 nm.

Microstructure of Films on Different Substrates

The film microstructures were characterized to determine their influence on the magnetic properties. XRD data (Fig. 2) and plan-view electron diffraction indicated that the film texture varied with substrate composition. The Bi-DyIG films on glass exhibited no preferred orientation from XRD and cross section electron diffraction. Plan view electron diffraction supported this, although in thin regions of the TEM sample (hence the top portion of the garnet film) some texturing was evident as the (420) spacing was weak and the (611) spacing was stronger than expected for a randomly oriented film. Considerable dispersion was found in the grain size, which varied from about 10 nm to 100 nm. No second phase lines were detected in the diffraction patterns. Cross-section TEM imaging revealed the garnet-glass interface to be fairly smooth both before and after (Fig. 3(a)) annealing.

For Bi-DyIG films grown on Si, a comparison of the XRD and magnetic data indicated that random orientation correlated with higher Θ_k. The first-run sample deposited on Si had H_c and Θ_k values comparable to the Bi-DyIG film on glass, and had a very similar XRD pattern (Fig. 2). The only difference in XRD data was the observation of a strong (842) reflection in the film on Si, probably due to orientation with the Si (400) planes at nearly the same spacing. Electron diffraction also indicated a fairly random orientation in that all reflections

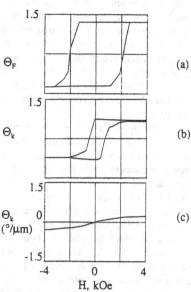

Figure 1: Faraday and Kerr hysteresis loops for Bi-DyIG films on (a) glass, (b) Si (100), and (c) GaAs (100).

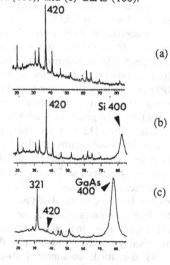

Figure 2: X-ray diffraction patterns of Bi-DyIG films deposited on (a) glass, (b) Si (100), and (c) GaAs (100).

were present; however, the diffraction rings were spotty even when a very large number of grains contributed to the pattern. A grain size of about 20 nm was observed, with less apparent variance than for the film on glass. For the second sample run, the normally prominent (420) reflection was weak and the (800) spacing was dominant. This sample also had a significantly reduced Θ_k indicating a definite connection between film microstructure and magnetic properties.

The Bi-DyIG samples on Si were also examined in cross section. Underneath the fine grain structure formed during annealing are larger crystals (about a micron in diameter) directly atop the Si (Fig. 4(b)). These are of the garnet phase as determined by selected area electron diffraction. The grain boundaries of the large crystals are not evident in plan view TEM unless the sample is ion milled briefly from film side, when the boundaries become easily visible (Fig. 5). The interface between these grains and the Si is fairly abrupt and smooth despite annealing. However there are two very thin (0.5-1 nm) layers, presumably one of which is an oxide, between the Si and garnet. Very faint second phase lines were observed in the diffraction pattern, which have not been definitely associated with any of the features observed in TEM.

In contrast to the Bi-DyIG films deposited on Si and glass, films on GaAs showed low coercivity and no perpendicular anisotropy. The film microstructure also contrasted with that of the Si and glass samples. XRD and electron diffraction data (Fig 2(c)) indicate strong (321) diffraction and very weak (420) diffraction. The grains were slightly smaller than for the other films (15 nm) and fairly uniform in size except for those grains at the Bi-DyIG / GaAs interface itself. No evidence for the mixed-sized grain structure seen in Si-grown films was present. Cross-section TEM did, however, reveal a much rougher interface in the Bi-DyIG / GaAs relative to the film interfaces with Si or glass. Electron diffraction also indicated second phase lines, conceivably reaction products with the substrate.

As deposited, un-annealed Bi-DyIG films on GaAs also were also analyzed. The un-annealed garnet/GaAs interface was different for the two sample runs on GaAs (100), which is not surprising since they had different magnetic properties when

(a) Glass

(b) Si (100)

(c) GaAs (100)   20 nm

Figure 3: Cross-section TEM images of Bi-DyIG films grown on glass, Si (100), and GaAs (100).

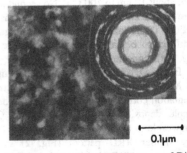

0.1μm

Figure 4: Plan-view TEM image of Bi-DyIG grown on Si (100). The electron diffraction pattern is shown in the inset.

annealed. The first run sample (before annealing) had a thin (10 nm) epitaxial garnet layer atop the GaAs, which was confirmed by plan view diffraction. On top of this were two amorphous layers, the contrast between them presumably due to a composition difference. On the other hand, neither an epitaxial layer nor pronounced contrast bands were seen in the un-annealed sample from the second run. As for films on Si, the variations in growth runs for films on GaAs are attributed to unintentional differences in substrate surface preparation, as well as sputtering and annealing conditions.

Compositional Analysis of Garnet Films

RBS was used to determine the compositional variation within the By-DyIG films before and after annealing. The results showed that after annealing, there was a difference in composition compared to as-deposited films. This variation in composition was most obvious in the Bi-DyIG / GaAs samples which showed changes in the Bi, Dy, and Ga levels. This data suggests that diffusion of species during the anneal is possible which could have resulted in interfacial reaction. As the film magnetic properties are sensitive to composition this may account for some of the reduction in Θ_k and H_c seen for Bi-DyIG films on GaAs.

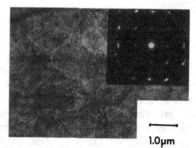

1.0μm

Figure 5: Plan-view TEM image of Bi-DyIG grown on Si (100). A portion of the film has been sputtered off, revealing a large grain structure superimposed on that showed in Fig. 4. The inset shows electron diffraction from one grain.

DISCUSSION

Observations from this study suggest three possible links between the Bi-DyIG / substrate interface and the film magnetic and MO properties. Grain orientation effects, which for films on GaAs exhibited by far the strongest deviation from randomness, are believed to be partly responsible for the reduced H_c of samples on this substrate. In addition GaAs has the closest thermal expansion coefficient to garnet (near room temperature); hence the stress induced during thermal processing may be lower than for Si or glass, reducing the magnetic anisotropy and resistance to domain rotation. Furthermore, in the GaAs case some degree of diffusion or intermixing of Ga may have lead to an effective doping of the garnet film which reduced the H_c or altered the compensation temperature.

In contrast, for Bi-DyIG films grown on Si, the three parameters discussed (film texture, stress, and composition inhomogeneity) were all more similar to the samples on glass, and the magnetic properties were much closer as well. Effects such as film texture [13] and movement of atoms to different lattice sites during processing [8] have been determined to affect magnetic properties for other material systems. Since several phenomena appear to be at work here, particularly for films on GaAs, isolating each one has thus far proved elusive. Variations between different growth runs on the same substrate underscore the sensitivity of the film to the growth surface.

Investigations are underway into use of diffusion barrier materials to reduce the sensitivity of the Bi-DyIG films to the substrate, in order to provide greater control in growing MO films with intact properties on Si and GaAs. This will also be important to realization of the integrated semiconductor devices whose operation would also be very sensitive to interface reactions.

CONCLUSION

The growth of Bi-DyIG films on Si and GaAs has been carried out along with the evaluation of some basic materials issues pertaining to garnet MO media / semiconductor integration. The garnet microstructure and magnetic properties are sensitive to the choice of substrate as well as the subtle variations in processing conditions. Some combination of reduced stress, grain orientation effects and possible interface reaction leads to low coercivities (about 0.03 kOe) in samples grown on GaAs. Films on silicon have comparable magnetic properties and microstructure relative to films on glass, but exhibit wide variations between growth runs. Samples on glass as expected have H_c above 1 kOe. Further work should lead to films with improved figures of merit on GaAs and improved process control for films on Si. As applications requiring MO elements integrated with circuitry and optoelectronics expand, this could be one of the implementations to consider.

ACKNOWLEDGEMENT

This work was supported by a NSF (DMR-PYI) award.

REFERENCES

1. S. Makio, S. Takeda, S. Sakano, N. Chinone, Elec. and Commun. in Jpn. Part 2 **74**, 50 (1991).
2. K. Okubo, M. Tsutsumi, Elec. and Comm. in Jpn. Part 2 **74**, 40 (1991).
3. M.N. Deeter, G.W. Day, R. Wolfe, V.J. Fratello, IEEE Tran. on Magn. **29**, 3402 (1993).
4. C.-J. Lin, Mat. Res. Soc. Symp. Proc. **150**, 15 (1989).
5. J. Cho, S. Santhanam, T. Le, K. Mountfield, D.N. Lambeth, D. Stancil, W.E. Ross, J. Lucas, J. Appl. Phys. **76**, 1910 (1994).
6. P. Hansen, C.-P. Klages, K. Witter, J. Appl. Phys. **60**, 721 (1986).
7. M. Okada, S. Katayama, K. Tominaga, J. Appl. Phys. **69**, 3566 (1991).
8. T. Suzuki, J. Appl. Phys. **69**, 4757 (1991).
9. K. Shono, H. Kano, S. Kuroda, Fujitsu Sci. Tech. J. **26**, 157 (1990).
10. J.H. Lee, W.K. Choo, Y.S. Kim, D.W. Yun, J. Appl. Phys. **75**, 2455 (1994).
11. W.R. Eppler, B.K. Cheong, D.E. Laughlin, M.H. Kryder, J. Appl. Phys. **75** 7093 (1994).
12. A. Azevedo, S. Bharthulwar, W.R. Eppler, M.H. Kryder, IEEE Trans. on Magn. **30**, 4416 (1994).
13. P.F. Carcia, M. Reilly, Z.G. Li, H.W. van Kesteren, IEEE Trans. on Magn. **30**, 4395 (1994).

Part II

Growth, Structure and Interfaces

Atomic Scale Engineering of Superlattices and Magnetic Wires

J. Camarero, J. de la Figuera, L. Spendeler, X. Torrellas*,
J. Alvarez*, S. Ferrer*, J.J. de Miguel, J.M. García,
O. Sánchez, J.E. Ortega, A.L. Vázquez de Parga,
and R. Miranda.
Dpto. de Física de la Materia Condensada, C-III
Univ. Autónoma, Cantoblanco, E-28049 Madrid, Spain
* E.S.R.F., B.P. 220, F-38043, Grenoble, France

1   Introduction.

In the past years artificially-structured materials have been grown with an increasing degree of sophistication due to steady progress in our ability to control growth processes down to the atomic level. These materials have yielded new physical properties due to the confinement of electrons in less than three dimensions. Thus, the confinement of electrons in two-dimensional (2D) metallic superlattices has resulted in oscillatory magnetic coupling with an associated oscillatory giant magnetoresistance (GMR). New properties are expected when the electrons are further confined to one dimension (1D) of free motion in the structures known as quantum wires. In this report we briefly describe two recent examples of atomic-scale engineering of materials. In the first case a surfactant is used to purposely modify the structure of magnetic/non magnetic superlattices. The second example illustrates a further reduction in dimensionality obtained by modifying the substrate onto which the growth takes place: the fabrication of 1D magnetic quantum wires on vicinal surfaces.

2   Experimental.

The experiments have been carried out in three different Ultra High Vacuum (UHV) chambers: a) A Molecular Beam Epitaxy (MBE) system equipped with Thermal Energy Atom Scattering (TEAS) and Low Energy Electron Diffraction (LEED), b) a system equipped with Scanning Tunneling Microscopy (STM) and LEED and c) a six-circle X-ray diffractometer placed at the ID3 beamline of the European Synchrotron Radiation Facility (ESRF) at Grenoble. The samples were Cu(111) flat and vicinal surfaces, cleaned in-situ by ion sputtering and annealing. In all cases, prior to the growth, the cleanliness was checked by Auger Electron Spectroscopy (AES) and the crystalline perfection by LEED, TEAS and STM respectively. The metals evaporated, Pb, Co, Cu and Fe were deposited from different electron bombardment and thermal sources at low rates (Å/min) under residual pressures of the order of 10^{-10} Torr.

49

Mat. Res. Soc. Symp. Proc. Vol. 384 ° 1995 Materials Research Society

3 Results and Discussion.

3.1 Surfactant effects on the growth of {Co/Cu}(111) super-lattices.

It has been shown that Co can be grown by MBE at room temperature (RT) in the metastable fct phase on an adequate substrate such as Cu(100) [1, 2]. The lattice mismatch being moderate ($\sim$ 2%), the Cu surface provides a template onto which the Co atoms lock, accommodating the forced lateral expansion with a contraction of their interlayer spacing. This structure can be maintained up to rather large thicknesses because the square symmetry of the Cu(100) face is very different from the hexagonal one of bulk hcp-Co, thus making the transition very unlikely [1]. Accordingly, Co/Cu superlattices grown along the (100) direction are strictly fcc and have shown antiferromagnetic (AF) coupling [3] and oscillations in the coupling as a function of the Cu spacer [4], reproduced by several groups [5, 6].

On the contrary, experimental studies of epitaxial films of Co grown on Cu(111) and (111)-oriented Co/Cu superlattices have been plagued with contradictory reports regarding not only their electronic and magnetic properties, but even their crystalline structure. Polycrystalline (111)-textured sputtered multilayers [7, 8] have shown MR oscillations and correlated oscillatory magnetic coupling, while samples grown by MBE either do not display oscillatory coupling [9, 10] or do not present AF coupling at all [11]. In the rare occasions in which oscillatory coupling was indeed observed, a significant fraction of the sample was ferromagnetically-coupled independently of the thickness of the Cu spacer layer [12].

Using a combination of different techniques (STM, LEED, and Surface X-ray Diffraction -SXRD-) we have performed a thorough characterization, of the structural evolution of the Co films in real time during growth. We have found that the origin of these contradictory results can be traced down to the existence of defects that result in the formation of channels between crystallites, thus preventing the formation of continuous spacer films [13]. Across these channels, magnetic bridges can develop providing a direct FM coupling between adjacent Co films [13]. It was further suggested that the Cu film grown on Co was composed of fcc-twinned crystallites [13]. This can be directly inferred from experimental data such as those depicted in Fig. 1. This figure shows LEED I-V curves measured on a clean Cu(111) substrate and on a trilayer formed by depositing 3 ML of Co on the substrate and then covering it with another 3 ML of Cu. Unlike the curves corresponding to the fcc substrate, which display the typical three-fold symmetry, those of the trilayer appear to be six-fold. Although this could be taken as an indication that the film is already hcp, a careful study of these I-V curves shows that they can be accurately reproduced just by incoherently adding the intensities of the two inequivalent curves of the clean substrate (the (1,0) and (0,1) families). This means that the film is composed of twin-fcc crystallites larger than the coherence length of a typical LEED instrument ($\leq$ 100Å), that cannot match and coalesce laterally. Therefore, this results provide an explanation for the lack of AF coupling observed in MBE-grown {Co/Cu}(111) multilayers. It seems reasonable to ascribe the different behavior of Co films grown on the Cu(100) and Cu(111) faces to the different symmetry of these two faces. In fact, the fcc-(111) face of Cu can only be distinguished from the (0001) basal plane of hcp-Co by the stacking sequence. It is therefore expected that the transition of Co from fcc to hcp happens very easily and has some influence on the twinning of the Cu spacers.

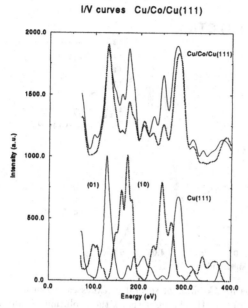

Figure 1: LEED I-V curves for a clean Cu(111) surface (lower panel) and a 3ML Cu/3 ML Co/Cu(111) trilayer (lower panel). The latter curve can be directly obtained by incoherently adding the curves of the two inequivalent beams ((1,0) and (0,1)) of the bare substrate, indicating that the Cu film consists of twin-fcc crystallites.

In order to study this problem we resorted to Surface X-Ray Diffraction (SXRD) experiments performed at the ID3-BL7 beamline of the ESRF. SXRD is a well-established technique for surface structure determination combining a very high sensitivity for structural defects with an easy interpretation of the experimental results, since kinematic scattering theory can be applied. Additionally, the very high photon flux available at the ESRF makes it possible to obtain a wealth of data *in real time* during deposition. This is crucial for a correct determination of all the relevant parameters involved in the growth process. Our experiments consisted of basically two types of measurements: first, monitoring the evolution of the diffracted intensity at selected points in k-space in real time during evaporation of the growing species, and second, measuring complete scans along the most relevant Crystal Truncation Rods (CTR's) at fixed coverages, after having stopped growth. By fitting the latter measurements with a suitable model, it is possible to accurately determine the crystallographic structure of the films, whereas the former ones allow us, by adequately choosing the point in reciprocal space where the intensity is going to be measured during the time-scans, to follow the time evolution of a given structural characteristic of particular importance; in our case, we were interested in studying the formation of stacking-faults (SF's) and twins in the epitaxial films of fcc-Co grown on Cu(111), and the transition from the fcc to the hcp phase as a function of the Co layer thickness.

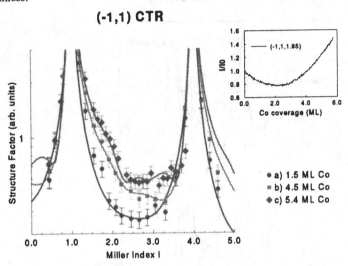

Figure 2: (-1,1) rod scans for Co films of several thicknesses grown at RT on Cu(111): (a) 1.5 ML; (b) 4.5 ML; (c) 5.4 ML. The inset shows the evolution of diffracted intensity at the twined fcc position

Fig. 2 shows scans measured on the (-1,1) CTR after depositing different amounts of Co. Bulk Bragg maxima in the CTR at Miller indexes l=1 and l=4 correspond to the fcc stacking while features at l=2 reveal the appearance of SF,s. Several aspects deserve being noted: first, as said before no SF's appear for the first two monolayers of Co. This information can be obtained from the fits to rod scans such as the one labeled (a), and also from timescans such as the one depicted in the inset. The initial decrease in intensity is due to the increasing disorder at the surface caused by the substrate etching and formation of bilayer islands with pure fcc stacking. At this stage a considerable amount of intermixing of Co into the Cu substrate takes place [14]. It appears therefore that the fcc structure of this bilayer is stabilized by the presence of Cu in the film. Above 2ML of Co, the segregation of Cu to the growing film is frozen and further growth occurs by single-layer islands that start to introduce SF's in the Co film leading to local hcp stacking. This is revealed by the asymmetric shape of the CTR in Fig. 2b and the increase in intensity in the inset above 2ML. Hence, on top of the faulted islands growth continues as hcp-Co. The transition to hcp of the remaining fcc fraction takes place in the same way over subsequent layers; an example of the structure of a thicker ($\sim$ 5.4 ML) Co film is shown in Fig. 2c. For this coverage, nearly 50% of the fourth layer is already hcp, whereas another 25% is forming an fcc-SF. The film therefore is not homogeneous, because different areas of the sample make their structural transition at different heights above the substrate surface. As a consequence, for instance, our fit indicates that all three possible hcp stackings (-ABAB-, -BCBC- and -ACAC-) are present in the film, thus preventing the formation of a continuous layer. Nucleation of Cu islands on top of hcp fractions with different stacking sequence seems thus to be the reason for the twinning of the Cu overlayers [15].

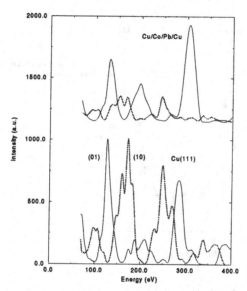

Figure 3: LEED I-V curves for selected diffraction spots of (a) the clean Cu(111) surface, and (b) a 3ML Cu/3 ML Co trilayer grown on a Pb-precovered Cu(111) surface. The three-fold symmetry of the latter curves indicates pure fcc-stacking, without detectable formation of twins.

Having determined successfully the origin of the problems found when working with this system is not enough if high-quality superlattices are to be grown: there remains the task of finding a method that allows us to avoid these undesirable results. To this end, we resorted to using Pb as a surfactant to try to modify the natural tendencies of both materials. Pb is a low-surface-energy metal, and is therefore likely to remain on top of the growing film without being incorporated into it, a basic condition that must be fulfilled by any surfactant agent to be of any practical use. In fact, we have observed that a monolayer of Pb deposited on top of the clean Cu(111) substrate remains on the surface after having subsequently deposited more than 20 ML of both Co and Cu. But most importantly, Fig. 3 shows how the presence of this Pb layer prior to the growth of a Cu/Co trilayer prevents the formation of fcc twins. The LEED I-V curves depicted are equivalent to those shown in Fig. 1 , but now it is evident that the trilayer grown with Pb has three-fold symmetry, with a structure rather similar to that of the Cu substrate although the shifts in the peak positions indicate relaxations of the interlayer spacing. In any case, no twin fraction is detected for trilayers grown using this method. This effect persists for additional Cu/Co trilayers, and can therefore be used to fabricate high-quality superlattices [15].

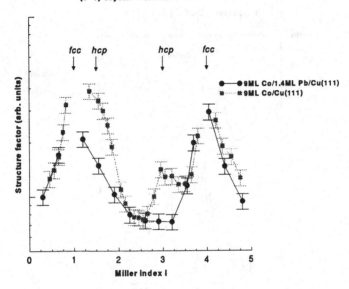

Figure 4: (0,-1) rod scans for Co films of 9.5 ML, grown on Cu(111) with and without using Pb as a surfactant. While in the former case the fcc structure is preserved, in the latter a large fraction of the film has already converted to hcp.

The mechanism of actuation of the surfactant at the atomic scale is not yet completely clarified, but we have determined that the main effect consists of delaying the transition of the Co film from fcc to hcp. Cu layers deposited on top of purely-fcc stacked Co films continue to grow with the correct stacking sequence, producing films of much higher structural quality. This effect of Pb is demonstrated by the SXRD data presented in Fig. 4: there, two scans of the same CTR are shown, corresponding to 9.5 ML Co films grown with and without precovering the Cu substrate with 1.4 ML of Pb. It is clear from the figure that while the film grown without Pb contains already a large fraction of hcp structure (indicated by the increasing intensity diffracted near the values 1.5 and 3 of Miller index l), the one grown with Pb has retained the fcc structure. In summary, these data clearly indicate that the use of surfactants for the growth of metallic superlattices is a valuable method that is bound to concentrate increasing attention in the near future.

3.2 Magnetic Quantum Wires and Lateral Superlattices.

The fabrication of quasi-1D materials is recently attracting strong interest. It has been reported that Ni and Co nanowires (600Å in diameter) exhibit perpendicular magnetization and enhanced coercivity [16]. Co/Cu multilayers shaped in the form of 400-800Å diameter nanowires show MR values of 15% at RT [17, 18]. Conventional vertical multilayers show GMR effects in both geometries: current in the plane and current perpendicular to the plane (CPP). For the first case the effect vanishes when the layer thickness surpass the electron mean free path ($\sim$ 100Å), while for the CPP geometry the number of interfaces sampled is much larger since the relevant length is now the spin diffusion length ($\sim$1000Å). Accordingly, a stronger GMR effect is expected for the CPP geometry. The measurements in this geometry, however, are difficult to perform. The growth of lateral superlattices on the surface of vicinal substrates may offer new possibilities in this respect, since an alternate array of magnetic/non magnetic stripes can be tested with a current in the surface plane but perpendicular to the stripes, that will sample all the interfaces. We have explored this approach [19] which is schematically illustrated in Fig. 5.

stepped surface

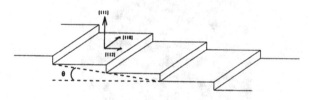

+ Fe, Co

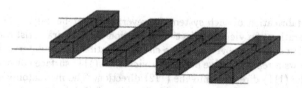

Figure 5: Schematic drawing of a lateral superlattice of magnetic wires grown by step flow during thermal deposition on vicinal substrates.

Vicinal substrates with varying terrace widths can be prepared by intentional misorientation of the crystal at selected angles out of the normal. For certain step orientations, terrace widths and deposition rates, evaporation of magnetic materials (Fe, Co) may result in preferential decoration of steps and formation of regular arrays of wires as discussed elsewhere [19].

Figure 6: STM images (270 Å wide) recorded with sample voltage, $V_s = -0.3V$ and tunneling current i=0.3 nA. Above: Clean Cu(111) 4.5° vicinal surface. Below: 0.2 ML of Fe deposited on Cu(111) 4.5°.

The actual fabrication of such system, however, may not be easy. The first task involves the preparation of a vicinal surface. Observation of the characteristic splitting of the LEED spots is not enough. A real-space characterization is mandatory. The upper panel in Fig. 6 shows a representative image of a vicinal Cu(111) surface cut nominally at 4.5° away from the (111) direction into the (112) direction. The monoatomic-high steps are aligned along [110] and display considerable "frizziness" [20] due to fast motion along the steps of atoms at the kinks. Images at smaller magnifications proof that the steps are straight over distances of microns, i.e., not pinned by impurities or curved by excessive annealing as commonly observed. Carefully prepared samples show a distribution of terrace widths that is a narrow Gaussian centered around a width of 22Å indicating that the actual miscut angle is 5.5°.

Deposition of 0.2 ML of Fe at RT onto such a vicinal surface produces images like the one reproduced in Fig. 6 (lower panel). Oblong islands 2-4Å-high, 15Å wide and 25-50Å long appear close to the steps edges. Frequently, bilayer-high steps are found. In many cases the islands are separated as a consequence of a significant etching of the terraces. The islands cover ~50% of the surface area. Since this is much larger than expected from the deposited amount of Fe (0.2ML), it indicates that the additional material removed from the clean Cu substrate is incorporated into the islands. Thus, the islands probably consist of a fcc-FeCu alloy. The islands are isolated from each other and have an average area of ~ 400Å², below the size of Fe islands on W(110) at the superparamagnetic limit (130 nm²) [21]. Therefore the islands probably behave magnetically as a dense ($\approx 1.4 \pm 0.2 \cdot 10^{13}$ islands·cm^{-2}) array of superparamagnetic "particles" lacking magnetic order, a system with potentially interesting properties.

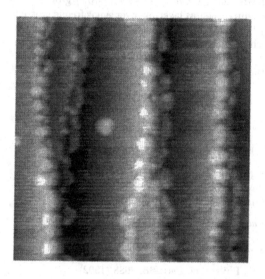

Figure 7: STM image (500Å wide) taken with $V_s = -0.6V$, and i=0.5 nA on a Cu(111) 2.5° surface containing multiatomic steps with 0.4 ML of Fe deposited.

We have observed evidence for deposition-induced etching of the substrate for RT-evaporation of Fe on most vicinal Cu(111) surfaces showing a regular array of monoatomic steps. Vicinal surfaces cut at smaller angles display larger terraces but, unfortunately have a tendency to yield partial step bunching, i.e. that the steps are grouped into step bunches including multilayer-high steps and separated by terraces larger that expected on the basis of the miscut angle. On these surfaces with partial step bunching the intermixing reaction seems to be severely reduced. Fig. 7 shows an STM image of a Cu(111) 2.5° surface onto which 0.4ML of Fe have been evaporated. Terraces ~100Å wide are separated by multiatomic steps grouped together in bunches of 3-4. Islands that almost coalesce decorate the steps. The islands are 4Å high and cover 20% of the area, i.e. they contain mainly the evaporated material.

Scanning Tunneling Spectroscopy (STS) can be used to identify the chemical nature of features on the surface. $\frac{dI}{dV}(V_s)$ curves show indeed that these islands and wires do not contain a significant fraction of Cu. Their magnetic properties, however, remain to be studied. In this respect it is worth mentioning that monoatomic-high Fe-stripes 200Å -wide grown of W(110) [21] show Curie temperatures, T_c, that scale inversely with their width in a manner similar to the thickness dependence of T_c in thin films [22].

In sumary, the fabrication of lateral superlattices is not yet as advanced as the production of vertical superlattices. Growth at steps takes place in the 1D analogs of the well known modes of growth at surfaces: intermixing and surface alloy formation, Stranski-Krastanov (islands) growth and layer-by-layer (row-by-row) growth. The latter, highly desiderable, may be uncommon just as layer-by-layer is not frequent on extended surfaces. Neverheless, the ability to grow artificial magnetic structures of reduced dimensions offers unique scientific and technological possibilities that are worth exploring in the next future.

This work has been financed by the CICyT under projects PB91-0929 and PB93-0271. JJM wishes to thank the Comunidad Autónoma de Madrid (CAM) for help with travel expenses.

References

[1] L. González et al., Phys. Rev. B **24**, 3245 (1981).

[2] J. R. Cerdá et al., J. Phys: Condensed Matter **5**, 2055 (1993).

[3] A. Cebollada et al., Phys. Rev. B **39**, 9726 (1989).

[4] J. J. de Miguel et al., J. Magn. Magn. Mat. **93**, 1 (1991).

[5] Z. Qiu. et al., Phys. Rev. B **46**, 8659 (1992).

[6] M. T. Johnson et al., Phys. Rev. Lett. **68**, 2688 (1992).

[7] S. S. P. Parkin, R. Bhadra, and K. P. Roche, Phys. Rev. Lett. **66**, 2152 (1991).

[8] D. H. Mosca et al., J. Magn. Magn. Mat. **94**, L1 (1991).

[9] M. T. Johnson et al., Phys. Rev. Lett. **69**, 969 (1992).

[10] G. R. Harp et al., Phys. Rev. B **47**, 8721 (1993).

[11] M. T. Kief and J. W. F. Egelhoff, Phys. Rev. B **47**, 10785 (1993).

[12] A. Schreyer et al., Phys. Rev. B **47**, 15334 (1993).

[13] J. de la Figuera, J.E. Prieto, C. Ocal and R. Miranda, Phys. Rev. B **47**, 13043 (1993).

[14] J. de la Figuera et al. (unpublished).

[15] J. Camarero *et al.*, Phys. Rev. Lett. **73**, 2448 (1994).

[16] T. M. Whitney *et al.*, Science **261**, 1316 (1993).

[17] L. Piraux *et al.*, Appl. Phys. Lett. **65**, 2484 (1995).

[18] A. Blondel *et al.*, Appl. Phys. Lett. **65**, 3019 (1995).

[19] J. de la Figuera *et al.*, Appl. Phys. Lett. **66**, 1006 (1995).

[20] M. Poensgen *et al.*, Surf. Sci. **274**, 430 (1992).

[21] H. J. Elmers *et al.*, Phys. Rev. Lett. **73**, 898 (1994).

[22] C. M. Schneider *et al.*, Phys. Rev. Lett. **64**, 1059 (1990).

NMR STUDIES OF BULK AND INTERFACE STRUCTURE IN CO BASED MULTILAYERS

P. PANISSOD*, J.P. JAY*, C. MENY*, M. WOJCIK** AND E. JEDRYKA**
* Institut de Physique et Chimie des Matériaux de Strasbourg, CNRS-ULP
23 rue du Loess, F-67037 Strasbourg, France
** Institute of Physics, Polish Academy of Sciences
Al. Lotnikow 32-46, 02668 Warsaw, Poland

ABSTRACT

Owing to the sensitivity of the hyperfine field to the topological and chemical environment of the probe nuclei, NMR spectra can be considered as histograms of the short range order ruling the structure of the material under investigation. Complementary to diffraction techniques this gives a local insight on the structure in the direct space. We review recent structural investigations of cobalt layers imbedded in Co/X multilayers and particularly of buried interfaces. Special attention has been given to the way intermixing takes place at the interfaces as its influence on the multilayer properties may be of considerable importance. Co/Cu multilayers, a case of weakly miscible elements, have been specially investigated owing to their GMR properties. But also cases of solid solution forming elements (Co/Ru or Co/Cr) or compound forming elements (Co/Fe) have been thoroughly studied. The latter case, which shows a stabilization of a bcc Co phase, will be discussed against the bulk alloy phase diagram.

I. INTRODUCTION

The influence of the detailed structure of metallic multilayers and sandwiches on their magnetic and transport properties is well accepted. Bulk structure of the individual layers and bulk defects, do influence magnetic and magnetotransport properties but even more critical is the structure of the interfaces. This makes structural studies of buried interfaces highly desirable since properties are strongly dependent on interface roughness or compositional intermixing. The sensitivity of NMR to the local environment of atoms (number and nature of the neighbors) and to the site symmetry can be used to study the local atomic structure of individual layers and the interface topology of metallic multilayers and superlattices. The structural information at atomic scale, as given by NMR, complements that of standard structural measurements like X ray diffraction and electron microscopy observations. In terms of probed distances and element selectivity NMR can be compared to the EXAFS technique: while, contrary to EXAFS, it does not yield quantitative information about distances, it provides much more detailed information about local chemical configurations. The NMR spectrum reflects the occurrence probability distribution of all nearest neighbor (NN hereafter) configurations in a sample (each configuration giving rise to a characteristic line in the spectrum). This distribution can be compared, in turn, to those which would result from various model structures of interfaces (from a perfect abrupt interface to strongly interdiffused interfaces or sharp interfaces containing monoatomic step defects or discontinuous, granular interfaces). From such a comparison, concentration profiles, densities of step defects, sizes of grains or islands can be evaluated thus characterizing the interface roughness at atomic distance scale. The method has been developed and applied mostly to cobalt based multilayers. We present here some studies which, in the bulk state, cover the cases of non miscible metals, solid solution forming metals and compound forming metals. Before we survey these examples, we

61

present first an introduction to the analysis of the NMR spectra and to various structural models for the interfaces that have been checked against the experimental results.

II. METHODS: SPECTRUM ANALYSIS AND INTERFACE MODELS

Typical Co NMR spectra observed in multilayers are shown in Fig.1. All spectra are normalized to the interface area of the samples so that the interface components of the spectra are superimposed, whenever the interface structure and composition are similar through samples, whereas contributions from the bulk increase proportionally to the Co layer thickness. Spectra can be coarsely separated into two parts: a main structure between 215 MHz and 230 MHz corresponding to bulk Co (fcc Co at 217 MHz, hcp Co at 225-228 MHz [0], stacking faults in between) and a set of lines (called satellites) or a tail below 200 MHz corresponding to nuclei lying in the interfacial regions (where Co and the other element are nearest neighbors). A perfect interface should yield, beside the main bulk line, a single satellite reflecting the unique environment of Co in the interface plane. This is obviously not the case here and several pieces of qualitative information can be readily drawn out of such spectra:

In the Co/Cu series shown, spectra are nearly superimposed in the low frequency range which corresponds to the interface. This shows that the interface topology does not depend on the Co layer thickness. It is often observed, in series of samples, that the interfaces of the thinnest Co layers look different, with a larger spectral intensity as compared to the other samples. The effect is weak here, though. It can be explained by misfit strains which may lead to a different structure for Co in thin layers. Another explanation may be that, from place to place, rough and/or diffused areas start to merge from both sides of the Co layer as it gets thinner: this modifies the Co environment distribution and the spectrum shape.

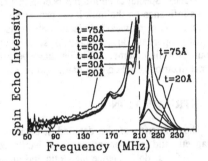

Fig.1. *Examples of Cobalt NMR spectra recorded in a series of Cu/Co$_t$ /Cu/NiFe/FeMn spin valves. The left part shows the interfacial component magnified 20 times with respect to the bulk components on the right.*

The example spectra exhibit a weak but well defined satellite line at 165 MHz, corresponding to Co with 3 Cu NN's at perfect (111) interfaces. However, the intensity of this satellite is weak and the extended tail which is observed at low frequency is typical of diffuse interfaces. This suggests that the interfaces are composed of a small number of islands with flat (111) oriented surfaces separated by rough and/or mixed areas.

Fig.1 also shows a significant increase in the spectral intensity between 200 and 210 MHz, as the Co thickness increases. Although this frequency range corresponds to Co atoms at the interface (with 1 Cu NN), the increase happens to be linear with t_{Co}. To explain this spectral feature, it is thus necessary to assume here an extra contribution from Co atoms in the bulk of the layers. Since the frequency of the extra intensity is largely away from fcc and hcp Co, it can only be attributed to Co located at the vicinity of bulk defects. Chemical analysis ruling out the possibility of bulk impurities, it is then concluded that the extra intensity arises from Co located at the boundaries of numerous thin columnar grains. The proportion of these sites yields the surface to volume ratio of the grains and, assuming smooth hexagonal columns, an average column diameter of 60 Å can be deduced. This compares well with the 100 Å irregular columnar grains observed on TEM cross section micrographs. In other samples, with much larger grain diameters, the NMR spectra show no thickness dependence of the signal intensity in this frequency range.

The last qualitative observation on the spectra concerns the bulk crystallographic structure of the Co layers. The high and sharp peak at about 217 MHz and the weak intensity at 228 MHz show that Co is primarily fcc but the considerable amount of intensity between 220 and 225 MHz indicates the presence of numerous stacking faults. Moreover the relative amount of Co in hcp environments (high frequency contribution) increases with the Co thickness. It is concluded that Co grows fcc first on fcc Cu but, with increasing Co thickness, it tends to recover its stable hcp phase through an increasing number of stacking faults.

One step further in the analysis is made by fitting the spectrum to a set of gaussian lines from which the amount of Co atoms in each specific spectral region is deduced. Such coarse analysis yields the proportions of Co involved in the two phases and the stacking faults of the bulk of the layers. The amount of Co in other spectral regions (i.e. in specific structural objects) are also quantified this way as for the grain boundaries mentioned above. Usually the thickness dependence of the various intensities is needed to discriminate between interfacial (thickness independent) and bulk quantities. From the remaining intensity in the low frequency range, the proportion of Co located in the interface regions of the multilayers is also deduced: for a perfect interface its value should correspond to 1 monolayer per interface, its usually larger value shows immediately the amount of admixture at the interfaces. In granular systems the value of the interface/bulk ratio is also used to estimate the average grain size [1].

Modeling of the interface spectral shape does provide more quantitative information about intermixing at the interfaces (short range roughness). The basis for the simulation of a spectrum is the shift of the Co NMR frequency resulting from the substitution of an alien element for Co in the NN shell [2]. This shift gives rise to a succession of satellites to the bulk line, corresponding to Co atoms with 1, 2, 3, ... alien neighbors, the relative intensities of which are the occurrence probabilities of such NN configurations. The spacing between satellites is generally regular but depends on the foreign element; it is estimated independently on reference bulk alloys of known structure (fully disordered and/or ordered). Hence, simulated spectra consist of a sum of L primary lines (L=13 in compact structures), arising from Co nuclei surrounded by L-n Co and n aliens in their NN shell. Lines are assumed gaussian like. Their spacing is primarily given by the reference study but a limited shift of the lines must be allowed since they are rarely exactly at the same frequency as in the references. This shift arises from differences between the electronic structure at the metal-metal interface and in the reference alloys: strain effects [3], or a different average magnetization in the mixed region, or the anisotropic neighbor distribution in an interface [4] may be responsible for the difference (such effects are illustrated in the next section). The width of the lines is treated as a secondary free parameter of the fitting procedure.

It is usually easy to fit the interface spectra with a set of independent intensities for the satellite lines but the resulting set of configuration probabilities that is obtained after refinement is not directly informative. An alternative way is to build a model (topological and chemical) of the interface structure where the basic parameters are distances and concentrations. From such a model, configuration probabilities (satellite intensities) are deduced which are used to reconstruct the spectrum. The spectral refinement procedure is applied to the structural parameters instead of the line intensities. The fitting procedure concludes about the applicability of the model and, if it can be accepted, its refined parameters give a direct insight on the short range order at the interfaces.

When the coarse analysis of the spectra suggests that interfaces are sharp (weak admixture between the two elements), the first interface model one can think of is a model where the interfaces defects or the interface roughness consist only in steps which have a monoatomic height [3,5]. Variants of such a model consider that the single mixed monolayer is built of patches of the two elements. Given the model, the probability of occurrence of the various neighbor configurations in the interface can be computed as function of a few characteristic lengths, diameter of patches or distance between steps. These lengths are the main parameters of the model which, if applicable, characterizes the nanostructure of the interface. In the illustrations below two parameters are used which describe, in plane, the density of steps (d: average spacing) and their straightness (l: average straight length). Monoatomic step models apply to situations where atoms of the two species are clustered in one mixed interface plane, deviations from the models become obvious as soon as there is a significant probability for atoms of any element to be isolated within a shell of the other species. Such event is more closely handled by the diffused interface model which is presented next. It must be pointed out here that NMR measurements provide information essentially about the distribution of NN configurations which means that the length scale at which the technique characterizes the interface roughness is about two atomic distances. As a consequence what is called here a monoatomic step interface includes any oblique or wavy interface as long as the slope is well below 1/2: to NMR this appears still as a sharp interface, monolayer thick, whereas X rays would feel it thicker.

When there is obviously much more than one mixed plane at each interface, one can use a general model which introduces a concentration profile through several monolayers [5,6]. The two elements are assumed distributed at random in the mixed atomic planes (i.e. the interfaces are built from successive two dimensional random alloys). Using a binomial law distribution of atoms in plane, the configuration probabilities and the spectral shape are fully defined by the concentrations in each atomic layer which are the free parameters of the model. The refinement against the experimental spectrum yields the interface profile with atomic resolution.

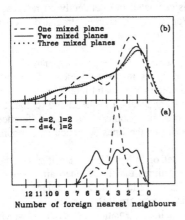

The number of parameters can be reduced by constraining the profile to an analytical function. For example, a linear profile has only one adjustable parameter: its slope or accordingly, the number of mixed planes within an interface region. Of course in real world the distribution of local configurations may differ from the binomial law which is used here but it is hard to develop such a model which would take into account both a concentration profile along the growth axis and the detailed in-plane structure of the interface. Such attempts are described below.

In Fig.2 are presented the expected shapes of the interfacial part of the NMR spectrum for these two models and several values of the pertinent parameters (d and l are expressed in atom spacing units) [5]. The

Fig.2. Simulations of the interface spectra using the step defect model (a) and the diffused interface model (b). Spectra normalized to unit area.

shapes are quite different between the step interface model and the diffused model which shows that the two types of interface structure can be easily discriminated. Among the diffused models, differences can be observed up to an interface thickness of 3 monolayers. For 3 mixed layers and above the spectral shape is not sufficient to discriminate between the models but, of course, the number of mixed layers is given by the relative intensity of the interfacial tail compared to that of the full spectrum (coarse analysis). Examples of spectral refinement using these models are given later in the paper.

These two basic models fail to describe the observed spectra when the interface admixture is inhomogeneous i.e. when parts of the interface are close to perfection or at least sharp, while others are strongly mixed. In such cases a correct distribution of signal intensities may be obtained assuming an interface model where islands of pure elements are surrounded by mixed, alloyed regions. The pure patches are assumed to have a perfect interface with the other element while the mixed parts are treated as an homogeneous alloy or like in the diffused model. This model combines simplified versions of the two first with a supplementary parameter which measures the surface area or volume ratio of the two regions (see Co/Fe).

III. EXPERIMENTAL EXAMPLES

Non Miscible Elements: Co/Cu multilayers and sandwiches

Here we summarize NMR studies of the local atomic structure of Co layers in Co/Cu multilayers and spin valves (1 Co layer) from various origins. Different fabrication techniques have been used to grow the multilayers (MBE or e-gun UHV evaporation, RF or DC, magnetron or diode sputtering), which result in quite different spectral shapes of the Co NMR. The differences arise from three main classes of structural differences between samples: an admixture of fcc and hcp phases in various proportions, a columnar growth with various grain diameters and various degrees of Co-Cu admixture at the interfaces. The bulk Co layer structure (stacking, stacking faults and other defects) as well as the Co/Cu interface topology was investigated. More details on samples and NMR results are given in references [3,5,7-9].

Concerning the bulk structure of the Co layers and to summarize the conclusions of the studies: (i) copper favors the fcc structure, upon Cu layers, Co grows initially with the fcc structure and, at constant Co thickness, the thicker the Cu layers the larger is the fcc Co content (a few exceptions of hcp growth of Co onto Cu have been observed though [10]); on the contrary, (ii) large Co thickness and strong (111) growth textures favor the most stable hcp stacking; (iii) large deposition rates favor numerous columnar grains of small diameters.

The results of the study about the interface structure of the layered systems are summarized in Fig.3. Significant differences between samples can be observed at first sight. These differences have been translated in terms of interface roughness by comparison with model spectra like shown on Fig.4. Samples are sorted, from top to bottom, in order of increasing interface roughness (amount of Co-Cu admixture):

a). A sample prepared by slow thermal evaporation of a single Co layer on a (111) oriented Cu single crystal. The purpose of the sample was to achieve the sharpest and cleanest interface between Co and Cu. The expected spectrum for a perfect interface should exhibit two narrow lines, one for the bulk part of the Co layer (at about the frequency of bulk Co) and one for Co atoms at the interface which have 3 Cu neighbors among their 12 NN's. Indeed the experimental spectrum is very close to the expected one both qualitatively (only one satellite line beside the main bulk one) and quantitatively (the satellite line intensity corresponds nearly exactly to one Co monolayer). The frequency for the interface satellite is here about 5% lower than expected from bulk references: this illustrates the modification to the hyperfine field that is due to strains and/or anisotropic chemical environment at interfaces. In other samples, where the chemical disorder is larger at the interface, the position of this main satellite is much closer, as expected, to the bulk position.

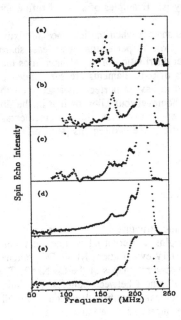

a) Nearly perfect interface (111) (1 monolayer). Single satellite line corresponding to Co atoms with 3 Cu neighbors. Traces of Co with 2 Cu neighbors.

b) The best multilayered structure observed (1.1 ML in the interface). The 3 Cu neighbors line is 30% of the interface spectrum. The average distance between monoatomic step defects is 4 atomic radii.

c) Sharp interfaces (<1.5 ML) but numerous monoatomic step defects: average distance 2 atomic radii.

d) Mixed interfaces (2.5 ML). Presence of flatter areas (3 Cu neighbors line resolved). Steep concentration profile.

e) Mixed interfaces (>3 ML.). Satellite lines of Co with 1 and 2 Cu neighbors show the presence of several planes containing a weak Cu concentration (3 to 8 %).

Fig.3. Examples of ../Cu/Co/.. interface spectra

b). A multilayered sample prepared by slow thermal evaporation of Co and Cu on a float glass substrate covered with a gold layer. Although the sample is polycrystalline in plane, the Au buffer layer presents an excellent surface flatness, large grains and good crystallinity; it favors an excellent (111) texture along the growth direction. The spectrum exhibits, as the previous one, the dominant satellite line of abrupt interfaces. The whole interface spectrum contains the equivalent amount of 1.1 monolayer of Co showing the very weak Co-Cu admixture. Interfacial defects can be described as monoatomic steps which are separated, on average, by 4 atomic distances of perfectly flat interface (75% perfect surface area). Fig.4 shows an MBE grown sample with similar features

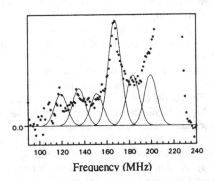

Fig.4. Reconstruction of a Co/Cu multilayer spectrum showing the contributions of Co neighbor configuration (from 1 to 6 Cu neighbors) involved in the interface according to a step defect model (d=2.5, l=1.6). Bulk contributions are not shown

and its reconstruction using the monoatomic step model.

c,d,e). Argon sputtered samples on Si substrates with different buffer layers designed more specifically for magnetoresistive properties. They exhibit to various degrees a significant interfacial admixture containing from 1.5 Co monolayer (still sharp interface but with a high density of steps) to 3 and more Co monolayers (several planes of Co containing Cu impurities and vice versa). Rather large series of such samples (5 to 10) with various Co or Cu layer thickness and various bilayer periods allowed a limited study of the influence of various parameters on the interface nanostructure: (i) High deposition rates (magnetron vs. diode sputtering) favor interface abruptness (typical spectrum c), (ii) at low deposition rates by diode sputtering, the columnar growth with large grains results in very rough interfaces at long distance (AFM study) and a weak texture (XRD), at short distance this is associated with the total absence of sharp interfaces and a strong interface admixture at atomic scale (typical spectrum e), (iii) the nature of the buffer layer deposited on the substrate prior to the multilayer growth has a striking effect on the interface quality: samples grown in the same conditions, except for the buffer layer, exhibit, on an Fe buffer, an interface spectrum characteristic of a sharp interface with monoatomic steps defects but, on a Cu buffer; an extended, structureless interface spectrum of disordered and mixed interfaces.

Miscible Elements: Co/Cr, Co/Ru multilayers

Chromium and ruthenium are elements which, in bulk alloys, form solid solutions with cobalt (the possibility of a miscibility gap in the CoCr phase diagram may be suspected though). Both elements favor the hcp stacking of Co. The NMR studies performed on Co/Ru and Co/Cr multilayers show very similar structures, as far as Co is concerned: they are two cases of large, alloy like admixture at the interfaces. Ru is hexagonal and in Co/Ru superlattices Co grows with the hcp stacking. In Co/Cr superlattices, thick Cr layers tend to impose their bcc structure to Co whereas thin Cr layers where found by RHEED analysis to grow as distorted (110) bcc planes with both NW and KS epitaxial relationships in order to mimic the hexagonal symmetry of the (0001) Co plane [4]. Only Co/Cr results are presented here.

Fig.5 shows typical spectra observed in these samples. Although XRD indicate a hcp stacking, the bulk line peaks around 222 MHz which is low for samples with in-plane magnetization [0]. This shows, in agreement with TEM cross section observations, that the Co layer contains a lot of stacking faults. This defective structure explains also the large line width as compared to the sharp fcc Co line observed in the Co/Cu samples above. The slight downshift observed for the thinnest Co layer may be associated with misfit strains but also, and more probably, with the overall decrease of the magnetization in the sample since Cr depresses very strongly

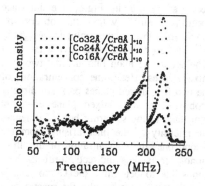

Fig.5. *Co NMR spectra observed in a series of Co/Cr multilayers.*

the magnetic moment of Co. The extended tail below the main line originates from Co atoms at the interfaces of the Co and Cr layers. The tails are superimposed in the three samples: the interfaces have the same topology. From the thickness dependence of the main Co line intensity, the amount of Co atoms alloyed with Cr at the interfaces is estimated to be 11 Å per Co layer. No NMR was observed for Co layers thinner than 10 Å because of the large admixture: they are alloyed to the core and they are too weakly magnetic to yield a significant signal.

Such cases of a large interfacial admixture are adequately analyzed using the diffused interface model. An example of spectrum reconstruction is given on Fig.6 which shows the contribution of each atomic plane in the interface. The reconstruction presented illustrates the possible difference between bulk alloys and alloyed interfaces. In bulk CoCr alloys the spacing between satellite lines is found to be 32 MHz. However, as observed in very diluted alloys [11], this value is an average between a 41 MHz and a 22 MHz spacing depending whether the Cr NN is in the same (0001) plane as Co or in the adjacent plane of the anisotropic hcp Co. The fits show clearly that the smallest spacing yields a better agreement than the average one which is not surprising since, in the interface, Co atoms have most of their Cr NN in the plane next to theirs. The concentration profile deduced from the fits is given in Fig.7 (the range of possible values corresponds to satellite spacings between 32 and 20 MHz). It shows that the mixed region is extended over five monolayers per interface. All concentrations do not have the same reliability. Indeed, as shown in Fig.6, the interface spectrum results, within the observation range, from three planes only: from right to left, the last full Co plane giving rise to a shoulder to the main line and the two first mixed planes. Hence the concentrations of the two first mixed planes only are directly probed. The third mixed plane does not contribute to the spectrum but it influences the intensity of the lines arising from the second one, the concentration of this third monolayer is thus indirectly probed. It is impossible to determine the Cr content of the two last planes because the spectra give neither direct nor indirect information about them. However their values are bounded in order to satisfy the constraints of a monotonous concentration profile and the amount of Cr deposited. Actually there is no pure Cr plane in

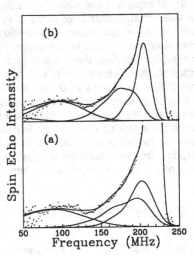

Fig.6. *Reconstruction of a Co/Cr multilayer spectrum showing the contribution of each plane involved in the interface. a) using the best satellite spacing (20 MHz); b) using the average bulk spacing (32 MHz) The bulk contributions are not shown.*

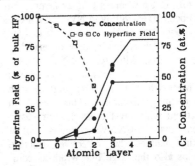

Fig.7. *Cr concentration and average hyperfine field (magnetization) profile in the interface. The monolayer at zero corresponds to the full Co plane which is in contact with the interface. The full lines show the minimum and the maximum Cr concentration values consistent with the interface model.*

the multilayer. From the average frequency of each plane subspectrum, it is possible to have an insight on the magnetization profile: although the hyperfine field is not fully proportional to the local magnetic moment, its average value provides a reasonable estimate of the average moment. Fig.8 displays the hyperfine field (magnetization) profile in the interface; it is much less sensitive to the choice of the satellite spacing because of its integral nature (centroid). From this profile it is possible to deduce that 1.8 Co layers are magnetically dead which compares well with bulk magnetization measurements.

Compound Forming Elements: Co/Fe superlattices

Cobalt can be stabilized in a bcc structure in Co/Fe multilayers [12]. In MBE superlattices, XRD and ion channeling experiments show the presence of the bcc phase only (no fcc or hcp Co phase) up to a critical Co thickness of 21 Å whereas, on the free Co surface, RHEED observations during the growth show that Co switches from the bcc ordering to the hcp one above 10 Å only [13-15]. NMR experiments have been carried out in these Co/Fe superlattices for Co thickness ranging from 5Å to 42Å. The purpose was to investigate the Fe-Co arrangement in the buried interfaces which leads to the doubling of the critical thickness. The effect of Fe thickness and samples deposited on MgO (100) substrate have also been studied which are not presented here. Observations in the (110) series are summarized in Fig.8. For Co thickness between 5 Å and 21 Å the spectra reveal a main peak at 198 MHz and a resolved satellite at 214 MHz. The 198 MHz line is well below the known NMR frequencies in the closed packed Co structures and have to be associated with the bcc structure as evidenced first in sputtered Co/Fe multilayers by combined NMR and EXAFS measurements[12]. Contrary to other elements, the vicinity of Fe atoms increases the Co NMR frequency (about 10 MHz/Fe in fcc and in bcc bulk alloys) [2,17], it is thus difficult from a stand alone NMR observation to decide whether the line at 214 MHz arises from a bcc Co/Fe interface or from a modified fcc structure. Actually XRD do not show the presence of any fcc or hcp Co phase: the 214 MHz line must be attributed to bcc Co at the Co/Fe interface. Above 21 Å Co a new peak appears suddenly at 222 MHz while XRD rocking curves and the lattice spacing evolution show abrupt discontinuities. This is associated with the onset of a distorted or misoriented hcp Co phase for a thickness which, as mentioned above, is twice the bcc thickness limit observed on an open surface by RHEED.

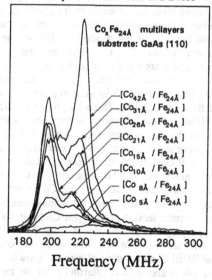

Fig.8. ^{59}Co NMR spectra at 1.6 K in $Co_xFe_{25Å}$ multilayers grown on GaAs (110), the integral intensity of each spectrum is set proportional to the Co layer thickness.

These observations show the role of the Fe overlayer to stabilize Co in a bcc structure and are discussed next.

Let us consider a perfect (sharp) interface between Co and Fe in bcc symmetry. In case of (110) growth 2 Fe NN are present in the Fe plane adjacent to the Co plane. From the study of bulk bcc alloys [17] a perfect Co/Fe (110) interface would exhibit a line at about 216 MHz; the well resolved satellite at 214 MHz is thus the indication of a considerable content of perfect interface. However the, as considerable, tail of the spectra up to 300 MHz shows also a non negligible amount of alloyed Co atoms with up to 8 Fe NN. The evolution of the spectra shows that the spectrum tails are superimposed only for two samples just before the onset of the hcp line. This means that continuous, pure Co planes are present only for 15 Å thick Co layers and above; in thinner Co layers interfacial regions from both sides are connected. The thinnest Co layers (5 Å) are even alloyed to core.

As in one of the Co/Cu examples above, both basic models fail to describe the spectra: the sharp interface model fails to explain the large interfacial intensity and particularly the high frequency part of the spectra while the diffused model cannot reproduce the dip between the pure bcc Co line and the line for 2 Fe NN (i.e. any monotonous concentration profile that gives the right amount of 2 Fe NN configurations gives also too many 1 Fe NN configurations). Two inhomogeneous models of the Co layers have then been built which can successfully explain the spectral shapes in the bcc range. In both of them the Co layer contains grains of pure bcc Co which give the spectral intensity at 198 MHz and partly at 214 MHz and a bcc CoFe alloy which gives a broad spectral background up to high frequencies. In the first of them Co grains are columns with a sharp Co/Fe interface on both sides whereas in the second one they are pyramidal with a sharp Co/Fe interface on the base side only. The first model has been discarded. Indeed, in order to agree with the experiment, the volume fraction of Co columns must be varied from 0 to 50% as the Co thickness increases; this is obviously unphysical since it implies that the column diameter is predetermined by the final layer thickness from the start of the deposition. The second model sticks more closely to what is expected from a 3D growth of Co islands onto Fe: each atomic plane is a patchwork of Co and CoFe areas which are piled up in such a way as to create Co pyramids in the third dimension. NMR spectra are thus simulated using two parameters for each atomic plane: (i) the area fraction of Co patches (from 0, fully alloyed plane, to eventually 1, full Co plane) and (ii) the Fe content of the alloy patches. These two topological/chemical parameters fix the configuration probabilities and the intensities of the corresponding NMR lines. The profiles of both parameters across the Co layer are found not to depend on the deposited Co thickness which shows the self-consistency of the model. These profiles are described below.

Fig.9 exemplifies the theoretical spectra calculated with the model. Three main components are shown for the spectrum: bcc Co peak and its perfect interface satellite (the pyramid base), the interface between bcc Co pyramids and CoFe alloy, and the CoFe alloy. All the results in the (110) series are summarized in the diagram presented in Fig.10 showing the model cross section of the open Co layer, as observed by RHEED and of the Co layer after capping by Fe, as determined from the NMR study of superlattices. It is worth noting that NMR cannot decide here what is top or bottom, the choice made on Fig.10 is suggested by the RHEED observation of a rough growth of Co onto Fe, particularly with the onset of the hcp phase above 10 Å. Starting from the top and across about 4 atomic planes Co is mostly diluted in an alloy the Fe content of which steeply drops from 100% to 25% (these planes contain most of the deposited Co in the thinnest case). Below these first planes, the Fe concentration in the alloy stays essentially constant (25-20 at% Fe) while the area fraction of bcc Co increases across the layer. For thick enough Co layers (21 Å) the pyramid bases eventually merge into up to 3 full bcc Co planes. Then hcp Co builds up in the layer. Out of 21 Å Co in a bcc phase, 10-11 Å are included in a bcc CoFe alloy. The remaining 11-10 Å of pure

Co are distributed in the following way: 3 full Co planes (6 Å) are formed on the flat Fe surface and the rest is clustered in the mixed planes, where it alternates with the rather homogenous CoFe alloy. In case of an open surface, the onset of the hcp structure is observed above 10 Å of deposited Co.

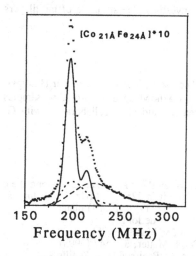

Fig.9. Experimental and model spectra in the thickest bcc sample from the (110) series, showing contributions to the calculated spectra. Solid line: bulk bcc Co and its sharp interface with Fe, dotted: interface between bulk Co and CoFe alloy, dashed: CoFe alloy.

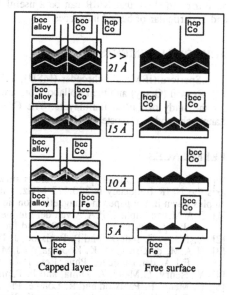

Fig.10. Cross section image of the Fe/Co/Fe superlattice as seen by NMR and RHEED (see text).

One main conclusion from this study is that 10 to 11 Å seems to be the stability limit for the bcc phase of pure Co when grown on Fe substrate, in case of an open Co surface as well as in superlattices. The reconstruction of the bcc structure up to 21 Å by the Fe overlayer in superlattices is due to the intermixing with the rough hcp Co open surface and formation of an alloyed CoFe layer. A rather astonishing fact is that, beside a transition interface which is fully alloyed, the pure bcc Co phase coexists with a rather homogeneous alloy containing 20 to 25 at% Fe. This Fe concentration in the alloyed part of the Co layer happens to be close to the stability limit for a bcc structure in the equilibrium phase diagram of bulk CoFe. The study shows that a comparable limit is preserved also in the non-equilibrium conditions of MBE growth and the system has a tendency to phase demixion between a CoFe alloy and Co. However under thin film growth conditions the bcc structure is preserved in all phases, below the critical thickness, whereas in bulk alloys the phase segregation occurs between the bcc CoFe alloy and an fcc, Co rich, phase. A similar conclusion is reached out of the study of bcc CoFe alloys prepared by codeposition in the same conditions as the multilayers [16]. It must be quoted that the study of these thin films suggests, for the 214 MHz line, another explanation than a (partly) sharp interface. Indeed, in the alloys, which are found bcc down to 14 at% Fe by XRD, the spectra reveal a chemical short range ordering of Fe which favors configurations of Co with 2 Fe neighbors like at a sharp (110) Co/Fe interface. Therefore it cannot be excluded that a similar ordering occurs in the superlattice interfaces.

IV CONCLUSION

Because it probes, in real space and at short distance, the atomic configurations, NMR complements efficiently standard diffraction techniques (and even diffuse scattering investigations) in case of absence of translation symmetry. Throughout this non exhaustive review we have tried to show that NMR can be a useful tool to investigate the structure of multilayers and in particular of buried interfaces.

ACKNOWLEDGMENTS

V.Speriosu (IBM), P.Schroeder (MSU), T.Valet (Thomson), F.J.A.den Broeder (Philips), J.P. Renard (Orsay) and their collaborators are gratefully aknowledged for providing samples for the study of Co/Cu multilayers. The Co-Fe system is studied in collaboration with G. Langouche and J. Dekoster (KU Leuven).

REFERENCES

[0] **Note**: In hcp Co the hyperfine field is anisotropic. When the magnetization is along the c axis the NMR frequency is 220-222 MHz, when in plane it is 225-228 MHz. In all the examples given in the paper the magnetization lies in the dense (111) or (0001) plane.
[1] E.A.M.van Alphen, P.A.A van der Heijden and W.J.M. de Jonge,
    J. App. Phys. **76**, 6607 (1994) and Phys. Rev. **B 51**, March (1995)
[2] C. Mény, E. Jedryka and P. Panissod, J. Phys: Cond. Matter, **5**, 1547 (1993)
[3] H.A.M. de Gronckel, K. Kopinga, W.J.M de Jonge, P. Panissod, J.P. Schillé and
    F.J.A. den Broeder, Phys. Rev. **B 44**, 9100 (1991)
[4] Y. Henry, C.Mény, A. Dinia, and P. Panissod, Phys. Rev. **B 47**, 15037 (1993)
[5] C. Mény, P. Panissod and R. Loloee, Phys. Rev. **B 45**, 12269 (1992)
[6] Y. Saito, K. Inomata, A. Goto and H. Yasuoka, J. Phys. Soc. Jap. **62**, 1450 (1993)
[7] T. Valet, P. Galtier, J.C. Jacquet, C. Mény and P. Panissod,
    J. Magn. Magn. Mat. **121**, 402 (1993)
[8] C. Mény, P. Panissod, P. Humbert, J.P. Nozières, V.S. Speriosu, B.A. Gurney and
    R. Zehringer, J. Magn. Magn. Mat. **121**, 406 (1993)
[9] C. Mény, J.P. Jay, P. Panissod, P. Humbert, V.S. Speriosu, H. Lefakis, J.P. Nozières
    and B.A. Gurney, in *Magnetic Ultrathin Films. Multilayers and Surfaces, Interfaces
    and Characterization,* edited B.T. Jonker et al (Mat. Res. Soc. Proc. **313**, Pittsburgh
    PA, 1993) pp 289-294
[10] P. Le Fèvre, D. Chandesris, H. Magnan and O. Heckmann,
    J. de Phys. (Paris) Col. **C9-4**, 159 (1994)
    see also N. Persat, A. Dinia, J.P. Jay, C. Mény and P. Panissod,
    to be presented at MML'95 Cambridge
[11] M. Kawakami, J. Phys. Soc. Jap.,**40**, 56 (1976)
[12] Ph. Houdy, P. Boher, F. Giron, F. Pierre, C. Chappert, P. Beauvillain, K. Le Dang,
    P. Veillet and E. Velu, J. Appl. Phys. **69**, 5667 (1991)
[13] J. Dekoster, E. Jedryka, C. Mény, and G. Langouche,
    J. Magn. Magn. Mater., **121**, 69 (1993) and Europhys. Lett. **22**, 433 (1993).
[14] J.Dekoster, E. Jedryka, M. Wójcik and G. Langouche,
    J. Magn. Magn. Mater. ,**126**, 12 (1993)
[15] J. P. Jay, E. Jedryka, M. Wojcik, J. Dekoster, G. Langouche, P. Panissod
    presented at ICM'94 (submitted to Phys. Rev. B)
[16] M. Wojcik, J. P. Jay, P. Panissod, E. Jedryka, J. Dekoster, G. Langouche
    presented at ICMFS'94 (to be published elsewhere)
[17] J. P. Jay, M. Wojcik and P. Panissod, presented at ICM'94 (to be published elsewhere)

GROWTH OF Fe/ZnSe MULTILAYERS ON GaAs (001) AND (111) BY MOLECULAR BEAM EPITAXY

H. ABAD*, B. T. JONKER, C. M. COTELL, S. B. QADRI and J. J. KREBS
Naval Research Laboratory, Washington, D. C., 20375-5343.
*National Research Council Postdoctoral Associate.

ABSTRACT

The growth of Fe/ZnSe/Fe multilayers on (001) and (111) GaAs substrates is reported. The samples were characterized *in-situ* by reflection high energy electron diffraction (RHEED), and *ex situ* by vibrating sample magnetometry (VSM), ferromagnetic resonance (FMR), cross sectional transmission electron microscopy (TEM), and x-ray diffraction. On the (001) surface, the quality of the layers deteriorated significantly with the growth of the first ZnSe spacer layer. In Fe/ZnSe/Fe trilayer structures, TEM revealed a well-defined layered structure, with a high density of defects in both the ZnSe spacer layer and the subsequent Fe layer. VSM and FMR clearly showed the presence of two Fe films with distinct coercive fields, with the higher coercive field attributed to the lower crystalline quality of the second Fe layer. θ-2θ x-ray diffraction measurements performed on samples grown on (001) GaAs substrates indicated that the ZnSe spacer layer (grown on (001) Fe) grew in a (111) orientation. Growth on GaAs(111) substrates produced better RHEED patterns for all layers with little deterioration in film quality with continued layer growth, so that the magnetic properties of the individual Fe layer could not be distinguished.

INTRODUCTION

The magnetic coupling between ferromagnetic thin films separated by several monolayers of a non-magnetic metallic element has generated a great deal of interest and activity in the past few years.[1] Such systems often exhibit giant magnetoresistance (GMR), with potential for practical applications in magnetic recording. Several example systems include Co/Cu, Fe/Ag, and Fe/Cr superlattices. The coupling in these metallic systems is thought to be mediated by the conduction carriers in the non-magnetic metallic spacer layer. In a recent paper, Mattson *et. al.*[2] reported on the coupling of ferromagnetic Fe films separated by thin layers of Si. The authors reported that the coupling at low temperature changed from ferromagnetic to antiferromagnetic upon illumination of the sample with visible laser light. Other reports on similar systems indicate that the coupling was induced by thermal excitation.[3,4] The prospect of extending magnetic interlayer coupling to include semiconductor spacer layers is exciting from both a scientific and a technological point of view. Such systems may allow the manipulation of the exchange coupling via the control of carrier densities in the spacer layer either by doping or by photo-excitation.

The Fe/ZnSe system offers some advantages over the Fe/Si system in this regard. Since ZnSe is a direct wide bandgap semiconductor, photo-induced effects may be more readily manifested and easily distinguished from thermally induced effects. From a materials growth standpoint, the lattice match is more favorable, with the mismatch between ZnSe and twice the lattice constant of Fe being 1.1% compared to 5.5% for the Fe/Si system. In addition,

73

extensive work has established that it is possible to grow Fe epilayers on ZnSe (001) with excellent magnetic and structural properties.[5] Auger electron diffraction and x-ray photoelectron spectroscopy studies have shown that Fe grows on the 2x1 reconstructed surface of ZnSe (001) in a layer-by-layer manner, with little interdiffusion or compound formation at the interface.[6] However, the growth of ZnSe on an Fe epilayer presents a new and different set of problems and is more challenging. The growth of a polar material on a non-polar one (*e. g.* the growth of GaAs or CdTe on Si)[7] is inherently difficult and can lead to structural defects such as antiphase domains. Although the optimum growth temperature of ZnSe on GaAs (001) is typically quoted as 300° C, a lower growth temperature for ZnSe on Fe is important to minimize interdiffusion at the ZnSe/Fe interface and to preserve the magnetic character of the Fe layer. Migration enhanced epitaxy (MEE)[8] allows the epitaxial growth of semiconductors at low temperatures and was utilized in the work reported to grow ZnSe epilayers on Fe.

RESULTS AND DISCUSSION

Details of the growth of ZnSe on Fe (001) films at low temperature (175° C) were reported recently.[9] To briefly summarize, in the present work all the ZnSe spacer layers were grown using migration enhanced epitaxy (MEE)[8] at 175° C. This method allows the growth to proceed at a low temperature compatible with the growth and preservation of the underlying Fe layer as well as the precise control of layer thicknesses. The growth of trilayers consisting of Fe/ZnSe/Fe structures was initiated by the deposition of a ZnSe buffer layer on the (001) GaAs substrate. For most structures the buffer layer was grown using standard codeposition at 300° C to thicknesses that varied from 1000Å to 1μm. The thickness of the Fe layers was varied from 25 to 125Å, while that of the ZnSe spacer was varied from 15 to 300Å. RHEED patterns for the second Fe layer were spotty[9], indicative of three dimensional growth and surface roughness but did not show rings which would be indicative of polycrystalline material. These RHEED patterns differed dramatically from those observed for the first Fe film (which were streaky), and presumably reflect the lower quality of the intermediate ZnSe spacer layer grown on the first Fe film, relative to the ZnSe buffer layer grown on GaAs.

Examination of the samples by transmission electron microscopy (TEM) confirmed that the quality of the spacer layer was much lower than that of the ZnSe buffer layer. Polished cross sections were thinned to electron transparency on a liquid nitrogen cold stage using 4.5kV Ar ions at a 15° angle of incidence. The samples were examined on a Philips CM30 TEM operating at 300kV. Figure 1 shows a bright field TEM photomicrograph of a sample consisting of the following layers: GaAs(001) substrate/1μm ZnSe (buffer)/ 118Å Fe/200Å ZnSe (spacer)/118Å Fe/2μm polycrystalline ZnSe cap layer. Observations made from this sample are consistent with previous results.[9] In general, the ZnSe (001) buffer layers grown directly on the GaAs substrates were of high crystalline quality with few defects. The Fe (001) layers grown on top of these ZnSe buffer layers were also relatively defect-free, although there was evidence for strain at the Fe/ZnSe buffer interfaces due to the 1.1% lattice mismatch. The interfaces between the first Fe layer and the ZnSe layers above and below it were reasonably sharp, suggesting that little interdiffusion took place between these layers.

Subsequently grown layers showed a high concentration of stacking faults. In Figure 1, the upper Fe layer shows a much higher concentration of defects than the first Fe layer, which

ZnSe cap layer
Fe layer
ZnSe spacer layer
Fe layer

ZnSe buffer layer
——— 200Å

Figure 1. Bright field cross sectional TEM micrograph for a structure composed of GaAs (001) substrate/ 1μm ZnSe (buffer)/ 118Å Fe/200Å ZnSe/118Å Fe/2μm polycrystalline ZnSe cap layer. The GaAs substrate is not shown. The wavy interface between the ZnSe buffer layer and the first Fe layer may be attributed to a surface perturbation which was propagated through the deposited layers.

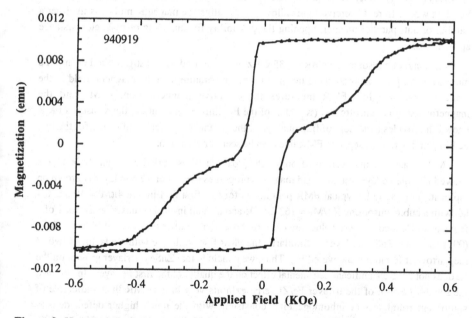

Figure 2: Hysteresis loop for the Fe/ZnSe trilayer shown in Figure 1. These data illustrate that the Fe layers have distinctly different coercive fields: the lower film has $H_c = 50$ Oe, while for the upper one H_c is approximately 200 Oe.

makes the interface between the ZnSe spacer layer and the upper Fe layer more difficult to image due to increased electron scattering. Figure 1 clearly shows that the stacking faults are nucleated in the ZnSe spacer layer and that all the layers above the first Fe layer are highly defective. The hillock or wave in the layers which appears to the left of center in the micrograph may be attributed to a perturbation on the GaAs substrate surface which was propagated through the deposited layers, although the ZnSe/GaAs interface was not imaged directly (due to sample thickness) to confirm this.

Vibrating sample magnetometer (VSM) measurements showed that the Fe layers buried in ZnSe remained ferromagnetic. Comparison of the measured magnetic moment and the moment calculated from the quantity of Fe deposited as measured using x-ray fluorescence spectroscopy (XRF) and assuming a bulk moment/atom showed that most of the Fe was incorporated in ferromagnetic form with a maximum net moment loss of 15%. Figure 2 shows the room temperature hysteresis loop as measured along the [100] direction (easy axis) for the sample shown in Figure 1. For this sample, XRF results and the measured magnetic moment indicate that all the deposited Fe was incorporated in ferromagnetic form. The figure shows clearly the presence of two Fe layers with distinctly different coercive fields. It is concluded that the layer with the larger coercive field is the top Fe layer as suggested by the poor crystalline quality revealed by the TEM and RHEED, and from previous studies of single Fe epilayers on (001) ZnSe, where much lower coercive fields were found.[5] The growth of two Fe films with different coercive fields allows one to align the magnetic moments in either a parallel or anti parallel manner, opening the possibility for tunneling or magnetoresistance applications.

Ferromagnetic resonance (FMR) at 35 GHz was also used to investigate the Fe/ZnSe/Fe multilayer samples. All data were taken at room temperature with the magnetic field in the plane of the sample. FMR measures the effective magnetization, $4\pi M'$, and the magnetocrystalline anisotropies (K_1, etc.) of the Fe films.[5] In addition, the measured FMR linewidth provides a measure of film quality. If the parameters differ significantly for the two Fe films in the trilayer, separate FMR lines can be seen for each film.

FMR measurements performed on single (001) Fe films buried between ZnSe layers showed that the Fe layer retains good magnetic properties even after a ZnSe layer is grown on top of it. For example, typical FMR parameters for this first Fe film are $4\pi M' = 19.3 \pm 0.4$ kG with a cubic anisotropy $K_1/M_s = 263 \pm 5$ Oe and a small in-plane uniaxial anisotropy of a few tens of Oersteds. These values are close to those found earlier for Fe (001) films on ZnSe (001) without a ZnSe top layer.[5] Similarly, the FMR linewidths are within a factor of two of the narrowest found for uncovered Fe. Thus, we conclude that ZnSe overlayer growth on the Fe does not have a significant detrimental effect on the quality of the first Fe layer. In contrast, the second Fe film of the trilayer Fe/ZnSe/Fe exhibits very broad FMR lines indicative of significant roughness or inhomogeneity, consistent with the much higher defect density observed in this layer with TEM. Typical linewidths are 90-180 Oe for the first film but 600-700 Oe for the second. The large difference in defect density is also the probable source for the

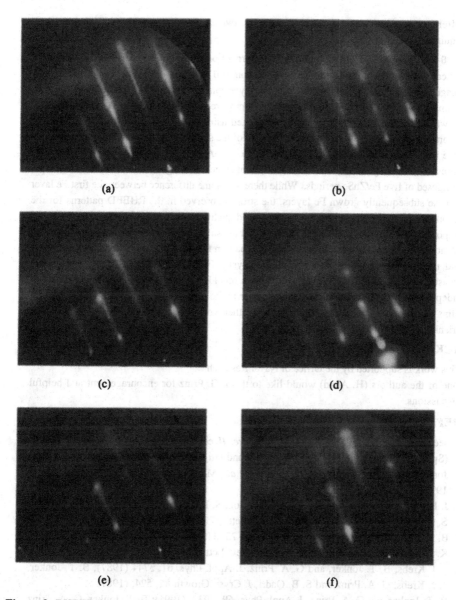

Figure 3: RHEED patterns for different stages of the growth of an (Fe/ZnSe)₅ /GaAs(111) sample. (a) Fe/ZnSe/GaAs(111) after 20 min. of Fe growth. (b) ZnSe/Fe/ZnSe/GaAs (111) after 9 periods of MEE ZnSe. (c) Second Fe layer after 24 min. of Fe growth. (d) 3rd ZnSe spacer layer, MEE period 9. (e) 4th Fe layer after 18 min. of growth. (f) 5th Fe layer after 25 min. of growth.

difference in coercive fields observed for the two Fe films, although additional work is required to quantify this contribution.

θ-2θ x-ray diffraction measurements performed on ZnSe/Fe/ZnSe/GaAs(001) structures indicated that at least some portion of the ZnSe on (001) Fe grows in a (111) orientation. TEM images also suggested that the ZnSe spacer layer contained more than one orientation. To examine this more closely, Fe/ZnSe multilayers were grown on GaAs(111) substrates. The growth of the ZnSe spacer layers was performed using MEE in a procedure similar to that reported in reference 9. The crystalline quality of the second and subsequent Fe layers in this case was markedly superior to that observed on GaAs(001) substrates. Figure 3 shows a series of RHEED pictures taken during different stages of the growth of a multilayer sample composed of five Fe/ZnSe periods. While there was some difference between the first Fe layer and the subsequently grown Fe layers, the streaks observed in the RHEED patterns for the second through the fifth Fe layers indicated good quality growth. This indicates that it should be possible to grow Fe/ZnSe superlattices on (111) surfaces. This contrasts sharply with the situation for (001) surface, where the RHEED pattern for the second Fe film becomes spotty[9] and the growth of more than two single crystal Fe films is very difficult. Magnetic measurements performed on multilayers grown on (111) GaAs did not show the presence of independent films. This could be due to either pinhole ferromagnetic coupling or to the Fe films having similar magnetic parameters. Further work to optimize the growth in the (111) orientation is underway.

ACKNOWLEDGMENTS

This work is supported by the Office of Naval Research.
One of the authors (H. Abad) would like to thank G. Prinz for encouragement and helpful discussions.

REFERENCES

1. See for example *Ultra Thin Magnetic Structures II*, edited by B. Heinrich and J. A. C. Bland (Springer-Verlaag, Berlin, 1994), chap. 2.; and *Magnetic Ultrathin Films*, edited by B. T. Jonker, S. A.. Chambers, R. F. C. Farrow *et al.* (Materials Research Society, Pittsburgh, 1993) vol. 313.
2. J. E. Mattson, Sodha Kumar, Eric E. Fullerton, S. R. Lee, C. H. Sowers, M. Grimsditch, S. D. Bader, and F. T. Parker, Phys. Rev. Lett. , **71**, 185, (1993).
3. B. Briner and M. Landolt, Phys. Rev. Lett. , **73**, 340, (1994).
4. K. Inomata, K. Yusu, and Y. Saito, Phys. Rev. Lett. , **74**, 1863, (1995)
5. J. J. Krebs, B. T. Jonker, and G. A. Prinz, J. Appl. Phys. 61, 3744 (1987); B. T. Jonker, J. J. Krebs, G. A. Prinz and S. B. Qadri, J. Cryst. Growth **81**, 524, (1987).
6. B. T. Jonker and G. A. Prinz, J. Appl. Phys. **69**, 2938, (1991); B. T. Jonker, G. A. Prinz and Y. U. Idzerda, J. Vac. Sci. Technol. **B 9**, 2437, (1991).
7. S. F. Fang, K. Adomi, S. Lyer, H. Morkoc, H. Zabel, C. Choi, and N. Otsuka, J. Appl. Phys., **68**, R31, (1990).
8. J. M. Gaines, J. Petruzzello, and B. Greenberg, J. Appl. Phys., **73**, 2835, (1993).
9. H. Abad, B. T. Jonker, C. M. Cotell and J. J. Krebs, J. Vac. Sc. Tech. **B 13**,716, (1995).

EPITAXIAL GROWTH OF (111) TbFe$_2$ BY SPUTTER DEPOSITION

C.T. WANG, R.M. OSGOOD III, R.L. WHITE and B.M. CLEMENS
Department of Materials Science and Engineering, Stanford University, Stanford, CA 94305-2205

ABSTRACT

The effect of in-plane strain in (111)-oriented epitaxial TbFe$_2$ films on the magnetization orientations was studied. Magnetocrystalline anisotropy, shape anisotropy, and magnetoelastic energy were calculated to determine the magnetization orientation in the presence of various in-plane strains. Theoretical considerations indicate that a compressive strain smaller than -0.065% can induce an out-of-plane magnetization. DC magnetron sputtering from Tb and Fe elemental targets was used to grow 50, 100, 200, and 400 Å thick epitaxial TbFe$_2$(111) films at 600°C on (11$\bar{2}$0)-oriented sapphire substrates with 1000 Å thick epitaxial Mo(110) buffer layers. The growth rate of the TbFe$_2$(111) films was 1.44 Å/sec. X-ray diffractometry, Rutherford backscattering spectrometry, and vibrating sample magnetometry were used to characterize the crystal structure, epitaxial orientation, composition, stress and strain state, and magnetic properties of the TbFe$_2$ films. The TbFe$_2$(111) films were epitaxial with twins rotated by 60° and were in tensile strain states with the resulting in-plane magnetization.

INTRODUCTION

TbFe$_2$ is a giant magnetostrictive material with the largest known room temperature magnetostriction[1]. Large magnetostriction is potentially useful for sensors, actuators or surface-acoustic-wave applications. In the magnetostrictive applications, the applied magnetic field changes the magnetization direction and then changes the dimensions of the material. On the other hand, we can use the inverse magnetostrictive effect to control the magnetization orientation by applying stress on the material.

The effect of film stresses on the anisotropy orientation have been studied in amorphous TbDyFe films with the result that tensile stresses induce in-plane anisotropy and compressive stresses induce out-of-plane anisotropy[2]. In this experiment we grow (111)-oriented epitaxial TbFe$_2$ films and study the effect of film strains on the magnetization orientation. Epitaxial TbFe$_2$(111) films with out-of-plane magnetization are potentially useful for magneto-optical recording applications.

The total magnetic anisotropy of (111)-oriented epitaxial TbFe$_2$ films in the presence of various in-plane strains is calculated to determine the orientation of the magnetization. These films have been successfully grown on (11$\bar{2}$0)-oriented sapphire substrates with 1000 Å Mo buffer layer at 600°C. The crystal structure, epitaxial orientation, composition, stress and strain state, and magnetic properties of the TbFe$_2$ films are investigated. Strain and magnetization data from this experiment are compared with those from calculation of the anisotropy.

MAGNETIC ANISOTROPY

Magnetocrystalline anisotropy, shape anisotropy, and magnetoelastic energy were considered to determine the magnetization orientations in the epitaxial TbFe$_2$(111) films. Using

79

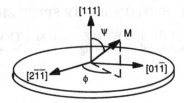

Figure 1: Film coordinate system for (111)-oriented TbFe$_2$ films.

Using the film coordinate system in Figure 1, we can write the magnetocrystalline aniostropy (E_{mc}) per unit volume of TbFe$_2$ as:

$$E_{mc} = K_1[\frac{1}{3}\cos^4\psi + (\sqrt{2}\cos\phi\sin^2\phi - \frac{\sqrt{2}}{3}\cos^3\phi)\sin^3\psi\cos\psi + \frac{1}{4}\sin^4\psi] \qquad (1)$$

where K_1 is the magnetocrystalline constant of -7600 x 10^4 erg/cm^3. The shape anisotropy (E_s) per unit vloume of TbFe$_2$ is:

$$E_s = 2\pi M^2\cos^2\psi \qquad (2)$$

where M is the magnetic moment of 800 emu/cm^3. For $\lambda_{100} \ll \lambda_{111}$ and equi-biaxial stress, the magnetoelastic energy (E_{me}) per unit volume of TbFe$_2$ can be written as[3]:

$$E_{me} = 9\lambda_{111}[\frac{c_{44}(c_{44} + 2c_{12})}{c_{11} + 2c_{12} + 4c_{44}}]\epsilon_{11}\cos^2\psi \qquad (3)$$

where λ_{111} is the magnetostrictive constant of 2460 x 10^{-6}, the c_{ij} are the elements of the elastic stiffness matrix, and ϵ_{11} is the in-plane strain. The values of the magnetocrystalline constant , magnetic moment, and magnetostrictive constant used in the calculation were measured from single crystal TbFe$_2$ by Clark[1].

Since the elastic stiffnesses of TbFe$_2$ are not known, materials with the same crystal structure and similar magnetic properties were surveyed to find typical elastic stiffness values. The crystal structure and magnetostrictive properties of Tb$_{0.3}$Dy$_{0.7}$Fe$_2$ were the same as those of TbFe$_2$[1]. Therefore, we use these elastic stiffnesses for the stress and strain calculations.

The ψ and ϕ dependence of the magnetocrystalline anisotropy are plotted in Figure 2(a). The anisotropy was calculated for ψ from 0 to 90° and ϕ from 0 to 120°. If only the magnetocrystalline aniostropy is considered, the two energy minima in Figure 2(a), one along the out-of-plane [111] direction and the other one along the nearly-in-plane [11$\bar{1}$] direction at $\psi = 70.2°$ and $\phi = 60°$, are equally favorable magnetization orientations as can be verified by minimizing Eq.(1) first with respect to ϕ and then with respect to ψ. However, if the shape anisotropy is also considered, the magnetization will prefer to lie along the nearly-in-plane [11$\bar{1}$] direction .

We calculate the total of the magnetocrystalline anisotropy, shape anisotropy, and magnetoelastic energy as a function of ψ at $\phi = 60°$ for various strains, as shown in Figure 2(b). At zero strain, the anisotropy energy along the out-of-plane [111] is a little larger than that along the nearly-in-plane [11$\bar{1}$]. A compressive strain of -0.065% makes the two directions equal in energy. A tensile strain of 0.5% makes the magnetization lie nearly in the film plane and a compressive strain of -0.5% makes the magnetization lie along the out-of-plane direction. Therefore, a modest compressive strain will induce out-of-plane magnetization in epitaxial TbFe$_2$(111) films.

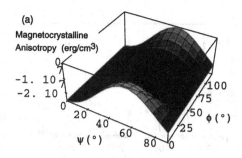 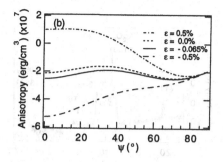

Figure 2: (a) Magnetocrystalline anisotropy of epitaxial $TbFe_2(111)$ films as a function of ψ and ϕ. (b) The total anisotropy energy as function of ψ at $\phi = 60°$ for strains of 0.5%, 0.0%, -0.065% and -0.5%.

EXPERIMENTAL

Epitaxial $TbFe_2(111)$ films with thickness of 50, 100, 200, and 400 Å were grown on $(11\bar{2}0)$ sapphire substrates with a 1000 Å epitaxial Mo(110) buffer layer using DC magnetron sputtering. The sapphire substrates were cleaned in solvents and then dried by high purity N_2. Prior to deposition, the substrates were first annealed at 650°C until a base pressure of 4×10^{-8} Torr was achieved. Mo buffer layers were grown at this temperature at a rate of 1 Å/sec in 3.0 mTorr Ar. The substrates were then cooled to 600°C. At this temperature, $TbFe_2$ films were grown by cosputtering from Tb and Fe elemental targets at a total rate of 1.44 Å/sec in 1.5 mTorr Ar. A 100 Å Mo layer was deposited over the $TbFe_2$ film at a temperature below 50°C to prevent oxidation.

The crystal structure and epitaxial orientation of the films were determined using a Philips XGR 3100 diffractometer and a Picker four circle x-ray diffractometer. Asymmetric and grazing incidence x-ray scattering (GIXS) with a wavelength of 1.4586 Å were performed at the Stanford Synchrotron Radiation Laboratory (SSRL) to measure the plane spacings at different ψ (where ψ is the angle between the normal of the diffracting planes and the film normal). The standard $\sin^2 \psi$ analysis for epitaxial (111) films was then used to determine the in-plane stress, in-plane strain and out-of-plane strain[4]. Three $TbFe_2$ {224} planes at different inclinations were selected for this measurement. Rutherford backscattering (RBS) energy spectra taken at 6° off film normal and at a tilt angle of 100° with incident He^+ of 2.2 MeV were used to determine the composition of these films. Vibrating sample magnetometry (VSM) with a maximum field of 14.0 kOe was used to measure the in-plane and out-of-plane hysteresis loops of these films.

RESULTS AND DISCUSSION

A symmetric XRD scan of a 400 Å film is shown in figure 3(a). The presence of strong $TbFe_2(111)$, (222), and (333) peaks and Mo(110) peak indicates that the $TbFe_2$ film is (111)-oriented and the Mo buffer layer is (110)-oriented. The two Tb (002) and (110) diffraction peaks come from a small amount of the Tb phase. The epitaxial quality and in-plane orientation of the Mo and $TbFe_2$ films were determined using asymmetric x-ray diffraction. Mo{121} at χ of 59.6° and $TbFe_2${111} at χ of 19.5° were picked for the ϕ scans, and the diffraction spectra are shown in Figure 3(b). The twofold symmetry of the

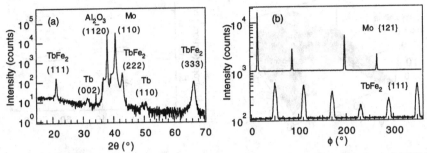

Figure 3: (a) XRD symmetric scan diffraction spectrum of a 400 Å TbFe₂ film. (b) XRD ϕ scan diffraction spectra of Mo{121} and TbFe₂{111} of the 400 Å TbFe₂ film.

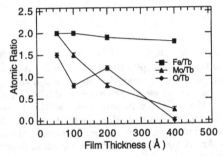

Figure 4: Averaged compositions of TbFe₂ films of 50, 100, 200, and 400 Å.

Mo {121} diffraction peaks indicates that the Mo buffer layer grows epitaxially. In the TbFe₂ film, six {111} diffraction peaks are found. For (111)-oriented single crystal epitaxial films, only three 111 peaks should be observed in the ϕ scan. Experimentally six TbFe₂ {111} peaks are observed because 60° rotation twins are formed in the epitaxial TbFe₂ films. The in-plane rocking curve FWHM of the 400 Å TbFe₂ film is about 5° indicating some mosaic spread in the in-plane orientation. From the XRD data we are able to determine that the orientation relationship between The Mo(110) buffer layer and the TbFe₂(111) film is $[\bar{1}\bar{1}2]_{TbFe_2}$ // $[001]_{Mo}$ (lattice mismatch 4.5%) and $[\bar{1}10]_{TbFe_2}$ // $[\bar{1}10]_{Mo}$ (lattice mismatch -14.3%).

The averaged compositions of TbFe₂ films of 50, 100, 200, and 400 Å are shown in Figure 4. The Fe/Tb atomic ratio is about 2/1 in these films as expected, however significant concentrations of Mo and O from the buffer layer and substrate respectively are found in these films as well. The Mo/Tb and O/Tb atomic ratios are 2.0 and 1.5 in the 50 Å film and are 0.25 and 0.0 in the 400 Å film. This implies that thicker TbFe₂ films have better averaged compositional quality. Since rare earth elements typically reacts with O from the sapphire substrate at high temperature, a refractory metal was chosen as a buffer layer to prevent reactions[5]. The diffusion of O through a Nb buffer layer on sapphire has been studied with the result that Nb buffer layer thicker than 1000 Å is needed to have a low O concentration at the surface for deposition temperatures higher than 500°C[6]. The data from Figure 4 might indicate that a Mo buffer layer thicker than 1000 Å is needed to prevent oxidation.

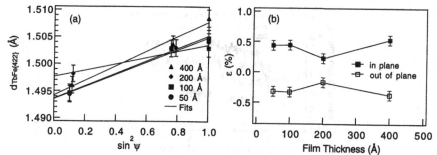

Figure 5: (a) d vs $\sin^2 \psi$ data. The lines are least-square fits to the data. (b) in-plane and out-of-plane strains in 50, 100, 200 and 400 Å TbFe$_2$ films as determined by $\sin^2 \psi$ analysis.

The stress and strain state in the TbFe$_2$ films were determined by using the d vs $\sin^2 \psi$ technique[4]. In (111) epitaxial films with equi-biaxial stress, the equation for film strain as function of ψ can be expressed as:

$$\epsilon(\psi) = \frac{d(\psi) - d_0}{d_0} = \sigma \left[\frac{2c_{44} - c_{11} - 2c_{12}}{3c_{44}(c_{11} + 2c_{12})} + \frac{1}{2c_{44}} \sin^2 \psi \right] \qquad (4)$$

where $\epsilon(\psi)$ and $d(\psi)$ are the strain and the measured plane spacing as a function of ψ, d_0 is the unstrained plane spacing, σ is the equi-biaxial stress, and the c_{ij} are the elements of the elastic stiffness matrix. From this we obtain the unstrained plane spacing d_0 and in-plane stress σ in each film from the least-square fits to the d vs $\sin^2 \psi$ data. Then, we can calculate the in-plane and out-of-plane strains in the film.

The plane spacing, $d(\psi)$, as a function of $\sin^2 \psi$ and least-square fits to the data for TbFe$_2$ films of 50, 100, 200, and 400 Å are shown in Figure 5(a). The error bars result from uncertainties in the energy of synchrotron radiation (energy error), 2θ (diffractometer error), and the position of the voigt peaks used to fit the diffraction data (fitting error). The data points agree with the linear fits within the error of the experiment. The positive slope of the fits indicates that the epitaxial TbFe$_2$(111) films are in in-plane tension. The largest tensile stress is in the 400 Å film and is about 0.88 GPa. A plot of the in-plane and out-of-plane strains of the TbFe$_2$ films is shown in Figure 5(b). The in-plane strains were calculated from the plane spacings at $\psi = 90°$ and the out-of-plane strains were calculated from the plane spacings at $\psi = 0°$. The tensile strains were likely caused by thermal strain from the sapphire substrate and lattice mismatch between Mo and TbFe$_2$. Since the thermal expansion coefficient of TbFe$_2$ ($12 \times 10^{-6}/°C$) is greater than that of sapphire ($8.8 \times 10^{-6}/°C$), we would expect the TbFe$_2$ film to develop a tensile stress when cooled from the deposition temperature.

VSM data for the 200 and 400 Å TbFe$_2$ films measured with the magnetic field parallel to the film plane are shown in Figure 6. Both films have an easy axis magnetization in plane. The total anisotropy in Figure 2(b) indicated that 0.5% tensile strain will cause the magnetization to lie in the film plane, which is consistent with the experimental data. The magnetic moments of the 50 and 100 Å TbFe$_2$ films are too weak to be measured with this apparatus. The coercivity (H$_c$) of the 200 Å film is about 10.5 kOe and the magnetic moment is approximately 273 emu/cm^3. The coercivity (H$_c$) of the 400 Å film is about 8.5 kOe and the magnetic moment is 486 emu/cm^3. The magnetic moment of single crystal TbFe$_2$ is about 800 emu/cm^3. The lower magnetic moment in the epitaxial TbFe$_2$ films is most likely caused by oxidation and structural defects in the films.

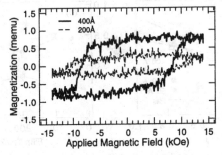

Figure 6: In-plane magnetic measurement of the 200 and 400 Å TbFe$_2$ films.

CONCLUSIONS

(111)-oriented epitaxial TbFe$_2$ films have been successfully grown on the (11$\bar{2}$0) sapphire substrates with 1000 Å (110)-oriented epitaxial Mo buffer layers using DC magnetron sputtering. XRD indicates the epitaxial TbFe$_2$ film has twins rotated by 60°. RBS data show the averaged concentrations of Mo and O in the TbFe$_2$ films decrease with the film thickness, indicating the averaged composition will be better in the thicker films. The TbFe$_2$ films are in tensile strain states and with in-plane magnetizations, consistent with our total anisotropy calculation. The tensile strains in the TbFe$_2$ films are caused by thermal strain from the sapphire substrate and lattice mismatch strain from the Mo buffer layer. In order to obtain compressive strains in the TbFe$_2$ films, the substrates with larger thermal expansion coefficients and buffer layer with compressive lattice mismatches should be used. MgO substrates and a Cu buffer layer would be good candidates.

ACKNOWLEDGMENTS

The authors would like to acknowledge the assistance of J. Xu, K.P. Fahey, J. Martinez, B.J. Daniels, M.C. Kautzky, and T. Ueda in the acquisition of synchrotron x-ray scattering data. Funding for this project was provided by NSF under contract No. DMR-91002920-A2. SSRL is funded by the U.S. Department of Energy (DOE), Office of Basic Energy Sciences.

REFERENCES

1. A.E. Clark, in Ferromagnetic Materials, edited by E. P. Wohlfahrt (North-Holland, Amesterdam, 1908), p. 531.
2. F. Schatz, M. Hirscher, M. Schnell, G. Filk, and H. Kronmüller, J. Appl. Phys. **76**, 5380 (1994).
3. S. Chikazumi and S.H. Charap, Physics of Magnetism (Kreiger , Malabar, Florida, 1986).
4. J.A. Bain, Ph.D. Dissertation, Stanford University, 1993.
5. J. Kwo, M. Hong, and S. Nakahara, Appl. Phys. Lett. **49**, 319 (1986).
6. C. Sürgers and H.V. Löhneysen, Thin Solid Films **219**, 69 (1992).

INTERFACE QUALITY AND MAGNETIC PROPERTIES OF
τ MnAl/Co SUPERLATTICES ON GaAs

C. Bruynseraede, J. De Boeck, W. Van Roy, G. Lauhoff*, , H. Bender, A. Van Esch, R. Mertens, J.A.C. Bland* and G. Borghs
IMEC, Kapeldreef 75, B-3001 Leuven, Belgium
* Univ. of Cambridge, Dept. of Physics, Cavendish Laboratory, Cambridge CB3 0HE, UK

Abstract

τ MnAl / Co superlattices are grown epitaxially on GaAs (100) substrates by molecular beam epitaxy. High angle XRD spectra are analysed and indicate that good structural quality is achieved. The behaviour of the superlattice upon annealing is described and compared with in-plane and out of plane magnetization data. Extraordinary hall effect measurements show low field switching comparable to MnAl thin films. Arguments on how structural properties are affecting the magnetic properties are discussed.

Introduction

τ MnAl and related intermetallic compounds, epitaxially grown on GaAs (001) have been studied for their interesting properties [1]. Epitaxy on the semiconductor substrate gives high quality films with the τ MnAl c-axis normal to the growth plane. This results in perpendicular magnetization due to the large magnetocrystalline anisotropy in τ MnAl. Polar magneto optic Kerr effect (MOKE) and extraordinary Hall effects (EHE) at moderate fields (± 10^4 Oe) have been described. Other Mn-based systems epitaxially grown on (001) GaAs include MnGa thin films [2] and multilayers of τ MnAl/NiAl [3].
Even more flexibility in tayloring the magnetic properties of the epitaxial heterostructures is anticipated from including a second epitaxial magnetic thin film in the heterostructures such as in the case of τ MnAl /Co [4]. In the following we describe growth experiments and sample analysis mainly by XRD and AGFM. We will concentrate on the interface quality and the effect of annealing on the structural and magnetic properties of the superlattices. We will discuss some phenomena that occur in these MnAl / Co heterostructures which demonstrate that excellent interface control will be essential in tuning the magnetic properties.

Experimental conditions

MBE growth includes the deposition of an AlAs buffer layer on GaAs (100) using standard conditions. After the deposition, at room temperature, of an amorphous template of 2.5 monolayers $Mn_{0.5}Al_{0.5}$, the sample is heated to initiate solid phase epitaxy. Upon the crystallized τ MnAl surface, the alternate deposition of MnAl and Co layers takes place at a substrate temperature of about 220 - 250 °C. RHEED is monitored during the superlattice growth and oscillations are recorded for establishing the deposition rate and for control of the completion of the τ MnAl monolayers.
We will show results of the sample analysis using X-ray diffraction (XRD), alternating gradient field magnetometry (AGFM) and extraordinary Hall effect (EHE).

Results

RHEED images during growth indicate the epitaxial relationship of the Co and the τ MnAl on GaAs (100). It is expected that the Co structure in these few monolayers thick films is dictated by the MnAl lattice. As a consequence, the often quoted forced bcc Co [5] phase should originate.
- Analysis of as grown superlattices:

85

X-ray diffraction analysis was used to study the crystal quality of the multilayers. It is well known that a high-angle XRD spectrum can yield detailed information on the quality of the interfaces as well as on the individual layer thicknesses, strain, etc. Furthermore, in comparison to a low-angle spectrum, it is less prone to systematic errors induced by small changes in the sample alignment. A high-angle XRD spectrum of a 6.5 x {τ MnAl(17Å)/Co(4Å)} superlattice is presented in Fig 1a, while Fig. 1b shows the spectrum of a 6.5 x {τ MnAl(18Å)/CoAl (3Å)/Co(4Å)/CoAl(3Å)}- superlattice, essentially the same structure whereby a CoAl unit cell is added at every MnAl-Co interface. Even for these small thicknesses, a clear multilayer structure can be observed. In both cases the central peak at 59° together with several satellite peaks and finite size peaks (between the satellite peaks) can be clearly observed and are indicative for a good superlattice structure. Higher order satellites at the right hand side of the central peak are hidden by the GaAs substrate peak at 66°.

The XRD spectra taken from the superlattices are fitted and best fits were obtained assuming a bcc Co structure. Fits to the spectra using the SUPREX program [6] are also shown in Fig.1. Excellent fits to the experimental data can be achieved. The analysis for the 6.5 x {τ MnAl(17Å)/Co(4Å)}-superlattice (nominal values) indicates that the true thicknesses are 16Å for the MnAl layer and 5.5Å for the Co layer. The τ MnAl and Co lattice spacing are respectively 1.6Å and 1.33Å, i.e. a little contracted in comparison with the theoretical values (Bulk τ MnAl: 1.75Å and "bcc" Co: 1.41Å). It should be noted that the spectrum can be best fitted when a intralayer disorder was included whereby an asymmetrical strain profile was assumed : we find a 0.02Å fluctuation on the MnAl lattice spacing , a contraction at the top τ MnAl-Co interface and an expansion in the lowest τ MnAl planes. For the Co-layers we find that the Co spacing for the Co layers growing on MnAl is contracted, while it is expanded for the Co layers closest to the next τ MnAl layer.

The fit to this spectrum also enables us to estimate the roughness on the individual layers. For crystalline/crystalline - systems, it is known that discrete fluctuations on the layer thicknesses lead to a broadening of the superlattice peaks in a XRD spectrum. However, one needs to be careful, since for an almost lattice matched system (e.g. Au/Ag with a 0.2% lattice mismatch) there is hardly any broadening and discrete thickness fluctuations cannot be easily inferred from the high-angle spectrum. It is therefore difficult to distinguish between continuous fluctuations on the interface width and discrete fluctuations on the crystalline layer thicknesses. In that case the high-angle spectrum only gives a lower limit for the discrete roughness and it is necessary to study the low-angle XRD spectrum . For our system, the theoretical mismatch is 1.8% and fits on the high-angle spectrum result in a neglectable continuous interface fluctuation width and a lower limit for the discrete roughness of about 1 monolayer for the τ MnAl layers with an error margin of 1 monolayer. This value of a one monolayer discrete roughness is confirmed by fits on low-angle XRD-spectra (not shown).

The same analysis has been made for the 6.5 x {τ MnAl(18Å) / CoAl(3Å) / Co(4Å) / CoAl(3Å)} superlattice and very comparable results emerge from this analysis. In this spectrum (Fig. 1b), more satellite peaks can be observed than in Fig.1a, due to the smaller modulation length of this superlattice and the slightly better layer quality - as indicated by fits - when a monolayer CoAl is inserted at every interface.

- Analysis of annealed superlattices:
While it has been observed [7] that the optimum temperature for obtaining bcc Co on GaAs is around 150 °C, MnAl magnetic quality is optimum for growth temperatures around 250 - 280 °C and was found to improve upon post growth annealing. This improvement is attributed to increased ordering in the τ MnAl lattice [8]. Attempts to achieve this ordering in situ by growing in an atomic layer epitaxy mode (alternate Mn / Al monolayer deposition) resulted in excellent RHEED patterns but poor low field magnetic properties. Hence it was decided to investigate the influence of post growth annealing on the superlattices.

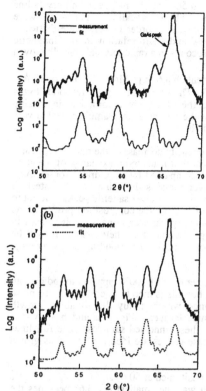

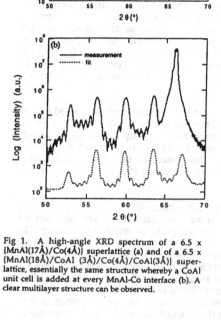

Fig 1. A high-angle XRD spectrum of a 6.5 x {MnAl(17Å)/Co(4Å)} superlattice (a) and of a 6.5 x {MnAl(18Å)/CoAl (3Å)/Co(4Å)/CoAl(3Å)} super-lattice, essentially the same structure whereby a CoAl unit cell is added at every MnAl-Co interface (b). A clear multilayer structure can be observed.

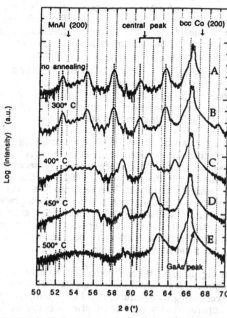

Fig. 2 High-angle XRD spectra of a 6.5x {MnAl(20Å) /CoAl(3Å) /Co(8Å)/ CoAl(3Å)} superlattice annealed (for 30 sec) at various temperature. A: As grown; B 300 °C; C: 400 °C; D: 450 °C; E: 500 °C.

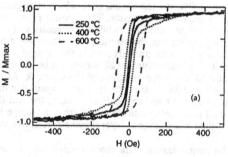

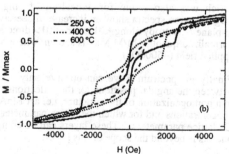

Fig. 3. A selection of alternating gradient field magnetometer data from annealed samples from the 6.5x {MnAl(20Å) /CoAl(3Å) /Co(8Å)/ CoAl(3Å)} superlattice (see also Fig 2). Curves a were taken with the field parallell to the plane, Curves b with the field perpendicular. (Note the horizontal scale difference (factor 10) between (a) in plane and (b) perpendicular.

In this section the effect of *ex-situ* annealing at 300° to 500° on the magnetic and crystalline properties of τ MnAl/Co superlattices is described. These annealing experiments were performed in combination with XRD and AGFM analysis in order to study interdiffusion phenomena occurring in these MnAl/Co- structures and their relation to the magnetic properties. The ex-situ annealing was done during 30 seconds in a rapid thermal annealing oven under forming gas atmosphere (N_2 - hydrogen mixture).

High-angle XRD spectra of such an annealing experiment on a 6.5x{τ MnAl(20Å) / CoAl(3Å) / Co(8Å) / CoAl(3Å)} superlattice are shown in Fig. 2. The upper curve A corresponds to the state of the sample before diffusion sets in, while the others are recorded at increasing annealing temperatures. A fitting on the spectrum of the non-annealed sample revealed very comparable quality features as dicussed for Fig.1a and 1b.

Roughness was shown to lead to a broadening and an intensity decrease of the multilayer peaks. Interdiffusion has another effect on the spectrum, because it leads to an exchange of the atom species (chemical disorder)which is not necessarily accompanied by structural disorder. As a result of this the modulation length in the multilayer remains the same but the interface becomes more diffuse. This will lead to a decrease of the multilayer satellite peaks, but not to broadening. In other words, interdiffusion gradually fades out the composition profile in the multilayer while the central peak of the average lattice parameter remains or increases in intensity. At a certain temperature, complete homogenization of the heterostructure will be achieved and the diffraction spectrum will only show a peak corresponding to the average lattice parameter of the materials involved.

In Fig. 2, curve B corresponding to an annealing temperature of 300°C appears to indicate an improved layer quality in comparison with curve A, but this is merely due to a reduced noise level during data acquisition. In this temperature range, which is only slightly above growth temperature, no significant effects occur in the superlattices. Fig. 3 a and b show the magnetization data, in-plane and perpendicular, for these annealed samples. The full lines show the data for samples annealed at 250 °C. These curves are identical to the as-grown situation.

From 400°C on, the central peak which is originally positioned at 60.6° shifts to higher angles and increases in intensity. This central peak, which was the smallest at 60.6° becomes the dominant one at 62.8°, while the intensities of the satellite peaks are steadily decreasing. The strong increase of the central peak reflects the creation of a mixed region in which the average lattice parameter prevails , while the layered structure is gradually lost by interdiffusion. The position of this central peak, which can be interpreted as the reflection from a random MnAlCo alloy, shifts in the direction of the Co peak. At the same time, hardly any broadening of the satellite peaks is observed, indicating again that interdiffusion is the prevailing mechanism and not interface roughening. Furthermore, the fact that the satellite peaks do not remain at their initial positions , indicate that there is a change in the modulation length. This effect can be ascribed to a change in layer thicknesses, or to a transformation of Co to an hcp-like structure.

The magnetization data show a reduction of coercivity in t MnAl (Fig 3 b) with annealing, which was also found for annealing of single τ MnAl thin films [8]. The in-plane magnetization spectra show an increase in coercivity for Co with the onset of remanence. No in-plane anisotropy along <011> or <001> directions of GaAs was found in these layers. The normalized quantity M/M Max is shown, where M Max is the magnetisation at the maximum applied field (~5000Oe).

Finally in spectrum E we can observe only a single peak which is located about halfway between the angular positions for the reflections of pure MnAl and Co. This peak corresponds to a homogenization of the structure, i.e. a τ MnAl-Co alloy with nearly equal τ MnAl and Co concentrations and for which the lattice parameter of the alloy is the average of the τ MnAl and Co lattice parameters. These layers can be compared with as grown MnAlCo thin films [9], but this is beyond the scope of this paper.

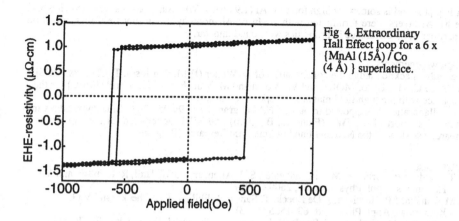

Fig. 4. Extraordinary Hall Effect loop for a 6 x {MnAl (15Å) / Co (4 Å) } superlattice.

Discussion

As mentioned above, no in-plane anisotropy is found for the magnetization in these τ MnAl/Co superlattices. This is in contrast with the observations in Co/CoAl epitaxial multilayers on GaAs [10] and with previous reports of Co on GaAs [11]. In the former case evidence of a faulted fcc structure was found, while in the latter case transformation to hcp was claimed. These results indicate a relation between the in-plane anisotropy and the Co crystal structure and are presently under further study.

Besides the good structural properties of the superlattices, they also reveal the perpendicular magnetization and hence the EHE -effect. The EHE curve in Fig.4 is taken from a MnAl / Co Superlattice and is very comparable to thin film MnAl EHE measurements [1]. Although, the signal magnitude is lower (factor of 4). Further, MOKE data on these samples show a smaller Kerr rotation compared to the MnAl thin films [12]. The coercivity of the measurement in Fig.4 is also smaller than what is regularly achieved in MnAl films and a better understanding of the interaction between structural and magnetic properties might lead to further optimisation of these films in perspective of applications such as MRAM.

A complete understanding of the magnetic properties of the superlattices cannot be achieved through low-field characterisation (to about 1T) only. High field MOKE experiments have been performed recently by G. Lauhoff and J.A.C.Bland [13] on these τ MnAl/Co superlattices. The MOKE measurements to 7 T show that the magnetisation is not saturated at the low field used in the AGFM characterisation. Hence, in Fig. 3, we normalize to M_{max} instead of the usual M_{sat}. The high field MOKE study has further revealed, for the first time, high field features that indicate coupling effects between the Co and τ MnAl layers in the superlattice. These features were later corroborated using EHE [14]. The lower magnitude of the Kerr effect and the EHE at low fields can also be attributed to the magnetic coupling effects occuring in the superlattice. These high field MOKE data and the interpretation will be published elsewhere [15].

Theoretical predictions by Van Leuken and de Groot [16] showed that coupling between τ MnAl and Co layers would be anti-ferromagnetic when a Mn-layer was terminating the τ MnAl. The predicted magnetic moment per Co atom is 1.67 and for Mn in MnAl is 2.7, which agree well with other reported values for Co (1.68 - 1.72, 1.76 μ_B for bcc Co [17], and 1.72 μ_B

for hcp [18] and is somewhat high for τ MnAl (1.94 μ_B [19]). Superlattices were grown where the MnAl layers were terminated with a Mn or Al monolayer. In low field magnetization experiments, no difference in properties was found thus far.

Acknowledgements
The authors acknowledge the collaboration of D. Weller (IBM, San Jose), T. Cheecks (UCSD) and T. Sands (UCB) for MOKE and VSM data and M. Van Hove and E. Carpi (IMEC) for their assistance with the high field magnetotransport measurements.
The collaboration is supported in part by NATO grant CRG 94 05 23. C.B. acknowledges the financial support of the IWT (Flanders, Belgium) and WVR acknowledges the support as a research assistant of the National Fund of Scientific Research (Belgium).

References
1. T. Sands, J.P. Harbison, M. L. Leadbeater, S. J. Allen, Jr. , G.W. Hull, R. Ramesh, and V.G. Keramidas, Appl. Phys. Lett. **57**, 2609 (1990)
2. M. Tanaka, J.P. Harbison, J. De Boeck, T. Sands, B. Philips, T.L. Cheecks, and V.G. Keramidas, Appl. Phys. Lett. **62**, 1565 (1993)
3. M. Tanaka, J.P. Harbison, T. Sands, B. Philips, T.L. Cheecks, J. De Boeck, L.T. Florez and V.G. Keramidas; Appl. Phys. Lett. **63**, 839 (1993)
4. J. De Boeck, C. Bruynseraede, H. Bender, A. Van Esch, W. Van Roy, G. Borghs, accepted for publication in J. Cryst. Growth (1995)
5. G.A. Prinz, Phys. Rev. Lett. **54**, 1051 (1985); F. Xu, J.J. Joyce, M.W. Ruckman, H.W. Chen, F. Bosherini, D.M. Hill, S.A. Chambers, and J.h. Weaver, Phys. Rev. **B35**, 2375 (1987)
6. E. Fullerton, I. K. Schuller, H. Vanderstraeten, and Y. Bruynseraede, Phys. Rev. **B45**, 9292 (1992)
7. S.J. Blundell, M. Gester, J.A.C. Bland, C. Daboo, E. Gu, M.J. Baird, and A.J. Ives, J. Appl. Phys. **73** 5984 (1993)
8. W. Van Roy, J. De Boeck, H. Bender, C. Bruynseraede, A. Van Esch, and G. Borghs, J. Appl. Phys. **78** (July 1, 1995) in print.
9. J. De Boeck, W. Van Roy, A. Van Esch, G. Borghs, P. Van Mieghem, R. O'Handly, J.A. del Alamo, M. Tanaka, J.P. Harbison, T. Cheeks, T. Sands, J. Electron Mater. 22 (7A) A50 (1993)
10. J. De Boeck, C. Bruynseraede, H. Bender, A. Van Esch, W. Van Roy, and G. Borghs, J. Appl. Phys. **76** 6281 (1994)
11. A. Moghadam, J.G. Booth, D.G. Lord, J. Boyle, and A.D. Boardman, IEEE Trans. on Magnetics, **30** 769 (1994) and ref [7]
12. D. Weller, T. Cheeks (Private communication)
13. G. Lauhoff and J.A.C. Bland (Private communication)
14. W. Van Roy, PhD Dissertation, K.U. Leuven 1995
15. G. Lauhoff, et. al., to be published
16. H. van Leuken and R.A. de Groot (Private Communication)
17. A.J. Freeman, A. Continenza, and C. Li, MRS Bulletin September 1990, pp. 27 - 33; and A.Y. Liu, and D. J. Singh, J. Appl. Phys. **73**, 6189 (1993)
18. D. Jiles "Introduction to magnetism and magnetic materials", (Chapman and Hall, London, 1990)
19. P.B. Braun and J.A. Goedkoop, Acta Cryst. **16**, 737 (1963)

GROWTH AND CHARACTERIZATION OF FePt COMPOUND THIN FILMS

M.R. VISOKAY and R. SINCLAIR
Department of Materials Science and Engineering, Stanford University, Stanford CA 94305

ABSTRACT

FePt alloy films were deposited at 50 and 490 °C onto amorphous SiO_2 and single crystal [001] MgO and [0001] Al_2O_3 using DC magnetron cosputtering, resulting in polycrystalline and [001] and [111] epitaxial films, respectively. High temperature deposition resulted in ordered films with the tetragonal $L1_0$ structure and out-of-plane magnetic easy axes while low temperature deposition yielded chemically disordered fcc alloys with in-plane easy axes. Significant modification of the magneto-optic Kerr spectrum is observed for ordered relative to disordered alloys for all orientations. The Kerr rotation has a strong orientation dependence for the ordered, but not disordered films.

INTRODUCTION

FePt and CoPt alloy thin films have received significant attention as possible magneto-optic (MO) recording media owing mainly to the existence of an ordered intermetallic phase with exceptional magnetic properties [1]. Near-equiatomic alloys adopt the tetragonal $L1_0$ crystal structure and have extremely strong magnetocrystalline anisotropy along [001], the unique c-axis [2-5]. Thin films with this structure have been found to have perpendicular magnetic anisotropy in the [111] [6,7] and [001] [7-10] orientations; in both cases the [001] easy axis is out of the film plane and therefore the magnetocrystalline anisotropy can be used to overcome that due to the thin film geometry, which favors in-plane magnetization. Several methods have been used to produce such films, including post-deposition annealing of both disordered alloys [6,11] and epitaxial multilayer precursors [8,9], which resulted in [111] textured and [001] epitaxial alloys, respectively. Simplification of this processing, namely direct formation of epitaxial ordered FePt and CoPt films by MBE [10,12] and sputtering [7] has also been reported. This paper describes the direct formation of polycrystalline and epitaxial FePt thin films by DC magnetron cosputtering and in particular the effect of both deposition temperature and substrate choice upon the resulting structural and magnetic properties.

EXPERIMENTAL

The alloys described here are equiatomic FePt produced directly by DC magnetron cosputtering from elemental sources. The film structure is: 0.5 nm Fe/5 nm Pt/100 nm FePt, which was produced simultaneously on [001] MgO, [0001] Al_2O_3 (sapphire) and amorphous SiO_2 substrates located within 1 cm of each other in the chamber. The Pt underlayer, which adopts the orientation relationships Pt{001} ∥ MgO{001} and Pt{111} ∥ Al_2O_3[0001], Pt(110) ∥ Al_2O_3(10$\bar{1}$0) [13,14], is present to allow epitaxial growth of the subsequently deposited alloy films at an arbitrary temperature. The Pt was formed at a substrate temperature of 490 °C while the alloys were deposited at 50 and 490 °C. Temperature was measured using a thermocouple in the heater head which was subsequently calibrated using an optical pyrometer. Actual, rather than nominal, temperatures are given here. The sputtering ambient was 3 mTorr Ar and the chamber base pressure was on the order of 1×10^{-8} and 1×10^{-7} Torr at room temperature and 490 °C, respectively. The composition was adjusted by changing the sputtering power of each source and the rates were measured using a quartz crystal monitor that was calibrated prior to film growth. The deposition rate was 0.1 nm/sec for the alloys and the substrates were rotated at approximately 20 rpm to insure compositional uniformity.

After deposition the film structure was characterized using transmission electron microscopy (TEM) and x-ray diffraction (XRD). TEM specimens were produced by standard grinding,

91

dimpling and ion milling techniques and the microscopy was performed using a Philips 430ST TEM operating at 300 kV. XRD experiments were performed using Philips 3100 two-circle and Philips MRD four-circle diffractometers, with CuK$_\alpha$ radiation in both cases.

Magnetic and magneto-optic properties were measured using vibrating sample, magneto-optic Kerr and torque magnetometry (applied field = 20 kOe for torque measurements). The variation of Kerr rotation with incident photon energy was determined in an applied field of 20 kOe using a Nihon Kagaku Kerr spectrometer for photon energies 1.24-3.54 eV (1000 nm ≤ λ ≤ 350 nm).

RESULTS AND DISCUSSION

Structural properties

The effect of substrate temperature upon the resulting film structure is shown in Figure 1 for 100 nm thick FePt films deposited at 50 and 490 °C onto [001] MgO. Symmetric XRD data show the presence of substrate and alloy (002) peaks in both cases, but strong alloy (001) diffracted intensity only for the high deposition temperature film. This peak confirms the formation of ordered FePt since it is forbidden for all the constituents as well as the disordered alloy. Plan-view selected area electron diffraction patterns (SADP) from the film deposited at 50 °C confirm that this alloy adopts a chemically disordered cubic structure, indicating that low temperature formation is analogous to quenching from above the ordering temperature in the bulk material. No orientations other than <001> are observed, indicating strong out-of-plane texture. For the film deposited at 490 °C the diffracted intensity on the low 2Θ side of the FePt (002) peak could be due to the Pt seed layer or either disordered or [100] oriented ordered FePt. Cross-section high resolution TEM (HRTEM) images, however, revealed that the Pt seed layer was not consumed during high temperature deposition and that no [100] FePt population was present. Symmetric XRD data from similar films grown with no Pt seed also show a complete absence of this diffracted intensity. The inset shows phi rocking curves from off-axis {111} planes for both the MgO substrate and FePt

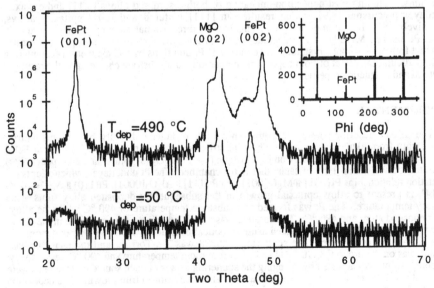

Figure 1. Symmetric x-ray diffraction data for 100 nm thick FePt alloy films deposited at 50 and 490 °C onto [001] MgO. The inset shows phi rocking curves from the {111} planes of the MgO substrate and FePt film deposited at 490 °C.

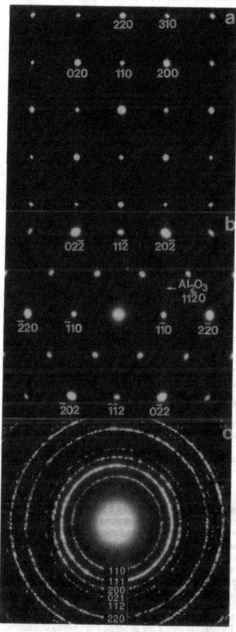

Figure 2. Plan-view SADP of 100 nm thick FePt deposited at 490 °C on a) [001] MgO, b) [0001] Al₂O₃ and c) SiO₂.

film deposited at 490 °C. The data show the four-fold symmetry expected for [001] materials, and that there is no relative in-plane rotation between the two, establishing the relationship FePt(110) ∥ MgO(110). The same orientation was observed for the film deposited at 50 °C.

The effect of substrate choice upon the resulting film structure is established in Figure 2 which shows plan-view SADP for 100 nm thick FePt films deposited simultaneously onto [001] MgO (Figure 2a), [0001] Al₂O₃ (Figure 2b) and amorphous SiO₂ (Figure 2c) at 490 °C. This is the same deposition for which data from the MgO sample are shown in Figure 1. Figure 2a was taken near the [001] zone axis and measurement of the (110) peak contained in this pattern and the (001) from the symmetric x-ray scan confirm the tetragonality of the alloy, with $c/a = 0.97$. No (001) intensity is observed, further demonstrating that no [100] out of plane material is present. As revealed by both plan-view and cross-section TEM imaging, the typical defect structures for these [001] films are orthogonally oriented [111] twins with a spacing on the order of 100 nm and a high density of anti-phase boundaries.

Deposition onto [0001] Al₂O₃ results in a strong [111] out of plane texture as observed using symmetric XRD. The [111] oriented SADP (Figure 2b) is consistent with this and shows six-fold symmetry of both the {2$\bar{2}$0} fundamental and (1$\bar{1}$0) superlattice spots, indicating both strong ordering and that there are at least 3 distinct orientation variants present since a singly oriented film would have only one pair of (1$\bar{1}$0) spots. Furthermore, phi rocking curves show six-fold symmetry of the off-axis {002} and (001) peaks for the Pt seed and FePt alloy layers, respectively, demonstrating that the Pt forms with two [13,14] and the FePt with six orientation variants. HRTEM images show that the microstructure consists of randomly distributed domains of each orientation variant that are as small as 10 nm across. Weak Al₂O₃ diffraction spots are also present in which show that each domain is epitaxially related to the substrate with an in-plane orientation relationship of the type FePt(1$\bar{1}$0) ∥ Al₂O₃ (10$\bar{1}$0).

The film deposited onto SiO₂ (Figure 2c) shows a ring pattern characteristic of a random in-plane orientation. Although all allowed spacings, both superlattice and fundamental, are observed, the (00ℓ) rings are extremely weak and were only observed for long exposure times and are not contained in the figure. A c/a ratio of 0.97 was derived from the measured radii. Thinner alloy layers (~40 nm) showed strong [111] out of plane and random in-plane texture; it is not known why the growth orientation changes as the film becomes thicker, but the overall result is a nearly randomly oriented alloy. The microstructure is polycrystalline with grain size on the order of 50-100 nm.

Magnetic properties

The magnetic properties of these alloy films are strongly affected by both the crystal structure and orientation. Most striking is the effect of deposition temperature, which was shown in the last section to determine whether the fcc or tetragonal $L1_0$ structure forms. The effect of deposition temperature upon the resulting magnetic properties is shown in Figure 3, the bulk of which contains the Kerr spectra (the variation of saturation Kerr rotation with incident photon energy) for FePt films deposited on [001] MgO at 50 and 490 °C, and are thus disordered and ordered, respectively. The inset shows the corresponding Kerr hysteresis loops (field applied perpendicular to the film plane) for an incident wavelength of 600 nm. The hysteresis loops clearly show that a switch from a hard to an easy magnetic easy axis out of the plane accompanies the formation of [001] oriented ordered FePt, as expected due to the strong magnetocrystalline anisotropy of this phase. The low coercivity is somewhat surprising considering the observed twins and antiphase boundaries, which would be expected to act as domain wall pinning sites. The hysteresis loops suggest that this is not the case, however.

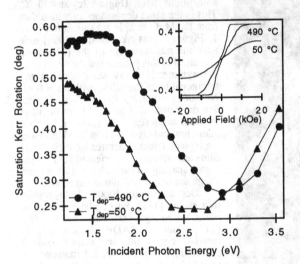

Figure 3. Magneto-optic Kerr rotation vs. incident photon energy for FePt films deposited onto [001] MgO at 490 and 50 °C. The inset shows the corresponding Kerr hysteresis loops taken using an incident wavelength of 600 nm.

The Kerr spectra show that a significant modification occurs upon ordering, and that ordered FePt has a higher angle of rotation for most of the energy range measured. This effect has also been observed for FePt formed by MBE [10,12] and by annealing epitaxial multilayer precursors [15], and recently reported ab-initio calculations show that this is consistent with the change in atomic environment associated with ordering [10]. For energies near 2 eV, the rotations obtained from these films (as high as 0.72°) compare quite well with competing structures such as Co/Pt multilayers. These results clearly show that high deposition temperatures that lead to ordered phase formation yield films with superior properties relative to those deposited at low temperature.

Owing to the strong anisotropy present in FePt with the $L1_0$ structure, it is reasonable to expect that the crystallographic orientation will also affect the resulting properties. This is shown to be the case in Figure 4, which contains Kerr hysteresis loops for ordered FePt films deposited simultaneously onto [001] MgO, [0001] Al_2O_3 and amorphous SiO_2. The [111] and randomly oriented films deposited onto Al_2O_3 and SiO_2 have extremely large coercivities, on the order of 9 kOe, compared to the [001] film on MgO, for which the coercivity is about 0.5 kOe. This difference can largely be attributed to the film microstructure, rather than any inherent anisotropy, however. The [001] alloy has no high angle grain boundaries or second phase particles that can act as domain wall pinning sites, while the other two films can be regarded as polycrystalline in nature, thus leading to higher coercivities. While deposition onto Al_2O_3 led to local epitaxy, six distinct in-plane orientation variants formed, resulting in an effectively polycrystalline structure. Since coercivity is largely controlled by microstructure, it is likely that the relatively poor magnetic properties of the [001] alloys can be overcome by proper processing and microstructural engineering: i.e. by introducing grain boundaries and forming [001] textured, rather than epitaxial, films.

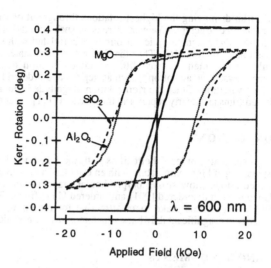

Figure 4. Magneto-optic Kerr hysteresis loops for ordered FePt films deposited simultaneously onto [001] MgO, [0001] Al_2O_3 and SiO_2.

A significant difference between the hysteresis loops that is related directly to the structural anisotropy is the magnitude of the saturation Kerr rotation, which is largely independent of microstructure. The [001] film clearly has a larger rotation than the other two for this wavelength. This is shown more completely in Figure 5 which contains Kerr spectra from two sets of films, one of which is ordered (deposited at 490 °C) (Figure 5a) and the other disordered (deposited at 50 °C) (Figure 5b). There is a substantial orientational variation for the ordered alloys, with the

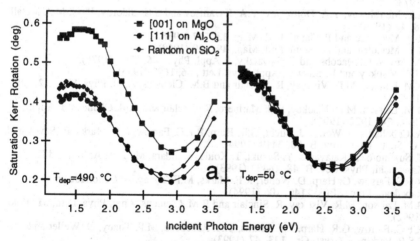

Figure 5. Magneto-optic Kerr rotation vs. incident photon energy for 100 nm thick FePt films deposited at a) 490 and b) 50 °C. For each temperature films were deposited simultaneously onto [001] MgO, [0001] Al_2O_3 and SiO_2. The scales are the same for both figures.

[001] film displaying the highest rotations throughout observed energy range. When the alloy adopts the fcc structure, however, there is no such orientation dependence. This is a manifestation of the fact that magneto-optic anisotropy is correlated with strong magnetocrystalline anisotropy present in a structure [16]. Thus the Kerr rotation for materials such as fcc Co or disordered FePt would be expected to show little or no orientation dependence, while those with strong magnetocrystalline anisotropy, such as hcp Co and ordered FePt, on the other hand, should show a strong variation. Since the Kerr rotation magnitude is microstructurally independent, the [001] oriented films are clearly superior to those with a [111] or random texture in this respect.

CONCLUSIONS

In summary, ordered FePt films with the $L1_0$ crystal structure can be formed directly by cosputtering at high temperature with an orientation determined by the choice of substrate. [001] oriented alloys show strong perpendicular magnetic anisotropy and enhanced Kerr rotations relative to both disordered [001] and ordered [111] films. The magneto-optic properties of the [001] ordered films make them attractive for recording applications, but the observed low coercivity is a significant drawback which requires further microstructural development.

ACKNOWLEDGMENTS

Funding from Kobe Steel U.S.A. under Stanford contract 9054 is gratefully acknowledged. Thanks are due to B.M. Lairson, B.M. Clemens, G.A. Bertero and T. Itoh for helpful discussions related to this work and to G. Temko for assistance with the magnetic measurements.

REFERENCES

1. T.B. Massalski, Binary Alloy Phase Diagrams, (American Society for Metals, Metals Park, (1986).
2. O.A. Ivanov, L.V. Solina, V.A. Demshina and L.M. Magat, Phys. Met. Metall., 35, 81 (1973).
3. A.Z. Men'shikov, Y.A. Dorofeyev, V.A. Kazantsev and S.K. Sidorov, Phys. Met. Metall., 38, 47 (1974).
4. R.A. McCurrie and P. Gaunt, Phil. Mag., 13, 567 (1966).
5. R.A. McCurrie and P. Gaunt, Phil. Mag., 19, 339 (1969).
6. D. Treves, J.T. Jacobs and E. Sawatzky, J. Appl. Phys., 46, 2760 (1975).
7. M.R. Visokay and R. Sinclair, Appl. Phys. Lett., 66, 1692 (1995).
8. B.M. Lairson, M.R. Visokay, R. Sinclair and B.M. Clemens, Appl. Phys. Lett., 62, 639 (1993).
9. B.M. Lairson, M.R. Visokay, E.E. Marinero, R. Sinclair and B.M. Clemens, J. Appl. Phys., 74, 1922 (1993).
10. A. Cebollada, D. Weller, J. Sticht, G.R. Harp, R.F.C. Farrow, R.F. Marks, R. Savoy and J.C. Scott, Phys. Rev. B, 50, 3419 (1994).
11. T. Sugimoto, T. Katayama, Y. Suzuki, T. Koide, T. Sidara, M. Yuri, A. Itoh and K. Kawanishi, Phys. Rev. B, 48, 16432 (1993).
12. R.F.C. Farrow, G. Harp, D. Weller, R.F. Marks, M.F. Toney, A. Cebollada and T.A. Rabedeau, Proc. SPIE, 2140, 106 (1994).
13. B.M. Lairson, M.R. Visokay, R. Sinclair and B.M. Clemens, Appl. Phys. Lett., 61, 1390 (1992).
14. R.F.C. Farrow, G.R. Harp, R.F. Marks, T.A. Rabedeau, M.F. Toney, D. Weller and S.S.P. Parkin, J. Cryst. Gr., 133, 47 (1993).
15. B.M. Lairson and B.M. Clemens, Appl. Phys. Lett., 63, 1438 (1993).
16. D. Weller, G.R. Harp, R.F.C. Farrow, A. Cebollada and J. Sticht, Phys. Rev. Lett., 72, 2097 (1994).

STRUCTURE AND MAGNETISM IN Mo/Co MULTILAYERS

C.L. FOILES, M.R. FRANKLIN AND R. LOLOEE
Department of Physics and Center for Fundamental Materials Research, Michigan State University, East Lansing, MI 48824

ABSTRACT

The magnetic and structural properties of Co in Mo/Co multilayers are an unresolved issue. Two independent studies report different structures for thicker Co layers in multilayers of comparable dimensions as well as significant differences in both the form and the layer thickness for a loss of ferromagnetism. Both studies base their structural claims on a limited number of Bragg lines. We have used x-ray diffraction, transmission electron diffraction, and EXAFS to probe the structure of Mo/Co multilayers having Co layer thicknesses from 4 to 63Å. Our structural and magnetic results for these samples are compared to those of the above studies and are contrasted with results for Mo/Fe multilayers having comparable dimensions.

INTRODUCTION

Non-equilibrium structures can be established in metallic multilayers and thereby alter macroscopic properties. The magnetic behavior of Co in Mo/Co is a possible but controversial example. A comparison of results in two published studies of sputtered Mo/Co multilayers produces limited points of agreement and significant differences. Reflection X-ray diffraction (XRD) in each study produces low angle peaks that confirm a layered structure. Both studies find highly disordered layers when the layer thickness is about 10Å or less and a near loss of magnetism occurs in this same layer thickness range. Agreement ends with these features and the studies report structural and magnetization results with fundamental differences for thicker layers. Sato used magnetron sputtering and his higher angle XRD data give a single dominant peak whose location is dependent upon the relative portions of Mo and Co in the bilayer unit [1]. For multilayers having equal amounts of Mo and Co with bilayer thicknesses of less than 40Å, this peak location is constant and by a thickness of 60Å it has split into Bragg peaks characteristic of the Mo and Co forming the bilayer. Wang, et al. used focused ion sputtering and they find very different higher angle XRD results [2]. Their data yield isolated Mo<110> and Co<111> peaks for multilayers with thicknesses comparable to those of Sato. In addition, a line from ε-Co_7Mo_6 is observed for many samples and the <002> peak for HCP Co is observed for thick Co layers. Transmission electron diffraction (TED) data reveal lines consistent with BCC Mo and 3 lines associated with ε-Co_7Mo_6. The dependence upon layer thickness for the room temperature saturation magnetization's of the two studies is also different. Consider the results for samples with equal amounts of Co and Mo in the bilayer. The studies agree that 10Å Co layers have a saturation magnetization less than 10% that of bulk Co but Sato reports a slow rise with increasing Co layer thickness, still less than 40% of bulk for 30Å Co layers, while Wang, et al. report a rapid rise that reaches >80% for 20Å Co layers and $\geq$ 95% for 30Å Co layers. The results for samples with differing amounts of Co and Mo in the bilayer vary somewhat in details but produce similar differences.

Mat. Res. Soc. Symp. Proc. Vol. 384 © 1995 Materials Research Society

Our earlier study of magnetron sputtered Mo/Fe multilayers had produced structural and magnetization patterns of behavior that are similar to a blend of those in these two Mo/Co studies [3]. A comparison of possible lattice mismatches in Mo/Co and Mo/Fe reveals interesting similarities and differences. Sputtered multilayers typically form with the most dense planes of its constituents parallel to the substrate. Bulk Mo and Fe have a BCC structure and the <110> plane is the most dense. The respective d-spacings for these planes are 2.225 and 2.027Å and this is a lattice mismatch of 8.9%. Whether Co is in its normal HCP phase, with the <002> plane being the most dense and having a d-spacing of 2.023Å, or in its FCC phase, with the <111> plane being the most dense and having a d-spacing of 2.046Å, the lattice mismatch in d-spacings is similar for both Mo/Fe and Mo/Co multilayers. However, the placement of atoms within the planes is clearly different for the different crystal structures. Comparing results for Mo/Fe and Mo/Co is one method of testing the importance of this in-plane atom arrangement. Given the conflicting results for the two Mo/Co studies using different sputtering techniques, a meaningful comparison between behavior in Mo/Co and Mo/Fe multilayers also requires additional data for Mo/Co. In the present paper we report our structural and magnetization results for magnetron sputtered Mo/Co multilayers and compare them to results for Mo/Fe multilayers, prepared in the same experimental system, and to results from the other studies of Mo/Co multilayers.

EXPERIMENTAL TECHNIQUES

The samples were prepared by DC magnetron sputtering in a chamber that can be evacuated to 8×10^{-8} Torr prior to sputtering. The sputtering environment was pure Ar at a pressure of 2.5 mTorr and 16 different samples could be made during each preparation run. A computer controlled substrate holder placed an unmasked substrate over the proper targets with a time pattern needed to obtain nominal thicknesses. The sputtering rates of the targets were determined by quartz crystal thickness monitors at the start of each preparation run and were remeasured several times during the run. A detailed description of this system has been published [4]. Sapphire and cleaved NaCl were used as substrates. The same pattern of individual layer thicknesses was used for both substrates but the total multilayer thicknesses differed. Samples on sapphire had a total thickness of about 2000Å while that for samples on NaCl substrates was about 500Å. Samples having the bilayer dimensions were prepared consecutively. Individual layer thicknesses were chosen to match an integral number of monolayers (ML) for the metals: 2.25Å and 2.05Å were used as the nominal ML thickness for Mo and Co, respectively.

Standard XRD using a rotating anode system with a Cu target and a graphite monochromator preceding the detector provided the initial structural characterization of the samples. Portions of the sapphire substrate samples were used for magnetization measurements in a SQUID magnetometer. These measurements were done at 5K and the field was typically applied parallel to the multilayer film. Portions of the multilayer films were floated off the NaCl substrates and onto Cu grids for TED studies. These were done in a field-emission STEM that permitted on-line examination of results and in a conventional TEM that recorded the results on film. For each sample in the TED study at least three different regions of the multilayer film were studied to confirm sample uniformity. EXAFS data were collected at the National Synchrotron Light Source on beam line X23A2 using a fluorescence detector.

RESULTS AND DISCUSSION

Low angle XRD data for our Mo/Co multilayers confirm that the samples are layered with the actual bilayer distances being within ±3% of the nominal values. The locations of all XRD peaks are independent of the substrate and the thinner samples give weaker signals as expected. As the bilayers of our Mo/Co samples vary from 5ML/5ML to 14ML/28ML the higher angle XRD data have a progression seen for numerous metallic multilayers [5]. As the following details indicate, the bilayer unit progresses from being highly disordered to being a well defined average crystalline unit and finally to being a crystalline unit whose constituent Bragg lines are resolved. The 5ML/5ML sample has a weak and broad line consistent with a d-spacing of between 2.135 to 2.156Å. The 7ML/7ML sample has a well defined line with a d-spacing of 2.145Å and four satellite lines, 2 below and 2 above. This d-spacing is consistent with a weighted composite line from the d-spacings of Co and Mo and is hereafter denoted as <D>. As the bilayer thickness increases this same pattern is maintained. For samples with equal amounts of Mo and Co the location of all <D> lines is consistent with 2.139 +/- 0.007Å and the satellite line locations shift consistent with the new bilayer distance. XRD data for the 14ML/14ML sample have this same pattern of locations although there is a slight change in the progression of intensities. For this last sample satellite lines are located very near the d-spacings for pure Mo and pure Co lines and the presence of pure element lines might account for the slight intensity change. Reliable quantification of this change is not yet possible. The higher angle XRD data for the 14ML/28ML sample are clearly different. Individual lines from Mo and Co are clearly resolved and these lines are flanked by satellite lines. The d-spacing for the Mo line is 2.208Å and the d-spacing for the Co line is 2.045Å. The relative error for our d-spacings is +/-0.003 Å. See Table I for a summary of results.

Given the possibility of strain in our multilayers, the preceding d-spacings are consistent with either a FCC<111> or HCP<002> line for Co and a BCC<110> line for Mo. Bragg lines having other indices are not observed in our XRD data and once again this is consistent with the high degree of texture that is typical of sputtered metallic multilayers [5]. Most of our samples with bilayer distances of 7ML/7ML or greater show second order XRD effects of the <D> line and satellites that confirm the interpretation given in the preceding paragraph. The 14ML/14ML sample has very poorly resolved second order effects and the 14ML/28ML sample has only the two pure element lines as its second order effect.

The TED data for our Mo/Co multilayers are consistent with the structural progression deduced from the preceding XRD data. All the TED patterns are dominated by rings. For the 5ML/5ML sample the pattern consists of one intense but broad ring and two weak, diffuse rings. This pattern is consistent with amorphous structure. The 5ML/7ML sample gives a TED pattern that suggests a mixture of crystalline and amorphous structures. For all samples with a bilayer distance of 7ML/7ML or greater, the TED patterns yield a sequence of strong Bragg lines that are consistent with a BCC structure. Typically 5 to 6 lines in the sequence are evident and we attribute these lines to the Mo layers. Evidence in the TED patterns for Co related lines is limited. The non-BCC lines are weaker, limited in number and do not always form complete rings. The 10ML/10ML sample shows 4 non-BCC lines: using bulk Mo parameters for the BCC lines as a standard, the d-spacings of the 4 lines are 2.05, 1.94, 1.06 and 0.81Å. Given the possibility of strain in the multilayers and the error limit of about +/- 0.04Å in our TED results for these lines, these values are consistent with Co_3Mo, the μ-phase of Co_7Mo_6 or HCP Co. In fact, with the intensity of this 1.94Å line being very weak, these results are also consistent with Co being a mixture of HCP and FCC structures.

Table I. XRD structural results for Mo/Co multilayers. <D> denotes a single line which we interpret as a composite line composed of the Mo<110> line and a Co line. The * denotes a weak, broad line which is more likely an indication of the nearest neighbor distances in an amorphous sample.

| Mo(ML)/Co(ML) | Bilayer Distance Å | | d-spacing of | # peaks from bilayer | |
	Nom.	Actual	<D> (Å)	satellites	low angle
5/5	21.4	21.8	2.135 to 2.156*	0	0
5/7	25.6	25.8	2.126	3	3
7/7	30.2	30.1	2.145	4	3
10/10	43.0	42.0	2.135	4	2
14/14	60.2	61.4	2.132	4	3
14/28	88.9	87.5	Mo<110> 2.208 Co<111> 2.045 or Co<002>	2 1	4

In comparing the present structural results with the earlier Mo/Co studies, a clear consistency with the results of Sato [1] emerges. The entire progression from amorphous, to crystalline with a dominant composite Bragg line (<D>), to Bragg lines for the individual elements as the bilayer distance increases is reproduced. Sato speaks of a dominant Bragg line in his XRD data and its d-spacings are consistent with those we observe for <D>. Only one major difference occurs. Our multilayers appear to maintain the composite structural behavior to a larger bilayer thickness than those of Sato. In his samples the individual element Bragg lines are evident for 60Å bilayer thicknesses while that does not occur in our samples until bilayer thicknesses of greater than 60Å. Our structural results are not consistent with those of Wang, et al. [2].

Comparing our structural results for Mo/Co and Mo/Fe reveals a similar structural progression as layer thicknesses increase but some of the details differ. As for the Mo/Co multilayers, XRD data for all Mo/Fe multilayers have low angle Bragg peaks that document layering. The structural behavior for very thin bilayers once again appears amorphous with XRD data giving a single broad peak at higher angles. In Mo/Fe, our data indicate the change from amorphous to crystalline structure occurs as the bilayer thickness increases by 2ML. A 3ML/3ML sample has a weak peak with a FWHM of greater than 4 degrees while a 4ML/4ML sample has a well defined composite <D> line with a FWHM of 0.8 degrees and a satellite line is observed. Using the FWHM to estimate a coherence length for this latter sample gives structural coherence over 7 bilayer units. As the bilayer distance for these Mo/Fe samples varies from 4ML/4ML to 15ML/15ML the location of the <D> remains essentially constant at 2.134 +/- 0.004Å. This abrupt onset of crystallinity seen in XRD data has a counterpart effect in the TED data. For the 6ML/6ML sample the TED data reveal only a single set of BCC Bragg lines. However, both the 10ML/10ML and 15ML/15ML samples have TED data that produce two well defined sets of BCC lines. Electron microscope focusing considerations prevent the determination of absolute spacings for these Bragg lines but the relative difference for the two sets of lines is determined precisely. That difference is constant within experimental error and has a value of 6.4 +/- 0.6%. The corresponding difference for bulk Fe and Mo is 8.9% and thus this result indicates that at least one (and more probably both) of the layers is (are) strained. The lack of a shift in the XRD composite <D> line is consistent with equal and compensating strains

Figure 1. Normalized saturation magnetization data for Mo/X multilayers with approximately equal amounts of Mo and X in the bilayer unit. The data connected by broken lines are room temperature results for Mo/Co from references [1], ◆, and [2] ■. The single ▲ is a 5K datum from [1]. The solid line data, ●, are 5K data from the present study and the ⊕ are 5K data for Mo/Fe from reference [3].

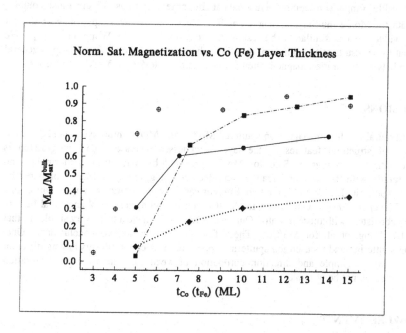

in the Mo and Fe d-spacings. This combination of a constant location for a composite <D> line in XRD data and the splitting of TED data from one common set of BCC lines into two sets was also observed in Fe/V multilayers [6].

Analysis of Co K-shell EXAFS data is still in progress and thus it is not possible to give any detailed results at this time. However, one qualitative feature of these data reinforces the validity of structure developing with increasing bilayer thickness as described above. The EXAFS data for the 5ML/5ML consists of a single decaying oscillation while data for the 5ML/7ML sample have some features of crystalline oscillations. Data for the 7ML/7ML sample have oscillations characteristic of crystalline Co. These data indicate the local structure as well as the average long range structure changes from amorphous to crystalline as the bilayer thickness is increased from 5ML/5ML to 7ML/7 ML in Mo/Co.

Figure 1 contains the magnetization results from a number of studies. All values have been normalized to the saturation magnetization at 5K for the pure ferromagnetic component. The points connected by broken lines are the room temperature saturation magnetization data for equal layer thickness Mo/Co samples from Sato [1] and Wang, et al. [2]. The single triangle datum is 5K data from Sato. The solid circles are data at 5K for the Mo/Co samples of the present study. The disagreement with our data and that from Wang, et al. is evident: their values approach pure bulk values at room temperature while ours remain substantially below that limit at 5K where the largest values occur. Any comparison of our data with that from Sato is complicated by the different temperatures for the measurements. Our data have a general dependence upon layer thickness comparable to that reported by Sato and our values would be expected to decrease at higher temperature. Such decrease should improve the agreement but, noting that his single datum at 5K is about 40% smaller than our corresponding result, it is clear that a believable comparison requires actual data at similar temperatures. Figure 1 also contains our 5K data for Mo/Fe multilayers, denoted as ⊕ . These data show a rapid increase to near pure bulk values that is similar to the results reported for Mo/Co by Wang, et al. [2]. We believe that these results raise significant questions about any model claiming a simple, general correlation between saturation magnetization and structure in Mo/Fe and Mo/Co multilayers.

CONCLUSIONS

Our structural results for magnetron sputtered Mo/Co and Mo/Fe multilayers yield a similar progression of structural features as the layer thicknesses increase. This progression is consistent with the findings of Sato for Mo/Co prepared by magnetron sputtering[1] and disagrees totally with the structural features reported by Wang, et al. for Mo/Co prepared by focused ion sputtering[2]. We observe two different variations of saturation magnetization as a function of layer thickness. Our results for Mo/Co conflict with those of Wang, et al but are reasonably consistent with those of Sato. Our results for Mo/Fe have a form very similar to that reported by Wang, et al. for Mo/Co. These findings lead us to two conclusions. One, magnetron sputtering and focused ion sputtering appear to produce multilayers having different structure. Two, a simple and direction correlation between such structure and saturation magnetization may not be possible.

ACKNOWLEDGMENTS

We gratefully acknowledge both the help of Viv Shull and H-K Sung in collecting some of these data, and support for this research from THE RESEARCH CORPORATION and the MSU Center for Fundamental Materials Research.

REFERENCES

[1] Noboru Sato, J. Appl. Phys. 63, 3476 (1988).
[2] Y. Wang, F.Z. Cui, W.Z. Li and Y.D. Fan, J. Magn.Magn.Mater.102, 121 (1991).
[3] H-K. Sung, "Structural Changes and Their Effects upon the Properties of Ultrathin Fe Layers", Ph.D. thesis, Michigan State University (1990); H-K. Sung, C.L. Foiles and T.I. Morrison, Bull. APS 36, 1045 (1991).
[4] J.M. Slaughter, W.P. Pratt, Jr., and P.A. Schroeder, Rev. Sci. Instrum. 60, 127 (1988).
[5] B.Y. Jin and J.B. Ketterson, Adv. in Phys. 38, 189 (1989).
[6] C.L. Foiles, Metallur. Trans. 23A, 1105 (1992).

MAGNETIC AND STRUCTURAL PROPERTIES OF IRON NITRIDE THIN FILMS OBTAINED BY ARGON-NITROGEN REACTIVE RADIO-FREQUENCY SPUTTERING

H. CHATBI, J.F. BOBO, M. VERGNAT, L. HENNET, J. GHANBAJA, O. LENOBLE, Ph. BAUER AND M. PIECUCH

LMPSM-URA-CNRS 155, Université Henry Poincaré BP 239 54506 Vandœuvre Cedex France

INTRODUCTION

Fe-N thin films have attracted considerable attention because they are potential candidates for magnetic recording with their large saturation magnetization and their good corrosion resistance[1]. It has even been demonstrated that saturation magnetization can be larger than the bulk iron one for low nitrogen contents[2]. The origin of the enhanced magnetic moment could occur from the metastable α'' $Fe_{16}N_2$ phase or from an expanded bcc FeN structure which is also called α FeN.

Several ways for obtaining iron nitride films have been investigated :

- thermal evaporation with a nitrogen partial pressure (either atomic or molecular nitrogen) : such a technique is not the most suitable for the preparation of nitrogen-rich Fe-N alloys, but several groups have successfully obtained α, α'' or γ FeN phases[3]. Let us also notice that MBE growth of α'' phase is possible, according to Komuro et al.[4].

- reactive sputtering : contrary to thermal evaporation, reactive sputtering allows to obtain a wide variety of iron nitrides. As underlined by Takahashi et al.[5] or Gao et al.[6], the use of adapted seed layers like (100) iron buffer grown on (100) MgO substrate accompanied with thermal annealing leads to α or α'' phases. The role of this thermal treatment is to re-order nitrogen atoms in the iron expanded lattice. From another point of view, Xiao and Chien[7] have sputtered all the iron nitrides (except α'') on unheated substrates and without any other treatment using ammonia reactive gas. These last authors suggest NH_3 is the best solution for growing single phase iron nitrides.

This work is devoted to preparation and study of as-deposited sputtered Fe-N films[8,9]. They have been prepared in a large range of nitrogen partial pressures and with substrate temperatures ranging from $\approx 40°C$ (unheated) up to 600°C. The different structural phases have been identified by X-ray diffraction and Mössbauer spectroscopy. These results are correlated with bulk magnetization measurements.

EXPERIMENTAL PROCEDURES

Iron nitride films are deposited on microelectronic-grade (100) Si wafers or carbon-coated TEM grids in an Alcatel SCM 650 automated sputtering set-up. The base pressure is 7.10^{-7} mb and the working pressure 3.10^{-3} mb. The iron target is 500 W RF-polarized (≈ 6.3 W/cm^2) and the deposition rate is close to 3 Å/s if substrates are located 10 cm above the target. Such sputtering conditions have been chosen because they provide high density and low roughness (110) textured Fe films in pure argon plasma sputtering. Note that this experimental context leads to a spontaneous (110) α Fe dense planes growth.

Nitrides have been obtained by introducing several controlled amounts of nitrogen in the main argon atmosphere, keeping the total pressure equal to the previous value of 3.10^{-3} mb. The nitrogen percentage x_{N2} in the gaseous flow ranges from 0 to 40%. The total thickness was 1250 ± 50Å for samples deposited on Si and 400Å for TEM grids. The substrates temperature is varied between room temperature (unheated substrate) and 600°C.

103

These samples were structurally characterized by X-Ray Diffraction (XRD) with a K_β-filtered Co K_α radiation (1.78892 Å) on a θ/2θ Philips goniometer operating with a Raytech Position Sensitive Detector. Crystallographic phases were deduced from comparison of experimental diffraction profiles with standard ones (JCPDS data). Some uncertainty is left for the determination of ε-$Fe_{2-3}N$ phase because of the relatively large variations of its crystalline parameters among the compositional domain where it exists.

Structure was also checked by Transmission Electron Microscospy (TEM) with a Philips CM20 microscope operating at 200 kV. Selected Area Electron Diffraction (SAED) results were consistent with XRD. Electron Energy Loss Spectroscopy (EELS) experiments were performed on this TEM fitted with a Gatan (model 666) spectrometer. We could measure the atomic abundances of iron and nitrogen from the respective intensities of their characteristic absorption edges, estimate the oxidation state of iron (L_2/L_3 intensity ratio) and get some qualitative informations about the atomic structure of the samples.

Local magnetic properties have been investigated by Conversion Electron Mössbauer Spectroscopy (CEMS). Mössbauer spectra were recorded at room temperature in the backscattering mode with a He (5% CH_4) gas flow proportional counter. This allows a non destructive study with a sampling depth of about 2500 Å, encompassing therefore the whole thickness of iron nitride films. The source drive and data storage were of usual design. The ^{57}Fe hyperfine pattern was fitted with standard routines where Lorentzian line shapes were assumed.

Bulk magnetization measurements have been performed with either a Vibrating Sample Magnetometer or a Quantum Design SQUID down to 5K. Room temperature Kerr rotation cycles have also been performed in both longitudinal and polar geometries, they give similar results than the VSM ones but faster and with a better accuracy for determining coercive fields.

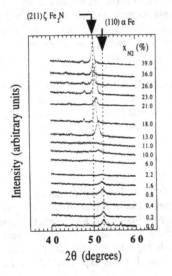

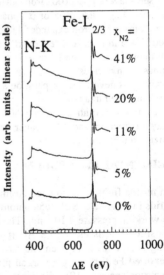

Figure 1 : XRD experiments performed on the series of Fe-N films deposited at room temperature as a function of the nitrogen concentration in the plasma.

Figure 2 : EELS spectra recorded for a series of samples deposited on unheated TEM grids and increasing nitrogen concentrations in the plasma ($0.00 < x_{N2} < 41\%$). The N/Fe atomic ratio evolves from 0 (Fe) to 0.5 (Fe_2N).

RESULTS

1. Crystallographic Structure and Stoichiometry

Structural properties of the films deposited on unheated substrates are illustrated in figure 1 which represents the evolution of XRD peaks as a function of the nitrogen partial pressure in the plasma. These results can be summarized as follows :
- At very low nitrogen concentrations, films only show the usual (110) diffraction peak of the bcc iron structure but with a small shift towards low angles. We attribute this to a lattice expansion of bcc Fe which is called α FeN. Lattice expansion reaches up to 0.7%.
- With increasing N_2 concentration, diffraction peaks become broader : it is the signature of an amorphous FeN alloy. This is the first observation, up to our knowledge, of an amorphous iron nitride phase.
- For x_{N2} equal to 0.08, an XRD peak appears at 51.4°, it could be the (111) peak of hexagonal ε-$Fe_{2-3}N$ phase, coexisting with amorphous Fe-N.
- For higher values of x_{N2} ($0.18 \leq x_{N2} \leq 0.24$), the (101) peak of hexagonal ε-Fe_3N phase is observed.
- Finally, for $x_{N2} \geq 0.26$, the diffraction spectrum only shows the (211) peak of the orthorhombic ζ-Fe_2N phase.

The samples deposited on heated substrates were better crystallized, mainly those which gave amorphous phase at room temperature. More precisely, increasing substrate temperature from 300°C to 600°C leads to a transformation into γ-Fe_4N of all the iron nitrides which could be obtained at room temperature. This behavior is easy to explain as the γ phase is the most stable in this range of temperatures. Of course, any combination of these situations could be found for intermediate temperatures. Finally, XRD indicates the presence of various iron nitride structures in our films. However, because of grain size or texture effects, we cannot accurately estimate their proportions in the samples. A summary of the identified phases vs. substrate temperature (T_S) and nitrogen atomic ratio (x_{N2}) is given in table I.

Figure 2 shows EELS spectra collected for a series of Fe-N samples deposited with increasing x_{N2}, spectra are presented after background substraction. One clearly sees the growth of the N K-edge contribution to the spectra. At the same time, the Fe L_2/L_3 intensity ratio remains characteristic of metallic iron[10] whatever the nitrogen concentration is. A quantitative estimation of the respective iron and nitrogen relative abundances is obtained by integrating the areas of N K-edge and Fe $L_{2/3}$-edge with a 60 eV integrating window. The nitrogen-to-iron atomic ratio in our films evolves from 0.00 for pure iron films up to 0.50±.05 for ζ-Fe_2N samples. It is therefore consistent with structural characterization.

Finally, we observe EELS oscillations spanning several eV above the edges. They can be interpreted, like EXAFS in X-ray absorption spectroscopy, to scattering effects of the incident electron with neighbours of the target atom. Remark that these oscillations are significantly reduced for amorphous Fe-N films, so it is proof that their interpretation is correct as they are structure-dependent. More investigations are being done in this topic.

2. Mössbauer Spectroscopy

Some experimental and calculated CEMS spectra are displayed in figure 3. They are representative of the evolution versus nitrogen concentration of the plasma (x_{N2}) and substrate temperature (T_S). The spectra analysis of crystallized phases was carried out mostly with superimposed discrete six line patterns (magnetic phases) and/or a quadrupole split doublet (non magnetic phases). For most of the magnetic nitrides, the intensity behaviour within the sextet indicates in-plane magnetization. Spectra of amorphous magnetic iron nitrides do not exhibit the discrete sextet of crystallized magnetic phases but a broad distribution, they have been analysed with a hyperfine field distribution according to the histogram method.

So, magnetic components have been attributed respectively to pure b.c.c. iron, to iron atoms with a slightly shifted hyperfine field (α expanded bcc iron), to amorphous magnetic Fe-N, to γ-Fe$_4$N multi-site compound and to ε-Fe$_{2-3}$N. According to their isomer shifts, paramagnetic components are found to be relevant to ζ-Fe$_2$N species. CEMS results have been found to be coherent with XRD along the main lines of our work with evidence for γ-Fe$_4$N at high T_S, ζ-Fe$_2$N for high x_{N2}, amorphous Fe-N for T_S close to room temperature and $x_{N2}\approx5\%$. However, some discrepancies exist for intermediate T_S and x_{N2}. In fact, while Mössbauer spectroscopy detects all the iron environments, only the best crystallized phases are revealed by X-ray diffraction when a mixture of various nitrides sets in.

Let us also notice that a small paramagnetic contribution is found in amorphous phase. This is due to the existence of non magnetic iron sites in amorphous Fe-N. Contrary to XRD, it has been possible to estimate the atomic abundances of the various phases from the analysis of the spectra.

600°C	α-Fe	γ-Fe$_4$N	γ-Fe$_4$N	γ-Fe$_4$N	γ-Fe$_4$N	γ-Fe$_4$N
400°C	α-Fe	α-Fe γ-Fe$_4$N ε-Fe$_{2-3}$N	α-Fe γ-Fe$_4$N ε-Fe$_{2-3}$N	γ-Fe$_4$N ε-Fe$_{2-3}$N	γ-Fe$_4$N	
200°C	α-Fe	α-Fe γ-Fe$_4$N ε-Fe$_{2-3}$N amorphous	γ-Fe$_4$N ε-Fe$_{2-3}$N	γ-Fe$_4$N ε-Fe$_{2-3}$N	ε-Fe$_{2-3}$N	ζ-Fe$_2$N
100°C	α-Fe	γ-Fe$_4$N ε-Fe$_{2-3}$N	γ-Fe$_4$N ε-Fe$_{2-3}$N amorphous	ε-Fe$_{2-3}$N	ε-Fe$_{2-3}$N	
unheated	α-Fe	amorphous	amorphous	amorphous ε-Fe$_{2-3}$N	ε-Fe$_{2-3}$N	ζ-Fe$_2$N
T_S $\quad x_{N2}$	*0.000*	*0.046*	*0.063*	*0.109*	*0.205*	*0.332*

Table I : "Phase diagram" of sputtered iron nitrides. Nitrogen to argon ratio increases along horizontal lines and deposition temperature increases along columns.

3. Magnetic properties

Concerning the magnetic properties of our samples, several studies have been led : magnetic anisotropy, saturation magnetization at room temperature and, lastly, ferromagnetic fluctuations in ζ-Fe$_2$N phase.

a. Saturation magnetization vs. x_{N2} of samples prepared at room temperature :

Figure 4-a reports saturation magnetization (Ms) dependence with x_{N2}. One clearly remarks the plateau at $\approx$1700 emu/cm3 for low x_{N2}. No enhancement of M_S can be observed as confirmed by CEMS spectra which show the usual 330 kOe hyperfine field sextet for these samples. For larger x_{N2}, M_S starts to decrease. Room temperature saturation magnetization sharply decreases to zero values for $x_{N2} > 0.24$, exactly when ζ-Fe$_2$N is the only phase in our samples. Therefore, ζ-phase is found to be paramagnetic at room temperature by magnetization measurements and Mössbauer spectrometry.

b. Magnetic anisotropy and coercivity of iron nitride films :

Hysteresis curves have been recorded for both in-plane and out-of-plane field geometries. In all the cases where nitrides are ferromagnetic (i.e. from α-Fe to ϵ-Fe$_3$N), in-plane magnetization curves saturate at low fields while several kOe are necessary to saturate the out-of-plane ones as a result of shape anisotropy. Figure 4-b shows the dependence with x_{N2} of the coercive fields (H$_C$) of iron nitride films deposited on unheated substrates. The increase of H$_C$ up to 140 Oe for $0.18 < x_{N2} < 0.24$ can be exactly correlated with the presence of ϵ-Fe$_3$N which is well crystallized. For lower nitrogen concentrations, H$_C$ has low values ($\approx$30 Oe), it is coherent with the poor crystallization of the corresponding films. The coercive field of our sputtered iron nitrides is low in all cases and therefore compatible with magnetic recording requirements.

c. Low temperature magnetic transition in ζ-Fe$_2$N films :

We have investigated low temperature magnetic behavior of iron nitride films prepared at room temperature with a nitrogen flow high enough for obtaining ζ-Fe$_2$N ($x_{N2} > 0.24$). The magnetic transition is characterized by Arrott plots : M^2(H,T) is plotted vs H/M at various temperatures. These curves are expected to be linear in a mean field model. Curie temperature is deduced from the M^2 vs H/M curve which passes by the origin. The good linear shape of these curves is a sign for the homogeneity of the samples. Our measurements show a decrease of the Curie temperature T$_C$ from 300K for $x_{N2} = 0.25$ down to 60K for $x_{N2} = 0.37$. These values are in agreement with those reported by Chen et al.[11] for bulk ζ-Fe$_2$N samples and their dependence vs x_{N2} is shown in figure 4-c.

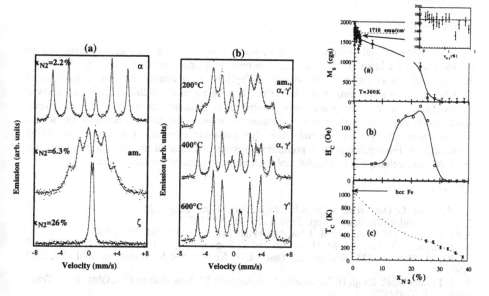

Figure 3 : CEMS spectra recorded for various preparation conditions of FeN films :
(a) effect of increasing x_{N2}, unheated substrates.
(b) influence of substrate heating for $x_{N2} = 6.3\%$.
The nature of the different species is reported.

Figure 4 : Summary of the magnetic properties of Fe-N films prepared at room temperature vs x_{N2} :
(a) saturation magnetization at 300K
(b) coercive field at 300K
(c) Curie temperature

DISCUSSION

Our results shed light on the problem of preparing iron nitride films by reactive sputtering : the use of nitrogen in reactive gas mixture is a good solution to obtain all the various Fe-N phases. We indicate the preparation conditions of single phase iron nitrides. Such phases can be obtained on unheated substrates except single phase γ'-Fe$_4$N wich requires substrate heating during sputtering. Concerning the synthesis of α''-Fe$_{16}$N$_2$, we did not find in this batch of samples the conditions to obtain it.

One of the most original results of our study is that we have obtained amorphous Fe-N alloy on unheated substrates. It presents a large saturation magnetization ($\sim \alpha$ Fe), soft magnetic properties (coercive field $\sim$25 Oe) and a rather large domain of existence vs the plasma composition (from $x_{N2} \approx 5\%$ up to 20%). This disordered phase transforms into Fe or Fe$_4$N when T_S is increased. Let us also notice that success in preparation of this amorphous phase is related to the total thickness t_f of the films : for t_f >5000 Å, amorphous iron nitride tends to transform into Fe or Fe$_4$N as well as for heated substrates. This behavior is attributed to a plasma heating of the substrates during the deposition duration ($\approx$1 to 5 hours).

Concerning low nitrogen concentrations, α Fe-N phase presents a non negligible lattice expansion. From naive band structure considerations, we would expect an enhanced magnetization in such a nitride compared to pure iron. We do not find any increase of M_S, so it proves that a structural expansion of iron nitride structure is not enough for an increase of magnetization.

CONCLUSION

Using conventional argon-nitrogen gaseous mixture, we have been able to produce iron nitride films in a large range of atomic concentrations and equilibrium phases : Fe, Fe$_4$N, Fe$_{2/3}$N, Fe$_2$N. By varying the stoichiometry of Fe$_2$N, we could control the Curie temperature of the films between room temperature and 60K. These preparation conditions have even allowed us to obtain expanded α Fe-N and a new amorphous iron nitride with soft magnetic properties. Furthermore, these results show that an increase of the lattice parameter is not sufficient for the stabilization of a higher spin material. Other effects related for instance to local order could play a crucial role. Finally, we present preliminary results of EELS on iron nitride thin films, we show that this technique is a powerful tool to check stoichiometry and local order.

References

1. S.F. Matar, G. Demazeau and B. Siberchicot, IEEE Trans. Magn., **26**, 60 (1990)
2. T.K.Kim and M.Takahashi, Appl. Phys. Lett., **20**, 492 (1972)
3. M. Takahashi, H. Shoji and M. Tsunod, J. Magn. Magn. Mater., **134**, 403 (1994)
4. M. Komuro, Y. Kozono, M. Hanazono and Y. Sugita, J. Appl. Phys., **60**, 5126 (1990) ; Y. Sugita, M. Mitsuoka, M. Komuro, H. Yoshiya, Y. Kozono and M. Hanazono, ibid., **70**, 5977 (1991)
5. M. Takahashi, H. Shoji, H. Takahashi, T. Wakiyama, M. Kinoshita and W. Ohta, IEEE Trans. Magn., **29**, 3040 (1993)
6. C. Gao and M. Shamsuzzoha, IEEE Trans. Magn., **29**, 3046 (1993)
7. J. Q. Xiao and C.L. Chien, Appl. Phys. Lett., **64**, 384 (1994)
8. J.-F. Bobo, M. Vergnat, H. Chatbi, L. Hennet, O. Lenoble, Ph. Bauer and M. Piecuch, J. Magn. Magn. Mater., **140-144**, 717 (1995)
9. J.-F. Bobo, H. Chatbi, M. Vergnat, L. Hennet, O. Lenoble, Ph. Bauer and M. Piecuch, J. Appl. Phys. (1995) in press
10. R.D. Leapman, L.A. Grunes and P.L. Fejes, Phys. Rev. B, **26**, 614 (1982)
11. G.M. Chen, M.X. Lin and J.W. Ling, J. Appl. Phys., **75**, 6293 (1994)

EPITAXIAL GROWTH OF (001)- AND (111)-ORIENTED PtMnSb FILMS AND MULTILAYERS

M.C. KAUTZKY AND B.M. CLEMENS
Dept. of Materials Science and Engineering, Stanford University
Stanford, CA 94305-2205

ABSTRACT

In this paper we report the successful growth of single-phase epitaxial PtMnSb films and multilayers by dc magnetron cosputtering, both in the (001) orientation on MgO(001) and W(001), and in the (111) orientation on $Al_2O_3(0001)$. Single-layer films in the thickness range $50\text{Å} \leq t \leq 1000\text{Å}$ were grown and characterized using x-ray diffraction (XRD), magneto-optic Kerr effect (MOKE), and vibrating sample magnetometry (VSM). The in-plane orientation relationships, as determined by asymmetric XRD, were PtMnSb[100]||MgO[110], PtMnSb[100]||W[100], and PtMnSb[$\bar{1}$01]|| $Al_2O_3[2\bar{1}\bar{1}0]$. The crystalline quality of the films was found to depend strongly upon the substrate, growth temperature, film thickness, and presence of a capping layer, but rocking curve widths of 1° or less were achieved on each substrate. Measurement of the in-plane strain showed that the films were almost entirely relaxed, with strains <1%. In-plane magnetization was observed in all cases, with moments and coercivities in the 400-500 emu/cm^3 and 100-200 Oe ranges respectively. Polar Kerr spectra showed large rotations (0.75° - 1.03°), whose peak wavelengths appear to depend on both film structure and optical interference effects.

INTRODUCTION

For years there has been considerable interest in the Heusler alloy compounds due to the wide variety in their magnetic properties and the ease with which these properties can be changed by elemental substitution. In recent years, much work has focused on PtMnSb, due to its unusual magneto-optical properties and electronic structure. This compound has the largest Kerr rotation of all known metallic systems at room temperature[1], making it attractive as a magneto-optical recording medium. It also has a calculated electronic structure predicting 100% spin polarization of the conduction band[2], which could produce very large giant magnetoresistance effects in spin valve structures. However, both of these applications have yet to be realized, in part because of problems with the structure and processing of PtMnSb films. The cubic crystal structure of PtMnSb and large demagnetizing field associated with its moment cause thin films to have in-plane magnetization, making them unsuitable for MO media. Similarly, chemical disorder, the use of polycrystalline samples, and the difficulty in producing a smooth unoxidized surface for band structure studies have all hindered verification of the predicted electronic structure[3]. Both areas might therefore benefit from the growth of high-quality epitaxial PtMnSb thin films and multilayers. We have calculated the strains necessary to produce perpendicular magnetization by inverse magnetostriction in PtMnSb films, and found that the levels required, +2.1% for the (001) case and +5.5% for the (111) case, can only be achieved by epitaxial strain. Alternatively, the possibility of growing high-quality, singly-oriented thin films offers a simple means to both probe the electronic structure of PtMnSb and investigate the effects of chemical disorder on it with angle-resolved photoemission.

With these applications in mind, we have investigated epitaxial growth of this alloy by magnetron sputtering. We previously reported the successful growth of epitaxial PtMnSb(001) on W(001) seed layers on MgO(001)[4]. In this paper, we report further successful epitaxial growth on oxide surfaces in multiple orientations, as well as in (001)-oriented multilayer structures with W.

EXPERIMENTAL

PtMnSb has the $C1_b$ structure, which consists of three interpenetrating fcc lattices originating at $(0,0,0)$, $(\frac{1}{4},\frac{1}{4},\frac{1}{4})$ and $(\frac{3}{4},\frac{3}{4},\frac{3}{4})$ of the 6.20Å unit cell. Using this as a basis, we have identified a number of candidate growth surfaces based on lattice matching in rotated and unrotated interfacial orientations. The most promising surfaces are listed in Table 1, with the orientation relationships calculated to produce the lowest coherency strain with the PtMnSb film. It is immediately obvious that good substrates exist for each orientation, with a variety of mismatches, and in particular that some of these mismatches are in moderate excess of the required tensile strain for perpendicular magnetization. Keeping in mind that the processing conditions needed to produce a well-ordered $C1_b$ phase usually involve elevated temperatures, the oxide ceramic and refractory metal substrates are then most attractive among this group, due to their high thermal stability and resistance to interdiffusion. They also have the lowest electrical conductivity of the substrates listed, a desirable quality if transport properties in PtMnSb-based film structures require a high fraction of electron conduction to take place through the PtMnSb layer. For this combination of reasons, we have used W(001), MgO(001), and Al_2O_3(0001) as growth surfaces.

Table I: Predicted orientation relationship and misfit strain for {001}-,{011}-, and {111}-oriented PtMnSb films on various substrates.

Orientation Relationship	$\epsilon_{misfit}(\%)$
PtMnSb(001)<100> ∥ W(001)<100>	2.1
PtMnSb(001)<100> ∥ Mo(001)<100>	1.5
PtMnSb(001)<100> ∥ MgO(001)<110>	-3.9
PtMnSb(011)<100> ∥ W(011)<100>	2.1
PtMnSb(011)<100> ∥ Mo(011)<100>	1.5
PtMnSb(011)<100> ∥ MgO(011)<0$\bar{1}$1>	-3.9
PtMnSb(011)<100> ∥ Al_2O_3(11$\bar{2}$0)<0001>	4.8,-6.0
PtMnSb(111)<$\bar{1}$01> ∥ Pt(111)<$\bar{1}$$\bar{1}$2>	9.6
PtMnSb(111)<$\bar{1}$01> ∥ Al_2O_3(0001)<2$\bar{1}$$\bar{1}$0>	8.5
PtMnSb(111)<$\bar{1}$01> ∥ Cu(111)<$\bar{1}$$\bar{1}$2>	1.0

All films were prepared by dc magnetron cosputtering from elemental targets for optimum compositional control. Prior to deposition, single-crystal 1 cm x 1 cm MgO and Al_2O_3 substrates were scrubbed for 2 min. in 1:1:1 $NH_4OH : H_2O_2 : H_2O$ and immediately dried with filtered nitrogen before insertion into the load-lock. The substrates were then annealed under UHV conditions for 1 hr. at 500-550°C before cooling to the deposition temperature, either 500°C or 350°C. The chamber base pressure at temperature was $\leq 5 \times 10^{-8}$ torr for all films. The PtMnSb films were sputtered onto MgO and Al_2O_3 in 1.5 mT Ar, while for films on W, both the W and PtMnSb layers were sputtered at 3 mT. The elemental rates were calibrated using a quartz crystal rate monitor to within ±.01 Å/s to insure good compositional control, while computer-activated shuttered sources and a rotating substrate were used to provide thickness control and uniformity. The deposition rate for all layers was in the range 0.5-1.0 Å/s. Following deposition the films were characterized using XRD, VSM, and MOKE.

RESULTS AND DISCUSSION

We observe that the films with the best structural quality are those sputtered at low Ar pressure, high temperature, and low rate. It appears, therefore, that surface diffusion is important in achieving good structural coherence. This observation is to be expected from general epitaxial growth considerations, but is particularly relevant for the $C1_b$ crystal structure, as the diffusion lengths required to bring adatoms to the correct sublattice sites may require several atomic jumps. For completeness, we also point out that continued chemical ordering via bulk diffusion should also play a role, since 500°C is higher than half the bulk melting temperature of PtMnSb[5].

Figure 1(a)-(d) show symmetric XRD scans and asymmetric phi scans, on a log scale, for 300Å films of PtMnSb grown on W(001), MgO(001), and Al_2O_3(0001) at 500°C. All four scans contain peaks from only the desired $C1_b$ phase, but while the films deposited on W(001) and Al_2O_3 have a single out-of-plane orientation, the one deposited on MgO has trace amounts of (111)-oriented PtMnSb ($2\theta=25°$) in addition to its strong (001) texture. Despite this, all films are epitaxial, as shown by the four- and six-fold symmetry of the phi scans. By comparing film peak positions in phi with those of the substrate we have established that the PtMnSb grows unrotated on W and Al_2O_3, but with a strain-energy driven 45° rotation between film and substrate on MgO. These orientation relationships match those predicted in Table 1 from lattice-matching arguments.

One unexpected feature in these results is an asymmetry of the phi scan for PtMnSb(111) on Al_2O_3. One would expect growth of the three-fold symmetric PtMnSb(111) surfaces in two equally-populated, but distinct, orientations on the six-fold symmetric $Al_2O_3(0001)$ surface. However, as Fig. 1(d) shows, we observe a very large difference in the population of the two orientations. Without a direct image of the film/substrate interface, it is difficult to interpret this inequality, but we submit that several mechanisms could account for our observation, for example a preferential nucleation of one orientation at ledges on the sapphire followed by a layer growth mode. Identifying the precise mechanism, however, will require further investigation.

Due to the success in growing (001)-oriented epitaxial films on W, we have also begun some exploratory work in growing epitaxial PtMnSb(001)/W(001) multilayers. Unlike the single-layer films, where the best quality PtMnSb is achieved at the highest temperatures, we have been unsuccessful to date in fabricating good-quality multilayers at any temperature above 350°C. Figure 1e shows structural data for a 10-repeat superlattice of bilayer spacing 85Å, which was grown on a 100Å W seed layer at 350°C. The low-angle x-ray diffraction pattern, which is sensitive to roughness and intermixing at the interfaces, is characterized by seven orders of primary Bragg maxima. The attenuation of these maxima is characteristic of interfaces whose thicknesses have been widened by interdiffusion - a reasonable expectation given the growth temperature. Despite this overall reduction in intensity, however, the width of the peaks remains constant in $|\vec{q}|$. This is usually a sign that there is no uncorrelated roughness between adjacent layers. Whether this interpretation is, in fact, correct is still in question, but additional information which is consistent with it can be found in the high-angle and asymmetric patterns. In the high-angle regime, the peaks are considerably broadened and attenuated, with only the fundamental superlattice line (58.7°), one set of satellite peaks (60.0°on high 2θ side of the fundamental), and the seed layer peak (57.8°) visible. We have fitted the measured data to Voigt profiles, and from the width of the fundamental peak, calculated a vertical correlation length of 120Å. Since this value indicates that coherent scattering is taking place over length scales larger than the bilayer period, we conclude that the interfaces must be crystalline. This idea is substantiated by the phi scan, which shows a clear epitaxial relationship which would not be present if the interfaces were amorphous. The increased width of the asymmetric peaks simply indicates that the in-plane mosaic spread of the multilayer is much larger than that of the single-layer film. Our preliminary conclusions about the structure of PtMnSb(001)/W(001) multilayers, then, are that they can be grown epitaxially, but with layers having a broad mosaic spread in-plane, and with interdiffused interfaces which do not appear to have uncorrelated roughness.

The crystalline quality of the single-layer films was measured as a function of temperature and thickness by rocking curves on the symmetric peaks. The results are shown in Table 2. The most obvious trend is that in all cases, the films grown at 500°C are of better quality than those grown at 350°C, consistent with improved surface and bulk diffusion. Another trend we observe, but which is not presented in Table 2, is that in uncapped films at a given deposition temperature, the FWHM of the rocking curves decreases with increasing film thickness. For example, samples grown at 500°C and 350°C both show reductions of 10-15% in the FWHM going from 100Å to 300Å. Capping may further enhance this effect, as the capped 1000Å samples show even larger reductions, between 30% and 70% relative to 300Å films. Conclusions based on this data, however, are questionable, since the samples were drawn from different experiments rather than from a systematic study. For example, since the deposition rate was held constant in all samples, it is still unclear whether this improvement in film quality is due to continued ordering in the bulk with increasing deposition time, or to a decrease in the fraction of film volume which is peak-broadened by oxidation. At this point we simply point out that growth of high-quality films, with rocking curves ≤1°, can be achieved on each surface using elemental cosputtering.

Strains in the films were calculated with the $\sin^2\psi$ technique, using the {400} and {422} families of planes in the <001> and <111> films respectively. A biaxial strain state in the films was assumed. Because the elastic stiffnesses C_{12} and C_{11} are not known for PtMnSb, the strains were calculated over the range $0.3\leq \frac{C_{12}}{C_{11}} \leq 0.8$[4]. The results, shown in Table 2, indicate that in all cases the strains are very small, consistent in sign and magnitude with essentially full relaxation of the films from their theoretical misfits. There are slightly larger strains in the films grown at 500°C, but this may be simply due to a larger strain contribution from thermal expansion differences between the film and substrate. We also observe slightly larger strains for films with thicknesses below those shown in Table 2, but even at 100Å the values are still less than 1%. From these results we conclude that epitaxial growth alone cannot produce strains of the magnitude required for perpendicular magnetization in PtMnSb films.

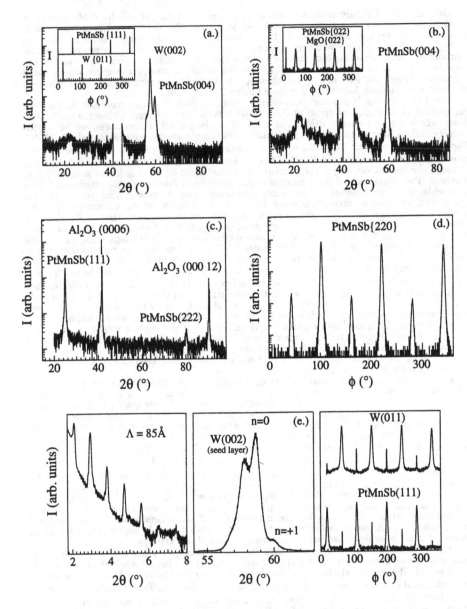

Figure 1: Symmetric and asymmetric x-ray diffraction scans, showing single-phase epitaxy, for 300Å PtMnSb films on (a.) MgO(001) (b.) W(001)/MgO(001) (c.),(d.) Al$_2$O$_3$(0001), and (e.) a 10*(45Å PtMnSb)(40Å W) multilayer on W(001)/MgO(001). The missing peaks in (a.) and (b.) are from the MgO substrate, as are the features at 22°-24°. (c.) and (d.) were taken from separate samples prepared under identical conditions.

Table II: Symmetric rocking curve FWHMs and in-plane strains for 300Å and 1000Å epitaxial PtMnSb films.

Substrate	Temp (°C)	Thickness (Å)	Cap	$\Delta\omega_\perp$ (°)	$\epsilon_\parallel$ (%)
W	500	300	-	0.89	0.3-0.5
W	350	300	-	1.12	0.2-0.3
MgO	500	300	-	2.14	<-0.1
MgO	500	1000	20Å Pt	0.66	<-0.1
Al$_2$O$_3$	500	300	-	1.49	0.2-0.3
Al$_2$O$_3$	350	300	-	4.23	<0.1
Al$_2$O$_3$	500	1000	20Å Pt	0.95	0.2-0.3

Figure 2(a) shows in- and out-of-plane MOKE hysteresis loops for a 300Å uncapped film deposited at 500°C on Al$_2$O$_3$. In this film and all others, the easy axis lies in the plane of the film, indicating shape anisotropy is the dominant contribution to the anisotropy energy. Values of the saturation moment in uncapped films are low relative to the bulk value of $\approx$550 emu/cm^3 - only 150 to 250 emu/cm^3. However, when oxidation in the films is eliminated by capping, the moments rise to between 400 and 500 emu/cm^3. These values, while consistent with other reported values for thin films of PtMnSb[6,7], are still somewhat less than the bulk value. This can be understood by considering the crystal structure of PtMnSb. Chemical disorder in the lattice and deviations from 1:1:1 stoichiometry during deposition will cause some fraction of the Mn atoms to sit on incorrect sites, with a smaller separation between neighboring Mn atoms than the 4.38Å of the perfect structure. Since large separation is a critical condition for ferromagnetic ordering in this class of compounds, this site disorder will result in paramagnetic ordering of the Mn atoms involved, and hence a lowering of the total moment. In our films, the magnitude of the residual moment is nonetheless still large enough to generate a sizable shape anisotropy, which, combined with the low levels of strain in the films produces in-plane magnetization. Comparison of the saturation fields out-of-plane, which is taken as the field required to produce 90% of the saturation moment, results in $\approx$8700 Oe and $\approx$6000 Oe for the 300Å films on W and Al$_2$O$_3$ respectively. A 20 kOe field was insufficient to saturate the film on MgO. The lower saturation field for the (111)-oriented film indicates that it has the largest perpendicular anisotropy component of the three surfaces studied. Since magnetoelastic anisotropy alone cannot account for the difference, the improved perpendicular component may be due to magnetocrystalline anisotropy, with <111> being the easy axis. Further work is needed to substantiate this point. The in-plane coercivities of the films are small, and like the moments, depend on the presence of oxidation. The 300Å uncapped films have coercivities of 250-300 Oe, while those of capped films are somewhat lower, in the 100-200 Oe range. Presumably the oxide impurities act to pin domain walls and restrict their motion, leading to the higher coercivities.

Finally, Fig. 2(b) shows comparisons of the Kerr rotation spectra for the uncapped 300Å samples, and for the multilayer sample described earlier. The films deposited on the bare oxides show identical behavior, with a peak rotation of 0.75° at 750 nm, but upon introduction of the W seed layer, there is a change in both the size of rotation and the peak wavelength, to 1.03° at 575 nm. Interestingly, the multilayer in which W is used as a spacer layer has the poorest magneto-optical properties, with a peak rotation of only 0.12° at 750 nm. These variations in rotation and peak position are due to a convolution of factors which are difficult to separate. On one hand, the changes may stem from differences in the structural quality of the films and multilayer. As evidenced by the rocking curves, the films on W grew with the best crystalline structure, and hence should have a band structure closest to the ideal. By the same turn, the multilayer had both the worst defect structure and the worst MOKE properties. This argument for a structural origin is supported even further by our observation that the magnitude of rotation for films on W increases with increasing film thickness in the range 50-300Å, for this correlates well with the decrease in rocking curve width over the same range. On the other hand, the changes in rotation magnitude that we see are also accompanied by shifts in the peak position, which suggests effects from interference between the film and metallic seed layer. As shown by others[8], changes in the Kerr spectrum of thin magneto-optic films are common when they are grown adjacent to nonmagnetic

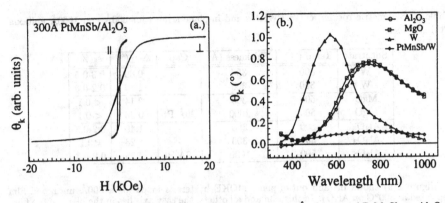

Figure 2: (a.)Polar ($\perp$) and longitudinal ($\parallel$) MOKE loops for 300Å of epitaxial PtMnSb on Al$_2$O$_3$. (b.)Polar Kerr spectra for epitaxial PtMnSb films and multilayers, measured at H$_{sat}$ = 20 kOe.

metal layers, and hence we might expect that optical effects could explain the shift for samples grown on W. The main inconsistency with this picture, however, is that enhancements of the rotation should happen in energy ranges where the real part of the complex index of refraction of the nonmagnetic layer is much smaller than that of the magneto-optic layer. The W seed layer, which is 1000Å thick and should therefore behave nearly as bulk W, has optical constants in the visible range[9] which are actually larger than those of PtMnSb[10,11]. At present, therefore, we are unable to give a coherent picture for the Kerr behavior in our films, but point out that our data has features which argue for contributions from both structural and optical sources.

CONCLUSIONS

We have demonstrated that high-quality epitaxial films of PtMnSb can be grown in the (001) and (111) orientations by elemental cosputtering on MgO(001), W(001) and Al$_2$O$_3$(0001). The optimum conditions for epitaxial growth of the C1$_b$ phase are high temperature, low pressure, and low rate. PtMnSb(001)/W(001) multilayers on W seed layers can also be grown epitaxially, but only at lower temperatures, and with considerable structural defects. The crystalline quality of single-layer films is sensitive to chemical ordering and oxidation, but symmetric rocking curves of 1° or less can be achieved on each surface. The strain in all films is consistently small, never exceeding 1%. Correspondingly, all films and multilayers show in-plane magnetization, with moments and coercivities in capped samples in the range 400-500 emu/cm^3 and 100-200 Oe respectively. Finally, polar Kerr spectra show large peak rotations (0.75° - 1.03°), whose peak wavelengths appear to depend on both film structure and interference effects.

REFERENCES

1. P.G. van Engen, K.H.J. Buschow, and R. Jongebreur, Appl. Phys. Lett. **42**, 202 (1983).
2. R.A. de Groot, P.G. van Engen, and K.H.J. Buschow, Phys. Rev. B **50**, 2024 (1983).
3. C. Park, private communication.
4. M.C. Kautzky and B.M. Clemens, Appl. Phys. Lett. **66**, 1279 (1995).
5. R.A. Laudise, W.A. Sunder, R.L. Barns, G.W. Kammlott, A.F. Witt, and D.J. Carlson, J. Cryst. Growth **102**, 21 (1990).
6. E. Attaran and P.J. Grundy, J. Magn. and Magn. Mtls. **78**, 51 (1989).
7. N. Sugimoto, T. Inukai, M. Matsuoka, and K. Ono, Jap. J. Appl. Phys. **28**, 1139 (1989).
8. L. Chen, W. McGahan, D. Sellmyer, and J. Woollam, J. Appl. Phys. **67**, 7547 (1990).
9. J.H. Weaver, D.W. Lynch, and C.G. Olson, Phys. Rev. B **12**, 1293 (1975).
10. P.M. van der Heide, W. Baelde, R.A. de Groot, A.R. de Vrooment, P.G. van Engen, and K.J. Buschow, J. Phys. F **15**, L75 (1985).
11. G.S. Bains, R. Carey, and B.W.J. Thomas, J. Magn. and Magn. Mtls. **104-107**, 1011 (1992).

LARGE IN-PLANE LATTICE EXPANSION IN NiAs-MnSb THIN FILMS INDUCED BY ns LASER RECRYSTALLIZATION

YUKIKO KUBOTA* **, GRACE L. GORMAN*
and ERNESTO E. MARINERO*
* IBM Research Division, Almaden Research Center, 650 Harry Road, San Jose, CA 95120-6099
**Department of Materials Science and Engineering, Stanford University, Stanford, CA 94305

ABSTRACT

Sputter deposited MnSb thin films were annealed utilizing KrF excimer laser pulses (16ns), and the resulting structural and magnetic changes investigated. These changes are compared to those observed when the samples are subjected to isothermal and rapid thermal annealing treatments. Isothermal and rapid thermal annealing induce significant lateral grain growth, whereas the laser treatment produces vertical grain size refinement with no appreciable lateral growth. Annealing is shown to increase the hexagonal c-axis, reaching an expansion value of 7% for the laser annealed samples. This c-axis expansion has a strong influence on the magnetic properties of the thin films. Mechanisms for the c-axis expansion are discussed.

INTRODUCTION

The ε-phase of Mn-Sb system is reported to exist for Mn atomic compositions ranging from 44 to 50% as shown in the phase diagram of figure 1. [1] This ε-phase has a hexagonal crystallographic structure and is classified as belonging to the NiAs type. [2] The lattice parameters and magnetic properties are reported to exhibit a strong composition dependence in this range. [2-6] The variations in magnetic moments and Curie temperatures are ascribed to changes in Mn-Mn atomic spacing resulting from changes in lattice parameter as a function of site occupancy. Similarly, Nagasaki et al [5] have studied the pressure dependence of the lattice parameters of MnSb and found dramatic changes in Curie temperatures.

In this work, we seek to induce lattice parameter changes by rapid resolidification and recrystallization of fixed composition MnSb thin films. We expect that volume changes due to recrystallization will result in lattice strain and seek to compare the corresponding changes to their magnetic and magneto-optic activity. To this effect we utilized three different time-temperature regimes of annealing; laser quenching (LQ), rapid thermal annealing (RTA) and isothermal furnace annealing (IA). Typical annealing times for each method are 16 ns, 120 sec. and 5 hours, respectively. A variety of structural and magnetic probes are employed to monitor the effect of annealing under this wide temporal range of conditions.

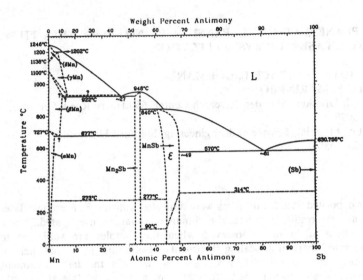

Figure 1. Mn-Sb phase diagram [1]. Note the compositional range for the ε-phase Ni-As formation.

EXPERIMENTAL TECHNIQUES

MnSb thin films were deposited by DC magnetron sputtering (base pressure < 1x10^{-7} Torr.) from elemental targets at a sputter pressure of 3 mTorr of Argon onto quartz substrates. The film layer structure is **substrate / SiN (20 nm) / MnSb (150 nm) / SiN (12.5 nm)** All layers were grown during the same pumpdown and the SiN was grown by reactive sputtering. To study the relation between the film structure and its magnetic and magneto-optic properties, we utilized the following analytical methods. X-ray microprobe for determining film composition. X-ray diffraction (XRD) and transmission electron microscopy (TEM) for microstructural analysis. Atomic force microscope (AFM) to measure film morphology. Magneto-optic Kerr Effect (MOKE), vibrational sample magnetometer (VSM) and torque magnetometer for characterizing magnetic activity.

Laser annealing was conducted within the pole pieces of a 20 kOe electromagnet which is part of the MOKE apparatus. The exposure was conducted with no applied magnetic field. Dielectric mirrors and beam combiners are utilized to guide both the excimer laser radiation and the probe He-Ne laser onto the same spot of the sample. The excimer laser irradiates a 6 mm diameter area (determined by the pole bore hole) whereas the He-Ne laser spot is approximately 1.5 mm in diameter and is fully confined within the irradiated area. In this fashion we are able to in-situ monitor the magnetic changes due to laser annealing. Irradiation in our work was conducted through the SiN/air interface. The laser is a Lambda Physik (102) excimer laser running on KrF (248 nm) and it generates 16 ns pulses of around 250 mJ in energy.

Control of the laser fluence is obtained by means of variable attenuator and the pulse energy is measured with a pyroelectric detector. To minimize pulse to pulse energy variations, a repetition rate of 2.5 Hz was employed. This also guaranteed that no thermal build up effects are present in our annealing experiments. The RTA (AG Associates 610) pulse heat profile in our studies was the following: 5 s hold at room temperature, ramp-up to the target temperature at 125°C/s, 120 s hold at the target temperature and the sample was removed from the instrument after cooldown below 100°C.

EXPERIMENTAL RESULTS

To ensure ε-phase structures, the nominal sample composition was targeted at $Mn_{56}Sb_{44}$. The x-ray microprobe revealed atomic % compositions of 55.6% Mn and 44.4% Sb within ±0.5% accuracy. Absolute atomic content determination was not undertaken and we note that impurity traces of C and O are present. Thus, the actual film composition is at best precise to 1 atomic%. XRD studies of as-deposited and annealed samples reveal that the films are single phase and strongly textured. Both as-deposited and annealed samples exhibit spectra dominated by the (110) diffraction peak. This is evident in the θ-2θ scan given in Figure 2 which corresponds to an RTA sample, annealed for 120 s at 550 °C. The only other orientation discernible in this spectrum (211) is 1:187 weaker than the (110) peak. The as-deposited and laser quenched samples showed also very weak (101) peaks whose intensity ratio compared to their (110) peaks are 1:94 and 1:30 respectively. The broad peak observed around 22°, is due to the quartz substrate background contribution. This growth orientation of the MnSb films leads us to conclude that the hexagonal c-axis is oriented in the film plane (//).

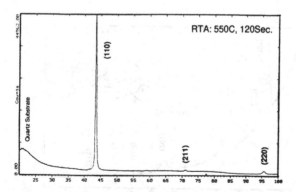

Figure 2. θ-2θ XRD spectrum of MnSb film (150 nm) deposited on quartz. Note the strong (110) texturing which is common for all samples studied. This particular spectrum corresponds to a film that was annealed for 120 sec at 550°C.

117

Although annealing does not change the texturing or number of diffraction features, its effect is to induce significant peak shifts. The effect of the various annealing treatments on the XRD spectra is shown in Fig. 3. We focus only on the (110) peaks in this figure to illustrate the changes in peak position, FWHM and intensity that occur when the as deposited samples are subjected to the various annealing treatments. The FWHM and peak areas are tabulated in the insert table in the figure. Such large peak shifts can be associated with lattice parameter changes in these films as no additional diffraction peaks are obtained that may indicate formation of secondary phases. To confirm that the peak shift is indeed due to lattice parameter changes, we utilized GIXS (grazing incidence x-ray scattering) to

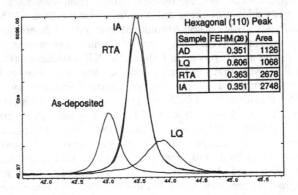

Hexagonal (110) Peak		
Sample	FEHM (2θ)	Area
AD	0.351	1126
LQ	0.606	1068
RTA	0.363	2678
IA	0.351	2748

Figure 3. θ-2θ scans around the (110) peak for AD, RTA, IA and LQ samples. Note the peak shifts, increments in peak intensity and FWHM which occur as a consequence of annealing.

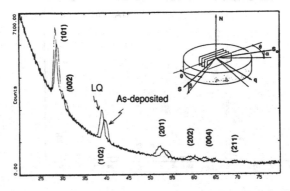

Figure 4. GIXS XRD spectrum of AD and LQ films. Peak shift is in the opposite sense as that seen in Fig 3. No in-plane texturing is observed and the spectra are assigned to the Ni-As phase of MnSb. The angle of incidence for the x-ray beam was ≤1°.

probe lattice planes perpendicular to the film surface. Thus, if the peak shift towards higher angles in the conventional geometry utilized in Figs. 2 and 3 is indicative of in-plane lattice expansion, the GIXS experiment should reveal an opposite trend for the same diffraction plane. GIXS scans for the as-deposited and the laser quenched samples are given in Fig. 4. The scattering geometry is also indicated in the figure. A larger number of peaks indexed to various crystallographic orientations are shown in the figure. This readily indicates that the polycrystalline film has no preferential in-plane growth orientation. The higher number of peaks observable, makes it possible to identify the crystal structure of these samples as hexagonal and the spectrum can be fitted to a Ni-As type of structure with lattice parameters a and c of 4.2Å and 5.77Å for the as-deposited films. These parameters compare well to those published for bulk $Mn_{56}Sb_{44}$ specimens of a = 4.21Å and c = 5.71Å. [4]

The only peak that is observed in both geometries is the (101) for AD and LQ samples. Thus, we utilized the peak position for this peak to ascertain the trend that one would expect concerning lattice parameter changes due to annealing. One clearly sees that the peaks in the LQ samples are distinctly shifted towards smaller angles, the opposite from the conventional geometry. Thus confirming that for the same atomic plane, lattice parameter expansion in the film plane is accompanied by lattice contraction in the vertical direction as a consequence of the laser annealing. In contrast as shown in Table 1, no such changes are observed for the AD samples in GIXS vs normal diffraction geometries. In other words, the effect of laser annealing is to contract the spacing between lattice planes in the $\perp$ direction, whereas those in // direction expand. In comparison, the AD (101) peak does not show any shift, indicating that the crystalline arrangement is essentially strain-free.

Figures 5 and 6 show plan and cross-sectional bright field images that compare the microstructure of the AD specimens to that of the annealed samples. Large lateral grain growth is observed after oven and RTA treatments, whereas comparable grain size is obtained following laser exposure. The average grain size following RTA and isothermal annealing increased from ~20-40 nm to ~50-100 nm. The cross sectional TEM images of AD and LQ films in Fig. 6 show that the grains in the AD films span the 150 nm thickness of the film, whereas after LQ, grain boundaries are clearly visible in the vertical plane, indicative that vertical grain size reduction or, grain

Sample	Lattice parameters				Tc
	a (Å) $\perp$	//	c (Å) $\perp$	//	(°C)
AD	4.20	4.20	5.77	5.77	170
LQ	4.13	4.18	5.88	6.17	275
RTA	4.16	N.A.	5.86	N.A.	220
IA	4.16	N.A.	5.84	N.A.	220
$Mn_{56}Sb_{44}$	4.215		5.71		100~160
Mn_2Sb_1	4.10		6.65		293

Table 1. Lattice Parameters and Magnetic Properties of MnSb films.

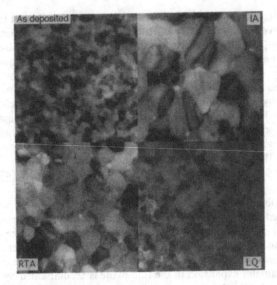

Figure 5. Plan view TEM images of MnSb samples showing the grain structure difference among AD, IA, RTA and LQ samples.

Figure 6. Cross sectional bright field and dark field TEM images obtained with (101) reflection: (a) AD film and (b) LQ film, fluence = 68.4 mJ/cm², 50 pulses. Grain refinement in the film thickness direction is observed in the LQ sample. The recrystallization front appears around the center of 150 nm thickness film in (b).

refinement, has occurred in this direction. The bright and dark field images of Fig 6 clearly show the columnar structure of the AD sample and the decrement to about half the original size after annealing the sample with 50 laser pulses of fluence = 68.4 mJ/cm².

DISCUSSION

Single phase MnSb films with NiAs hexagonal crystalline structure have been produced by sputtering. Subsequent annealing (oven, rapid and ns-laser annealing) lead to grain growth and refinement (LQ). However, the single-phase nature of the films is retained. Remarkable lattice expansion of the c-axis is observed upon annealing. Lattice parameters, for perpendicular and in-plane geometries (where applicable) are tabulated in Table 1 in which for comparison, we also give published literature values for $Mn_{56}Sb_{44}$ and Mn_2Sb_1 [7]. The table also contains measured Curie temperatures for the AD and the annealed samples. Utilizing our measured lattice parameters and Nagasaki's analysis [5], we calculate also the predicted Curie temperatures based on the Mn-Mn atomic spacing for the AD and LQ samples for which we have accurate determination of the in-plane lattice parameters.

The a and c lattice parameters for the AD samples are in good agreement with published values for $Mn_{56}Sb_{44}$. T_C, on the other hand, is 68 °C lower than the calculated value. We note that the work of Teramoto et al and Nagasaki et al [4,5] on the impact of Mn composition and lattice parameter changes on magnetic properties were conducted in bulk, powder samples with grains significantly larger than in our thin films. As determined in our TEM studies, the grain size of the AD samples is ~20-40 nm. It is possible that at such scales, dimensional effects may also lead to decrements in Curie temperature as discussed by Schaefer et al. [8] Isothermal and rapid thermal annealing yield hexagonal structures with virtually the same lattice parameters and the c-lattice expansion with respect to the AD sample (for out-of-plane hexagonal lattices) is 1.5%. Although the FWHM (figure 3) did not change upon RTA and IA, the peak and integrated intensity for the (110) peak increased by ~2.4. Large lateral grain growth was observed in the TEM analysis of the RTA and IA samples. Nevertheless, such growth cannot account for the large intensity increment. However, stronger (110) texturing as consequence of additional crystallization would explain such intensity change. Rocking curve analysis is in progress to determine this possibility. The c-axis expansion upon RTA and IA coupled with the crystal growth observation, can be interpreted in terms of grain boundary annihilation associated with the lateral grain growth and the development of in-plane tensile stress evolution due to the difference in thermal expansion coefficients between MnSb and the substrate. [9]

In the case of the LQ samples, we note from Fig 3 that the FWHM has increased almost by a factor of 2X as compared to the AD case. The ratio of the FWHM between AD and LQ (110) peaks is 0.53 (Fig. 3). In Fig 6, we showed by cross-sectional TEM that upon laser annealing, the columnar grains are reduced in size to approximately 1/2 the film thickness. Thus, this increment in FWHM for the (110) peak is consistent with reduction in grain size in the vertical direction as expected from the diffraction geometry (lateral grain size changes do not contribute to changes in FWHM). This reduction in vertical grain size is explained by recrystallization upon thin film melting. This grain size refinement is a consequence of the directionality of the melt front propagation and the limited time for nucleation and growth. As shown in Figure 5, the plan view TEM analysis reveals that no lateral grain growth occurs for laser annealing.

Comparison of GIXS data for AD and LQ specimens reveals that if the crystal structure following laser quenching is still NiAs, the diffraction spectrum fitting requires a 7% expansion of the c-axis to fit the diffraction peak positions after LQ.

Such a large degree of expansion could be induced by the melting/resolidification process and accompanying volume changes. In the time scales involved in the resolidification process, there is no time to relax the resulting strain. Therefore, the unit cells are frozen with a high degree of lattice strain.

We have also considered, the possibility of phase stabilization from the melt of the tetragonal Mn_2Sb_1. As shown in the table 1, the c/a ratio of the LQ film has an intermediate value between hexagonal MnSb and tetragonal Mn_2Sb. According to the phase diagram, this phase can be produced at or near the melting temperature [1] and the fast quenching could potentially stabilize this phase. We note however, that although the Curie temperature compares well with our measured value for the LQ sample, the measured saturation magnetization of the LQ sample is more than 2X larger than the published value for Mn_2Sb_1.[7] In addition, we do not observe new peaks in the GIXS spectrum of LQ, the single distinguishing feature following the transient annealing is a systematic peak shift for all spectral peaks shown in Fig. 4.

CONCLUSION

We have investigated the effect of various annealing process on MnSb thin films and its impact on magnetic properties. The films exhibit strong (110) texture with the hexagonal c-axis oriented in the plane of the film. All annealing methods induced large in-plane lattice expansion in c-axis. RTA and conventional annealings, induce lateral grain growth which we attribute as the origin of the ~1% in the lattice distortion. No lateral grain growth was observed in LQ. Although two possibilities are likely to explain the large changes in the GIXS spectra, our experimental results are consistent with the very large 7% expansion resulting from melting and resolidification of the thin film leading to grain size refinement and the development the large strain values observed.

ACKNOWLEDGEMENTS

This work was partially supported by the NSIC/ARPA contract number MDA972-93-1-0009.

REFERENCES

1. T. B. Massalsky, in Binary Alloy Phase Diagrams, 2nd. edition (1990)
2. C. Guillaud, **Ann. Phys., 4** 671, (1949).
3. T. Okita and Y. Makino, **J. Phys. Soc. Jpn., 25**, 120, (1968)
4. I. Teramoto and A. M. J. G. van Run, **J Phys. Chem. Solids, 29**, , 347 (1968).
5. H. Nagasaki, I. Wakabayashi and S. Mimomura, **J. Phys. Chem. Solids, 30**, 329 (1969).
6. W. Reimers, E. Hellner, W. Treumann and P. J. Brown, **J. Phys. Chem. Solids, 44**, 195, (1983).
7. F. J. Darnell, W. H. Cloud and H. S. Jarrett, **Phys. Rev., 130** , 647 (1963).
8. H. -E. Schaefer, H. Kisker, H. Kronmuller and R. Wurschum, **NanoStruct. Mat., 1**,, 523, (1992).
9. M. F. Doerner and W. D. Nix, in <u>CRC Critical Reviews in Solid State and Materials Science,</u> vol. 14, Issue 3, pp.225 - 268, (1988).

ADJACENT LAYER COMPOSITION EFFECTS ON FeTbCo THIN FILM MAGNETIC PROPERTIES

MICHAEL B. HINTZ
Imaging and Electronics Sector Materials Application Laboratory, 3M Co., St. Paul, MN 55144

ABSTRACT

The magneto-optical (MO) layer in current rare earth-transition metal (RE-TM) based MO recording media is typically 20 nm to 60 nm thick. It has been suggested, however, that media structures employing a multiplicity of thinner MO layers may be advantageous, e.g., for multi-level recording applications [1] or media noise reduction [2]. As magnetic layer thickness is reduced, interactions among magnetic layers and adjacent materials can have an increasingly large influence on magnetic properties; in many instances, these interactions can dominate the observed magnetic behavior.

As a means of studying MO layer - adjacent layer interactions, we have used thin ($\approx$3 nm) films of several materials to separate single 24 nm thick ion-beam-deposited FeTbCo layers into N thinner layers of 24/N nm thickness (N x 24/N). As N increases, the FeTbCo magnetic properties generally change; however, the relative magnitude of the changes is strongly dependent upon the adjacent layer composition. Magnetization (Ms), energy product (MsHc) at 30 C and Curie temperature data for 1 x 24 nm structures and 6 x 4 nm structures are compared and discussed for specimens employing SiC_x, SiN_x, YO_x, HfO_x, Si and SiO_x adjacent layer materials.

INTRODUCTION

Current magneto-optical media constructions are typically four-layer thin film stacks which are formed by successively depositing a first dielectric layer, a rare-earth-transition metal (RE-TM) magneto-optical layer, a second dielectric layer, and a metallic reflector layer on a transparent substrate. The magneto-optical (MO) layer exhibits perpendicular magnetic anisotropy. RE-TM layers are highly reactive and must be protected from the ambient environment to achieve acceptable stability. The dielectric layers on either side of the MO film serve as barriers between the MO material and the surrounding environment. In addition to environmental protection requirements and acceptable optical properties, these adjacent dielectric layers must exhibit low reactivity with the RE-TM magneto-optical film.

The magneto-optical layer in current rare earth-transition metal based MO recording media is typically 20 nm to 60 nm thick. It has been suggested, however, that media structures employing a multiplicity of thinner MO layers may be advantageous, e.g., for multi-level recording applications [1] or media noise reduction [2]. Furthermore, there is considerable interest concerning the fundamental properties of very thin magnetic films [3]. As MO layer thickness is reduced, the ratio of MO layer - adjacent layer interfacial area to MO layer volume increases; consequently, interfacial reactivity and/or interactions therefore have an increasingly large influence on MO layer magnetic properties.

As a means of studying MO layer - adjacent layer interactions, thin ($\approx$3 nm) films of several materials have been used to separate single 24 nm thick ion-beam-deposited FeTbCo layers into N thinner layers of 24/N nm thickness (N x 24/N). As N increases, the FeTbCo film's magnetic properties generally change; however, the relative magnitude of the changes is strongly dependent upon separator/adjacent layer (S/AL) composition. The present study compares and discusses the magnetic properties of 1 x 24 nm structures and 6 x 4 nm structures employing Si, SiC_x, SiN_x, SiO_x, HfO_x, and YO_x S/AL materials. The magnetic properties monitored include saturation

123

magnetization (Ms), the product of Ms and the coercivity, Hc (energy product, MsHc) at 30 C, the temperature at which the remanent perpendicular magnetization rapidly decreases (Td), and the Curie temperature (Tc).

EXPERIMENTAL PROCEDURE

Figure 1 schematically illustrates the dual ion beam system used to prepare specimens for this study. A pentagonal prism target carousel allows sequential exposure of up to 5 target materials to the primary beam for deposition. Computer control of the primary ion beam and target carousel facilitates flexible and reproducible thin film deposition. A shutter/shield assembly close to the substrate planetary permits target pre-cleaning prior to deposition and reduces substrate bombardment by specularly reflected primary beam components. The planetary enables uniform coating of four 130 mm diameter disk substrates. A secondary ion source can be used to clean substrates prior to deposition and/or to modify properties of the growing films. The system is cryopumped and has a base pressure in the low 10^{-8} torr range.

All specimens were deposited using a 900 eV Xe+ primary beam; a primary beam current of 225 mA results in an average deposition rate of 3-6 nm/min. Xe working gas pressure in the deposition chamber was 2.5 x 10^{-4} torr. Polished Si (100) wafers are used as substrates. The 1 x 24 nm specimen structure consists of a ≈ 24 nm thick FeTbCo film sandwiched between two 40 nm thick layers of a particular separator/adjacent layer material. For the otherwise identical 6 x 4 nm structure, the 24 nm FeTbCo layer is divided into six ≈ 4 nm thick sheets by ≈ 3 nm thick separator material layers. The The FeTbCo films were sequentially deposited from an Fe-5.3 at.% Co alloy target and an elemental Tb target. The Fe-Co and Tb deposition times were adjusted to produce a layered structure with the desired composition and a nominal repeat periodicity of 1 nm; the 24 nm thick layers therefore consist of 24 FeCo - Tb layer pairs, while each of the 4 nm thick layers contains 4 FeCo - Tb layer pairs. For a given deposition process, the 1 nm repeat period roughly corresponds to the maximum in MsHc vs. layer periodicity for FeTbCo materials [4, 5]. Si, SiC, Si_3N_4, SiO, HfO_2 and Y_2O_3 were used as targets for the

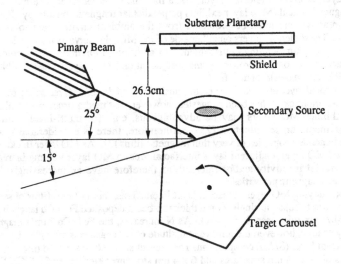

Figure 1:
Schematic illustration of dual ion beam deposition system geometry.

separator materials. No reactive gases were employed during deposition; consequently the resulting films from compound targets comprising volatile constituents (i.e. O and N) are likely somewhat deficient in O or N relative to the starting target composition. As the exact stoichiometry is unknown, the S/AL materials are referred to as SiN_x, HfO_x, etc. All but the Si S/AL material targets were formed from hot pressed powder compacts. The Si target was dense polycrystalline material.

A Digital Measurement Systems model 1660 VSM was used to measure magnetic properties. M-H hysteresis loops at 30 C were used to determine Ms and MsHc perpendicular to the film plane. The Curie temperature, Tc was determined by measuring perpendicular magnetization as a function of temperature in an applied field of 5,000 Oe. While the large applied field is likely to increase the apparent Tc value due to induced magnetization effects, it was sufficiently large to exceed $4\pi M$ of all specimens, thereby enabling measurement of specimens which would otherwise have demagnetized at temperatures well below Tc. For the purpose of this paper, the "demagnetization temperature", Td, is defined as the temperature at which the remanent magnetization, Mr, becomes less than half Ms at the same temperature. Td was determined by saturating the specimens in the perpendicular direction at 30 C and monitoring the remanent magnetization while ramping up the temperature in 5 C increments; noting the temperature at which remanent magnetization suddenly decreased and comparison with the M - T data used to determine Tc enabled estimation of Td. Film chemistries were determined using X-ray florescence (XRF).

RESULTS AND DISCSUSSION

A set of 1 x 24 nm and 6 x 4 nm structures were fabricated using identical deposition parameters for the FeTbCo layers. All specimens exhibit perpendicular anisotropy at 30 C. Excluding the HfO_x specimens for which Hf florescence strongly interferes with the Co peak, XRF measurements result in an average film composition of Fe-19.4 at.% Tb - 3.4 at.% Tb. The standard deviation of the Tb concentration is 0.5 at.%, while that for the Co is 0.2 at.%.; some of the variation is probably due to differing adsorption and fluorescence characteristics among the separator materials. For a given separator material, the variation in Tb concentration between the 1 x 24 and 6 x 4 structures is typically ≤ 0.3 at.%. The magnetic properties of the 1 x 24 nm and 6 x 4 nm structures are summarized in Table I. Td and Tc are rounded to the nearest 5 degrees, the size of the temperature steps in the measurement. The compensation temperature of all specimens is well below 30 C, the temperature at which Ms and MsHc were determined.

It is well documented that the microstructure [6] and magnetic properties [3] of thin films are influenced by the substrates and/or underlayers upon which they are grown, so some differences among the specimens is expected. Inspection of the 1 x 24 nm data reveals that Ms varies by less than ± 20 % from a mean value of 89 emu/cm3 for all S/AL materials; this observation is consistent with the composition measurements and indicates that the S/AL composition has a relatively minor effect on Ms for the 1 x 24 nm specimen structure. The largest relative changes are seen in MsHc, with the SiN_x specimen exhibiting the largest value and the SiC_x specimen the smallest. The maximum variation from the mean of 4.2 x 10^5 ergs/cm3 is just slightly more than 20 %. Thus, while the S/AL composition has some effect on the 1 x 24 nm specimen magnetic properties, the effects are not dramatic.

Comparison of the 6 x 4 nm and 1 x 24 nm specimen data shows that S/AL composition has a much more pronounced effect on magnetic properties as the magnetic layers become thinner. MsHc, which drops by an order of magnitude or more for five of the six S/AL materials, is the most strongly affected. The 6 x 4 nm YO_x specimen exhibits exceptional behavior this regard, as MsHc decreases by only about 40% relative to the 1 x 24 structure. Ms, which increases by as much as a factor of three, is also strongly influenced by S/AL composition.

TABLE I.

Ms, MsHc, Td and Tc of 1 x 24 nm and 6 x 4 nm films structures for several separator/adjacent layer materials.

Construction/ Separator ID	Ms (emu/cm^3)	MsHc (ergs/cm^3)	Td (C)	Tc (C)
1 x 24 SiC$_x$	98	3.3 x 10^5	150	160
1 x 24 SiN$_x$	89	5.1 x 10^5	160	165
1 x 24 SiO$_x$	100	4.8 x 10^5	165	165
1 x 24 HfO$_x$	74	4.6 x 10^5	160	165
1 x 24 YO$_x$	100	3.7 x 10^5	155	165
1 x 24 Si	62	4.8 x 10^5	150	160
6 x 4 SiC$_x$	111	7.9 x 10^3	<50	120
6 x 4 SiN$_x$	198	2.7 x 10^4	<45	145
6 x 4 SiO$_x$	291	1.8 x 10^4	<30	170
6 x 4 HfO$_x$	166	4.0 x 10^4	60	140
6 x 4 YO$_x$	131	2.3 x 10^5	130	150
6 x 4 Si	80	1.6 x 10^4		115

Several investigators [7, 8] have observed that RE-TM films undergoing oxidation characteristically exhibit decreases in compensation temperature and MsHc. As Tb in the films reacts with O, it becomes less effective at compensating the TM moment, and the compensation temperature, Tcomp, drops; this is consistent with the observed increases in Ms for many of the 6 X 4 nm specimens. Tc and Td are also reduced to varying degrees relative to the 1 x 24 structures. The MsHc decrease has been correlated with decreases in magnetic layer anisotropy as oxidation proceeds [7].

However, while oxidation can can cause the observed behavior, it does not appear to be necessary; very similar magnetic property changes are observed for specimens with S/AL materials which contain no oxygen. It is proposed that the present observations can be accounted for by considering simple layer intermixing in addition to previously proposed interactions such as Tb oxidation. In fact, a primary effect of the 6 x 4 nm structure is to reduce the kinetic barriers to intermixing of the magnetic layer and S/AL materials by increasing interfacial area and reducing the average transport distances. Closer inspection of the Table I data lends support to both the Tb oxidation and intermixing hypotheses. Tc is substantially reduced for Si and SiC$_x$ S/AL materials, but remains relatively high for SiN$_x$, SiO$_x$ and YO$_x$. The increase in Ms for the SiN$_x$, SiO$_x$ and HfO$_x$ specimens is considerably larger than that occurring for the remaining S/AL materials. YO$_x$ is the only S/AL material for which the changes in all measured properties, including MsHc, are relatively small.

For the SiO$_x$ and HfO$_x$ specimens, it is plausible that the increases in Ms result from Tb oxidation. The SiN$_x$ specimen contains no significant O, but Tb - N compounds do exist and exhibit very low curie temperatures, so the effect of N on Ms could be similar to that of O. Alternatively, it is not inconceivable that incorporation of N into the magnetic layer could increase the Fe magnetization [12]. The relatively small change in Ms exhibited by the oxygen-bearing YO$_x$ specimen can be rationalized by examination of Table II, which shows the Gibbs free

TABLE II.

Gibbs free energy of formation for the stable oxides of Tb and the S/AL materials [10,11]

Oxide Compound	ΔG @ 298K (KJ/mol oxide)	ΔG @ 298K (KJ/mol O_2)
Co_3O_4	-774	-387
Fe_3O_4	-1015	-508
SiO_2	-856	-856
HfO_2	-1088	-1088
Tb_2O_3	-1776	-1184
Y_2O_3	-1817	-1211

energies of oxide formation per mol of the most stable oxide product and per mol of oxygen reactant for the S/AL and magnetic layer constituents. Inspection of table II reveals that only Y_2O_3 is thermodynamically stable relative to Tb_2O_3. The argument that simple intermixing of the magnetic and S/Al materials strongly contributes to the observed magnetic property changes is most easily made for the Si 6 x 4 nm structure. The low and comparatively stable value of Ms for both the 1 x 24 and 6 x 4 nm Si specimens is consistent with a low reactivity among Si and the magnetic layer constituents. However, Si has ≈ 25 at.% solid solubility in bulk Fe and decreases Tc of resulting alloy by more than 200 C at the solubility limit relative to pure Fe [11]; similar effects may be anticipated for Si additions to FeTbCo. Furthermore, while the origins of perpendicular anisotropy in RE-TM alloys is still debated, it seems clear that the magnetic anisotropy must result from some structural anisotropy introduced during film growth [13]. Consequently, perturbations of the magnetic layer structure due to intermixing can be expected to influence magnetic anisotropy, and consequently may also affect MsHc.

Some of the differences observed among the various S/AL materials can also be rationalized by considering various trade-offs between reaction to form a stable compound such as Tb_2O_3 and intermixing. Formation of a continuous Tb_2O_3 layer at the interface between the materials could serve as a kinetic barrier to further intermixing of the materials and thereby reduce the relative decreases in Tc and MsHc. It is interesting to note that the SiC_x, for which an interfacial oxide reaction barrier is not expected, exhibits an MsHc and Tc decrease similar to Si.

Unlike the other S/AL materials, Y_2O_3 is more stable than any other known compound among the S/AL and magnetic layer constituents; consequently, there is neither much free O to react with Tb and increase magnetization nor is there much of any other free constituent to diffuse into the magnetic layer material and decrease MsHc or Tc. Diffusion rates for a stable compound such as Y_2O_3 are on average much lower than would be expected for elemental constituents. Consequently, the magnetic properties of the 6 x 4 nm YO_x specimen are comparatively unaffected.

CONCLUSIONS

The magnetic properties of FeTbCo films are comparatively insensitive to adjacent layer composition for an FeTbCo layer thickness of 24 nm, but adjacent layer interaction effects dominate the observed behavior of 4 nm thick films. The observed magnetic property changes are consistent with two types of magnetic layer - adjacent layer interactions. The first is a simple intermixing of the layer materials, which reduces magnetic layer energy product and can weaken exchange interactions sufficiently to reduce the magnetic layer Curie temperature. The second type of interaction appears to be reaction of MO layer Tb with adjacent layer constituents such as O and N; such reactions reduce the average Tb moment for the layer, causing a drop in layer compensation temperature and an increase in magnetization. The comparatively small changes observed between the 1 x 24 nm and 6 x 4 nm structures for the YO_x specimens are attributed to the thermodynamic stability of Y_2O_3.

REFERENCES

1. N. Saito et. al., Jap. J. Appl. Phys. **28**, supplement 28-3, 343 (1989).
2. C. -J. Lin, Appl. Phys. Lett. **62**, 636 (1993).
3. L. M. Falicov et. al., J. Mater. Res. **5**, 1299 (1990).
4. N. Sato, J. Appl. Phys. 59, 2514 (1986).
5. D. R. Callaby, R. D. Lorentz and S. Yatsuya, J. Appl. Phys. **75**, 6843 (1994).
6. J. A. Thornton, Ann. Rev. Mater. Sci. **7**, 239 (1977).
7. T. C. Anthony, J. Brug, S. Naberhuis, and H. Birecki, J. Appl. Phys. **59**, 213 (1986).
8. M. M. Yang and T. M. Reith, J. Appl. Phys. **71**, 3945 (1992).
9. L. Eyring, Handbook on the Physics and Chemistry of Rare Earths, edited by K. A. Gschneidner and L. Eyring (North-Holland Publishing, 1979), p. 368.
10. Handbook of Chemistry and Physics, 74th ed, edited by D. R. Lide (CRC press, 1993).
11. R. M. Bozorth, Ferromagnetism (Van Nostrand Company, 1978) p 71.
12. Y. Hoshi and M. Naoe, J. Appl. Phys. **69**, 5622 (1991).
13. F. Hellman and E. M. Gyorgy, Phys. Rev. Let. **68**, 1391 (1992).

Part III

Interlayer Coupling

BLS STUDIES OF EXCHANGE COUPLING IN THE IRON WHISKER/Cr/Fe SYSTEM

J.F.COCHRAN, K.TOTLAND, B.HEINRICH, D.VENUS*, AND S.GOVORKOV
Simon Fraser University, Department of Physics, Burnaby, BC,
Canada V5A 1S6
*Permanent address: Physics Dept., McMaster University, Hamilton,
Ontario, Canada

ABSTRACT

Brillouin light scattering (BLS) and magneto-optic Kerr effect
have been carried out on a series of specimens consisting of a
Fe(001) whisker substrate upon which thin Cr(001) and Fe(001)
layers have been deposited by means of molecular beam epitaxy in
ultrahigh vacuum. The Fe film was 20 monolayers (ML) thick, and
the Cr(001) films were grown having various thicknesses. It is
demonstrated that BLS thin film frequencies measured in the
saturated magnetic state with the thin film magnetization parallel
with the applied magnetic field can be used to obtain the exchange
coupling strength between the thin film and whisker magnetizations
both for antiferromagnetic coupling and for ferromagnetic
coupling, provided that the ferromagnetic coupling is not too
strong. It is also shown that the coupling strength is extremely
sensitive to the quality of the chromium growth: a small
deterioration in the growth conditions has been found to reduce
the exchange coupling by nearly a factor of two.

INTRODUCTION

In a beautiful series of experiments Unguris[1] et al, Pierce[2,3]
et al, and Stroscio[4] et al have used scanning electron microscopy
with polarization analysis (SEMPA) and scanning tunneling
microscopy (STM) to investigate the growth and magnetic structures
in thin Cr(001) and Fe(001) layers grown by means of molecular
beam epitaxy (MBE) on very clean and perfect Fe(001) whisker
substrates. A trilayer structure was used to investigate the
exchange coupling between the iron whisker substrate and a thin
iron film, 5-10 monolayers thick (ML), separated by a Cr(001)
interlayer whose thickness varied from 0 to 40 ML over a distance
of ~0.5 mm. They showed that the coupling alternated between antiferromagnetic coupling (AF) and ferromagnetic coupling (FM)
depending on the thickness of the chromium interlayer, and that
the pattern of changes from AF to FM coupling depended on the
substrate temperature at which the chromium growth took place. The
smoothest growths, corresponding to a layer-by-layer growth mode,
occurred when the substrate temperature was held at 350C; the
resulting magnetization patterns could be described as the
superposition of two oscillatory terms[2]- a short wavelength
component having a period of 2.105±.005 ML and a long wavelength
component having a period of 12±1 ML. Unfortunately, the SEMPA

131

technique can not be used to provide a quantitative measure of exchange coupling strengths, and it can only be used in very small applied magnetic fields. One must therefore turn to other techniques such as Brillouin light scattering (BLS) or magneto-optic Kerr effect (MOKE) in order to obtain a quantitative measure of the exchange coupling between iron layers separated by a chromium interlayer.

We are interested in the use of BLS to measure the strength of the exchange coupling between a thin Fe(001) film separated from a bulk whisker substrate by a Cr(001) interlayer. The principle of the method is simple. The magnetization in an isolated thin Fe(001) film when subjected to an in-plane applied magnetic field, H, applied along a cubic axis oscillates about equilibrium with a circular frequency, ω_m, given by[5]

$$\left(\frac{\omega_m}{\gamma}\right)^2 = \left(H + 2K_1/M_s\right)\left(H + 2K_1/M_s + 4\pi M_{eff}\right) \qquad (1)$$

where γ is the spectroscopic splitting factor (for Fe γ=1.8379x10^7 /Oe corresponding to g=2.09), M_s is the saturation magnetization (=1.706 kOe at 295K for bulk iron[6]), K_1 is the cubic anisotropy constant (=4.76x10^5 ergs/cc for iron[7] at 295K), and

$$4\pi M_{eff} = 4\pi M_s - 2K_u/dM_s \qquad (2)$$

for a film d cm thick. The last term in (2) is derived from a surface anisotropy energy term of the form

$$E_s = -K_u \left(m_z/M_s\right)^2 \text{ ergs/cm}^2. \qquad (3)$$

In eqn. (3) m_z is the component of magnetization normal to the film plane; $K_u > 0$ corresponds to a tendency for the magnetization to orient itself normal to the film. Oscillations of the magnetization about its equilibrium direction are thermally excited at any finite temperature, and they may be observed by means of light scattering[8]. A small fraction of the light incident at frequency ω_0 is scattered at the frequency ω_s shifted by the frequency of the oscillating magnetic system, $\omega_s = \omega_0 \pm \omega_m$. If the thin magnetic film is allowed to interact with a second magnetic system, the iron whisker in the present instance, the resonant frequency of the film will be altered; the shift in frequency increases with the strength of the coupling between the thin film and the substrate. A FORTRAN program has been developed[9] to calculate the dependence of the intensity of frequency shifted light on the frequency shift for a thin magnetic film exchange coupled to a bulk magnetic substrate through an energy having the form[5]

$$E_c = -J_1\cos(\Delta\phi) + J_2\cos^2(\Delta\phi) \text{ ergs/cm}^2, \qquad (4)$$

where Δφ is the angle between the thin film magnetization and the bulk surface magnetization. The intensity of the scattered light exhibits peaks at frequency shifts corresponding to magnetic excitation frequencies; one of the peaks can be identified as the thin film resonant frequency. The calculation is an obvious extension of the theory described by Camley and Mills[10]. The spatial variation of the magnetic excitation amplitude in the bulk magnet is taken into account along with dipole-dipole coupling between the thin film and the bulk magnet. The calculation is only valid when the thin film and the bulk static magnetizations are parallel with the applied magnetic field. In order to simplify the calculations it has been assumed that the amplitude of the precessing thin film magnetization is independent of position across the film thickness: this approximation limits the applicability of the model to films less than ~30 ML thick. It will be shown below that this model calculation provides a good description of BLS frequencies measured at fields sufficiently large to saturate the specimen, i.e. to align the thin film and bulk static magnetizations along the applied field direction.

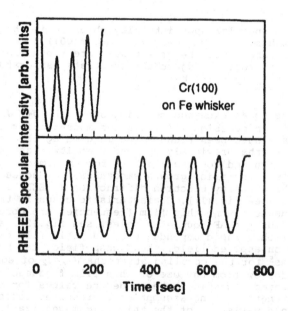

Fig.(1) RHEED specular spot intensity observed for a good growth of 12 ML Cr(001) on an iron whisker substrate. The substrate temperature was adjusted for optimum growth[5], approximately 300C. The lower part of the figure is a continuation of the growth shown in the upper part of the figure with the zero of time reset to zero.

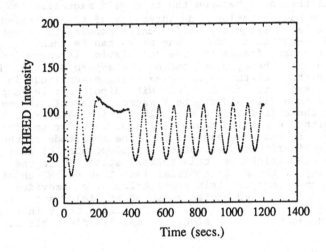

Fig.(2) RHEED specular spot intensity observed in the 11 ML region for the growth of a 11 ML- wedge- 12 ML Cr(001) interlayer on an iron whisker substrate. The growth was interrupted between 200 and 400 seconds in order to adjust the substrate temperature for optimum growth[5], approximately 300C.

Values of the total exchange coupling strength, $J_T = J_1 - 2J_2$, deduced from thin film precessional frequencies measured for the saturated magnetic state have been found to be in good agreement with coupling strengths previously deduced from lower and upper cusp fields (Heinrich and Cochran[5], p.599) for specimens that exhibited a superior growth smoothness as deduced from reflection high energy electron diffraction (RHEED) intensity oscillations (see fig.(1)). Characteristic cusps in the magnetic field dependence of the thin film frequencies are observed at low magnetic fields for AF coupled samples, see fig.(3), and are associated with a rapid dependence of the thin film magnetization orientation on applied field. These cusp fields can be used to obtain values[5] for the coupling strengths J_1, J_2 of eqn.(4). Two advantages follow from the use of thin film frequencies measured in the saturated magnetic state to deduce values for the exchange coupling strength: (1) no assumption is required concerning the magnetic field variation of the thin film magnetization orientation; (2) coupling strengths can be measured for both positive and negative values. It is possible to obtain FM coupling strengths because the thin film peaks retain their identity even when those frequencies overlap with the bulk spin-wave frequencies. The thin film frequencies do become very broad as the FM coupling strength increases so that in practice the method is not useful for coupling strengths J_T greater than ~0.5 ergs/cm^2.

We have used the BLS method, along with MOKE, to investigate the exchange coupling in a recently prepared series of specimens having the structure Whisker(001)/NCr(001)/20Fe(001)/20Au(001), where N is the number of Cr monolayers. The best Cr growths achieved for these samples were not as good as the best growths previously attained; typical RHEED intensity oscillations are shown in fig.(2). The pattern of fig.(2) indicates that the Cr(001) growth proceeded in a good layer-by-layer manner and the RHEED specular spot profiles for the growths of figs.(1) and (2) were very similar. However, the Cr surface for the growth of fig.(2) was definitely not as smooth as surfaces characterized by the RHEED pattern shown in fig.(1). As will be shown below, this rather slight variation in surface quality had a profound effect on the exchange coupling between the 20 ML Fe film and the whisker. The bilinear coupling strength, $|J_1|$ in eqn.(4), was most affected: it was reduced by approximately one half. The biquadratic term (J_2 in eqn.(4)) was little affected. These observations suggest that the coupling strength is very sensitive to the details of the formation of the first Fe/Cr interface. It may very well be that the exchange coupling is very sensitive to the formation of interfacial ordered compounds as suggested by Stoeffler and Gautier[11].

GROWTHS

Well prepared Fe(001) whiskers represent the best available metallic templates and are characterized by atomic terraces whose dimensions[4] are in excess of 1μm. The cross-section of a typical Fe(001) whisker is rectangular, approximately 0.15 to 0.20 mm square, and a typical whisker length lies between 7-15 mm. The whisker surfaces are bounded by {100} planes.

The Cr was grown at elevated substrate temperatures. The choice of substrate temperature is very crucial. We found that the growth of Cr proceeds properly only if the substrate temperature lies within a very narrow range; see Heinrich and Cochran[5] for the details. For sufficiently high substrate temperatures the first RHEED intensity oscillation always shows a very cuspy dependence on time, see fig.(1). For low temperatures the RHEED oscillations exhibit a clearly deteriorating behavior. The best growth was achieved when the RHEED oscillations maintained their intensity minima close to that of the first minimum (0.5ML) and the RHEED intensity maxima were comparable to, or larger than, the specular spot intensity for the bare Fe substrate, see fig.(1). The oscillations in that case were clearly cuspy, and that cuspy behaviour does not deteriorate with the number of deposited atomic layers. The specular spot line scans at the RHEED intensity maxima showed narrow lines whose linewidths were very nearly the same as those for the Fe substrate, and therefore the Cr layers at that point were nearly as smooth as the uncovered Fe whisker substrate. When a new atomic layer started to nucleate the intensity decreased and the line scans started to exhibit a noticeable broadening away from the specular spot along the direction of the (0,0) reciprocal rod. The broadening reached a maximum for a half filled atomic layer. In fact the line scan for a half filled atomic layer showed a clear splitting (two separated maxima) and this indicated that the mean separation between deposited atomic islands was well defined. The mean separation between atomic islands was ~700-800Å (at half ML coverages). Tunneling microscope studies by Stroscio[4] et al at optimum growth temperatures have confirmed that the average distance between nucleated islands is ~700-800 Å in agreement with the RHEED studies, and that growth proceeds in a layer by layer manner.

Specimens for magnetic studies were prepared by depositing the desired number of Cr(001) monolayers on the clean iron whisker following the above prescription. 20 ML of iron were deposited at room temperature, followed by 20 ML of gold also deposited at room temperature. The gold film provided a protective overlayer which prevented the iron from oxidizing when the specimens were removed from the vacuum chamber in order to carry out BLS measurements.

Specimens were also prepared for which the best RHEED patterns obtained displayed amplitude oscillations that were approximately half as large as those shown in fig.(1); see fig.(2) for a typical example. Such specimens displayed a reduced exchange coupling between the 20 ML thin film and the iron whisker. In order to investigate the possibility that the reduced coupling was due to termination of the growth at less than optimum smoothness, a specimen was prepared that contained a thickness wedge between 11 and 12 ML of Cr(001). The RHEED pattern for this case is shown in fig.(2). The chromium growth was carried out with the help of a movable shutter so that two thicknesses of Cr(001) could be grown joined by a wedge shaped ramp. The entire whisker (~10 mm long) was exposed to the flux of Cr atoms until 11 ML had been deposited according to the RHEED oscillations. The growth was then stopped and the shutter moved so as to cover approximately 4mm of the whisker. The Cr growth was resumed and the shutter slowly advanced so that at the conclusion of the growth the first 3 mm of the whisker was covered by 12 ML. At the conclusion of the Cr growth, 20 ML of Fe(001) and 20 ML of Au were deposited as above. BLS and MOKE studies carried out on this specimen are described below.

BLS and MOKE

BLS measurements were carried out using a standard multi-pass Fabry-Perot interferometer in the back-scattering configuration[12,13]. The angle of incidence of the 5145Å laser light was 45°; the intensity of the incident light at the specimen varied between 100 and 200 mWatts. Data were typically collected using a 60 GHz free spectral range divided into 256 channels, with a collection time of 1 second per channel.

Magneto-optic Kerr effect data were carried out using approximately 5 mWatts of 6328Å laser light in the longitudinal configuration. The apparatus was similar to that described by Bader and Erskine[14].

EXPERIMENTAL RESULTS

Examples of the dependence of magnetic excitation frequencies on applied magnetic field are shown in fig.(3) for a 20 ML Fe(001) thin film antiferromagnetically (AF) coupled to a whisker(001)

substrate through 11 ML of Cr(001), and in fig.(4) for a 20 ML
Fe(001) film weakly ferromagnetically coupled (FM) to a whisker
through 10 ML of Cr(001). In each case the largest frequency for a
given applied magnetic field is the bulk surface mode[12]. The lowest
lying frequencies in fig.(3), labeled TF, are associated with the
thin film mode. Note the cusps in frequency vs. field at H_1=1.25
kOe and at H_2=4.0 kOe. These cusps indicate that the orientation of
the static magnetization in the 20 ML Fe(001) film is changing
rapidly with applied field strength. For fields less than H_1 the
magnetization in the thin film lies antiparallel with the applied
field direction. For fields larger than H_2 the thin film static
magnetization and the bulk static magnetization are oriented
parallel with the applied field direction. For fields between H_1
and H_2 the thin film magnetization has turned away from the applied
field direction and takes up an orientation more nearly
perpendicular to the applied field direction. This behaviour can
be understood as a consequence of the minimization of the total
magnetic free energy[15,16]; the exchange energy terms of eqn.(4) act
to turn the magnetization away from the applied field direction in
competition with the term $-\mathbf{M}\cdot\mathbf{H}$ which acts to align the

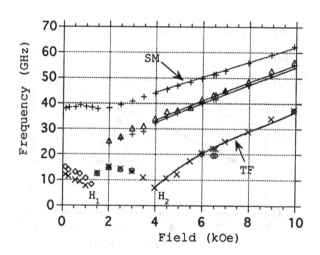

Fig.(3). Magnetic excitation frequencies measured for a 20 ML
     Fe(001) film antiferromagnetically coupled to a Fe(001)
     whisker through 11 ML of Cr(001); this specimen exhibited
     RHEED intensity oscillations similar to those shown in
     fig.(1). SM- bulk iron surface mode frequencies; TF- thin
     film frequencies. The intermediate frequencies correspond
     to the spin-wave manifold edges for bulk iron. The solid
     lines for fields larger than 4.0 kOe were calculated
     using $4\pi M_s$=21.44 kOe, K_1=4.76x10^5 ergs/cm^3, and J_T= -1.65
     ergs/cm^2.

magnetization with the field direction. The frequency splitting observed for fields less than H_1 are a consequence of dipole-dipole coupling between the thin film and the bulk coupled with a non-zero spin-wave wave-vector in the plane of the film. The frequency splitting between up-shifted and down-shifted scattered light modes becomes particularly pronounced for a two component magnetic system for which the static magnetizations are antiparallel. The small but relatively rapid decrease with applied field observed for the surface mode for fields near H_1 is presumably a second manifestation of dipole-dipole coupling between the thin film and the bulk; it is associated with a rotation of the thin film magnetization away from a direction that is antiparallel with the bulk magnetization.

The intermediate frequencies shown in fig.(3) are due to light scattered from bulk spin-wave modes[10,12]. A peak in the bulk spin-wave density of states in the long wavelength limit results in well defined maxima in the scattered light intensity; these intensity maxima are referred to as the bulk edge modes. For the light scattering geometry used to measure the data shown in fig.(3) these edge frequencies lie within ~0.3 GHz of the frequencies that would be measured in a ferromagnetic resonance experiment[17].

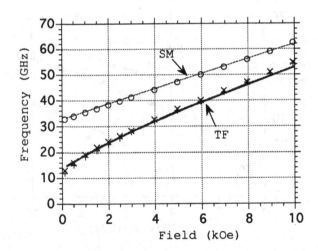

Fig.(4) Magnetic excitation frequencies measured for a 20 ML Fe(001) film coupled to a Fe(001) whisker through 10 ML of Cr(001); this specimen exhibited RHEED intensity oscillations similar to those shown in fig.(1). SM- bulk iron surface mode frequencies; TF- thin film frequencies. The solid lines were calculated using $4\pi M_s$= 21.44 kOe, K_1= 4.76x10^5 ergs/cm^3, and J_T= 0.

Fig.(4) illustrates a case for which the 20 ML Fe(001) film is weakly ferromagnetically coupled to the bulk substrate. In this case the thin film frequencies are nearly equal to the bulk edge frequencies. The thin film scattered light intensity dominates the bulk edge mode intensity with the result that the edge frequencies cannot be observed. The magnetizations in the thin film and in the bulk remain parallel with the applied field direction over the entire field range from 0.2 to 10 kOe consequently both bulk surface mode and thin film frequencies exhibit a monotonic dependence on the field.

The solid lines shown in figs.(3) and (4) have been calculated using the modified Camley-Mills theory mentioned above. The saturation magnetization and cubic anisotropy parameters for both the thin film and the substrate have been taken to be those for bulk iron[6,7] at 295K: $4\pi M_s= 21.44$ kOe, and $K_1=4.76\times10^5$ ergs/cm^3. We have in addition assumed a surface energy parameter $K_{uB}=0.5$ ergs/cm^2 between the bulk iron surface and the chromium (see eqn.(3)). There appears to be no data in the literature from which a value of the surface energy can be obtained. However, surface energy parameters for Fe(001) in contact with transition metals tend to lie between 0.5 and 0.8 ergs/cm^2 (de Jonge[18] et al). The calculated frequencies are insensitive to the value used for K_{uB}; they typically vary by less than 0.3 GHz if K_{uB} is changed from 0.5 to 1.0 ergs/cm^2. We have also included a term of the form of eqn.(3) to represent the surface energies associated with the thin film interfaces between the Fe film and the chromium and gold layers. The thin film frequencies are sensitive to the value chosen for the surface energy parameter K_{uA}; the choice of this parameter is equivalent to the choice of a value for the thin film effective magnetization, eqn.(2). The thin film frequency is also sensitive in the saturated state to the total coupling parameter $J_T= J_1-2J_2$. In effect one has available two parameters to fit the field dependence of the thin film frequencies. We have used the data of fig.(4) to determine an appropriate value for the parameter K_{uA}. For a given K_{uA} the parameter J_T was chosen to yield the observed frequency at 6.0 kOe. Calculated frequencies were then compared with observed frequencies over the entire field range. Values of K_{uA} ranging between 0 and 0.5 ergs/cm^2 gave an adequate representation of the data; $K_{uA}=1.0$, $J_T=0.5$ ergs/cm^2 resulted in calculated frequencies at low fields that were ~4 GHz larger than the observed frequencies. We have chosen to use $K_{uA}=0.5$ ergs/cm^2 because this choice results in better agreement with J_T obtained from the cusp fields for a number of AF coupled specimens, see Table(I). The value $K_{uA}=0.5$ ergs/cm^2 corresponds to an effective magnetization $4\pi M_{eff}=19.4$ kOe. As can be seen from Table(I), values of the total exchange coupling, J_T, deduced from the 6.0 kOe thin film frequencies are generally a little larger than values deduced from the cusp fields H_1 and H_2; however, the discrepency does not exceed 0.23 ergs/cm^2. It can be concluded tha thin film frequencies measured in the saturated magnetic state can be used to obtain the total exchange coupling strength, J_T.

BLS and MOKE measurements were carried out on the wedged specimen whose RHEED intensity oscillations are shown in fig.(2).

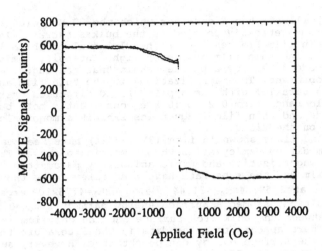

Fig.(5) MOKE signal measured at position A of fig.(6) for the specimen containing a wedge between 11 and 12 ML of Cr(001), and whose RHEED pattern is shown in fig.(2). The Cr layer was 11 ML thick at position A.

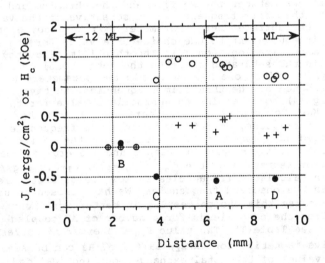

Fig.(6) Exchange coupling, J_T, and MOKE critical fields measured as a function of position along the wedged specimen whose RHEED intensity oscillations are shown in fig.(2). ●- J_T, total coupling deduced from the BLS high field data. o- upper critical field H_2 from MOKE. +- lower critical field H_1 from MOKE. A,B,C,D are positions along the specimen at which BLS measurements were carried out.

Fig.(5) shows the magnetic field dependence of the MOKE signal at
a point 6.4 mm from the end of the whisker in the 11 ML Cr region
(at position A in fig.(6)). The MOKE signal saturates for fields
between 1.3 and 1.5 kOe; this corresponds to the saturated state.
A small plateau is visible for fields less than 0.3-0.38 kOe; this
corresponds to the thin film magnetization oriented antiparallel
to the applied field. Let the field at which the antiparallel
state becomes unstable be H_1. Let the field at which the MOKE
signal saturates be H_2. The fields H_1 and H_2 have been plotted in
fig.(6) at various positions along the specimen. It is clear from
the MOKE signal that in the 12 ML CR(001) region the 20 ML Fe(001)
film did not exhibit AF coupling with the substrate. This
observation was confirmed by BLS measurements taken at position B
of fig.(6). The BLS frequency data could be well fit using J_T=+0.07
ergs/cm^2: ie a weak ferromagnetic coupling. In the region outside
the 12 ML region the coupling was found to be antiferromagnetic.
The strength of the AF coupling was found to be J_T~-0.6 ergs/cm^2
from the BLS data. The coupling strengths estimated from the MOKE
critical fields H_1= 0.36 kOe and H_2= 1.45 using a minimum energy
principle[15,16] were found to be J_1=-0.41 ergs/cm^2 and J_2=0.17
ergs/cm^2 corresponding to J_T= J_1-2J_2= -0.75 ergs/cm^2. Coupling
strengths calculated from the BLS data were found to be
consistently smaller than coupling strengths estimated from the
MOKE data. The reason for this discrepancy is not known. We plan

TABLE (I)

Exchange coupling constants deduced[16] from the cusps in the
magnetic field dependence of the thin film frequencies observed
for a 20 ML Fe(001) film coupled to a Fe(001) whisker substrate
through N ML of Cr(001); these specimens correspond to good Cr
growths characterized by RHEED intensity oscillations similar to
those shown in fig.(1). J_1 and J_2 are the bilinear and biquadratic
coupling coefficients of eqn.(4): see Heinrich and Cochran[5] Table6,
p599. Values of the total exchange coupling, J_T= J_1-2J_2, deduced
from the cusp field data are compared with J_T deduced from the thin
film frequency measured for an applied field of 6.0 kOe.

N	Coupling from Cusps			Coupling from 6.0 kOe Frequency		
(ML)	J_1 (ergs/cm^2)	J_2 (ergs/cm^2)	J_T=J_1-2J_2 (ergs/cm^2)	Freq.@6 kOe (GHz)	J_T (ergs/cm^2) K_{uA}=0	J_T (ergs/cm^2) K_{uA}=0.5
5	-0.78	0.44	-1.66	19.5	-1.75	-1.69
6	-0.59	0.17	-0.93	32.5	-0.81	-0.70
7	-0.70	0.22	-1.14	25.0	-1.39	-1.31
8	-0.28	0.11	-0.50	32.2	-0.84	-0.72
9	-0.71	0.23	-1.17	25.0	-1.39	-1.31
11	-1.04	0.24	-1.52	20.0	-1.74	-1.65
13	~-0.5	~0.2	~-0.9	28.0	-1.18	-1.09

further investigations. Finally, two cusps in frequency vs magnetic field were observed from BLS measurements at position A of fig.(6). These cusps occurred at $H_1=0.3\pm0.1$ kOe and at $H_2=0.8\pm0.1$ kOe. These cusps were expected to occurr at the fields $H_1=0.36$ and $H_2=1.45$ kOe measured using MOKE. The BLS cusp fields correspond to coupling parameters $J_1=-0.27$ ergs/cm^2, $J_2=0.09$ ergs/cm^2, and $J_T=J_1-2J_2=-0.45$ ergs/cm^2 assuming static magnetic configurations that minimize the free energy. This value of J_T is smaller than the value -0.75 obtained from the MOKE data, but agrees with the value J_T deduced from the BLS cusp fields within the range of uncertainties observed for the specimens listed in Table(I). Unfortunately, no MOKE data was available for the specimens listed in Table(I).

DISCUSSION

It is quite clear that the exchange coupling between two iron layers separated by a chromium interlayer is supersensitive to the morphology of the chromium growth. The relatively small change in the character of the RHEED intensity oscillations between the growths of figs.(1) and (2) resulted in quite different coupling strengths for Cr(001) layers 11 ML thick. We suspect that the different results obtained using "good growths", fig.(1), and those obtained using "less good growths", fig.(2), can be attributed to the quality of the interface between the iron whisker surface and the first Cr layer. It is very likely that mixing of the iron and chromium atoms at the interface can have a profound effect on the nature of the exchange coupling between the whisker and a thin iron overlayer. Further investigation is required.

We found one qualitative feature of the results shown in fig.(6) to be unexpected. The transition from AF coupling to weak FM coupling occurred abruptly and very near a chromium thickness of 12 ML. We had expected a gradual decrease of coupling strength over the 2.5 mm interval occupied by the chromium wedge.

ACKNOWLEDGEMENT

The authors would like to thank the Natural Sciences and Engineering Research Council for grants that supported this work.

K.T. gratefully acknowledges support from the Swiss National Science Foundation.

REFERENCES

(1) J.Unguris, R.J.Celotta, and D.T.Pierce, Phys.Rev.Lett.**67**,140 (1991); ibid,**69**,1125 (1992).

(2) D.T.Pierce, J.A.Stroscio, J.Unguris, and R.J.Celotta, Phys.Rev.**B49**,14564 (1994).

(3) D.T.Pierce, J.Unguris, and R.J.Celotta in Ultrathin Magnetic Structures II, edited by B.Heinrich and J.A.C.Bland (Springer-Verlag, Berlin, 1994), p.117.

(4) J.A.Stroscio, D.T.Pierce, J.Unguris, and R.J.Celotta, J.Vac.Sci.Technol.**B12**,1789 (1994).

(5) B.Heinrich and J.F.Cochran, Advances in Physics,**42**,523 (1993).

(6) A.S.Arrott and B.Heinrich, J.Appl.Phys.**52**,2113 (1981).

(7) P.Escudier, Ann.Phys.**9**,125 (1975): H.Gengnagel and U.Hofmann, phys.stat.sol.**29**,91 (1968).

(8) J.F.Cochran in Ultrathin Magnetic Structures II, edited by B.Heinrich and J.A.C.Bland (Springer-Verlag, Berlin, 1994), p222.

(9) J.F.Cochran. Unpublished.

(10) R.E.Camley and D.L.Mills, Phys.Rev.**B18**,4821 (1978).

(11) D.Stoeffler and F.Gautier, Phys.Rev.**B44**,10389 (1991); J.Magn.Magn.Mater.**104-107**,1819 (1992).

(12) J.R.Sandercock in Topics in Applied Physics Volume 51: Light Scattering in Solids III, edited by M.Cardona and G.Güntherodt (Springer-Verlag, Berlin, 1982), p.173.

(13) P.Grünberg in Topics in Applied Physics Volume 66: Light Scattering in Solids V, edited by M.Cardona and G.Güntherodt (Springer-Verlag, Berlin, 1989), p.303.

(14) S.D.Bader and J.L.Erskine in Ultrathin Magnetic Structures II, edited by B.Heinrich and J.A.C.Bland (Springer-Verlag, Berlin, 1994), p.297.

(15) W.Folkerts and S.T.Purcell, J.Magn.Magn.Mat.**111**,306 (1992).

(16) J.F.Cochran, J.Magn.Magn.Mat. To be published.

(17) B.Heinrich in Ultrathin Magnetic Structures II, edited by B.Heinrich and J.A.C.Bland (Springer-Verlag, Berlin, 1994), p.195.

(18) W.J.M. de Jonge, P.J.H.Bloemen, and F.J.A.den Broeder in Ultrathin Magnetic Structures I, edited by J.A.C.Bland and B.Heinrich (Springer-Verlag, Berlin, 1994), Table2,7, p.79.

MAGNETIC PHASE TRANSITIONS IN EPITAXIAL Fe/Cr SUPERLATTICES

Eric E. Fullerton,* K. T. Riggs,*† C. H. Sowers,* A. Berger,** and S. D. Bader*
*Materials Science Division, Argonne National Laboratory, Argonne, IL 60439, USA
**Department of Physics and Institute of Surface and Interface Science, University of California-Irvine, Irvine, CA 92717

Abstract

The surface spin-flop and Néel transitions are examined in Fe/Cr superlattices. The surface spin-flop, originally predicted by Mills [Phys. Rev. Lett. **20**, 18 (1968)], is observed in Fe/Cr(211) superlattices with antiferromagnetic interlayer coupling and uniaxial in-plane anisotropy. The Néel transition (T_N) of Cr is observed in Fe/Cr(001) superlattices, for which the onset of antiferromagnetism is at a thickness t_{Cr} of 42Å. The bulk value of T_N is approached asymptotically as t_{Cr} increases and is characterized by a three-dimensional shift exponent. These T_N results are attributed to finite-size effects and spin-frustration near rough Fe-Cr interfaces.

Introduction

Fe/Cr superlattices exhibit the intriguing magnetic properties of oscillatory interlayer coupling [1,2] and giant magnetoresistance [3]. Growth of epitaxial Fe/Cr superlattices allows the interlayer coupling and magnetic anisotropy to be tailored in order to probe additional, rather subtle, magnetic transitions. We discuss two such transitions, the surface spin-flop transition in Fe/Cr(211) superlattices [4] and the Néel transition of thin Cr layers in proximity with Fe in Fe/Cr(001) superlattices. The surface spin-flop transition is a first-order, field-induced phase transition in antiferromagnets with uniaxial magnetic anisotropy and the magnetic field applied along the easy axis. It was first predicted over 25 years ago,[5] but not realized experimentally until the appearance of Ref. 4. In Fe/Cr(100) superlattices, the antiferromagnetic ordering of the Cr spacers results in anomalies in a variety of physical properties. The Néel temperature (T_N) is strongly dependent on the Cr thickness. A transition-temperature shift exponent is extracted from the data in the thick Cr regime (<160Å) and discussed in terms of a combination of finite-size effects and spin-frustration near rough Fe/Cr interfaces. Results are compared with mean-field calculations of a two-dimensional Ising system at finite temperature to get rudimentary insight into the problem. The work provides auxiliary demonstrations of the value of epitaxial superlattices grown via sputtering, and of the versatility of a recent magneto-optic simulation formalism to handle different and arbitrary magnetization orientations in each ferromagnetic layer within a multilayer structure.

Fe/Cr(100) and (211) superlattices were epitaxially grown by d.c. magnetron sputtering onto single-crystal MgO(100) and (110) substrates, respectively. The epitaxial relations are Fe/Cr[011] // MgO[001] for Fe/Cr(100) and Fe/Cr[01̄1] // MgO[001] for Fe/Cr(211). The growth procedure and structural characterizations are provided elsewhere [6]. The magnetic properties were measured by means of a SQUID magnetometer and the surface magneto-optic Kerr effect using p-polarized, 633-nm light. Transport properties were measured using a standard four-terminal d.c. technique with a constant current of 10 mA and the applied field H in-plane.

Mat. Res. Soc. Symp. Proc. Vol. 384 © 1995 Materials Research Society

Surface spin-flop transition

In a MnF_2-type antiferromagnet (AF), a magnetic field parallel to the easy axis induces a first-order phase transition from the spin sublattices being antiparallel along the easy axis direction to the spin-flop phase in which the spin sublattices reorient almost $90°$ from the field direction but canted toward it. The spin-flop transition occurs at a field H_{SF} given by

$$H_{SF} = \sqrt{2H_E H_A - H_A^2} \quad , \tag{1}$$

where H_E and H_A are the exchange and anisotropy fields, respectively. For a thin AF film, surface effects strongly influence the magnetic response of the system [4,5,7,8]. The lower coordination of the surface spins allow them to respond more easily to external fields. If the surface spin is pointed antiparallel to the external field, the surface is expected to undergo a surface spin-flop transition at a lower field of $\approx H_{SF}/\sqrt{2}$.

Experimental searches of the MnF_2 system for the surface spin-flop transition at the time of the original theoretical prediction were unsuccessful. In the work of Ref. 4 Fe/Cr(211) superlattices were utilized. In these structures, the spin configuration is predominantly governed by: (i) the Zeeman interaction of the Fe with the external field, (ii) the AF interlayer coupling across the Cr spacers, and (iii) a uniaxial, in-plane anisotropy for the Fe. Therefore, these superlattices are isomorphic to the MnF_2-class antiferromagnets with a (100) surface.

Consider a superlattice which contains N layers of Fe. If N is even, then the two terminal Fe layers will point antiparallel to each other. Therefore, in an applied field, one of the surface layers will be antiparallel to the external field and will undergo the surface spin-flop transition. If N is odd, then the two terminal Fe layers will be parallel and align with the field, and only a bulk spin-flop transition is observed. Figure 1 shows both theoretical [4] and experimental results for superlattices with an even number of magnetic layers.

Shown in Fig. 1a is the calculated magnetization curve as a function of applied field for an N=16 superlattice with H_A=0.5 kOe and H_E=2.0 kOe. For small values of H, the Fe layers are antiparallel and aligned along the easy axis. As H increases the system undergoes a surface spin-flop transition at 0.93 kOe which is less than the expected spin-flop field of 1.5 kOe predicted from Eq. (1). For higher fields, the surface spin-flop transition evolves in a quasi-continuous manner into the bulk spin-flop transition, as originally predicted by Keffer and Chow [9] and described in detail in Refs. 4 and 7.

Shown in Fig. 1b are the magnetization curves for a (211)-oriented [Fe(40 Å)/Cr(11 Å)]$_{N=22}$ superlattice. The magnetization curves were measured using SQUID magnetometry and Kerr effect from H = -4 kOe to +4 kOe with H along the easy axis. As was observed in the theoretical calculations, the system undergoes two transitions upon going from the plateau at low fields into the canted state at higher fields. These two transitions can be seen more vividly in Fig. 1c which shows a static susceptibility plot obtained from differentiating the magnetization data. Two peaks are observed and identified in the figure as S and B for the surface and bulk spin-flop transitions, respectively. Similar measurements on superlattices with an odd number of layers show only the bulk transition as expected.

To confirm that the transition at the lower field results from the surface, we compare the SQUID results with Kerr-effect measurements. The Kerr-effect is surface sensitive, by virtue of the optical skin depth, which should reflect itself as a stronger intensity for the surface transition. In the Kerr intensity measurements (Fig. 1b) for H<0, the bulk spin-flop transition is reduced and the surface spin-flop transition is enhanced relative to their strengths of the SQUID magnetization measurements. At the surface spin-flop transition, the Kerr intensity switches from negative to positive, which indicates that the top Fe layer is oriented in opposition to H for small negative fields. As H crosses over to positive fields, the top Fe

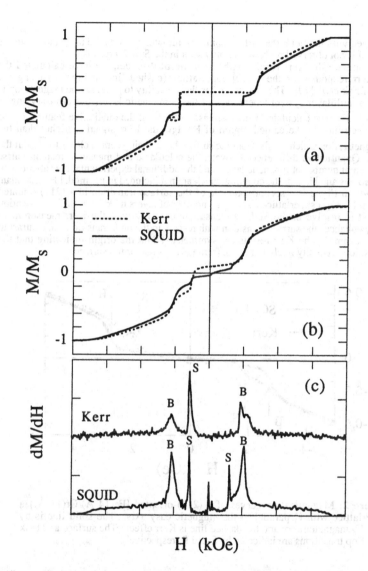

Figure 1: (a) Magnetization results for a model 16-layer superlattice structure. The solid line is the magnetization and the dashed line is the calculated longitudinal magneto-optic Kerr effect. (b) Magnetization curve of a [Fe(40 Å)/Cr(11 Å)]$_{22}$ superlattices from -M$_s$ to M$_s$ with the applied field parallel to the magnetic easy axis. The solid line is measured by a SQUID magnetometer and the dashed line is measured by longitudinal magneto-optic Kerr effect. (c) The numerical derivative of the measured curves in (b). S and B refer to the surface and bulk spin-flop transitions, respectively.

layer is aligned with the field; the surface spin-flop initiates from the Fe layer closest to the substrate, and is not observed by Kerr and only seen in the SQUID results.

To compare the Kerr results with the theoretical model, we have calculated the expected Kerr response for the model superlattice (dashed line in Fig. 1a) using the formalism of Zak *et al.* [10]. The formalism has the versatility to generate the magneto-optic response of a multilayer system where each of the N magnetic layers has arbitrary angles with respect to $\vec{H}$. The calculated Kerr response exhibits all the qualitative features of the measured spectrum. In the canted region of $H > H_{SF}$, the Kerr signal is higher than the SQUID magnetization which results from the surface layers being canted closer to $\vec{H}$ than the bulk layers. Quantitative differences between the calculated and measured response arise from the different number of magnetic layers and the additional experimental contributions of cubic anisotropies and biquadratic coupling which are not included in the model Hamiltonian.

Shown in Fig. 2 is a second example of the surface transition in a (211)-oriented [Fe(14 Å)/Cr(11 Å)]$_{44}$ superlattice. The large number of layers makes the surface transition very difficult to resolve in the SQUID measurement. However, the Kerr measurement dramatically enhances the surface transition with respect to the bulk transition. In contrast to the previous example, the Kerr results are symmetric about the origin indicating that the surface transition is equally likely to initiate from the substrate or top surface.

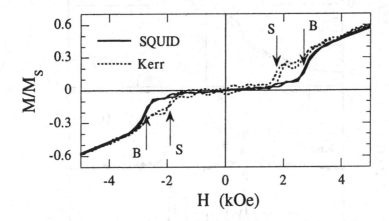

Figure 2: Magnetization curves of a (211)-oriented [Fe(14 Å)/Cr(11 Å)]$_{44}$ superlattice with H parallel to the magnetic easy axis. The solid line is by SQUID magnetometry and the dashed line is Kerr effect. The surface and bulk spin-flop transitions are indicated by S and B, respectively.

This study provided the first experimental observation of the surface spin-flop transition. The surface contributions to the magnetic response of the superlattice is elucidated by the combination of SQUID which measures the total magnetization and Kerr effect which is sensitive to the vicinity of the illuminated surface. This example highlights the rich magnetic phases possible in coupled magnetic superlattices.

Cr Néel transition

An outstanding problem in understanding the interlayer coupling in Fe/Cr superlattices is the role of the magnetic ordering within the Cr spacers. Bulk Cr is an itinerant AF with $T_N = 311$ K. An incommensurate spin-density wave (SDW) is formed which is characterized by a wave vector Q determined by the nested feature in the <100> direction of the Cr Fermi surface. At high temperature, the Cr spin sublattices S are transverse to Q ($S \perp Q$), while below the spin-flip transition at 123K S rotates 90° to form a longitudinal SDW with $S \| Q$ [11].

For the case of Cr(100) spacers, two periods in the interlayer coupling have been observed, a short period [two monolayers (ML)] and a long period (18Å) [2,12-14]. The short-period oscillations results from the same nesting responsible for the SDW of bulk Cr. The long period has also been observed in (110) [2] and (211) [6] oriented films, which suggests that it is not related to the nesting but, results instead from a short spanning-vector associated with the relatively isotropic 'lens' feature of the Fermi surface [15]. In general, only the long-period oscillation is observed in superlattices. Atomic steps in the Fe-Cr interface are sufficient to suppress the short period coupling. However, the magnetic ordering of the Cr in the presence of a stepped or rough interface and its role on the interlayer coupling are not known. Slonczewski [16,17] predicted that fluctuations in the short-period interactions can give an additional non-oscillatory, biquadratic coupling term in which the magnetization orientation between adjacent Fe layers is 90°, rather than 180° or 0°. This extrinsic biquadratic coupling can become a prominent characteristic of the thick Cr-spacer regime in which the long-period coupling is weak.

In the present work, the AF ordering of Cr spacer layers in sputtered, epitaxial Fe/Cr(001) superlattices is considered and we report a number of new observations. Firstly, we find that the AF order is suppressed for Cr spacers of thickness $t_{Cr} < 42$Å. Secondly, for $t_{Cr} > 42$Å, T_N initially rises rapidly, asymptotically approaches the bulk value for the thickest spacers studied (165Å), and exhibits a transition-temperature shift exponent $\lambda = 1.4 \pm 0.3$ characteristic of three-dimensional (3D) Heisenberg or Ising models. The overall T_N-$vs.$-t_{Cr} behavior can be understood in terms of a combination of finite-size and spin-frustration effects. Thirdly, the AF ordering of the Cr spacers results in anomalies in a variety of physical properties, including the interlayer coupling, remanent magnetization (M_r), coercivity (H_c), resistivity (ρ) and magnetoresistance (MR). Finally, the biquadratic coupling of the Fe layers observed for $T > T_N$ vanishes below T_N.

Transport measurements are often used as a probe of the AF ordering in Cr and Cr alloys, where ρ is enhanced above its extrapolated value as the temperature T decreases through T_N [11,18]. This anomaly in ρ, attributed to the formation of energy gaps opening on the nesting parts of the Fermi surface, is commonly used to locate T_N. Shown in Fig. 3 are T-dependent transport results for an [Fe(14Å)/Cr(70Å)]$_{13}$ superlattice. Figure 3(a) shows ρ $vs.$ T for H=500 Oe, which is a sufficient field to align the Fe magnetization. An anomaly in ρ below ~200K is observed as an increase above its expected linear behavior, as shown by the dotted line. The difference between the measured ρ and the linear extrapolation ρ_{lin} is plotted in Fig. 3(b). The 7% enhancement in ρ at 190K is consistent with similar measurements in bulk Cr and Cr(001) films [11,19]. The reduced value of T_N=195 K is determined by the point of inflection of ρ $vs.$ T [see Fig. 3(c)].

The AF-ordering of the Cr dramatically alters the magnetic properties of the superlattices. Shown in Fig. 4 is the T dependence of M_r, H_c, the saturation field (H_s), and the MR of the same superlattice as in Fig. 3. All four of these quantities exhibit anomalous behavior which is directly related to the Néel transition of the Cr. The M_r shows a transition at

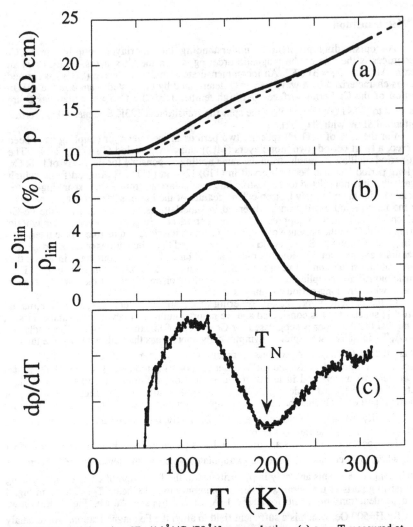

Figure 3: Resistivity of an [Fe (14Å)/Cr(70Å)]$_{13}$ superlattice. (a) ρ *vs.* T measured at H=500 Oe. The dashed line is a linear extrapolation of the data above 280 K. (b) The difference between the measured ρ and the linear extrapolation ρ_{lin} normalized to ρ_{lin}. (c) Derivative of ρ smoothed for clarity. The minimum in dρ/dT locates T$_N$.

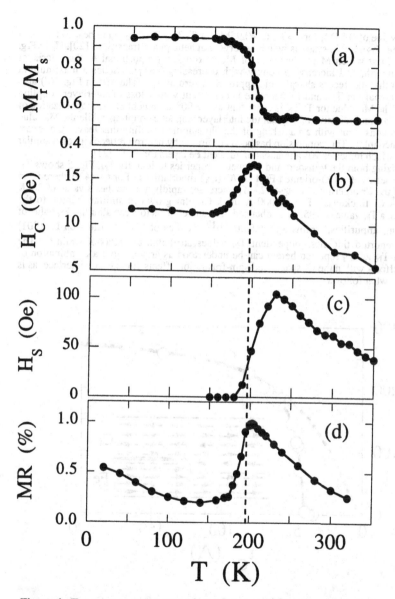

Figure 4: Temperature dependent magnetization results for the superlattice shown in Fig. 3. (a) Squareness ratio M_r/M_s; (b) coercivity, (c) saturation field defined at 90% of M_s, and (d) the magnetoresistance. The vertical dashed line locates T_N of the Cr interlayers.

T_N from a value of $0.53M_s$ for $T>T_N$ to $\approx 0.95M_s$ for $T<T_N$. The H_c value peaks at T_N in a manner often observed in systems which undergo magnetic phase transitions [20]. H_s in Fig. 3(c), the field at which M reaches 90% of M_s, is roughly proportional to the interlayer coupling strength, and increases strongly with decreasing $T>T_N$, reaches a maximum at ~230K, and then decreases sharply and approaches zero at T_N. The MR in Fig. 4(d) also shows an anomaly at T_N, and its decrease is consistent with a loss of interlayer coupling below T_N. The M_r value for $T>T_N$ is consistent with a 90° alignment of the magnetization of adjacent Fe layers, indicative of biquadratic interlayer magnetic coupling, while the M_r value for $T<T_N$ is consistent with a vanishing of the biquadratic coupling that leaves the layers relatively uncoupled. This conclusion has been confirmed by neutron reflectivity on a similar superlattice which indicates 90° alignment of adjacent Fe layers for $T>T_N$ [21].

Utilizing both the transport and magnetic properties to identify T_N, Fig. 5 shows T_N vs. t_{Cr} for a series of (001)-oriented Fe(14Å)/Cr(t_{Cr}) superlattices. For $t_{Cr}<42$Å there is no evidence of the Cr ordering. For $t_{Cr}>42$Å T_N increases rapidly and reaches a value of 265K for a 165-Å Cr thickness. For a 3000-Å thick Cr film grown in similar fashion to the superlattices, a T_N value of 295K was obtained. A number of factors can alter the T_N value of Cr, including impurities, strains and defects.[11,19] Studies of Cr(001) films on LiF(001) substrates reported thickness dependent T_N values attributed to epitaxial strain.[19] The behavior of T_N vs. t_{Cr} reported herein can be understood as arising from a combination of finite-size effects within the Cr spacer and spin-frustration effects at the Fe-Cr interface, as is discussed in what follows.

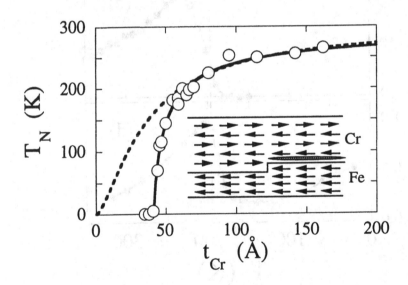

Figure 5: T_N for a series of [Fe(14Å)/Cr(t_{Cr})]$_N$ superlattices vs. Cr thickness. The dashed and solid lines are fits to Eq. (2) and (3), respectively. The inset shows a possible spin configuration of Cr on a stepped Fe surface in which the region of spin frustration at the Fe-Cr interface is shown schematically by the shaded ellipse to the right of the atomic step.

In thin films, magnetic properties are altered due to the surface contribution to the free energy [22]. Since this contribution is generally positive, the magnetic order is weakened at the surface and the ordering temperature is reduced. Scaling theory predicts that T_N should have the form:

$$\frac{T_N(\infty) - T_N(t_{Cr})}{T_N(t_{Cr})} = b\, t_{Cr}^{-\lambda} \quad , \tag{2}$$

where $\lambda = 1/\nu$ is the shift exponent, b is a constant and ν is the correlation-length exponent for the bulk system. The theoretically expected λ values are $1/0.7048 = 1.419$ [23] and $1/0.6294 = 1.5884$ [24] for the 3D Heisenberg and Ising models, respectively. Fitting the data for $t_{Cr} > 70$Å to Eq. (2) is shown by the dashed line in Fig. 5, where $T_N(\infty) = 295$K has its thick-film value. These data are well represented by $\lambda \approx 1.4 \pm 0.3$, which is in agreement with expectation from scaling theory. To fit the complete data set, we use the empirical expression:

$$\frac{T_N(\infty) - T_N(t_{Cr})}{T_N(t_{Cr})} = b\, (t_{Cr} - t_0)^{-\lambda'} \quad , \tag{3}$$

where $t_0 = 42.3$Å represents the zero offset in the Cr thickness and $\lambda' = 0.8 \pm 0.1$ as shown by the solid line in Fig. 5. The sharp drop in the value of T_N near $t_{Cr} \sim 50$Å and the nonuniversal value of λ' indicates the presence of an additional effect for the thinnest Cr spacers that we will identify below as being due to spin frustration.

For an ideal Fe/Cr(001) interface, $T_N(t_{Cr})$ should *increase* with decreasing Cr thickness due to the Fe-Cr exchange coupling since the Fe Curie temperature is much higher than T_N for Cr. This agrees with the experimental observations of Ref. 25 that the surface-terminated ferromagnetic layer of Cr on an Fe(001) substrate oscillates in its magnetization orientation relative to that of the Fe with a ≈ 2-ML period, consistent with the SDW AF anticipated for Cr. The oscillations are observed well above the bulk T_N, which suggests that the substrate, a relatively perfect Fe whisker, stabilizes the AF spin structure of the Cr. In the present study, however, we find $T_N(t_{Cr})$ *decreases* for thin Cr layers. We believe that this behavior arises from spin-frustration effects in the vicinity of the rough Fe-Cr interfaces. Such interfaces contain atomic steps as shown in the Fig. 5 inset. The interfacial exchange energy can be minimized only locally, and frustration of the interfacial spins will occur if the Fe and Cr magnetically order long-range. In the Fig. 5 inset, excess magnetic energy is schematically located at the Fe-Cr interface to the right of the step where the Fe and Cr moments are forced to align ferromagnetically. For a superlattice, assuming random monatomic steps at both the Fe and Cr surfaces, 25% of the Cr layer will be frustrated at both interfaces, 50% will be frustrated at one interface, and 25% will match with the Fe layers. The value of T_N, therefore, should be influenced by a balance between the energy gained from long-range AF ordering of the Cr and the energy cost due to magnetic frustration at the Fe-Cr interfaces. For thin Cr layers, the frustration energy is sufficiently high to suppress long-range ordering of the Cr. As t_{Cr} increases, the system overcomes the frustration energy and begins to order. The crossover thickness for the present samples is 42Å of Cr.

To further understand the mechanism for suppressing the Cr ordering, we have performed mean-field calculations at finite temperatures for an AF layer sandwiched between two ferromagnetic (F) layers as shown in Fig. 6. The model utilizes periodic boundary conditions, and the F, interface and AF exchange coupling constants are given by $J_F = -1.88 J_I = -3.35 J_{AF}$. The ferromagnetic layers are given by the open squares and the AF layer is represented by the circles. The circle are gray-scaled to represent the moment of the layer. Each ferromagnetic layer has two atomic steps to induce spin-frustration in the system.

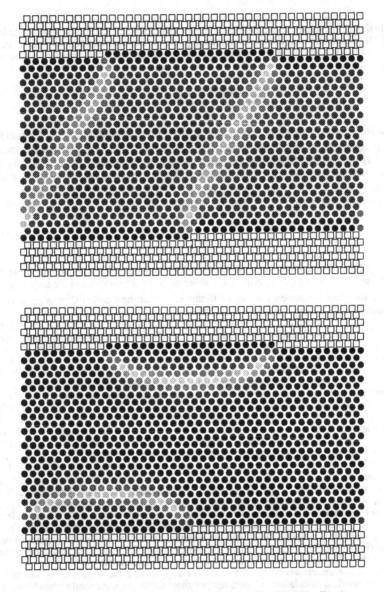

Figure 6: Finite-temperature mean-field calculations of a F/AF/F trilayer structure. The circles and open squares are the AF and F layers, respectively. The circles are gray-scaled to represent the moment; filled and open correspond to full and zero moment, respectively.

Two stable magnetic configurations are identifies in the calculations and shown in Fig. 6. The striped regions in the AF layer represent walls which separate AF domains which are 180° out of phase. In one configuration (upper panel), the domain walls go across the AF layer connecting steps on adjacent ferromagnetic layers. In the other stable configuration (lower panel) the domain walls are parallel to the interface and connect steps on the same ferromagnetic layer. In this configuration, the center of the AF layer exhibits homogeneous long-range order. For thin AF layers, the upper configuration is the lowest energy configuration. For thicker layers, the lower panel represents the lowest energy configuration. For this model structure, a crossover at 18 ML is shown in the phase diagram of Fig. 7. Below 18 ML layer thicknesses the upper-panel configuration is lowest energy for all temperatures. Above 18 ML, the lower panel is lowest energy and T_N approaches T_N for a free standing film for increasing AF layer thicknesses.

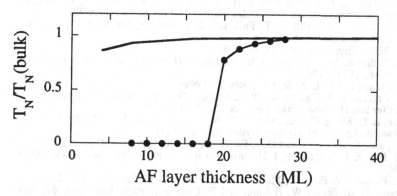

Figure 7: Néel temperatures for AF-layer (solid circles) calculated for the model shown in Fig. 6 versus AF layer thickness. The solid line is a similar calculation for a free standing AF film.

Although this model is too simplistic to quantitatively describe the Fe/Cr system, it is able to reproduce salient qualitative features of our measurements and highlights the important role roughness plays in the magnetic ordering of the Fe/Cr(001) system. More realistic electronic structure calculations [26] of Fe/Cr/Fe trilayers determine that the energy cost in suppressing the Cr moment (at 0 K) is only 0.8 meV/atom, as compared to 200 and 80 meV/atom for Fe and Mn, respectively. Theoretical calculations of diffuse or stepped Fe/Cr interfaces demonstrate that the presence of frustrated Fe-Cr bonds can strongly suppress the Cr moment over extended distances [26-28]. Thus, Cr is highly sensitive to its local environment with local distortions being capable of causing strong moment reductions. This again supports the idea that roughness plays a dominant role in the magnetic ordering.

In summary, we have investigated the AF-ordering of Cr spacer layers in epitaxial Fe/Cr(001) superlattices. AF order is suppressed for Cr spacers of thickness $t_{Cr} < 42$Å and is attributed to spin-frustration effects. For $t_{Cr} > 42$Å T_N initially rises rapidly, and then asymptotically approaches the bulk value for the thickest spacers studied (165Å) with a transition-temperature shift exponent characteristic of 3D Heisenberg or Ising models. Finally, the AF ordering of the Cr spacer layers dramatically alters the interlayer coupling in the sense that the biquadratic coupling of the Fe layers observed for $T > T_N$ vanishes below T_N.

We thank J. Mattson, D. Stoeffler, R. Wang and D. Mills for helpful discussion and R. Wang and D. Mills for supplying the spin-configurations of their model calculations. Work supported by the U.S. Department of Energy, Basic Energy Sciences-Materials Sciences, under contract No. W-31-109-ENG-38. One of us (A.B.) gratefully acknowledges support from the Alexander von Humboldt-Stiftung through a Feodor Lynen Research Fellowship.

References

†   Permanent address: Stetson University, DeLand, FL 32720

1. P. Grünberg, R. Schreiber, Y. Pang, M. B. Brodsky, and C. H. Sowers, Phys. Rev. Lett. **57**, 2442 (1986).
2. S. S. P. Parkin, N. More, and K. P. Roche, Phys. Rev. Lett. **64**, 2304 (1990).
3. M. N. Baibich, J. M. Broto, A. Fert, F. N. VanDau, F. Petroff, P. Etienne, G. Creuzet, A. Friederich, and J. Chazelas, Phys. Rev. Lett. **61**, 2472 (1988).
4. R. W. Wang, D. L. Mills, E. E. Fullerton, J. E. Mattson, and S. D. Bader, Phys. Rev. Lett. **72**, 920 (1994).
5. D. L. Mills, Phys. Rev. Lett. **20**, 18 (1968).
6. E. E. Fullerton, M. J. Conover, J. E. Mattson, C. H. Sowers, and S. D. Bader, Phys. Rev. B **48**, 15755 (1993).
7. R. W. Wang and D. L. Mills, Phys. Rev. B (1994).
8. A. S. Carrico, R. E. Camley, and R. L. Stamps, Phys. Rev. B **50**, 13453 (1994).
9. F. Keffer and H. Chow, Phys. Rev. Lett. **31**, 1061 (1973).
10. J. Zak, E. R. Moog, C. Liu, and S. D. Bader, Phys. Rev. B **43**, 6423 (1991).
11. E. Fawcett, Reviews of Modern Physics **60**, 209 (1988).
12. J. Unguris, R. J. Celotta, and D. T. Pierce, Phys. Rev. Lett. **67**, 140 (1991).
13. D. T. Pierce, J. A. Stroscio, J. Unguris, and R. J. Celotta, Phys. Rev. B **49**, 14564 (1994).
14. S. T. Purcell, W. Folkerts, M. T. Johnson, N. W. E. McGee, K. Jager, J. ann de Stegge, W. B. Seper, W. Hoving, and P. Grünberg, Phys. Rev. Lett. **67**, 903 (1991).
15. D. D. Koelling, Phys. Rev. B **50**, 273 (1994).
16. J. C. Slonczewski, Phys. Rev. Lett. **67**, 3172 (1991).
17. J. C. Slonczewski, J. Magn. Magn. Mater. (in press).
18. E. Fawcett, H. L. Alberts, V. Y. Galkin, D. R. Noakes, and J. V. Yakhmi, Reviews of Modern Physics **66**, 25 (1994).
19. J. Mattson, B. Brumitt, M. B. Brodsky, and J. B. Ketterson, J. Appl. Phys. **67**, 4889 (1990).
20. S. D. Bader, D. Li, and Z. Q. Qiu, J. Appl. Phys. **76**, 6419 (1994).
21. S. Adenwalla, G. P. Felcher, E. E. Fullerton, and S. D. Bader, unpublished.
22. K. Binder, in *Phase Transitions and Critical Phenomena*, edited by C. Domb & J. L. Lebowitz (Academic Press, London, 1983), p. 1.
23. K. Chen, A. M. Ferrenberg, and D. P. Landau, Phys. Rev. B **48**, 3249 (1993).
24. A. M. Ferrenberg and D. P. Landau, Phys. Rev. B **44**, 5081 (1991).
25. J. Unguris, R. J. Celotta, and D. T. Pierce, Phys. Rev. Lett. **69**, 1125 (1992).
26. D. Stoeffler and F. Gautier in *Magnetism and Structure in Systems of Reduced Dimension*, edited by B. Dieny R. F. C. Farrow M. Donath, A. Fert, B. D. Hermsmeier (Plenum Press, New York, 1993), p. 411.
27. D. Stoeffler and F. Gautier, Phys. Rev. B **44**, 10389 (1991).
28. D. Stoeffler and F. Gautier, J. Magn. Magn. Mat. (submitted).

BLOCKING OF MAGNETIC LONG RANGE ORDER BETWEEN THE MONOLAYER AND THE DOUBLE LAYER OF Fe(110) ON W(110)

HANS-JOACHIM ELMERS*, JENS HAUSCHILD*, GUOHUI LIU*, HELMUT FRITZSCHE*, ULRICH KÖHLER** AND ULRICH GRADMANN*
*Physikalisches Institut, Technische Universität Clausthal, D 38678 Clausthal-Zellerfeld
**Institut für Experimentalphysik, Universität Kiel, D 24098 Kiel, Germany

ABSTRACT

In extension of a recent study on submonolayer magnetism of Fe(110) on W(110) [1], we observed the interplay of morphology and magnetic order in Fe(110)-films prepared on W(110) at 300 K, in a range of coverages Θ between the pseudomorphic monolayer (Θ = 1) and the pseudomorphic double layer (Θ = 2), using a combination of STM, SPLEED, CEMS and TOM. Whereas the Curie-temperatures of the monolayer and the double layer are given by $T_c(ML)$ = 230K and $T_c(DL)$ = 450K, respectively, we observe in an interval of $1.20 < \Theta < 1.48$ ML a gap of magnetic long range order, for temperatures down to 115K. In CEMS, we observe superparamagnetic fluctuations for $T > T_c(ML)$, but magnetic short range order for $T < T_c(ML)$. The surprising blocking of long range order in the gap can only be explained from a quasi-antiferromagnetic indirect coupling between double-layer islands

1. INTRODUCTION

Ultrathin films of Fe(110) on W(110) in the monolayer regime have been used widely as model systems for monolayer magnetism [2, 3, 4, 5, 6, 7,], because of the thermodynamic stability of the monolayer [4]. However, even in this structurally well defined model system, dramatic influences of the micro-morphology on the magnetic properties are observed. The combination of STM imaging with magnetic measurements provides the opportunity to obtain new detailed insight in the relation between magnetism and morphology. This has been shown recently in a study of submonolayer magnetism of Fe(110) on W(110) using STM in combination with spin polarized electron diffraction (SPLEED) [1]. In extending these studies beyond the monolayer, we were surprised to find, for films prepared at RT, an interval of coverages $1.20 < \Theta < 1.48$ (in pseudomorphic ML), in which long range order was completely blocked. In the present paper, we report on experiments about this new phenomenon using STM, SPLEED, Mößbauer spectroscopy and torsion oscillation magnetometry, and on our conclusions.

2. GROWTH AND MORPHOLOGY

Our Fe-films were grown in UHV at room temperature (RT) on atomically clean W(110)-substrates. Film thickness was controlled by quartz oscillator monitors, structure was checked by LEED and AES. The growth mode for coverages between Θ = 0.8 and 1.9 (pseudomorphic) ML is shown by STM images in Figure 1. At Θ = 1.0 the 2nd monolayer nucleates on a nearly complete 1st monolayer. Coalescence of the double layer patches proceeds between 1.4 and 1.7 ML. For $\Theta \geq 1.7$ one observes substantial 3rd layer contributions and the incipient relaxation of the misfit by dislocations. In this paper, we

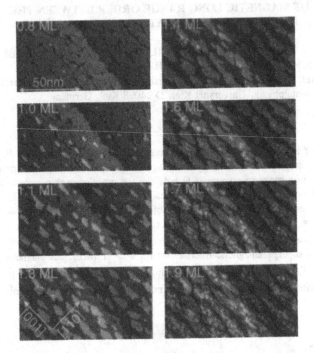

Figure 1: STM images of Fe-films prepared on W(110) at 300K. Coverages Θ in units of pseudomorphic monolayers are indicated. The easy axis [1Ī0] is at right angles with the axis of the elongated islands.

focus on the range Θ ≤ 1.6 where the films were pseudomorphic with the W-substrate (strained by 10.4% in the plane to accomodate the misfit $f_{FeW} = -9.4\%$), and composed of monolayer and double layer patches.

3. SPIN POLARIZED ELECTRON DIFFRACTION (SPLEED)

Our main magnetic probe was SPLEED. We used scattering geometries as described previously [1], with the quantization axis of the electron spin polarization along the easy axis [1Ī0] of the magnetic film. We measure the exchange asymmetry A_{ex} [1], the properly normalized difference in electron reflectivities for spin parallel or antiparallel to [1Ī0]. A_{ex} is roughly proportional to the magnetization [8]. External fields up to 200Oe could be applied along [1Ī0], generated by currents through the W-substrate. SPLEED was possible in fields up to 2Oe. Immediately below T_c, we then obtained, even in these low fields, standard easy axis loops, with coercivities down to 0.1Oe, and perfect saturation in the remanent state, e.g. the inset in Figure 2. This is an important consequence of the strong uniaxial in-plane anisotropy of the samples [9], which enabled us to restrict the measurements at lower temperatures to the remanent state, and to take remanence as saturation. After cooling down the samples to 115K under the action of periodic field pulses

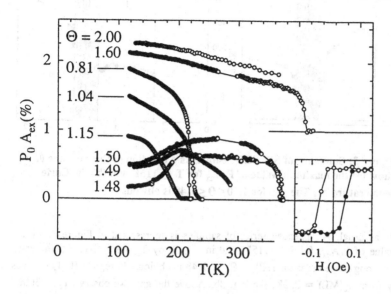

Figure 2: Exchange asymmetry P_0A_{ex} versus temperature T for Fe(110)-films on W(110), with coverage Θ as a parameter. The inset shows a magnetization loop for a film $\Theta = 1.48$, 10 K below T_c.

of 200 Oe·20 msec, we followed the remanent value of P_0A_{ex} (primary beam spin polarization $P_0 = 20\%$) versus T during warming up. Results are shown by some samples in Fig. 2. The temperature dependence was reversible. With respect to $A_{ex}(T)$, four regimes of coverage Θ can be distinguished. In regime I, $\Theta \leq 0.58$, the films are nonmagnetic, see [1]. In regime II, $0.58 < \Theta < 1.2$, represented in Fig. 2 by 3 samples, finite values of A_{ex} indicate long range remanent order roughly up to the Curie-temperature of the monolayer, $T_c(ML) = 230$ K [1]. Near the upper limit of this regime, A_{ex} rapidly decreases with increasing Θ. The following regime III, $1.20 < \Theta < 1.48$, forms the gap of long range order, where A_{ex} disappeared completely, for $T \geq 115$K. Finite values of A_{ex} could neither be observed in remanence nor in external fields up to 2 Oe, for a large number of samples in the gap. Long range order returned quite abruptly in the following regime IV, $\Theta \geq 1.48$. The Curie-temperatures, see Figure 3b, were now always above 300 K. The ideal easy axis loop of the inset in Figure 2, taken from a film $\Theta = 1.48$ just above the gap, contrasts sharply with the inability to obtain any magnetic signal for the last film in the gap ($\Theta = 1.47$). The dependence $A_{ex}(T)$ is non-monotonic for films immediately above the gap ($\Theta = 1.48, 1.49, 1.50$). With decreasing temperatures, A_{ex} first increases as expected, but then decreases again below $T_c(ML)$. For thicker films ($\Theta = 1.6, 2.0$), $A_{ex}(T)$

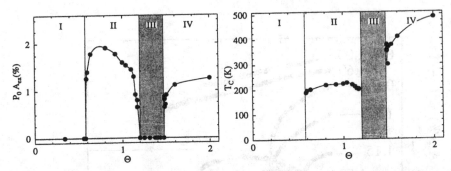

Figure 3: Summary of SPLEED results. As a function of the coverage θ, we show (a) the maximum value of P_0A_{ex} for $T \geq 115K$ and (b) the Curie temperature T_c. The gap for $1.20 < \theta \leq 1.48$ is obvious.

is monotonic as usual. An extended series of samples is represented in Figure 3a by the maximum value of P_0A_{ex}, for $T > 115K$, and in Fig 3b by T_c, respectively, both versus θ. The gap of long range order for $1.20 < \theta < 1.48$ is obvious. In regime II, T_c deviates only slightly from $T_c(ML) = 230K$, see Fig. 3b. Above the gap, we observe $T_c > 300K$. Because the upper end of the gap coincides with the coalescence of the double layer patches, see Figure 1, we conclude that the group of T_c-values of about 375K at this upper end belongs to an interconnected double layer network. The Curie-temperature of the double layer is roughly estimated as $T_c(DL) = 450K$.

The bare SPLEED data could be explained in a model where the double layer islands, prepared at 300K, that means above $T_c(ML)$, nucleate as superparamagnetic particles with statistical distribution of the moments along [1$\bar{1}$0] and [$\bar{1}$10], respectively, which magnetically polarize some monolayer surroundings. At $\theta = 1.2$, the system of monolayer haloes spreads over the hole sample, which then becomes macroscopically nonmagnetic, being however composed microscopically by up and down magnetized double layers islands with attached monolayer haloes. At $\theta = 1.48$, the breakthrough of long range order apparently is induced by direct exchange coupling between the islands by coalescence bridges. The Mößbauer data of section 4 support this model, the magnetometric measurements of section 5 contradict it.

4. CONVERSION ELECTRON MÖSSBAUER SPECTROSCOOPY (CEMS)

CEMS-spectra of a film $\theta = 1.3$, in the center of the gap, are shown in Figures 4a,b. At 250K (Fig.4a), only 30K above $T_c(ML)$, the spectrum shows a quadrupole doublet, the signature of superparamagnetic fluctuations. At 150 K (Fig.4b) one observes a magnetic pattern. Quantitative analysis shows that the 3 sixline components can be attributed to the monolayer and the first or second atomic layer of the double layer patches. Reasonably, the whole system including the double layer islands is frozen with freezing of the surrounding monolayer sea at $T_c(ML)$, different from the usual blocking of superparamagnetic islands by intrinsic thermal slowing down of their fluctuations.

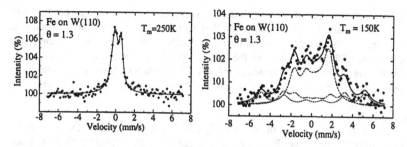

Figure 4: CEMS-spectra of a pure ^{57}Fe-film on W(110), $\Theta = 1.3$, prepared at 300K, measured in zero field (a) at 250K and (b) at 150K.

5. TORSION OSCILLATION MAGNETOMETRY (TOM)

Straightforward estimates using the island sizes of Figure 1 show that the superparamagnetic patches of the model in section 3 should be easily magnetized in moderate fields of the order of 0.1Tesla, at least at temperatures above 250K, where the rapid fluctuations shown by Figure 4a clearly exclude any hysteresis effects. SPLEED is impossible in fields of this order. We therefore checked the samples by TOM [10], which could be done in situ in fields up to 0.4Tesla. Results are shown in Figure 5. Note that we measure in TOM a magnetic torque constant R, and that for the case of a magnetically saturated sample R is connected with the saturation moment m_0 by $R/H = m_0/(1 + H/H_L)$, with an out-of-plane anisotropy field H_L that usually is of the order of some Tesla. This is just the signature of the thick films above the gap ($\Theta = 2.7$, 2.2 and 1.9). The film in the center of the gap, $\Theta = 1.35$, behaves completely different, showing only negative values of R/H, without any indication of saturation. Straightforward analysis of TOM shows that negative values of R/H must be explained by a quasi-antiferromagnetic system of uniaxial

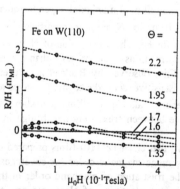

Figure 5: TOM of Fe(110)-films on W(110), at 300K. R/H (torque constant R per magnetic field H), in units of m_{ML}, the moment of a pseudomorphic monolayer with 2.17 μ_B per atom, versus H, applied along [1$\bar{1}$0].

islands with equal or comparable abundance of moments parallel and antiparallel to H, respectively. The frustration of saturation for R/H therefore must be read as frustration of magnetic saturation. We conclude that an interaction between the double layer patches must be active even at 300K which stabilizes the macroscopically nonmagnetic state even in the rapidly fluctuating system. This interaction induces a frustration of long range order in the double-layer island system which cannot be removed by external fields up to 0.4 Tesla.

6. DISCUSSION

The most important result of our investigation is the resistance of the films in the gap against magnetization, see Figure 5, which can be only explained from a quasi-antiferromagnetic interaction between the double layer islands which frustrates long range ferromagnetic order in this system of fluctuating single domain double layer islands. Simple estimates show that the magnitude of the magnetizing field (0.4Tesla) in Fig. 5 excludes magnetostatic interactions as cause of the frustration. Instead, we suggest that the postulated interactions are of indirect electronic origin. Our data thus provide experimental evidence for an (antiferromagnetic) **lateral indirect interaction between islands**, mediated by the interspaces consisting of monolayers plus their W substraste. The details of this interaction remain to be investigated. It is an analogue to the (vertical) indirect interaction between two ferromagnetic films by a nonmagnetic spacer, which forms the highlight of thin film magnetism of the last decade [11]. At the upper end of the gap, $\Theta = 1.48$, the increasing direct ferromagnetic coupling by coalescence bridges overcomes the frustrating interaction and long range order breaks through. The delicate balance of competing strong interactions makes the abrupt transition from a frustrated to an ordered system of single domain islands reasonable, to some degree. The frustrating interaction provides a natural explanation for the drop of A_{ex} in Figure 2 for films near the upper end of the gap, below $T_c(ML)$. Apparently, the strength of the interaction increases below $T_c(ML)$. Because of the tight analogies with spin-glasses, we might address our films in the gap as super-spinglasses, and the films just above the gap as superferromagnets [12].

Our results are related to recent data of Back et al. [7], who investigated Fe-films on W(110) which apparently were just above the gap; they observed giant susceptibilities $\chi = 3 \cdot 10^5$ near T_c. The paramagnetic susceptibility of a uniaxial super-ferromagnet above T_c, in a molecular field approximation, is given, for N atoms per particle, by $\chi = C/(T - T_c)$, with a Curie-constant $C \approx 2N$ Kelvin for Fe. The mean value of N at the upper end of the gap is 7000. Having in mind that correlations result in an enhancement of χ above the molecular field value, the susceptibilities observed by Back et al. are easily explained as super-ferromagnetic. Moreover, the sensitive dependence of T_c on coverages by several metals, which was observed by Weber et al. [6] for films just above the gap, is easily understood from the sensitive balance between frustrating and bridge interactions, and by the electronic nature of the former.

In conclusion, we have shown that clean Fe(110)-films prepared on W(110) at 300 K show a gap of ferromagnetic long range magnetic order for coverages between 1.2 and 1.47 pseudomorphic monolayers. The frustration of magnetic order in this gap apparently results from a virtually antiferromagnetic interaction of electronic origin between double-layer islands, mediated by the surrounding monolayer sea and its W-substrate. The nature of this interaction remains to be explained in detail. The films in the gap can be considered as super-spinglasses. At the upper end of the gap, they switch to super-ferromagnets when the ferromagnetic coupling by bridges overcomes the frustrating interaction. Much remains to be done for a complete understanding of this new type of interaction and its evidence in ultrathin films. This work was supported by the Deutsche Forschungsgemeinschaft.

References

1   H.J. Elmers, J. Hauschild, H. Höche, U. Gradmann, H. Bethge, D. Heuer and U. Köhler, Phys. Rev. Lett. 73, 898 (1994)

2   M. Przybylski and U. Gradmann, Phys. Rev. Lett. 59, 1152 (1987)

3   H.J. Elmers, G. Liu and U. Gradmann, Phys. Rev. Lett. 63, 566 (1989)

4   U. Gradmann, M. Przybylski, H. J. Elmers and G. Liu, Appl. Phys. A 49 , 563 (1989)

5   U. Gradmann, Magnetism in Ultrathin Transition Metal Films, in K.H.J. Buschow (ed), *Handbook of Magnetic Materials* Vol. 7/1, p. 1-96, Elsevier Science Publishers, Amsterdam 1993

6   W. Weber, D. Kerkmann, D. Pescia, D. A. Wesner and G. Güntherodt, Phys.Rev.Lett. 65, 2058 (1990)

7   C.H. Back, C. Würsch, D. Kerkmann and D. Pescia, Z. Phys. B 96, 1 (1994)

8   W. Dürr, M. Taborelli, O. Paul, R. Germar, W. Gudat, D. Pescia and M. Landolt,, Phys.Rev.Lett. 62, 206 (1989)

9   H. Fritzsche, H.J. Elmers and U. Gradmann, J. Magn. Magn. Mat. 135, 343 (1994)

10 R. Bergholz and U. Gradmann, J. Magn. Magn. Mat. 45, 389 (1984)

11 A. Fert, P. Gruenberg, A. Barthélémy, F. Petroff and W. Zinn, J. Magn. Magn. Mat. 140-144, 1 (1995)

12 S. Morup, M.B. Madsen, J. Franck, J. Villadsen and C.J.W. Koch, J. Magn. Magn. Mat. 40, 163 (1983)

MAGNETIC EXCHANGE COUPLING IN ASYMMETRIC TRILAYERS OF Co/Cr/Fe

K. THEIS-BRÖHL, R. SCHEIDT, TH. ZEIDLER, F. SCHREIBER, H. ZABEL
Ruhr-Universität Bochum, D-44780 Bochum, Germany
TH. MATHIEU, CH. MATHIEU, and B. HILLEBRANDS
Universität Karlsruhe, Engesser Str.7, D-76128 Karlsruhe, Germany

ABSTRACT

We present first results of anisotropy and exchange coupling studies of a system with two different magnetic layers (Fe and Co) separated by a nonmagnetic Cr spacer. For the magnetic measurements we used the longitudinal magneto-optical Kerr effect and ferromagnetic resonance. The hysteresis data obtained from the trilayer were fit to a theoretical model which contains both bilinear and biquadratic coupling. The in-plane anisotropy was found to be four-fold with the same easy-axis orientation for both the Fe and the Co layers. An analysis of the easy-axis hysteresis loops indicates long period oscillatory coupling and also suggests a short period coupling.

INTRODUCTION

Oscillatory exchange coupling has been discovered for different magnetic materials and a wide variety of nonmagnetic (NM) spacer materials over the last several years (for references see [1]). The period of the oscillations can be understood in terms of the electronic structure of the non-magnetic spacer layer material [2]. However, some spacer materials like Cr or Mn are not completely "non-magnetic". In the case of Cr an antiferromagnetic (AF) spin structure is formed below the Neel temperature together with a spontaneous incommensurate collinear spin density wave. Therefore the exchange interaction between the Cr atomic planes should affect the coupling behaviour of the adjacent magnetic layers as well. For the case of Fe/Cr(001), tight binding calculations [3] give AF interactions and, in the case of bcc Co/Cr(001), predict stable states for either ferromagnetic (FM) or AF alignment.

Experimental investigations on high quality Fe/Cr(001) trilayers with short 2 monolayer (ML) period oscillations find the strongest AF maxima for even Cr spacer atomic layer (AL) numbers. These are 8 ML reported by Purcell [4] and $\approx$4 ML reported by Demokritov [5]. However, if the magnetic layers are different on either side of the Cr spacer, then the AF maxima are expected at odd AL numbers.

EXPERIMENTAL

To study the coupling behavior of a system similar to Fe/Cr(001), but with a different second FM/NM interface, we have grown a (3.0 nm Co)/Cr/(5.6 nm Fe) trilayer with a wedge-shaped (0.5 – 3.0 nm thick) Cr spacer by MBE method; the growth temperature was 300°C. To grow this system epitaxially we employed MgO(001) substrates with Cr(001) buffers. The sample surface was covered with a Cr cap layer (for protection) subsequent to growth.

The Co layer grows in the hcp phase in the (11$\bar{2}$0) orientation with the c-axis laying in the film plane. Since the uniaxial (11$\bar{2}$0) Co structure is grown on bcc Cr(001) with a four-fold crystallographic symmetry, there are two equivalent orientations for the Co to grow: with the c-axis parallel to either the [110] and [$\bar{1}$10] axes of the Cr, respectively. This results in a twinned crystallographic domain structure [6].

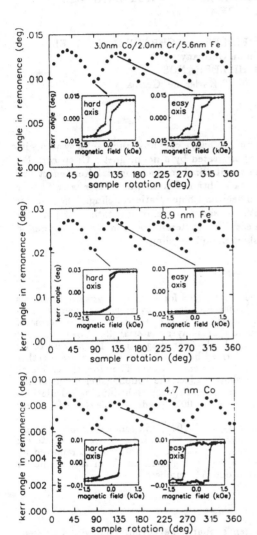

Figure 1. Scans of the Kerr angle in remanence as a function of the sample rotation. Data from the MOKE hysteresis loops were taken during a complete sample rotation. The insets show the hard axis (at minimum remanence) and easy axis (at maximum remanence) hysteresis loops of the (Co/Cr-wedge/Fe)-trilayer for 2.05 nm Cr layer thickness (a), a single 8.9 nm thick Fe-layer (b) and a 4.7 nm thick Co-layer (c). Both single layers are embedded in Cr.

We have also grown superlattices with 10 periods of the sequence [Co/Cr/Fe/Cr] and samples with individual Co or Fe layers, respectively. In the latter case the magnetic films are embedded in Cr layers. The layer thicknesses for all samples are chosen to be thick enough such that the magnetization is in the film plane.

To study the magnetic properties of the Co/Cr/Fe trilayer sample we measured hysteresis loops with the magneto-optical Kerr effect (MOKE). We measured in the longitudinal configuration as a function of the Cr interlayer thickness; the 100 μ-diameter laser spot was moved along the wedge shaped sample. For each Cr thickness we performed a complete in-plane sample rotation to determine the easy and hard axes. By plotting the Kerr angle in remanence as a function of the sample orientation, indicated by the angle Φ_H of the external field with respect to the Cr[001] in-plane direction, a four-fold anisotropy was obtained.

In Fig. 1a, we present such a plot for a Cr thickness of 2.0 nm. The hysteresis loops, however, show a nearly uncoupled behavior (insets of Fig. 1a). We show also plots of an individual 8.9 nm thick Fe layer (Fig. 1b) and an individual 4.7 nm thick Co layer (Fig. 1c). Notice the much higher coercitivity of the Co-layer.

The maxima of the Kerr angles in remanence indicate the in-plane easy axis. For Fe films they correspond to the [100] and [010] axes; this reflects the well-known result of a positive first order cubic anisotropy parameter K_1^{cub} [7] in [001] oriented films.

In the case of Co, a four-fold anisotropy was measured as well. The hard axes are *parallel* to the Co c-axes.

This behavior was found recently for Co(11$\bar{2}$0) on Cr(001)/Nb(001)/Al$_2$O$_3$(1$\bar{1}$02) with FMR measurements and was explained as an effect of the higher order uniaxial anisotropy parameter K_2^{hcp} [8].

To explain the four-fold anisotropy of the Co in these samples (grown on Cr(001) with a MgO(001) substrate) FMR measurements in addition to the MOKE measurements were carried out. We measured the superlattices and the individual Co layers. Only one resonance line at each easy axis orientation was found within the FMR spectra. This indicates that the crystallographic Co-domains are much smaller than the magnetic domains. For this case, and with a nearly 1:1 proportion of the domains (found for Co on Cr(001/MgO(001)), the resulting anisotropy energy is $F_{ani}^{eff} = \frac{1}{2}F_{ani}^{eff}(\Phi) + \frac{1}{2}F_{ani}^{eff}(\Phi + 90°)$ with the uniaxial anisotropies $F_{ani}^{eff}(\Phi)$ and $F_{ani}^{eff}(\Phi + 90°)$. The resulting in-plane anisotropy is four-fold and has the same shape as that for cubic in-plane anisotropy. The uniaxial anisotropy constant K_2 is in this case the relevant parameter for the anisotropy expression. The easy axis will then be found *between* the two c-axes. Therefore, neglecting the out-of-plane Θ–dependence, we can write the anisotropy energy (as for the Co case as in [8]) with Φ as the in-plane angle of the direction of magnetization measured against the in-plane Cr[001] axis:

$$F_{ani}^{eff} = \frac{1}{8}K_2 \cos(4\Phi) = -K_2 \sin^2\Phi \cos^2\Phi + const. \tag{1}$$

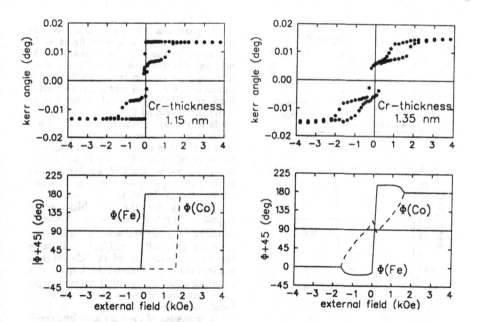

Figure 2. The MOKE hysteresis loop for a Cr thickness of 1.15 nm measured at 45° (easy axis configuration), showing a weak AF behavior (above); and the field dependence of the angles Φ_{Fe} and Φ_{Co} obtained from the fit (not shown) plotted for increasing field strength (below).

Figure 3. The MOKE hysteresis loop for a Cr thickness of 1.35 nm measured at 45° (easy axis configuration), showing a strong AF behavior (above); and the field dependence of the angles Φ_{Fe} and Φ_{Co} obtained from the fit (not shown) plotted for increasing field strength (below).

DISCUSSION

To analyze the MOKE hysteresis loops of the wedged-shaped Co/Cr/Fe sandwich (see Figs. 2–4) we fit our easy axis loop data. We assumed the following phenomenological expression for the free energy:

$$F_{mag} = \left(-\mu_0 M^{Fe} H \cos\left(\Phi_{Fe} - \Phi_H\right) + K_1^{Fe} \sin^2 \Phi_{Fe} \cos^2 \Phi_{Fe}\right) t_{Fe}$$
$$+ \left(-\mu_0 M^{Co} H \cos\left(\Phi_{Co} - \Phi_H\right) + K_2^{Co} \sin^2 \Phi_{Co} \cos^2 \Phi_{Co} - const.^*\right) t_{Co} \quad (2)$$
$$- 2A_{12} \cos\left(\Phi_{Fe} - \Phi_{Co}\right) - 2B_{12} \cos^2\left(\Phi_{Fe} - \Phi_{Co}\right).$$

Notice, that the anisotropy energy of the Co-layer was transformed to the Fe coordination system. These fits to the data points (not shown) are in reasonable agreement mainly for those parts of the loops at which no domain processes do determine the magnetization behavior. Taking into consideration the different thicknesses (5.6 nm for Fe and 3.0 nm for Co), the product of the saturation magnetization (assuming bulk-like behavior) and the layer thickness ($M_s \cdot t$) in the external field term of the free energy is about twice as high for the Fe layer as for the Co layer. Therefore the behavior of the Fe layer is expected to be influenced much stronger by the external field than of the Co layer. This is demonstrated by the results of the fits for Φ_{Fe} and Φ_{Co} of the easy axis hysteresis loops (see Figs. 2–4).

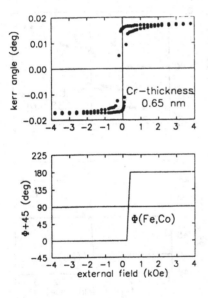

Figure 4. The MOKE hysteresis loop for a Cr thickness of 0.65 nm measured at 45° (easy axis configuration), showing a FM behavior (above); and the field dependence of the angles Φ_{Fe} and Φ_{Co} obtained from the fit (not shown) plotted for increasing field strength (below).

In particular, the Fe spins flip already at small (reverse) magnetic fields while the spins in the Co layer remain in the original direction. Apparently the magnetization of the Co layer follows the rotation of the Fe layer and not the external field. In addition the much higher coercitivity of individual Co layers (see Figs. 1) leads to the conclusion that Co must undergo a much more complicated domain structure. Due to the different magnetic behavior of Fe and Co (see Figs. 1b-c), a nearly uncoupled Co/Cr/Fe trilayer shows hysteresis loops which are close to those shown in Figs. 1a for t_{Cr}=2.0 nm.

The fits for the Cr thickness of t_{Cr}=2.0 nm reveal the absence of bilinear coupling (A_{12}=0) and a weak biquadratic coupling (B_{12}=-0.01 mJ/m^2). Because the high coercitivity of the Co layer is not included in our model this biquadratic coupling value of B_{12}=-0.01 mJ/m^2 might represent the Co coercitivity and not a "real" biquadratic coupling.

The shapes of the easy axes hysteresis loops for a weak AF coupling differ from those with a nearly uncoupled behavior, mainly by a slightly longer step in the loop for the former case.

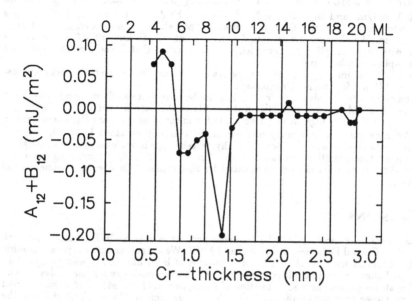

Figure 5. Results of our fits to the data for $A_{12}+B_{12}$ as a function of the Cr spacer thickness.

In Fig. 2 we present an easy axis hysteresis loop for weakly AF coupled Co and Fe layers at a Cr thickness $t_{Cr}=1.15$ nm (with the fit results $A_{12}=-0.03$ mJ/m^2, $B_{12}=-0.01$ mJ/m^2). In comparison, for a weak FM coupling strength the length of the step decreases or the step nearly vanishes. It is interesting to note that the length of the step is very sensitive to the coupling behavior.

In all cases with weak coupling characteristics, the Co spins first remain in their original spin alignment after the Fe spin flip occurs. With increasing field then domain and/or spin rotation processes occur.

For a stronger AF coupling the Co spins change their alignment before the Fe spin flip occurs (see Fig. 3). For a Cr thickness of 1.35 nm the strongest AF coupling constants were found ($A_{12}=-0.12$ mJ/m^2 and $B_{12}=-0.08$ mJ/m^2). Note that a high biquadratic coupling constant B_{12} (compared to A_{12}) is obtained. For Cr thicknesses between 0.85 nm and 1.05 nm, the shapes of the hysteresis loops again suggest a strong AF coupling. For this region we found coupling constants of ($A_{12}+B_{12}=-0.05$ ... -0.07 mJ/m^2). In this case the biquadratic coupling constant B_{12}, is much higher than the bilinear one, A_{12}.

For a strong FM coupling we find hysteresis loops without any steps and a clear FM shape. In this case the Co layer couples so strongly to the Fe with FM spin alignment that the magnetization processes of both materials occur together. Such loops can be obtained at $t_{Cr}=0.55$ nm and $t_{Cr}=0.75$ nm. At $t_{Cr}=0.65$ nm the shape of the hysteresis loops suggests that the Co spins do not reach complete saturation after the spin flip (see Fig. 4). For all three Cr layer thicknesses the fits give a positive coupling constant $A_{12}=0.07-0.09$ mJ/m^2.

In Fig. 5 we present the results of our simulations for the coupling constants $A_{12}+B_{12}$ as a function of the Cr spacer thickness. A long period coupling oscillation with a period of 10–11 ML is deduced from the two AF maxima at ≈ 9 ML and ≈ 20 ML and from the

FM maxima at 4–5 ML and $\approx$15 ML. The second long period AF maxima at t_{Cr}=2.85 nm (A_{12}=-0.01 mJ/m^2 and B_{12}=-0.01 mJ/m^2) also can be deduced clearly from the shapes of the hysteresis loops in this Cr region. Also a 2 ML short period coupling might exist in this system. We obtained two strong AF maxima at $\approx$7 ML and $\approx$9 ML, separated by a region with weak coupling characteristics. But in other Cr-thickness regions no short period coupling can be found.

The odd Cr atomic layer number suggests a FM-like Co/Cr interface exchange, assuming an AF one for the Fe/Cr interface.

In the case of Fe/Cr, biquadratic coupling can be observed mostly in a region between the AF and the FM maxima (see [9]). But our fits to the data for the Co/Cr/Fe trilayer system suggest that the highest negative biquadratic coupling constant B_{12} occurs at the strong AF maxima. As previously mentioned the coupling constant B_{12}=-0.01 mJ/m^2 found (from the simulations) for most of the hysteresis loops in the case of a weak coupling may represent the coercitivity of the Co layer. The strong biquadratic coupling constants found at the AF maxima should be reduced by this value. This does not influence the main behavior.

CONCLUSIONS

In conclusion we have studied the magnetic properties of (Co/Cr/Fe) samples, as well as individual Co and Fe layers, by MOKE and FMR. We suggest coupled crystallographic (11$\bar{2}$0)Co domains with an almost 1:1 distribution of both domains causing a four-fold anisotropy behavior of the Co layer, with similar characteristics as for the Fe layer. The MOKE hysteresis loops show an indication of a long period (10-11 ML) and a short period (2 ML) exchange coupling oscillation, with the AF maxima at odd numbers of AL of the Cr spacer. The strongest AF maximum (with a coupling constant of -0.20 mJ/m^2) was found at 9 ML. The fits to the data also reveal a high biquadratic constant whenever the AF coupling shows maxima.

The authors wish to thank K. Ritley, W. Oswald, J. Podschwadek and P. Stauche for technical help. We wish also to thank Z. Frait (Prague) for the use of his FMR spectrometer for additional measurements. The work was supported by the Deutsche Forschungsgemeinschaft through SFB 166. One of us (KTB) acknowledges C. P. Flynn for hospitality and acknowledge partial support from the Physics Department at the Univ. of Illinois at Urbana-Champaign.

REFERENCES

1. M.D. Stiles, Phys. Rev. **B48**, 7738 (1993).
2. K.B. Hathaway in Ultrathin Magnetic Structures II, ed. by B. Heinrich and J.A.C. Bland.
3. J.H. Hasegawa, Phys. Rev. **B42**, 2368 (1990); Phys. Rev. **B43**, 10803 (1990).
4. S.T. Purcell, W. Folkerts, M.T. Johnson, N.W.E. McGee, K. Jager, J. Aan de Steege, W.P. Zeper and P. Grünberg, Phys. Rev. Lett. **B42**, 67, 903 (1991).
5. S. Demokritov, J.A. Wolf, and P. Grünberg and W. Zinn, Mat. Res. Soc. Symp. Proc. Vol. **231**, 133 (1992).
6. W. Donner, N. Metoki, A. Abromeit and H. Zabel, Phys. Rev. B,**48**, 14745, (1993).
7. Th. Mhüge, Th. Zeidler, Q. Wang, Ch. Morawe, N. Metoki, and H. Zabel, J. Appl. Phys. **77** 1055 (1995).
8. F. Schreiber, Z. Frait, Th. Zeidler, N. Metoki, W. Donner, H. Zabel and J. Pelzl, Phys. Rev. B,**51**, 2920, (1995).
9. M. Rührig, R. Schäfer, A. Hubert, R. Mosler, J.A. Wolf, S. Demokritov and P. Grünberg, Phys. Stat. Sol. (a) **125**, 635 (1991).

Studies of Exchange Coupling in Fe/Cu/Fe(001) "Loose Spin" Structures

M. Kowalewski, B. Heinrich, K. Totland*, J.F. Cochran, S. Govorkov, D. Atlan**, K. Myrtle
Simon Fraser University, Physics Department, Burnaby, Canada; and
P. Schurer, Royal Roads Military College, Victoria, Canada.

Abstract:
The interlayer exchange coupling has been studied in two trilayer structures:
(a) 5.7Fe/5Cu/1Fe_cCu_{1-c}/5Cu/10Fe(001), where c=0.0, 0.1, 0.2, 0.45 0.60
(b) 5.7Fe/5Cu/1Cr_cCu_{1-c}/5Cu/10Fe(001), where c=0.1, 0.45, 0.8 and 1.0.
The intention of these studies was to identify the role of Fe and Cr atoms in the alloyed Fe_cCu_{1-c} and Cr_cCu_{1-c} layers on the direct interlayer coupling which is facilitated by the Cu valence electrons. FMR, BLS and MOKE studies were used to determine the interlayer exchange coupling. Mossbauer spectroscopy was used to identify the magnetic state of the Fe atoms in the alloyed layer. The results showed that the presence of foreign atoms inside the Cu spacer significantly decreased the bilinear antiferromagnetic coupling between the Fe layers. In the low concentration limit the Fe and Cr atoms behaved in a similar manner. A significant difference was found in the high concentration limit where the Fe atoms start to be partially magnetically ordered.

Introduction:

We have studied extensively the exchange coupling in structures which consisted of two Fe(001) ferromagnetic layers separated by a bcc Cu(001) non-ferromagnetic spacer (trilayers) [1-6]. Magnetic trilayers represent simple systems in which the magnetic behavior can be adjusted by careful control of the epitaxial growth. The magnetic properties were measured using Ferromagnetic Resonance (FMR), Brillouin Light Scattering (BLS), Magneto-Optical Kerr Effect (MOKE), and by Mossbauer spectroscopy (using a single atomic layer of ^{57}Fe). The FMR measurements were carried out in the temperature range 77-400K. The BLS, MOKE and Mossbauer studies were performed at room temperature.
Recently we have investigated the role of the "loose spins" of Fe atoms on the exchange coupling between ferromagnetic layers [1,5,6]. A single additional monolayer (ML) of Fe_cCu_{1-c} was inserted inside the Cu spacer. The Cu spacer was surrounded by two Fe layers, Fe1 and Fe2, which were 5.7 and 10ML thick. In this paper we present the results of our investigation which was directed towards the study of the role of Fe and Cr atoms, inside the Cu spacer, on the direct exchange interlayer coupling facilitated by Cu itinerant electrons. The following structures have been investigated:
5.7Fe/5Cu/Fe_cCu_{1-c}/5Cu/10Fe(001), where c=0.0, 0.1, 0.2, 0.45 and 0.60;
5.7Fe/5Cu/Cr_cCu_{1-c}/5Cu/10Fe(001), where c=0.1, 0.45, 0.8 and 1.0.
The integers describe the number of MLs. The samples were grown on Ag(001) substrates held at room temperature (RT) using MBE. The layers were then covered by 20ML thick Au(001) to provide a protective layer for the ambient measurements.
The total thickness of the interlayer in the above samples is 11ML. The exchange coupling through bcc Cu(001) interlayers was extensively investigated in our previous studies. It was found that the interlayer exchange interaction across 11ML of Cu reaches its maximum antiferromagnetic coupling. With a decreasing Cu thickness the interlayer exchange interaction changes rapidly to ferromagnetic coupling. It crosses zero coupling around 8 MLs of Cu.

Results and discussion:

All deposited layers were terminated at the RHEED intensity maxima, see Fig.1. Alloyed layers were prepared by co-depositing Fe or Cr atoms together with the Cu. The RHEED intensity oscillations were not affected during the deposition of alloyed layers, see Fig.1, and therefore it is

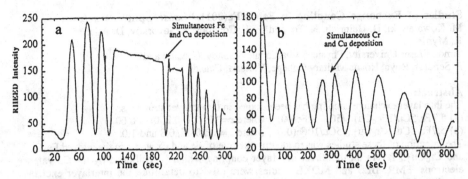

Fig.1: RHEED intensity oscillations of the specular spot during the growth of the Cu spacer. The arrows point to the growths of the alloyed layers: 1(a) 45% Fe - 55% Cu, 1(b) 45% Cr - 55% Cu. The angle of incidence of the RHEED electron beam corresponds to the first anti-Bragg condition.

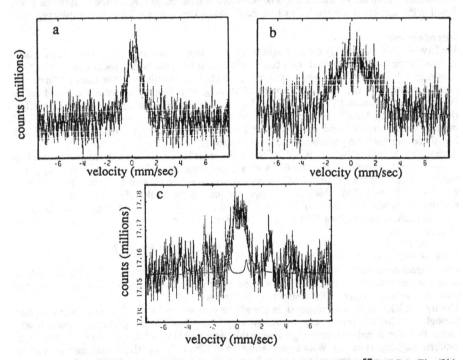

Fig.2: Mossbauer spectra for the following samples: (2a) 5.7Fe/5Cu/1^{57}Fe/5Cu; Fig.(2b) 5.7Fe/5Cu/1^{57}Fe/5Cu/10Fe; (2c) 5.7Fe/5Cu/1^{57}Fe$_{0.5}$+Cu$_{0.5}$/5Cu(001)/10Fe. The isomer shifts in (2a) and (2b) are 0.14, 0.17 mm/sec respectively. The isomer shift for Fe surrounded by Cu is expected to be 0.22 mm/sec. The samples (a) and (b) were grown using a ^{56}Fe source for layers Fe1 and Fe2. The spectrum in (2b) is broadened by the presence of inhomogeneous hyperfine fields. The sample in (c) was grown using a natural Fe source for the growth of Fe1 and Fe2. The central doublet in (2c) is similar to that in (2a). The presence of satellites in (2c) is due to abundance (2%) of ^{57}Fe in natural Fe.

reasonable to assume that the formation of alloyed layers followed the growth mode of the Cu spacer.

The measured isomer shifts (i.s.) in Mossbauer spectra in samples 5.7Fe/5Cu/1^{57}Fe/5Cu and 5.7Fe/5Cu/1^{57}Fe/5Cu/10Fe, see Fig.2a,b, showed that the ^{57}Fe atoms were surrounded on average by ~70% of Cu. In a perfect bcc lattice each Fe atom in the Fe layer would be surrounded only by Cu atoms; the above result implies that during the growth of the alloyed layer there is some tendency for Fe atoms to move vertically. The Mossbauer spectra for 5.7Fe/5Cu/1^{57}Fe/5Cu and 5.7Fe/5Cu/1^{57}Fe$_{0.5}$+Cu$_{0.5}$/5Cu(001)/10Fe show a single central doublet, see Fig.2a,c. The central doublet is broadened by the distribution of isomer shifts due to the different atomic surrounding of the ^{57}Fe atoms. The central broad peak in sample 5.7Fe/5Cu/1^{57}Fe/5Cu/10Fe, see Fig.2b, indicates that the second Fe2 layer leads to a partial ferromagnetic ordering of ^{57}Fe. However a relatively narrow doublet in sample 5.7Fe/5Cu/1 ^{57}Fe$_{0.5}$+Cu$_{0.5}$/5Cu(001)/10Fe, see Fig.2c, shows that the Fe atoms dispersed in smaller concentrations between Cu atoms have fluctuating magnetic moments at RT and therefore can be considered as "loose spins".

Slonczewski recently proposed a model based on the concept of "loose spins" [7]. "Loose spins" contribute to the total exchange energy. The exchange energy of "loose spins" due to the RKKY field of the surrounding Fe layers can be expressed as

$$E= (U_1^2+U_2^2+2U_1U_2\cos(\Theta))^{0.5}$$

where U_1 and U_2 describe the exchange energy between a "loose spin" atom and ferromagnets Fe1 and Fe2. Θ is the angle between the magnetic moments of the surrounding ferromagnetic layers Fe1 and Fe2. The free energy of the "loose spins" is then used to evaluate the contribution of "loose spins" to the bilinear and biquadratic exchange coupling. The exchange coupling between ferromagnetic layers separated by a non-magnetic spacer can be described by

$$E= -J_1.\cos(\Theta) + J_2.\cos^2(\Theta)$$

where J_1 (bilinear) and J_2 (biquadratic) exchange couplings are measured in ergs/cm^2 and Θ is the angle between magnetic moments in two iron films.

The exchange coupling between Fe layers was measured using FMR, BLS and MOKE. FMR studies were carried out from 77 to 373 K, BLS studies and MOKE studies were performed at RT only. The results of our FMR studies are shown in Fig.3. Values of the total exchange coupling, $J_{tot}(T)=J_1-2J_2$, were obtained from the absorption peak fields corresponding to the acoustic and optical FMR modes, see details in [2,3]. The field positions of cusps in the BLS measurements, see Fig.4a, and the fields corresponding to saturated and antiferromagnetic configurations of the magnetic moments in MOKE measurements, see Fig.4b, allow one to determine the individual values of J_1 and J_2 [2,3]. The exchange coupling in Fe/Cu/Fe(001) samples having a pure Cu spacer was studied extensively in our previous work [4]. The exchange coupling through a pure 11ML Cu layer is strongly antiferromagnetic and reaches its maximum value at this thickness. For Cu thicknesses less than 8 ML the coupling is ferromagnetic and rapidly increases with a decreasing Cu thickness. It is tempting to assume that the coupling U_1 and U_2 between a "loose spin" and the surrounding Fe layers Fe1 and Fe2 is nearly the same as that between ferromagnetic layers separated by a Cu spacer of an equivalent thickness. Our choice of samples was guided by this simple assumption. One expects that in our structures $U_1=U_2$ since the "loose spins" were surrounded by two Cu layers having equal thicknesses. An extrapolation of the thickness dependence of the exchange coupling in Fe/Cu/Fe samples suggests that the strength of U_1 and U_2 should not significantly exceed 6K. For this case the Slonczewski "loose spin" model predicts a small ferromagnetic bilinear coupling (with J_2~0) and a 1/T dependence on temperature, see Fig.3a. The measured exchange coupling in 5.7Fe/5Cu/Fe$_c$Cu$_{1-c}$/5Cu/10Fe(001) samples decreases rapidly with an increasing concentration, c, of Fe, see Fig.3a. This is in qualitative agreement with the Slonczewski's model of "loose spins". The measured temperature dependence of $J_{tot}(T)$ can be fitted with a combination of linear and $1/T$ terms. However, the $1/T$ terms are significantly larger and more importantly they have opposite sign to that expected from the Slonczewski's model, see Fig.3a. Therefore, the $1/T$ terms in $J_{tot}(T)$ are not likely caused by "loose spins". It is more probable that the presence of Fe atoms inside the Cu spacer decreases the direct bilinear exchange coupling, $J_1(T)$. The Fe impurity atoms in the middle of the Cu spacer create a local electronic potential which affects the spin dependent reflectivity of the Cu itinerant electrons at the Fe1/Cu and Cu/Fe2 interfaces which facilitate the direct exchange coupling [8].

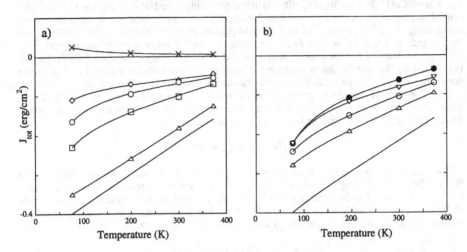

Fig.3a: The total exchange coupling J_1-$2J_2$ in series 5.7Fe/5Cu/1Fe$_c$Cu$_{1-c}$/5Cu/10Fe(001), where c= 0.1 ($\triangle$), 0.2 ($\square$), 0.45 ($\bigcirc$), 0.60 ($\Diamond$). The bottom solid line corresponds to a pure 11 ML thick Cu spacer. ($\times$) symbols represent the calculated values of the "loose spin" bilinear coupling using U1=U2=6K.

Fig.3b: The total exchange coupling J_1-$2J_2$ in series 5.7Fe/5Cu/1Cr$_c$Cu$_{1-c}$/5Cu/10Fe(001), where c=0.1 ($\triangle$), 0.45 ($\bigcirc$), 0.8 ($\triangledown$) and 1.0 ($\bullet$). The bottom solid line corresponds to a pure 11 ML thick Cu spacer.

The solid lines through points for all the "loose spin" samples are fits of the form a+b*T+c/T, where T is temperature in Kelvin, and a, b, c are constants.

In high concentration limit of Fe, c > 0.6, the interlayer exchange coupling becomes ferromagnetic [1] and is weakly dependent on temperature. These results show that a partial ferromagnetic ordering as evidenced by Mossbauer spectra, see Fig.2b, strongly enhances the coupling between the ferromagnetic layers and results in a noticeable ferromagnetic interlayer coupling. The exchange coupling between the Fe1 and Fe2 ferromagnetic layers (5.7 and 10MLs) is then facilitated through the partially ordered middle layer ("loose spin" layer). Since the coupling through 5ML thick Cu spacer layer is ferromagnetic the resulting exchange coupling in 5.7Fe/5Cu/Fe$_c$Cu$_{1-c}$/5Cu/10Fe samples becomes increasingly more ferromagnetic when c →1.

In this paper we present our recent measurements in which the "loose spin" Fe atoms were replaced by Cr atoms. The main idea in this study is to replace the Fe atoms inside the Cu spacer with the Cr atoms which have similar valence bands (3d, 4sp) to those of Fe atoms, but which do not possess a long range ferromagnetic order. In high concentration limit the Cr magnetic moments order antiferromagnetically, but their Neel ordering temperature is expected to be well below LN2 temperatures; therefore the Cr atoms should maintain their "loose spin" character better than the Fe atoms.

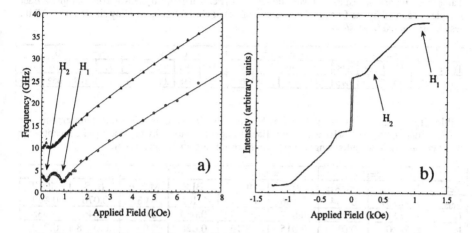

Fig.4a: BLS peak frequencies against applied magnetic field for sample 5.7Fe/5Cu/1Cr$_{0.45}$Cu$_{0.55}$/5Cu/10Fe(001). The upper and lower branches correspond to the acoustic and optical precessional modes. The solid lines represent theoretical fits with J$_1$=0.074 ergs/cm^2 and J$_2$=0.015 ergs/cm^2; 4πM$_{eff}$=4.25 kG, 13.55 kG, 2K$_1$/M$_s$=0.018, 0.24 KOe for the Fe layers 5.7 and 10ML thick, respectively. The upper cusp corresponds to the field H$_1$ at which the magnetic moments start to turn away from the applied field, the lower cusp corresponds to the field H$_2$ at which the magnetic moments orient antiparallel.

Fig.4b: The magnetization curve of sample 5.7Fe/5Cu/1Cr$_{0.45}$Cu$_{0.55}$/5Cu/10Fe(001) obtained by MOKE measurements. The position of fields H$_1$ (saturation state) and H$_2$ (antiparallel state) were obtained using J$_1$=0.077 ergs/cm^2 and J$_2$=0.015 ergs/cm^2 with the same magnetic properties of the individual Fe layers as given in the caption for Fig.4a.

The exchange coupling in 5.7Fe/5Cu/1Cr$_c$Cu$_{1-c}$/5Cu/10Fe samples was measured by FMR, BLS and MOKE. The results of FMR, BLS and MOKE studies are summarized in Table I and II. The temperature dependence of J$_{tot}$(T) is shown in Fig3b. The results of FMR, MOKE and BLS measurements are in good agreement. The exchange coupling again decreases with an increasing concentration of Cr atoms. The room temperature measurements of J$_1$ and J$_2$, see Table II, showed that the observed decrease in the exchange coupling is mostly caused by a decrease in the bilinear coupling, J$_1$. The biquadratic coupling, J$_2$, in 5.7Fe/5Cu/1Cr$_c$Cu$_{1-c}$ /5Cu/10Fe samples is very close to that in a pure 11ML thick Cu interlayer grown at RT and to those in 5.7Fe/5Cu/Fe$_c$Cu$_{1-c}$/5Cu/10Fe(001) samples [6] . However there are some noticeable differences. In a low concentration limit the exchange coupling in the Cr samples decreases with an increasing concentration of Cr more rapidly than in the Fe samples, Fig.3. This trend changes in high concentration limit. The decrease in the antiferromagnetic exchange coupling slows down with an increasing concentration of Cr, see c=0.8 and 1.0 samples. In fact even 1 ML of Cr maintains antiferromagnetic exchange coupling through the Cu spacer. Again as in the Fe$_c$Cu$_{1-c}$ samples, the temperature dependence of the exchange coupling, J$_{tot}$(T), can be fitted by linear and 1/T terms. However, no direct dependence between the strength of the 1/T terms and the appropriate concentrations of Cr and Fe atoms was found. This behavior combined with the incorrect sign clearly indicates that the 1/T terms are not caused by "loose spin" contributions as envisioned by Slonczewski's model.

Table I: Summary of results of FMR studies on 5.7Fe/5Cu/1Cr$_c$Cu$_{1-c}$/5Cu/10Fe samples. Each sample is denoted by the fractional concentration (c) of chromium in the alloy single layer. All the results are quoted in ergs/cm^2.

	c=0.1				c=0.45				c=0.8				c=1.0			
T(K)	77	195	295	375	77	195	295	375	77	195	295	375	77	195	295	375
-J$_{tot}$	.278	.192	.134	.095	.244	.153	.104	.070	.223	.118	.082	.056	.223	.108	.062	.035

Table II: Summary of results of BLS and MOKE studies at room temperature on 5.7Fe/5Cu/1Cr$_c$Cu$_{1-c}$/5Cu/10Fe samples. Each sample is denoted by the fractional concentration (c) of chromium in the alloy single layer. All the results are quoted in ergs/cm^2.

	c=0.1		c=0.45		c=0.8		c=1.0	
method	BLS	MOKE	BLS	MOKE	BLS	MOKE	BLS	MOKE
J$_{total}$	-0.137	-0.134	-0.106	-0.104	-0.105	-0.082	-0.076	-0.062
J$_1$	-0.127	-0.098	-0.077	-0.074	-0.050	-0.048	-0.040	-0.038
J$_2$	0.005	0.018	0.015	0.015	0.028	0.017	0.0178	0.012

Conclusions:

The above measurements indicate that the interlayer exchange coupling is decreased by modifying the spin dependent reflectivity of the spacer electrons by the inner electronic potential of the Fe or Cr atoms inside the Cu spacer. In high Fe concentration limit, c →1, the character of the exchange coupling starts to be strongly affected by the onset of a long range ferromagnetic order in the Fe$_c$Cu$_{1-c}$ layer. The Cr atoms in the Cu spacer also decrease the antiferromagnetic coupling. However, the exchange coupling remains antiferromagnetic even for c=1.

Acknowledgments:
The authors would like to thank the Natural Sciences and Engineering Research of Canada for grants that supported this work.
*K. Totland gratefully acknowledges support by the Swiss National Science Foundation.
**D. Atlan would like to express his gratitude for a Lavoisier Fellowship granted by the French Ministry of Foreign Affairs .

References:
[1] B. Heinrich and J.F. Cochran, Advances in Physics, **42**, 523 (1993).
[2 B. Heinrich, J.F. Cochran, M. Kowalewski, J. Kirschner, Z. Celinski, and A.S. Arrott, Phys.Rev.**B44**, 9348 (1991).
[3] B. Heinrich, Z. Celinski, J.F. Cochran, A.S. Arrott, K. Myrtle, and S.T. Purcell, Phys.Rev.**B47**, 5077 (1993).
[4] Z. Celinski and B. Heinrich, J.Magn.Magn.Mater., 99,L55 (1990).
[5] B. Heinrich, Z. Celinski, L.X. Liao, M. From, and J.F. Cochran, J.Appl.Phys. **75**, 6187
[6] B. Heinrich, M. From, J.F. Cochran, M. Kowalewski. D. Atlan, Z. Celinski, and K. Myrtle, Proceedings of ICM-94, J. Magn.Magn.Mater., in press.
[7] J.C. Slonczewski, J.Appl.Phys., **73**, 5957 (1993).
[8] M.D. Stiles, Phys.Rev.**B48**, 7238 (1993).

Interlayer Coupling in Magnetic/Pd Multilayers

Zhu-Pei Shi and Barry M. Klein

Department of Physics, University of California, Davis, CA 95616

Abstract

The Anderson model of local-state conduction electron mixing is applied to the problem of interlayer magnetic coupling in metallic multilayered structures with palladium (Pd) spacer layers. An oscillation period of 5 spacer monolayers and the tendency towards ferromagnetic bias of the interlayer magnetic coupling that we obtain are consistent with the experimental data.

The discovery of oscillatating interlayer magnetic couplings between ferromagnetic layers separated by a nonmagnetic metallic spacer [1] and of the related giant magnetoresistance effect [2], has stimulated a lot of experimental and theoretical activity. It has been shown that the periods of the coupling are related to the topology of the Fermi surface of the spacer layers. This interpretation has been confirmed by model and first-principles calculations, and is also supported by experiments [3, 4]. There are, however, other aspects of the coupling, e. g., the bias (ferro- or antiferro-magnetic) of the interlayer magnetic coupling, which have not been fully explained.

For Fe(001) layers separated by Pd(001) spacers of thickness between 4 and 12 ML the interlayer magnetic coupling is observed to have a strong ferromagnetic bias as seen in the experiments [5]. Above a 13 ML thickness of the Pd spacer the coupling begins to be antiferromagnetic. Metallic Pd is believed to be near the threshold of becoming ferromagnetic. The non-relativistic calculations of Moruzzi and Marcus [6] and of Chen *et al.* [7] predicted the onset of ferromagnetism in fcc palladium with a 5% expanded lattice. In a recent publication the ferromagnetic bias of the coupling in magnetic multilayer structures with a Pd spacer is explained in terms of the Pd as an almost ferromagnetic media [3]. Alternatively, in this paper, we interpret this ferromagnetic bias to be a consequence of a competition between RKKY-like and superexchange couplings, with RKKY coupling being dominant.

The RKKY-like coupling comes from intermediate states which correspond to spin excitations of the Fermi sea. States corresponding to electron-hole pair production in the Fermi sea, with an attendant spin-flip, contribute to the RKKY coupling as [8];

$$j_{RKKY}(\mathbf{q}) = \sum_{n_1,n_2,k} \left[\frac{|V_{n_1 k}|^2 |V_{n_2 k'}|^2}{(\varepsilon_{n_2 k'} - \varepsilon_+)^2} \frac{\theta(\varepsilon_F - \varepsilon_{n_1 k})\theta(\varepsilon_{n_2 k'} - \varepsilon_F)}{\varepsilon_{n_2 k'} - \varepsilon_{n_1 k}} + c.c. \right], \qquad (1)$$

where θ is a step function, ε_F is the Fermi energy, $\mathbf{k}' = \mathbf{k} + \mathbf{q} + \mathbf{G}$, $\mathbf{G}$ is a vector of the reciprocal lattice, ε_+ is the energy of the local impurity state, and V_{nk} represents the strength of the $s - d$ mixing interaction [9].

177

Mat. Res. Soc. Symp. Proc. Vol. 384 ⊚ 1995 Materials Research Society

The superexchange coupling arises from charge excitations in which electrons from local states are promoted above the Fermi sea (one from each layer) providing a second contribution to the coupling in addition to the RKKY coupling [8]:

$$j_S(\mathbf{q}) = - \sum_{n_1,n_2,k} \left[\frac{|V_{n_1 k}|^2 |V_{n_2 k'}|^2}{(\varepsilon_{n_2 k'} - \varepsilon_+)^2} \frac{\theta(\varepsilon_{n_1 k} - \varepsilon_F)\theta(\varepsilon_{n_2 k'} - \varepsilon_F)}{\varepsilon_{n_1 k} - \varepsilon_+} + c.c. \right] . \qquad (2)$$

The real space coupling between two sheets of spins can be obtained by Fourier transforming Eqs. (1) and (2) [8], with the coupling in the real space given by,

$$J_t(z) = \frac{a}{2\pi} \int_0^\infty dq_z \, j(q_z) \, \cos(q_z z) \quad , \qquad (3)$$

where a is a lattice constant, and z is in the direction perpendicular to the magnetic layers. The sign is chosen so that positive $J_t(z)$ signifies ferromagnetic coupling.

Using the Slater-Koster parameters [10], one can easily diagonalize small matrices (9x9 for a typical transition metal) to obtain the energy bands and density of states (DOS) for fcc Pd. The electron wave function $| n, \mathbf{k} >$ is a Bloch state belonging to band n and wave vector $\mathbf{k}$, and is expressed as linear combinations of localized orbitals:

$$| n, \mathbf{k} > = \frac{1}{\sqrt{N}} \sum_\nu e^{i\mathbf{k}\cdot\mathbf{R}_\nu} \sum_i a_{ni}(\mathbf{k}) \, u_i(\mathbf{r} - \mathbf{R}_\nu) \quad , \qquad (4)$$

where N is the number of cells in the material considered, $\mathbf{R}_\nu$ is a lattice vector, $u_i(\mathbf{r} - \mathbf{R}_\nu)$ is the ith orbital basis function, and $a_{ni}(\mathbf{k})$ is a (real) normalized eigenvector component determined by diagonalization of the single-particle Hamiltonian. We use a plausible approximation $V_{n_1 k} V_{n_2 k'}^* = V^2 M_{n_1 k, n_2 k'}^*(\mathbf{q})$ [8], where the matrix element is defined as $M_{n_1 k, n_2 k'} = < n_1 \mathbf{k} | e^{i\mathbf{q}\cdot\mathbf{r}} | n_2 \mathbf{k}' >$. The explicit expression for the matrix element is

$$M_{n_1 k, n_2 k'} = \sum_\nu e^{i\mathbf{k}\cdot\mathbf{R}_\nu} \sum_{i,j} a_{ni}(\mathbf{k}) \, a_{n_2 j}(\mathbf{k}') \int d\mathbf{r} \, u_i(\mathbf{r}) e^{i\mathbf{q}\cdot\mathbf{r}} u_j(\mathbf{r} - \mathbf{R}_\nu) \quad . \qquad (5)$$

The essential conditions for the simplication of this matrix was already discussed by Callaway et al. [11]. $u_i(\mathbf{r})$ are approximated as Clementi wave functions for the d states [11], and $a_{ni}(\mathbf{k})$ can be related to the Slater-Koster parameters in Ref. [10].

We consider one local level below ε_F for simplicity and set $\varepsilon_+ = \varepsilon_F - E_h$, where E_h is the energy required to promote an electron from an occupied local magnetic impurity level to the Fermi level. Based on the band structure of bulk paramagnetic Pd, we have calculated the couplings $j_{RKKY}(q_z)$ and $j_S(q_z)$ with $E_h = 0.08 \, Ry$, as shown in Fig. 1.

The couplings in real space are plottted in Fig. 2. The dashed line, dotted line and solid line are for RKKY-like, superexchange, and RKKY + Superexchange couplings, respectively. We see that the superexchange interaction gives a small contribution to the coupling, and the total coupling has a strong ferromagnetic bias. This tendency for a ferromagnetic bias resembles the experimental observation in Fe/Pd(001) trilayered structures [5]. The 5 ML oscillation period in the calculated interlayer magnetic coupling $J(z)$, as shown in Fig. 2, corresponds to the peak at $q_z \simeq 0.4 \frac{2\pi}{a}$ in $j_{RKKY}(q_z)$. It agrees with the experimental period of $4 - 5 \, ML$ in Fe/Pd/Fe(001) for trilayered structures [5].

We explain the result of ferromagnetic bias for multilayers with Pd spacers as being due to the structure of the Pd DOS and the location of the Fermi level, as shown in

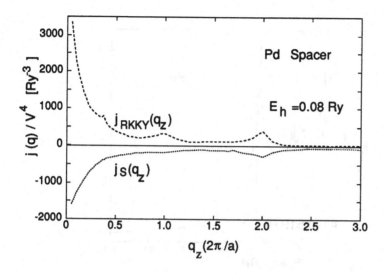

Figure 1: RKKY-like coupling, $j_{RKKY}(q_z)$, and superexchange coupling, $j_S(q_z)$, for Pd(001) spacers in reciprocal space calculated with $E_h = 0.08\ Ry$. V is expressed in rydbergs.

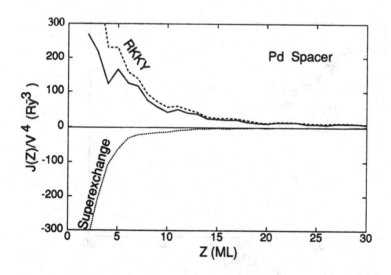

Figure 2: Interlayer magnetic coupling in the real space. The dashed line and dotted line are for RKKY-like and superexchange couplings, respectively. The solid line is for the total coupling (RKKY+superexchange).

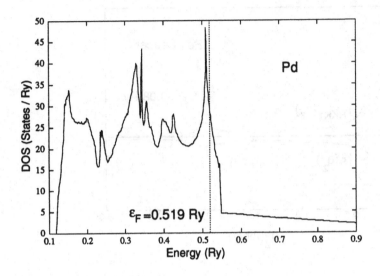

Figure 3: Density of states for Pd.

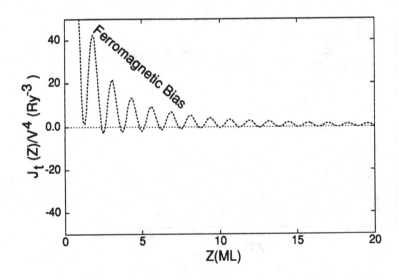

Figure 4: The total interlayer magnetic coupling for a densities of states with a peak below, but near, the Fermi surface. Ferromagnetic bias appears in the coupling.

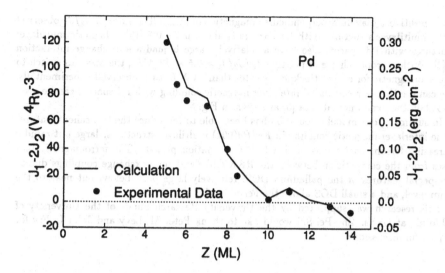

Figure 5: Interlayer magnetic coupling as a function of Pd spacer thickness. The solid line and filled circles are our calculations and the experimental data observed at $T = 77\ K$ [13], respectively. The theoretical (experimental) results are refered to the left (right) scales.

Fig. 3. In particular, the fact that the Fermi level falls *above* a peak in the DOS followed by a relatively smooth and structureless DOS, results in a relatively small superexchange contribution above $\sim 4\ ML$, and leads to a ferromagnetic bias driven by the large RKKY coupling.

To confirm our explanation of the cause of the ferromagnetic bias we use a free-electron gas model which enables us to obtain analytic results for the couplings. In a previous study [12], we showed that in the free-electron gas approximation, RKKY + Superexchange coupling resembles pure RKKY coupling, but without any magnetic bias. To illustrate the effect of a peak in the density of states on top of a free electron-like background, we use a "toy model" calculation by adding a Lorentzian shaped peak to the DOS of the free electron gas,

$$D(E) = \sqrt{E} + \frac{\sqrt{E_F}}{(E/\varepsilon_F - p)^2 + h^2} \ , \qquad (6)$$

where the position of the peak is at $E_p = p\ \varepsilon_F$, and h adjusts height of the peak (small h corresponds to a large peak). For example, by fixing the position ($p = 0.9$, below the Fermi level) and increasing the height of the peak (*e.g.*, $h = 0.3$), ferromagnetic bias occurs in the coupling, as shown in Fig. 4. We noted that in multilayer structures with a Cr spacer, RKKY + superexchange coupling gives an antiferromagnetic bias due to the structure of the DOS, with a peak above, but near to the Fermi level ε_F [12]. This is also confirmed in our "toy model" calculations.

The coupling observed in the experiments contains a bilinear exchange interaction J_1 and a biquadratic exchange interaction J_2. In conventional notation, $J_{exp} = J_1 - 2J_2$ [3]. Here, RKKY-like and superexchange couplings are contained in the bilinear coupling, J_1, and a positive J_2 favors a perpendicular magnetic coupling. The ratio of J_2/J_1 observed in the multilayer structures with Cr spacers is about 0.3 – 0.5 [13]. Magnetic multilayer structures with Pd spacers also have a relatively large biquadratic exchange interaction J_2 [3]. With a proper choice of positive J_2 ($J_2/J_1 \approx 0.5$ at 13 ML), the bias can switch to antiferromagnetic for spacer thickness greater than 13 ML, as is observed experimentally. One can see that our calculated interlayer magnetic coupling with Pd spacers can be used to explain the experimental data [5] as shown in Fig. 5.

In summary, our model calculation has been able to reproduce the two salient features of the interlayer magnetic coupling in $Fe/Pd(001)$ multilayer structures: large but rapidly decreasing ferromagnetic bias, and a 5 ML oscillation period. The ferromagnetic bias arises from the competition between the RKKY-like and superexchange couplings due to the special features of the palladium DOS: relatively large peak below, but near to the Fermi level, and a small DOS above the Fermi level.

This research was supported by the University Research Funds of the University of California at Davis. Zhu-Pei Shi would like to thank Peter M. Levy and John L. Fry for very useful discussions.

References

[1] S. S. P. Parkin, N. More and K. P. More, Phys. Rev. Lett. **64**, 2304 (1990).

[2] M. N. Baibich et al., Phys. Rev. Lett. **61**, 2472 (1988).

[3] B. Heinrich and J. F. Cochran, Adv. Phys. **42**, 523 (1993); and references therein.

[4] K. B. Hathaway, in *Ultrathin Magnetic Structures*, Vol. II, edited by B. Heinrich and J. A. C. Bland (Springer & Berlin, 1994), Chap. 2.

[5] Z. Celinski, B. Heinrich and J. F. Cochran, J. Appl. Phys. **70**, 5870 (1991).

[6] V. L. Moruzzi and P. M. Markus, Phys. Rev. B **39**, 471 (1989).

[7] H. Chen, N. E. Brener and J. Callaway, Phys. Rev. **40**, 1443 (1989).

[8] Z. P. Shi, P. M. Levy and J. L. Fry, Phys. Rev. Lett. **69**, 3678 (1992).

[9] P. W. Anderson, Phys. Rev. **124**, 24 (1961).

[10] D. A. Papaconstantopoulos, "Handbook of the Band Structure of Elemental Solids", pp159-160, Plenum Press, New York (1986).

[11] J. Callaway et al., Phys. Rev. B **28**, 3818 (1983).

[12] Z. P. Shi, P. M. Levy and J. L. Fry, Europhys. Lett. **26**, 473 (1994); Phys. Rev. B **49**, 15159 (1994).

[13] B. Heinrich et al., Mat. Res. Soc. Sym. Proc., **313**, 119 (1993).

STRUCTURAL COHERENCE AND MAGNETIC COUPLING IN Fe/Si

C.L. FOILES, M.R. FRANKLIN AND R. LOLOEE
Michigan State University, Department of Physics and Center for Fundamental Materials
Reasearch, East Lansing, MI 48824

ABSTRACT

A number of studies have inferred the presence of an Fe-silicide in Fe/Si multilayers. Our transmission electron diffraction data provide direct evidence for the presence of an Fe-silicide. Despite similarities in structural coherence and saturation magnetization behavior for Fe/Si and Fe/{FeSi}, direct evidence for Fe-silicide only occurs for the Fe/{FeSi} multilayers.

INTRODUCTION

The properties of Fe layers separated by Si spacer layers are a topic of current interest. Prior to the early 1990's studies had involved thick Si layers, 35Å or greater, with varying Fe layers thicknesses and a consistent pattern of properties had emerged. Fe layers >20Å were isolated crystalline units having a reduced magnetization and there was no magnetic coupling between these layers. As the layer thickness dropped below 20Å these Fe layers became amorphous and near a thickness of 12Å they ceased to be magnetic [1]. Recently a number of studies have found quite different properties when the spacer layer is thin. Toscano, et al. found evidence of an oscillating magnetic coupling in evaporated FeSiFe trilayers when the Si layer thicknesses were in the range of 7 to 30Å [2]. A group at Argonne reported both a non-oscillating, antiferromagnetic coupling and a structural coherence between Fe layers for sputtered Fe/Si multilayers with thin Si spacer layers [3]. The magnetic coupling was a maximum for 14Å Si layers and structural coherence was lost as the Si layer thickness increased above 17Å. The Argonne group subsequently reported antiferromagnetic coupling through a thicker spacer layer and structural coherence up to spacer layers of 40Å when the spacer was a mixture of Fe and Si [4]. There are independent reports that this magnetic coupling can be modified optically [3,5]. More recently, Inomata, et al. studied electron transport in sputtered Fe/Si multilayers and reported a magnetoresistance (MR) that is negative and has a significant change in its temperature dependence as the Si layer exceeds 15Å [6]. They also claim a change in coupling from ferromagnetic to antiferromagnetic at room temperature for their samples.

The structure of these thin Si spacer layer samples is not well determined. Toscano, et al. explain the magnetic coupling in the FeSiFe trilayers by claiming that the Si layer is an amorphous semiconductor [2]. The Argonne group use a number of properties to infer that the coupling and structural coherence are related to the formation of Fe-silicides [3,4]. Inomata, et al. use the temperature dependence of magnetic properties to infer that their Si spacer layers are a narrow gap Fe-silicide, ε-FeSi, when the Si layer is less than 15Å and is a combination of this silicide and amorphous Si when the Si spacer thickness is greater than 15Å [6].

In an earlier study we used transmission electron diffraction (TED) in an attempt to obtain direct evidence for Fe-silicides in sputtered Fe/Si and found no such evidence [1]. In the present paper we report the extension of our TED study to Fe/{FeSi} multilayers: samples for which the spacer layers are a mixture of Fe and Si, {FeSi}. We also report the results for reflection X-ray diffraction (XRD) studies done using a rotating anode and a synchrotron source.

Mat. Res. Soc. Symp. Proc. Vol. 384 ° 1995 Materials Research Society

EXPERIMENTAL PROCEDURES

The samples were prepared by DC magnetron sputtering using conditions similar to those used by the Argonne group [3,4]. The sputtering environment was pure Ar at a pressure of 2.5 mTorr and 16 different samples could be made during each preparation run. A computer controlled substrate holder placed an unmasked substrate over the proper targets with a time pattern needed to obtain nominal thicknesses. The sputtering rates of the targets were determined by quartz crystal thickness monitors at the start of each preparation run and were remeasured several times during the run. A detailed description of this system has been published [7]. The Fe layers in all samples were nominally 28.7Å thick, approximately 14 monolayers, and pure Si spacer layers varied from 10 to 30Å in thickness. {FeSi} spacer layers were formed by sputtering about 2Å of Si and then 2Å of Fe and repeating this sequence to obtain total spacer thicknesses varying from 12 to 40Å. Sapphire, crystalline Si and cleaved NaCl were used as substrates. The total multilayer thickness on all substrates was about 500Å. The only significant change in the preparation of the Fe/Si multilayers was the total thickness for samples on sapphire and Si; these multilayers had a total thickness of about 2000Å.

Standard θ-2θ XRD using a rotating anode system with a Cu target and a graphite monochromator preceding the detector provided the initial structural characterization of the samples. Subsequently, XRD data for a number of samples was collected at NSLS on beamline X3B1. Portions of the sapphire and Si substrate samples were used for magnetization measurements in a SQUID magnetometer. These measurements were done at 5K and the field was typically applied parallel to the multilayer film. A number of measurements were done in perpendicular fields as a consistency test. Portions of the multilayer films were floated off the NaCl substrates and onto Cu grids for TED studies in a field-emission STEM that permitted online examination of results. These TED data provided in-plane diffraction results and at least three different regions of each multilayer film were studied to test sample uniformity.

RESULTS AND DISCUSSION

Although there are variations in the absolute intensity for samples of comparable thickness on different substrates, the XRD Bragg peak locations are independent of the substrate used to within experimental error. Low angle XRD data confirm that all samples have a layered structure. The Fe/{FeSi} samples with thicker spacer layers typically have more than 2 peaks and 5 was the maximum number of peaks observed. The corresponding numbers for low angle peaks with our Fe/Si samples were 3 and 8 [1]. The higher angle XRD data have a Bragg peak consistent with the <110> line of Fe. For the entire range of {FeSi} spacer thicknesses the FWHM of this line lies between 0.51 and 0.68 degrees and indicates a structural coherence extending over 3 bilayer distances for all but the thickest {FeSi} spacer layer. This thickest spacer, 40Å thick, has structural coherence over 2 bilayer distances. This result differs from our Fe/Si results where structural coherence occurred only for Si spacer layers thinner than about 15Å. Our results for both Fe/Si and Fe/{FeSi} are consistent with the results reported by the Argonne group [3,4].

The variation in location of these <110> lines is shown as d-spacings in Figure 1. The patterns are different for Fe/Si and Fe/{FeSi} and are consistent with the differences in structural coherence found for these samples. The d-spacing for the <110> line of bulk BCC Fe is 2.027Å and the d-spacing we measure for a 450Å film of pure Fe on a NaCl substrate is 2.025Å. Alloying Si into Fe causes a decrease in d-spacing proportional to the Si content [8].

When two different crystalline materials having similar d-spacings are used to form the bilayer unit of a multilayer, XRD data for the layer thickness range used in the present study produce a single d-spacing that is a weighted average of two d-spacings. Thus, the gradual decrease in d-spacing with increasing spacer thickness for the Fe/{FeSi} samples is consistent with the presence of an increasing amount of crystalline Fe-Si alloy in the bilayer unit. Since the {FeSi} is formed under non-equilibrium conditions, any speculation on its structure should be limited. Therefore, we simply note that the entire range of d-spacings in Figure 1a is within the range of 2.027Å and 2.006Å which are the respective d-spacings of pure Fe and the solubility limiting Fe(23.4%Si) alloy. In Figure 1b the Fe/Si samples having thin Si layers and structural coherence show an even greater decrease in this d-spacing. For Fe/Si samples with thicker Si layers the structural coherence is lost and both the d-spacing and the FWHM of the observed <110> line are consistent with the structural coherence being limited to one bilayer unit.

The presence of satellites surrounding the <110> line are additional confirmation of layering. In the XRD data obtained with the rotating anode system, observation of one weak satellite is typical although 2 satellites are observed for the two thickest {FeSi} spacer layers. For the XRD data obtained using the synchrotron source at NSLS, observation of 2 satellites is typical. The satellite on the lower angle side of the <110> peak is always stronger. Figure 2 shows the strengths of these satellite peaks, normalized to the strength of the <110> peak to remove texture effects, as a function of {FeSi} spacer thickness. With the exception of the

Figure 1. Location of the <110> peaks in X-ray data as a function of spacer layer thickness. This location is given as a d-spacing.

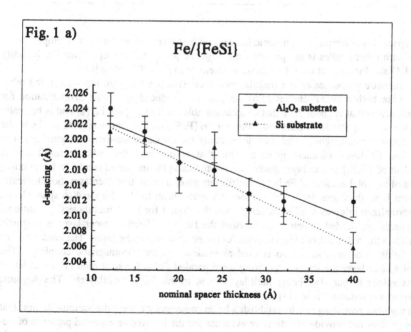

Fig. 1 a)

Fe/{FeSi}

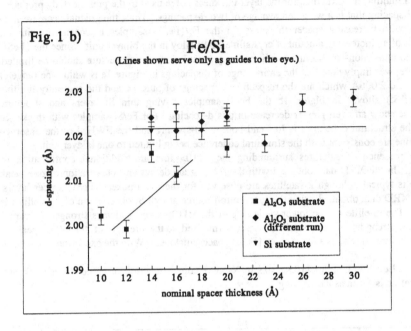

Fig. 1 b)

Fe/Si

(Lines shown serve only as guides to the eye.)

d-spacing (Å)

- ■ Al₂O₃ substrate
- ◆ Al₂O₃ substrate (different run)
- ▼ Si substrate

nominal spacer thickness (Å)

thickest spacer layer sample, a monotonic increase with spacer thickness occurs. Comparison of this data with other studies is not possible: the Argonne group did not report satellite intensity results [3,4] and Inomata, et al. did not observe satellites in their XRD data [6].

The magnetic properties of our multilayers are consistent with those reported in the other studies. Like both previous studies [3,6] we observe a reduced saturation magnetization for Fe/Si multilayers that is about 70% that of a comparable amount of pure Fe [1] and is basically independent of Si spacer layer thickness. For our Fe/{FeSi} multilayers, assuming the Fe in the {FeSi} spacer is non-magnetic and then normalizing the measured saturation magnetization to the nominal Fe layer thickness gives a value that agrees with the bulk Fe result and is independent of {FeSi} spacer layer thickness. Normalizing the measured results to the nominal thickness of the total amount of Fe in the multilayers gives a result that decreases monotonically with spacer layer thickness: from 85% for the 12Å spacer layer to 58% for the 40Å spacer layer. Either normalization gives a different result than that found for Fe/Si multilayers. Additional measurements are needed to determine whether the Fe in the {FeSi} spacer layer is magnetic and/or affects the reduction of magnetization for the nominally pure Fe layer. The need for large parallel fields to achieve saturation is used as evidence of antiferromagnetic coupling. The region of spacer layer thicknesses requiring large parallel fields for saturation occurs at larger thickness values in our Fe/{FeSi} multilayers than in our Fe/Si multilayers. The Argonne group report a comparable result [4].

Although the preceding results establish a basic consistency among the various studies, that consistency does not provide any decisive evidence for the inferred or assumed presence of Fe-silicides. Since studies of single Fe-Si interfaces are numerous and produce evidence for silicide

Figure 2. Satellite peak intensities as a function of spacer layer thickness. All values are normalized to the intensity of the <110> peak for their respective samples.

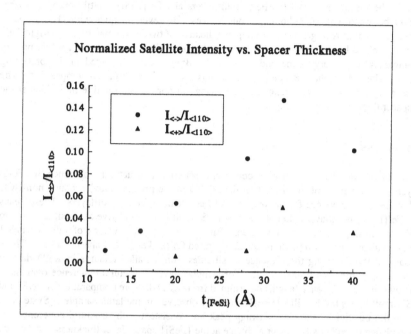

formations that are highly dependent upon growth method and conditions [9,10], reliable extrapolation of such results to repetitive interfaces in sputtered multilayers is difficult. Direct evidence for Fe-silicide in multilayers is needed and TED data for our Fe/{FeSi} samples give that direct evidence. All samples show 10 or more lines that are consistent with BCC structure and "most" samples show one additional line which we attribute to an Fe-silicide. If a very weak line in only one region of a sample is accepted as evidence, "most" becomes all. No sample shows more than 2 non-BCC lines and, in fact, only two samples have 2 lines. Neither the number of these Fe-silicide lines nor their strengths scale in any simple manner with the {FeSi} layer thickness. The 16Å and 40Å {FeSi} spacer layer samples are the multilayers having 2 non-BCC lines. Using the bulk Fe spacings for the BCC lines, the d-spacings of these 2 non-BCC lines are 3.08Å and 2.51Å.

With only 2 different lines being observed and the majority of samples giving only a single line, identification of the Fe-silicide is suggestive but not definite. The non-BCC line most frequently observed is the one with a d-spacing of 2.51Å. This value is nearly 0.1Å smaller than that of the <111> line in ε-FeSi. The less frequently observed line with a 3.08Å d-spacing is a comparable shift from a <110> line. The <210> and <321> lines of ε-FeSi are much stronger than these <110> and <111> lines in the x-ray powder patterns for this silicide but these higher index lines have d-spacings that are consistent with them being hidden under the observed BCC lines. Thus, we suggest that identification of the silicide as ε-FeSi is plausible but remind the reader that the non-equilibrium conditions associated with preparation of multilayers need not produce equilibrium silicides. The weakness of these lines as well as possible crystal texture effects prevent any estimate for the amount of silicide present.

The electron microscope images of our samples provide additional information about structure in the multilayers. All the Fe/Si multilayers and Fe/{FeSi} multilayers with spacer thicknesses below 26Å produce homogeneous images. However, samples with {FeSi} spacer layer thicknesses of 28Å or greater give images indicative of two phase material: the bright field images are dominantly dark (one phase) with small lighter regions (inclusions of another phase). The dominant phase is crystalline and exhibits the Bragg peaks discussed in the preceding paragraphs. The lighter phase is amorphous and has a yet to be determined composition. The formation of two phases may be related to the change in normalized satellite intensities at the largest spacer thickness shown in Figure 2.

CONCLUSIONS

X-ray diffraction data establish a consistent pattern of structural coherence in Fe-Si multilayers. However, in-plane diffraction data (TED in the present study) produce non-BCC peaks that are direct evidence for the presence of Fe-silicide in Fe-Si multilayers. These peaks occur in Fe/{FeSi} multilayers, do not occur in Fe/Si multilayers, and give a plausible match to the peaks for a single equilibrium silicide. Electron microscope images of our Fe/{FeSi} multilayers indicate that a two phase material is formed for thicker {FeSi} spacers layers. These results suggest that inferring the presence of silicides from detailed features in XRD data or macroscopic properties has limited value. For samples having structural coherence over more than one bilayer, the coherence lengths determined from XRD data are comparable for Fe/Si and Fe/{FeSi} multilayers but Fe-silicide peaks are only observed in the latter samples. Systematic changes in the position of the <110> Bragg peak are consistent with structural coherence but give no evidence for either silicides or a change at the {FeSi} spacer layer thicknesses for which electron microscope images show evidence of two phases with different structures. Changes in structural coherence and sample homogeneity produce no clear effects in saturation magnetization.

REFERENCES

[1] C.L. Foiles, M.R. Franklin and R. Loloee, Phys. Rev. B50,16070 (1994).
[2] S. Toscano, B. Briner, H. Hopster and M. Landolt, J. Magn. Magn. Mater. 114,L6 (1992).
[3] J.E. Mattson, Sudha Kumar, Eric E. Fullerton, S.R. Lee, C.H. Sowers, M. Grimsditch, S.D. Bader, and F.T. Parker, Phys. Rev. Lett. 71,185 (1993).
[4] J.E. Mattson, Eric E. Fullerton, Sudha Kumar, S.R. Lee, C.H. Sowers, M. Grimsditch, S.D. Bader, and F.T. Parker, J. Appl. Phys. 75, 6169 (1994).
[5] B. Briner and M. Landolt, Z. Phys. B92, 137 (1993).
[6] K. Inamoto, K. Yusu, and Y. Saito, Phys. Rev. Lett. 74, 1863 (1995).
[7] J.M. Slaughter, W.P. Pratt, Jr., and P.A. Schroeder, Rev. Sci. Instrum. 60, 127 (1988).
[8] W. B. Pearson, Handbook of Lattice Spacings and Structure of Metals and Alloys, Volume 2, Pergamon Press (1967).
[9] J. Alvarez, A.L. Vazquez de Parga, J.J. Hinarejos, J. de la Figuera, E.G. Michel, C. Ocal and R. Miranda, Phys. Rev. B47, 16048 (1993); J. Vac. Sci. Technol. A11, 929 (1993).
[10] Papers in Silicides, Germanides, and Their Interfaces, (Mater. Res. Soc. Proc. 320, 1994).

REDUCING INTERGRANULAR MAGNETIC COUPLING BY INCORPORATING CARBON INTO Co/Pd MULTILAYERS

WENHONG LIU*, JONATHAN MORRIS*, ALEX PAYNE** AND BRUCE LAIRSON*
*Materials Science and Engineering, Rice University, Houston, TX 77251
**Censtor Corporation, San Jose, CA 95126

ABSTRACT

The ideal magnetic switching mechanism for many types of data storage, including hard disk recording, is isolated domain coherent rotation (Stoner-Wohlfarth switching). However, in typical Pd/Co multilayers with high coercivity, the dominant switching mechanism is domain wall motion, which causes noise in the read back signal. We show that the proper addition of elements, such as carbon, into Pd/Co multilayers reduces the coupling between adjacent magnetic domains. The reduction of magnetic coupling reduces the length scale over which incoherent switching occurs.

Introduction

Pd/Co multilayers with high perpendicular anisotropy are prospective candidates for perpendicular recording media[1]. In typical Pd/Co multilayers with high coercivity, the dominant switching mechanism is domain wall motion, which causes transition noise[2].

In thin film recording media, when the linear density is over 100,000 bits per inch, noise is a prominent factor which impedes further increases in the linear density. One way to decrease the transition noise is to reduce the coupling between adjacent domains, effectively increasing the number of statistically independent particles involved in storage of a bit.

Carbon was selected to achieve grain isolation because it commonly adopts interstitial lattice positions and is relatively immiscible with Co and Pd, with a solid solubility limit of less than 1%.

Experimental

Multilayers were grown in a high vacuum deposition system equipped with a 14" APD cryopump and three 1.3" sputter deposition sources. Substrates were baked at 100°C prior to deposition to remove environmental deposits. Pd was RF sputtered at a power of 25W, Co was DC sputtered at 20W, and C was DC sputtered at 40W. The carbon/cobalt volume ratio was expected to be 10% based on the relative sputter yields[3]. Measured lattice parameter changes upon the incorporation of carbon indicated that the cobalt layer initially contained 8 at% carbon, assuming that the carbon in the lattice existed at interstitial sites. Measurement of the total film thickness increased by 10% upon incorporation of carbon, indicating that about 20% of the deposited carbon was initially dissolved in the Co layer.

Pd/Co multilayers were grown on single crystal Si(100) substrates coated with 500Å of silicon nitride. 100Å thick polycrystalline Pd seed layers were deposited prior to the multilayers to remove initial layer effects such as poor texture and islanding. Two sets of "as grown" samples will be discussed, once consisting of 20x(Pd(11Å)/Co(4Å)) and the other of 20x(Pd(11Å)/CoC(4Å)), with the second set incorporating carbon by co-sputtering with cobalt.

Magnetic Energy Model

In Pd/Co multilayers, the Pd layer is polarized for thicknesses less than approximately 20Å[4]. We therefore assume that within a domain, the multilayers are uniformly magnetized throughout the thickness of the film. This assumption is consistent with the measurement of equal coercivities from the front and back sides of the film using MOKE. Given the domain repetition length d and film thickness T, Suna[5] gives the demagnetized magnetostatic energy density for a stripe domain configuration as

$$\varepsilon_m = \sum_n (8/\pi^2) M_s^2 (d/T)(2n+1)^{-3}[1 - \exp(-(2n+1)\pi T/(2d))] \quad (1)$$

For 0.01<d/T<1000, a good approximation is

$$\varepsilon_m = \frac{2\pi d}{d + 7.38T} M_s^2 \quad (2)$$

Assuming that domain walls exist at grain boundaries, the domain wall energy density is

$$\varepsilon_w = 2\frac{C\sigma_w}{d} M_s^2 \quad (3)$$

where C represents coupling coefficient across the grain boundary. In a uniform film without grain boundaries, C=1. The total energy density is $\varepsilon_{tot} = \varepsilon_m + \varepsilon_w + \varepsilon_{ext}$. In equilibrium, $\frac{\partial \varepsilon_{tot}}{\partial d} = 0$. Therefore,

$$d_{equ} = \frac{7.38T\sqrt{C\sigma_w}}{(4.82\sqrt{T} - \sqrt{C\sigma_w})} \quad (4)$$

Equation (4) shows that the repetition length falls monotonically as the coupling coefficient C is reduced to zero, illustrating the value of reducing intergranular coupling. It is therefore desirable to reduce the coupling across grain boundaries. A minimum in the equilibrium domain repetition length also occurs at a critical thickness $T_0 \approx 0.054 C\sigma_w$.

Results and Discussion

Out-of-plane X-ray diffraction scans were performed on all of the annealed samples, from which the calculated lattice parameters are shown in Figure 1. Upon annealing, the Pd/CoC multilayer (111) spacing decreases, approaching the lattice parameter of the as-deposited Pd/Co multilayers. This suggests that upon annealing, carbon diffuses out of Co interstitial sites into grain boundaries and into the Pd layer. However, as shown in Figure 2, Transmission Electron Microscopy (TEM) shows no change in film morphology. High resolution cross-sectional TEM showed an large increase in the fraction of the film which produced lattice fringes, however. The as deposited film incorporating carbon showed only small regions within a grain which produced lattice fringes, while high resolution TEM on the annealed sample shown in Figure 2 produced large regions of continuous lattice fringes in high resolution imaging.

(a) As deposited

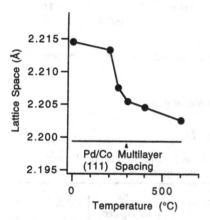 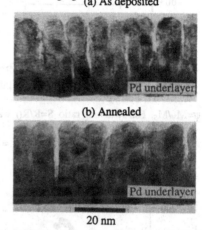

(b) Annealed

20 nm

Fig. 1 Pd/CoC multilayer (111) spacing vs annealing temperature.

Fig. 2 TEM cross-sectional images of Pd/CoC multilayers, (a) as deposited, (b) annealed.

Annealing increased the coercivity and saturation moment of Pd/CoC multilayers, as shown in Figure 3(a). Further, the hysteresis loop of post annealed Pd/CoC multilayers is more sheared than that of Pd/Co multilayers (Fig. 3(b)). The increase in shearing is a desirable result for magnetic data storage applications, since sheared loops result from decoupling of magnetic domains in the film. However, shearing also occurs due to other effects, such as inhomogeneity. Therefore measurements of decoupling must be made to establish if the increased loop shearing shown in Figure 3(b) is due to a reduction in the intergranular exchange coupling in the film. This question was addressed by making measurements of the switching radius ratio $S=R/R_0$, where R_0 is the radius required to achieve Stoner-Wohlfarth rotation.

Transverse magnetization measurements M_y were used to determine the anisotropy energy. Fig. 4 plots the transverse magnetization M_y vs ϕ, the angle between sample normal and the applied field at constant H. The equation

$$\frac{H_k}{2H}\sin\psi = \sin 2(\phi - \psi) \qquad (5)$$

was used to determine anisotropy energy K_u[2], where the anisotropy field is $H_k = 2K_u/M_s$, and

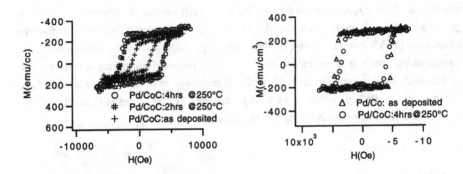

Fig.3 (a) Hysteresis loops of Pd/CoC. (b) Hysteresis loops of Pd/Co and post annealing Pd/CoC.

$\sin\psi = M_y/M_s$. The switching ratio $S = R/R_0$ was calculated using [6]

$$S^2 = 1.08\, H_k / H_c \qquad (6)$$

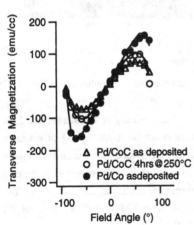

Fig 4. Transverse magnetization vs angle ϕ for H=16.3kOe.

Table 1. Anisotropy energy Ku and switching ratio S.

Sample	Ku 10^6ergs/cc	S
Pd/Co ML	2.5	1.95
Pd/CoC ML	1.7	1.56
CoCrTaPt	0.31	1.3
CoCrTa	0.58	1.2
Stoner-Wohlfarth	------	1.04

Measured values of K_u and S for Pd/Co and Pd/CoC are listed. For comparison, K_u and S on CoCrTa and CoCrTaPt of perpendicular alloy media from previous results[2] are also listed in Table 1. The decrease of S for annealed Pd/CoC compared to Pd/Co means that incorporation of carbon reduces the switching length scale, while the multilayers retain the high anisotropy that make them attractive media candidates.

Another way to evaluate switching length scale is to measure intergranular coupling using Kelly-Hankel plots. Isothermal remanence (IRM) curve and dc demagnetization (DCD) curve were measured (Fig. 5(a,b,c))). Isothermal remanence (IRM) curves are the remanence obtained by the progressive magnetization of an initially demagnetized sample. The dc demagnetization (DCD) curve is obtained by the application of a progressively increasing positive field to an initially negatively saturated sample, again measuring the remanent moment after each field application. For comparison, we also measured IRM and DCD curves on oxidized Pd/Co multilayers. Oxidation was performed in a 1 mTorr oxygen atmosphere at 250°C for 1hr. Following Kelly et al[7], We define

$$\Delta M(H) = 2 I_r(H) - I_d - 1 \qquad (7)$$

where $I_r(H) = M_r(H) / M_r(\infty)$, $I_d(H) = M_d(H) / M_d(\infty)$. Mr and Md are the remanent moments on the IRM and DCD curve, respectively.

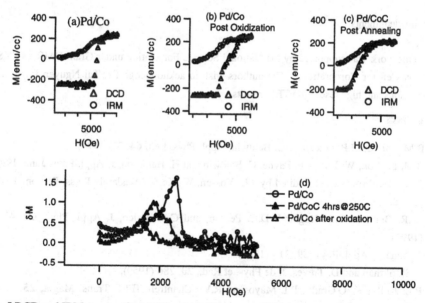

Fig. 5 DCD and IRM curves of (a) Pd/Co, (b) Pd/Co post oxidization, (c) P/CoC post annealing, (d) ΔM-H plots for Pd/Co, Pd/CoC post annealing and Pd/Co post oxidation samples.

For a noninteracting Stoner-Wohlfarth system it can be shown that $\Delta M(H) = 0$ [8]. The larger the magnetic interaction between particles in the film, the larger the ΔM peak in ΔM-H plot[7]. From equation (2) and (4) we obtain, for d much greater than the grain diameter,

$$\varepsilon_m = 1.3 \sqrt{\frac{C\sigma_w}{T}} \, M_s^2 \qquad (8)$$

e.g., in equilibrium the self-field energy decreases with decreasing coupling across grain boundaries $C\sigma_w$.

Fig. 5(d) shows that ΔM-H plots for Pd/Co, Pd/CoC post annealing and Pd/Co post oxidation samples. We get the largest peak in ΔM-H plot for Pd/Co multilayers, and smallest peak for Pd/CoC multilayers. The ΔM-H plots show that the coupling across the grain boundaries in the Pd/CoC post annealed sample is smaller than that of Pd/Co sample.

Conclusions

Carbon was incorporated into Pd/Co multilayers and subsequently phase segregated to achieve domain isolation. Annealing at 250°C resulted in a change in the Pd/CoC lattice parameter toward that of Pd/Co, with no change in film morphology observed by TEM. Radius ratios (S) and Kelly-Hankel plots (ΔM vs H) show a significant reduction in coupling between adjacent domains upon annealing in Pd/CoC multilayers.

Acknowledgments

This work was supported by the National Science Foundation under Grant DMR-9419684 and by Censtor Corporation. The authors wish to acknowledge Dr. Tai Nguyen for useful discussions and high resolution TEM results.

References

1. B.M. Lairson, J.P. Perez and C. Baldwin, Appl. Phys. Lett. **64**, 23 (1994).
2. B.M. Lairson, W. Liu, A.P. Payne, C. Baldwin and H. Hamilton, J. Appl. Phys. June, 1995.
3. Thin Film Processes II, edited by J.L. Vossen, W. Kerm (Academic Press, Boston, 1991) p. 758.
4. W.R. Bennett, C.D. England, D.C. Person, and C.M. Falco, J. Appl. Phys. **69**, 4384 (1991).
5. A. Suna, J. Appl. Phys. **59**, 313 (1986)
6. S. Shtrikman and D. Treves, J. de Phys. et Rad., **20**, 286 (1959).
7. P.E. Kelly, K. O'Grady, P.I. Mayo and R.W. Chantrell, IEEE Trans. Magn., **25**, 3881 (1989).
8. E.P. Wohlfarth, J. Appl. Phys., **29**, 595 (1958).

CRYSTAL STRUCTURE DEPENDENCE OF ANTIFERROMAGNETIC COUPLING IN FE/SI MULTILAYERS

R. P. MICHEL, A. CHAIKEN, M. A. WALL
Materials Science and Technology Division, Lawrence Livermore National Laboratory,
P. O. Box 808, Livermore CA 94551

ABSTRACT

Recent reports of temperature dependent antiferromagnetic coupling in Fe/Si multilayers have motivated the generalization of models describing magnetic coupling in metal/metal multilayers to metal/insulator and metal/semiconductor layered systems. Interesting dependence of the magnetic properties on layer thickness and temperature are predicted. We report measurements that show the antiferromagnetic (AF) coupling observed in Fe/Si multilayers is strongly dependent on the crystalline coherence of the silicide interlayer. Electron diffraction images show the silicide interlayer has a CsCl structure. It is not clear at this time whether the interlayer is a poor metallic conductor or a semiconductor so the relevance of generalized coupling theories is unclear.

INTRODUCTION

The magnetic coupling of adjacent ferromagnetic layers separated by a broad range of non-magnetic metal spacer layers oscillate from anti-ferromagnetic to ferromagnetic as the spacer layer thickness increases[1]. The variation of the coupling can result in oscillations in easily measured quantities such as the saturation field and the magneto-resistance as a function of interlayer thickness.

Most features of the oscillating exchange coupling have been successfully explained by applying RKKY type interactions to the layered geometry and using the Fermi surface characteristics of the interlayer metal[2]. Recent experimental observations of anti-ferromagnetic coupling in Fe/Si multilayers[3, 4] have motivated generalization of models of interlayer exchange to include systems without well defined fermi surfaces such as semiconductors and insulators[5, 6, 7]. Among the most pronounced predicted differences between metallic and non-metallic interlayer systems is the strong temperature dependence of the coupling in multilayers with a non-metallic interlayer due to the thermally activated nature of the carriers which carry the exchange.

In this paper we present data describing structural and magnetic characteristics of Fe/Si multilayers deposited using ion beam sputtering and discuss their significance to the theories of Bruno[6] and Zhang[5]. Consistent with previous studies we find that increasing the Si interlayer thickness from 14Å to 20Å, while keeping the Fe thickness fixed at about 30Å, has a dramatic effect on the magnetic properties and the morphology of the multilayer. We find that for Si layers around 14Å thick, the multilayer maintains crystalline coherence in the growth direction through more than one bilayer period and magnetically the Fe layers are anti-ferromagnetically coupled resulting in a high saturation field. For slightly thicker Si layers (around 20Å) the crystalline coherence in the growth direction is only as thick as a single bilayer, and the saturation field is small consistent with either ferromagnetically coupled or uncoupled Fe layers. Further, for Fe layers sufficiently thin, crystalline coherence is not achieved in the multilayer. We find that even for 14Å thick Si interlayers, disordered ferromagnetic Fe layers are either ferromagnetically coupled or uncoupled. Our TEM study reinforces the assertion of Fullerton

195

et. al.[3] that the crystalline iron-silicide that forms in the interlayer may be the CsCl structure. This silicide structure is likely stabilized in the multilayer because it is closely lattice matched to BCC Fe. We discuss the possibility that the crystallinity of the interlayer is crucial to produce AF interlayer coupling.

EXPERIMENTAL DETAILS

Our films were grown in a ion beam sputtering(IBS) systems described in detail elsewhere[8]. Briefly, four targets can be rotated in front of the 3 cm ion gun which sputters material up through a circular aperture in a stationary liquid nitrogen (LN) cooled Cu tray. A rotating tray above the Cu tray has positions for four substrates. The substrate to target distance is approximately 30 cm. The target carousel and ion beam voltage are computer controlled, and the layers thicknesses are monitored by a calibrated quartz crystal oscillator. The base pressure of the system is $1-2 \times 10^{-8}$ torr. We sputter in 2.5×10^{-4} torr partial pressure of UHP Ar which is about an order of magnitude lower than typical magnetron sputtering. With a beam voltage of 1kV and a beam current of 20mA the deposition rates are around 0.2Å/sec. IBS is unique in that at a fixed deposition voltage, the deposition rate can be independently adjusted by changing the beam current. As a result, the energetics of ion beam sputtering deposition can be quite different than those of thermal evaporation or magnetron sputtering.

We use glass and Si substrates and find no dependence in the magnetic or structural properties of the Fe/Si multilayers (MLs). We deposit at nominal room temperature (RT) or, by bringing the substrate tray into contact with the cooled Cu tray, at nominal LN temperature. Thick Si films sputtered under typical conditions are amorphous, and thick Fe films are BCC, polycrystalline, and textured in the (110) close packed direction.

The structural properties of the MLs were probed using a Rigaku rotating anode x-ray machine with a reflected beam monochromator and CuK_α radiation. Low angle θ-2θ scans reveal properties of the ML in the growth direction, and high angle scans measure the crystalline coherence. In addition, selected films were studied using TEM. RT magnetic characteristics of the MLs were measured using a vibrating sample magnetometer, and low temperature magnetic measurements were performed on a SQUID magnetometer.

RESULTS AND DISCUSSION

Figure 1 illustrates the effect of changing the interlayer thickness on the magnetic properties of Fe/Si multilayers. Film A ([Fe30Å/Si14Å]x50) (30/14 ML) has a low remanent magnetic moment (M_r/M_S=0.4) and a high saturation field H_S=1.7 kOe (H_S is define to be the field at which M(H) reaches 90% of its saturated value: i.e. $M(H_S)$=0.9M_S.). When the Si layer thickness is increased to 20Å, the magnetic behavior changes dramatically. As seen in figure 1, film B ([Fe30Å/Si20Å]x50) (30/20 ML) behaves like a single thick film of iron with a high remanence and a low saturation field. The magnetization of all of the films (1100-1200 emu/cm^3) is reduced from that expected if each iron atom had its bulk magnetization (1710 emu/cm^3 at RT). One expects in a perfectly layered system that roughly one monolayer of Fe at each interface would have a reduced moment due to Si nearest neighbors. The reduction we observe indicates more extensive interdiffusion. Our results are consistent with those of Fullerton et al on films deposited using magnetron sputtering[3]. The high saturation field and low remanence indicate the Fe layers are antiferromagnetically coupled through the interlayer. We calculate an AF coupling energy density $A_{12}=M_SH_St_{Fe}/2$= 0.25 erg/cm^2 at RT. In film B, the Fe layers are either ferromagnetically coupled or uncoupled and are thus easily aligned in a small applied field.

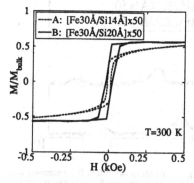

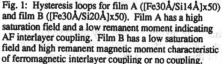

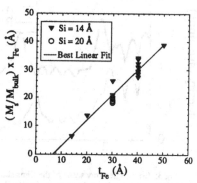

Fig. 1: Hysteresis loops for film A ([Fe30Å/Si14Å]x50) and film B ([Fe30Å/Si20Å]x50). Film A has a high saturation field and a low remanent moment indicating AF interlayer coupling. Film B has a low saturation field and high remanent magnetic moment characteristic of ferromagnetic interlayer coupling or no coupling.

Fig. 2: Estimate from magnetization data of the fraction of Fe which is interdiffused into Si for [FexÅ/Si14Å] MLs. (M/MBulk) is the fraction of the expected bulk Fe magnetization which is experimentally observed. The best linear fit to the data gives (M/M$_{bulk}$) x t$_{Fe}$ (Å) = $0.9t_{Fe}$ - 6Å, while an ideal multilayer would have (M/M$_{bulk}$) x t$_{Fe}$ (Å) =t$_{Fe}$. An approximately constant thickness of Fe (6- 8Å) becomes non-magnetic due to interdiffusion into the Si layer.

The magnetic properties of Fe/Si MLs also show strong dependence on the Fe layer thickness. Figure 2 shows the saturation moment normalized to the bulk magnetization times the Fe layer thickness versus Fe layer thickness in MLs with 14Å thick Si interlayers. The linear dependence indicates the fraction of iron that is non-magnetic due to interdiffusion into the Si layer in independent of Fe thickness. Assuming Fe atoms either have the full bulk atomic moment or are non magnetic, the intercept shows 6-8Å of the Fe layer is lost into the Si layer. This is a low estimate of the total degree of interdiffusion because it has been shown that Fe atoms with 3-5 Si nearest neighbors retain a reduced but non-zero moment. [9] Notably we find the magnetic moment of MLs with 20Å of Si also shown in fig 2, are generally reduced from their 14Å counterparts. Thus it is reasonable to picture the entire Si interlayer interdiffused to some degree with Fe. The AF coupling energy is approximately constant for thick Fe layers but between 20Å and 15 Å A$_{12}$ drops to zero. High angle x-ray scans indicate this drop may be a result of the thin Fe layer remaining amorphous. It may be that when the Fe layers are amorphous, the strain energy necessary to stabilize the crystalline silicide structure is not present.

Low angle θ-2θ x-ray scans of films A and B(fig.3) show 4 strong reflections indicating the ML are well layered. The ML peaks for the film B are narrower than those of film A indicating a reduced degree of roughness at the interfaces in the 30/20 ML. The bilayer periods derived from the positions of the ML peaks are reduced from the nominal periods by 4-8Å consistent with the reduced magnetization. Analysis of θ-2θ high angle x-ray scans reveals that most of the films are textured with Fe(110) perpendicular to the film plane. Using the well known Scherrer formula, the FWHM of the Fe(110) peaks can be used to approximate the range of the crystalline coherence in the growth direction. For the MLs that show square magnetic loops, the crystalline coherence extends over less than one bilayer period. On the other hand, in the AF coupled MLs, the crystalline coherence propagates typically through 2-3 bilayer periods. Thus the deposition of thinner Si layers allow crystalline coherence to reach from one Fe layer to adjacent layers and implies that the interlayer is itself crystalline. In addition, the AF coupled ML often show a strong Fe(200) reflection indicating a change in the preferred growth orientation in these films. This is the first report of a change in texture in Fe/Si MLs and may be

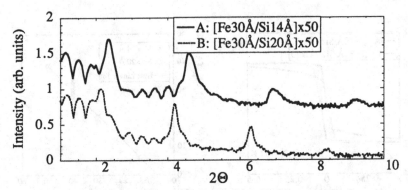

Fig 3: Low angle x-ray scans of the films shown in fig. 1. The ML peaks in film B are narrower than those of film A indicating better layering in the film that is not AF coupled. The interlayers in film B are amorphous while those in film A are crystalline.

unique to the energetics of IBS deposition. The change in texture is probably the result of an interaction between the Fe layer and the crystalline silicide interlayer that forms. A search in x-ray diffraction for Bragg reflections characteristic of the known iron rich silicide phases was unsuccessful.

In order to better correlate the magnetic characteristics of the Fe/Si multilayers to their structural properties and possibly identify the interlayer alloy phase, we carried out a detailed TEM comparison of two MLs. The first, [Fe40Å/Si14Å]x50, (40/14), showed AF interlayer coupling and the second, [Fe30Å/Si20Å]x50, (30/20), showed no evidence of interlayer coupling. Figure 4a,b show real space images of the two MLs. Both films are well layered, consistent with the low angle x-ray scattering results, but the grain structure in the growth direction of the two films is dramatically different. The 40/14 has grains that appear to reach from the substrate all the way through the ML stack. High resolution images confirm the crystalline coherence of Fe layers and the silicide interlayers in this ML. These results are consistent with the long coherence lengths derived from high angle x-ray scattering. In figure 4b no such extended crystalline coherence in observed. Instead the grain size is limited to

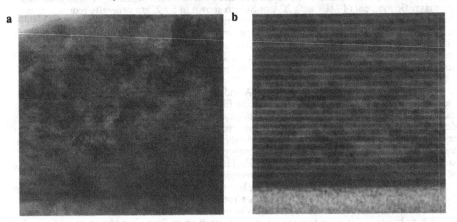

Fig. 4a,b: Real space TEM images of the ML shown in fig 1. Substrate is at bottom. Both ML are well layered. Film A which shows AF interlayer coupling has crystalline grains that reach through the entire stack. In contrast the grains in film B are limited to about the size of one bilayer thickness in the growth direction.

approximately one bilayer thickness. High resolution images confirm that the interlayer of the ML with 20Å of Si is amorphous. The long continuous layering observed in fig 4b are similar to Mo/Si MLs used for x-ray mirrors[10, 11].

Selected area electron diffraction images reinforce the marked difference in the structure of the two multilayers seen in the real space images. Figure 5b shows one nearly continuous ring consistent with the Fe(110) planes. Each Fe layer in this film consists of small grains with random in-plane orientation, textured in the (110) direction. On the other hand figure 5a shows more extensive crystalline order. The 6 bright spots on the (110) ring indicate much larger in plane grain size. The Fe(200) positions are the brightest reflections in the growth direction, indicating a (200) texture consistent with the x-ray scattering results. Significantly, intensity at the Fe (100) position is also evident. The (100) reflection is forbidden in the BCC structure of Fe and its presence in figure 5a is a clue to the interlayer crystalline phase. Both the CsCl and the Fe$_3$Si phases would produce (100) reflections in this position. The Fe$_3$Si phase is unlikely since it is ferromagnetic and would produce direct exchange coupling of the Fe layers. The CsCl structure is simple cubic with Fe at the corners and Si at the body center positions, and is closely lattice matched to BCC Fe. Fullerton et al have proposed that the spacer was either CsCl[3] or the epsilon phase[12]. The presence of a (100) reflection in figure 5a is direct evidence that the interlayer crystalline structure is CsCl. The equilibrium bulk binary phase diagram[13] shows the CsCl structure stable at RT for Si concentrations between 10 and 22 at. %. However, von Kanel has shown[14] that strain energy can stabilize the CsCl structure over much broader concentrations for silicide layers grown epitaxially.

In order to test the hypothesis that it is the loss of crystalline coherence and not the increased thickness of the silicide interlayer that results in the loss of AF interlayer coupling, we attempted to disrupt the crystalline coherence by growing the ML at a reduced substrate temperature. Two ML with the same nominal thicknesses were grown, the first on a nominal RT substrate and the second on a LN cooled substrate. [Fe40Å/Si14Å]x40 grown at RT consistently showed a high saturation field and low remanence. In contrast the LN cooled ML showed a square magnetic loop consistent with uncoupled Fe layers. Comparison of the low angle x-ray scans shows the LN cooled growth produces higher quality interfaces, and a bilayer period (52Å) much closer to the nominal period than the RT growth (49Å). Comparison of the FWHM of the high angle structural peaks in the two ML shows the crystalline coherence of the LN grown ML is limited to less than a single bilayer period, while that of the RT grown ML

a
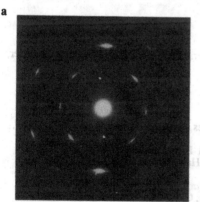

b
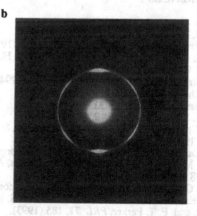

Fig 5a,b: Electron diffraction images of the MLs in fig 4 reinforce the conclusions drawn from the TEM images. (Growth direction along long axis of page.) The texture of film A is predominantly in the (200) direction in contrast to that of film B whose texture is in the usual (110) close packed direction. A faint (100) spot is visible in the image from film A indicating the crystalline silicide interlayer is in the CsCl crystal structure.

extends over 3 bilayer periods. The x-ray results indicate the low deposition temperature affects the crystal structure of the interlayer and that the loss of crystalline coherence through the interlayer eliminates the AF coupling.

CONCLUSIONS

Our results show that in the Fe/Si ML system, it is not straightforward to measure interlayer magnetic coupling as a function of the layer thickness. Unlike most metal/metal ML systems, the gross crystalline structure of Fe/Si MLs changes dramatically with changing layer thicknesses, and strongly affects the magnetic characteristics. We have shown that when the silicide interlayer is in the CsCl structure and is around 14Å thick, neighboring Fe layers are coupled anti-ferromagnetically. When the nominal Si layer thickness is increased to 20Å, the silicide interlayer is amorphous and the AF coupling disappears. It is possible, however, if the growth conditions can be adjusted to maintain the crystalline coherence of the silicide interlayer for larger layer thicknesses, then oscillations in interlayer coupling in Fe/silicide MLs may be measured that are similar to those observed in metal/metal MLs. We plan to try elevated deposition temperatures and post annealing of the MLs to explore this possibility. The exact stoichiometry of the silicide interlayer that produces AF coupling and whether it is a poorly conducting metal or a semiconductor is not clear at this point. The fact that when the Fe layers are amorphous, the AF coupling goes away is strong evidence that the AF coupling depends on the crystal structure of the silicide interlayer and not necessarily on its stoichiometry.

The relevance of the Bruno[6] and the Zhang[5] theories of magnetic interlayer coupling in ferromagnet/insulator or ferromagnet/semiconductor MLs to Fe/Si MLs is uncertain at this time. Future measurements of the conductivity of these MLs in the current-perpendicular-to-plane geometry may reveal whether the silicide interlayer is metallic or semiconducting.

We would like to thank Troy Barbee Jr., Tim Weihs and Patrice Turchi for helpful discussions. Also we thank Michael Lane, Ben O'Dell, Eric Honea and Sam Torres for experimental assistance. Part of this work was performed under the auspices of the U.S. Department of Energy by LLNL under contract No. W-7405-ENG-48.

REFERENCES

1. S. S. P. Parkin, *PRL* **67**, 3598 (1991).
2. E. Bruno and B. L. Gyorffy, *PRL* **71**, 181 (1993).
3. E. E. Fullerton, J. E. Mattson, S. R. Lee, C. H. Sowers, Y. Y. Huang, G. Felcher, S. D. Bader, and F. T. Parker, *JAP* **73**, 6335 (1993).
4. B. Briner and M. Landolt, *PRL* **73**, 340 (1994).
5. S. Zhang, *Unpublished* , (1994).
6. P. Bruno, *PRB* **49**, 13231 (1994).
7. J. C. Slonczewski, *PRB* **39**, 6995 (1989).
8. A. Chaiken, E. C. Honea, W. S. Rupprecht, S. Torres, and R. P. Michel, *Rev. Sci. Instr.* **65**, 3870 (1994).
9. C. Dufour, A. Brunson, G. Marchal, B. George, and P. Mangin, *JMMM* **93**, 545 (1991).
10. N. M. Ceglio, D. G. Stearns, D. P. Gaines, A. M. Hawryluk, and J. E. Trebes, *Opt. Lett.* **13**, 108 (1988).
11. D. G. Stearns, R. S. Rosen, and S. P. Vernon, *J. Vac. Sci. Technol. A* **9**, 2662 (1991).
12. J. E. Mattson, S. Kumar, E. E. Fullerton, S. R. Lee, C. H. Sowers, M. Grimsditch, S. D. Bader, and F. T. Parker, *PRL* **71**, 185 (1993).
13. *Binary Alloy Phase Diagrams*, Edited by T. B. Massalski, (American Society for Metals International, Materials Park OH, 1986), vol. 2, pp. 1108.
14. H. von Kanel, N. Onda, H. Sirringhaus, E. Muller-Gubler, S. Goncalves-Conto, and C. Schwarz, *Applied Surface Science* **70/71**, 559 (1993).

Part IV
Magnetic Anisotropy

PROCESS-INDUCED UNIAXIAL MAGNETIC ANISOTROPY IN EPITAXIAL Fe and $Ni_{80}Fe_{20}$ FILMS

J.R. CHILDRESS*, O. DURAND**, F. NGUYEN VAN DAU**, P. GALTIER**, R. BISARO** AND A. SCHUHL**.
* Department of Materials Science and Engineering, University of Florida, Gainesville, FL 32611.
** Laboratoire Central de Recherches-Thomson-CSF, Domaine de Corbeville , 91404 Orsay, France.

ABSTRACT

The occurrence of a weak in-plane uniaxial magnetic anisotropy in Fe thin films grown by molecular beam epitaxy onto (001)-oriented MgO substrates has been previously reported. We explain the occurrence of this anisotropy by measuring the in-plane tetragonal distortion of the cubic Fe lattice in a 800Å-thick film. The analysis of the full x-ray diffraction spectrum reveals a 0.1% difference between the two in-plane parameters. This small difference is sufficient to fully explain the observed anisotropy (≈ 20 Oe) using a standard magnetoelastic model. Although it is established that the uniaxial anisotropy results from the angle of incidence of the Fe atomic flux during deposition, the relationship between angle of incidence and in-plane tetragonalization is still unexplained. However, this anisotropy is shown to also occur in other epitaxial systems such as $Ni_{80}Fe_{20}$ on (111)Si. Control of this effect can help design epitaxial multilayer films with specific and reproducible magnetic states.

INTRODUCTION

Recently, there has been reports of a weak uniaxial magnetic anisotropy occurring in epitaxial thin-films and multilayers prepared by molecular beam epitaxy.[1,2] In the case of (100)Fe films grown on (100) MgO, the usual 4-fold easy axes of magnetization in the plane are (100)-oriented, with an anisotropy field of about 500 Oe with respect to the (110) axes. For highly uniform films, grown while rotating the substrate in its plane, the films display the expected soft magnetic properties of bulk Fe in the (100) directions, with a low coercivity of about 5-10 Oe. However, we have found that when the sample is deposited without substrate rotation, an additional uniaxial magnetic anisotropy breaks the symmetry between the two in-plane easy axes. The result of this effect on the hysteresis loops in the <100> and <010> directions is shown in Fig.1. While the hysteresis loop is one direction (labeled <100>) is square as expected, the loop performed with the field in the <010> direction is split, with a very small magnetization in the low-field region. We have shown conclusively, using magneto-optical Kerr microscopy, that this region of low magnetization is the result of a spontaneous nucleation and growth of magnetic domains oriented 90° from the field direction, i.e., along the <100> direction.[1] The splitting of the hysteresis loop in the <010> direction gives the uniaxial anisotropy field, and the value observed here (about 20 Oe) is typical of that observed for this system. Of course, the existence of a uniaxial anisotropy term is obvious only because the intrinsic coercive field of these films (which can be obtained from the width of the split half-loops) is only 5-10 Oe. If the film had a coercive field greater than the uniaxial anisotropy field, this effect would be observed simply as a (relatively small) difference between the coercivities in the two directions. Nevertheless, it is clear that such an effect can have significant impact on the magnetic properties of soft magnetic film and the investigation of interlayer coupling effects in multilayers.

ORIGIN OF UNIAXIAL ANISOTROPY

The uniaxial anisotropy component is found to occur only when the films are grown without substrate rotation. Consequently, any anisotropies in the substrate material, the vacuum environment, and the geometry of deposition may be responsible for this effect. However, only one factor, namely the off-axis position of the Knudsen cells with respect to the normal of the film, has been found to cause this anisotropy in our case. Factors which have been definitely excluded are substrate morphology (such as a stepped surface), film thickness gradients, and residual magnetism of the growth chamber.[3] We have found that it is possible to predict to occurrence and direction of the uniaxial anisotropy simply from the angle of incidence of the incoming Fe flux with respect to the surface.

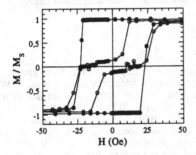 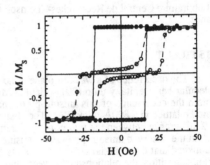

Figure 1: Hysteresis loops at T=15K for (a) a $Fe_{(18Å)}/Pd_{(21Å)}$ superlattice and (b) a 800Å-thick Fe film deposited on a (100)MgO substrate. In each case, the stepped loop (open circles) is obtained with the field applied along the cubic axis closest to the direction of the Fe atom flux projected in the plane of the substrate.

Uniaxial anisotropy occurs when the projection of the Fe flux in the plane of the substrate is directed preferentially towards one of the in-plane (100) axes, which we call the <010> axis. This axis is consequently always found to be the "hard" (100) axis, i.e., the axis which yields the stepped loop in Figure 1(a). In our MBE system, the Fe flux is 30° away from normal incidence, and its projection in the plane makes a 15° angle with the <010> axis, and a 75° angle with the <100> axis. Thus, the origin of the uniaxial anisotropy component is the angle of incidence of the Fe atom flux with respect to the substrate. Because this effect was first observed in Fe/Pd superlattices with very thin Fe layers (<20Å), it was speculated that an anisotropy may arise because the roughness at the Fe/Pd interface (about 5Å) can have an anisotropic morphology due to the angle of incidence of the Fe atoms. Although this explanation cannot be rejected *a priori* for multilayer samples, we have subsequently found that a similar anisotropy could be induced in single, relatively thick Fe films, as shown in Fig.1(b). This sample is an 800Å-thick Fe film grown directly on MgO with a 15Å Pd cap layer, and the most important conclusion is that an anisotropy with a magnitude similar to that found in multilayer is observed. This indicates that the origin of the anisotropy resides not in an interfacial effect, but rather is a "bulk" property of the iron layer, since an anisotropy of interfacial origin would necessarily result in an inverse dependence of the anisotropy field on the layer thickness.

To investigate a structural origin to the uniaxial anisotropy, we have undertaken a detailed structural characterization of this 800Å-thick Fe layer, using a four-circles x-ray diffractometer which allowed us to probe in-plane as well as out-of-plane average lattice spacings in our film.[3] Consistent with the 4% lattice mismatch between bcc Fe ($\sqrt{2}a_0 = 4.06$ Å) and fcc MgO ($a_0 = 4.20$ Å), we find, by measuring the position of the (002) Fe peak, an average contraction of the Fe lattice in the growth direction of 0.56%. The width of this peaks also yields the crystallographic coherence length of our epitaxial film along the growth direction (≈ 400Å). A measurement of the position of off-axis diffraction peaks (-22-2) and (22-2) yield the average lattice spacing in the plane ($a_0 = 2.8498$Å), which is therefore expanded by 0.53%. This expansion in-plane and contraction out-of-plane are consistent with each other using classical Poisson's ratio analysis.

The more surprising fact, however, comes from the analysis of the (013) and (103) peaks, which yield the relative magnitude of the two in-plane lattice parameters a and b. As shown in Fig.2, the peaks corresponding to these two planes are not found at exactly the same value of 2θ, which means that there is a difference between the two in-plane lattice parameters a and b. From the shift in peak position, this difference is calculated to be $(a-b)/a = 1.2 \pm 0.3 \times 10^{-3}$, i.e., about 0.1%. It is well known that a tetragonal distortion of a cubic lattice will lead to a uniaxial magnetic anisotropy whose amplitude and direction is determined by the magnetoelastic constants of the material. This calculation,[3] using the bulk magnetoelastic constants of Fe at 300K yields an expected uniaxial anisotropy field of $H_a = 25\pm8$ Oe, in excellent agreement with our measured value of the anisotropy field at 300K of 18 ± 5 Oe.

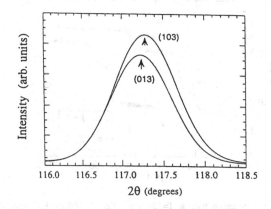

Figure 2: Fitted position of diffraction peaks corresponding to the (103) and (013) planes for a 800Å-thick Fe layer deposited on (100)MgO.

Although it is not possible to perform a detailed characterization of the Fe structure in the multilayer case (because of the superposition of Fe and Pd contributions) the similarity between the hysteresis loops for single and multilayer sample is a clear indication that the origin of the uniaxial anisotropy term is the same in both cases. Indeed, we have not observed any systematic variation in the magnitude of the uniaxial anisotropy with Fe layer thickness.

The possibility of inducing a uniaxial anisotropy in epitaxial Fe films has several important practical purposes. Most multi-source MBE growth environments are similar in that the the atom beam has a significant angle of incidence with the substrate, and therefore the effect described above is likely to occur, to varying degrees, in many growth configurations. Although rotation of the substrate is an obvious and easy remedy, it is not always performed, usually because of the need to monitor the crystalline structure during growth with *in situ* probes such as RHEED, or because a wedge-shaped magnetic layer is being grown for thickness-dependent studies. In those cases, the possible occurrence of growth-induced uniaxial anisotropy must be checked when studying the magnetic properties of epitaxial films and superlattices with weak intrinsic anisotropies. In particular, the "stepped" loop of Figure 1 is remarkably similar to those obtained in the case of weak antiferromagnetic indirect exchange coupling between magnetic layers, making the investigation of true coupling effects at these low fields more difficult.

APPLICATION OF UNIAXIAL ANISOTROPY IN EPITAXIAL LAYERS

Nonetheless, the ability to induce controlled magnetic anisotropies during a process is a potentially useful tool to fabricate magnetic superlattice structures with complex hysteretic behavior, which may be used for devices such as multi-state magnetic memories. Fig.3, for example, shows the hysteresis loop obtained for a structure containing 2 Fe layers, where the substrate has been rotated 90° between deposition of the two layers. The axis of uniaxial anisotropy for the two Fe layers are now 90° from one another, which means that the hysteresis loop for the whole structure is a combination of the two loops in Fig.1(a). In that case, there is a region at low applied field where the two Fe magnetizations are directed 90° from one another. As shown in Fig.3, this structure has, therefore, four distinct magnetic states.

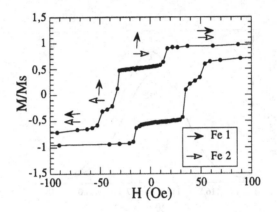

Figure 3: Magnetic hysteresis loop at T=15K for a $Fe_{(20Å)}/Pd_{(150Å)}/Fe_{(20Å)}/Pd_{(50Å)}$ multilayer. The two Fe layers have different uniaxial anisotropies, resulting in a 90° orientation at low fields.

Another example where a uniaxial anisotropy in an epitaxial film might be useful is a magnetic field detector based on the planar Hall effect, which can detect a 90° rotation of the magnetization using the anisotropic magnetoresistance effect, by measuring the transverse voltage in a planar Hall geometry (i.e., the magnetic field is applied in-plane). Such a device[4] requires a magnetic film whose uniaxial easy axis is directed in the direction of the current. The transverse voltage is then used to detect a rotation of the magnetization resulting from a small applied transverse field. To maximize the signal, one must use a magnetic material in which the anisotropic magnetoresistance is large, such as $Ni_{80}Fe_{20}$ permalloy. Fig.4 shows stepped hysteresis loops obtained for a Fe/Pd superlattice, both before and after a subsequent deposition of a 60Å epitaxial permalloy layer. The superlattice is deposited as described previously on a fixed substrate, while the permalloy film is deposited while rotating the substrate, to avoid any additional anisotropic contributions. It is evident that only the magnitude of the total magnetization is changed after deposition of the permalloy film, while the hysteresis loop maintains its stepped shape. This means that the uniaxial anisotropy has been transferred to the NiFe layer by exchange coupling to the Fe/Pd superlattice. Therefore, a multilayer structure can be constructed where the various magnetic layers play their own separate roles, namely, the Fe layers carry the uniaxial anisotropy which is determined by the deposition conditions, while the permalloy layers are responsible for the large signal obtained in a anisotropic magnetoresistance experiment. It is worthwhile noting that epitaxial permalloy has been shown[5] to have conduction properties equivalent to bulk NiFe, contrary to ultrathin polycrystalline permalloy in which the resistivity increases significantly with decreasing layer thickness below 100Å.

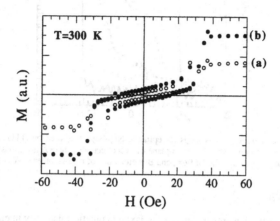

Figure 4: Magnetic hysteresis loops along the <010> direction (see text) for a $Fe_{(20Å)}/Pd_{(5Å)}$ structure (a) before and (b) after deposition of an additional isotropic 60Å-thick $Ni_{80}Fe_{20}$ layer. The uniaxial anisotropy in the Fe layers has been transferred to the NiFe layer by exchange coupling. The change in the magnetization is consistent with the bulk magnetic moment of $Ni_{80}Fe_{20}$.

UNIAXIAL ANISOTROPY INDUCED ON (111)Si SUBSTRATES

The experiment described by Figure 3 has already established that the MgO/Fe interface is not a factor in determining the occurrence of uniaxial anisotropy, since the direction of anisotropy could be changed in subsequent layers by rotating to substrate to a new position. Likewise it is important to consider whether this effect can also be observed in a totally different system, such as $Ni_{80}Fe_{20}$ (permalloy) deposited epitaxially on (111)Si. In that case, the permalloy film is grown from two *separate* atom beams of Ni and Fe, at a substrate temperature of 50°C, and it is the direction of the Ni flux (presumably because of the higher Ni concentration) that determines the orientation of the uniaxial anisotropy. Also, 40Å-thick buffer layer of (111)Ag is first grown on the Si, to prevent the formation of silicide compounds at the interface. In the plane of a (111) substrate, there are 3 equivalent [1-10] directions (referred to as A, B and C) which would normally be the easy axes. In our experiment, the in-plane projection of the Ni flux is first directed midway between two of these three axes (namely B and C). The resulting hysteresis loops obtained with the field applied along axes A and B are shown in Fig.5 (the hysteresis loop obtained along C is identical to that of B). The result is that axis A is now an easy direction of magnetization, while axes B and C are comparatively harder. Similarly to what was described earlier, this uniaxial anisotropy can be eliminated by rotating the substrate during growth. Note that the coercive field in this example is significantly higher to that expected for NiFe, due to relatively poor structure of the Ag buffer layer. Similar results with much thinner Ag buffers (6 Å) and consequently much narrower loops ($H_c \approx 3Oe$) will be presented in a forthcoming paper. In any case, the general conclusion is the same as in the case of Fe on MgO, namely that the axis (or axes) closest to the in-plane projection of the flux direction are comparatively harder magnetically. X-ray diffraction analysis is now underway to determine whether a structural origin to this uniaxial anisotropy can be found for the NiFe layer, by analogy to the results for Fe described above.

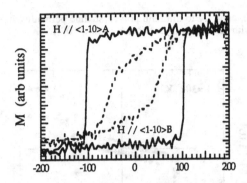

Figure 5: Magnetic hysteresis loops for epitaxial $Ni_{80}Fe_{20}$ deposited on (111)Si, at T=300K. The field is applied along two different in-plane <1-10> directions, where A is the axis perpendicular to the in-plane projection of the Ni flux , and B is either one of the two other <1-10> axes.

CONCLUSIONS

We have observed a process-induced uniaxial magnetic anisotropy in epitaxial Fe and NiFe thin films deposited by MBE onto (100)MgO and (111)Si substrates, respectively. This relatively weak uniaxial anisotropy is found to be determined uniquely by the geometrical relationship between the crystalline axes of the substrate and the direction of the incident magnetic atom flux. In both cases, the normally-easy axis closest to the projection of the flux in the plane becomes relatively "harder" magnetically. In the case of Fe on (100)MgO, a detailed x-ray structural analysis has revealed a 0.1% tetragonal distortion of the in-plane cubic mesh, with a contraction in the direction closest to the flux direction in the plane. This distortion fully explains the magnitude of the observed uniaxial anisotropy ($H_a \approx 20$ Oe). However, the mechanism that relates the incidence of the atom flux to the in-plane distortion has not been determined. One may speculate that the mobility of the atom immediately after adsorption may be greater along the incident direction, but the consequence on the in-plane lattice parameters is still unclear. An anisotropy in the structure of defects in the volume layers, which would be determined or amplified by the atom flux directions, has not been observed to this date. However, because the direction of the anisotropy can be modified <u>during</u> growth, it is clear that defects at the substrate/film interface are not the determining factor.

REFERENCES

[1] J.R. Childress, R. Kergoat, O. Durand, J.-M. George, P. Galtier, J. Miltat and A. Schuhl, J. Mag. Mag. Mat. **130**, 13 (1994).
[2] Y. Park, E.E. Fullerton and S.D. Bader, Appl. Phys. Lett., in press, (1995).
[3] O. Durand, J.R. Childress, P. Galtier, R. Bisaro and A. Schuhl, J. Magn. Magn. Mat. **145**, 111 (1995).
[4] A. Schuhl, F. Nguyen Van Dau and J.R. Childress, Appl. Phys. Lett., in press (1995)
[5] A. Schuhl, O. Durand, P. Galtier, J.R. Childress and R. Kergoat, Appl. Phys. Lett. **65**, 913 (1994).

SURFACE ANISOTROPY IN EPITAXIAL FE(110)/MO(110) MULTILAYERS

R.M. OSGOOD III, R.L. WHITE and B.M. CLEMENS
Department of Materials Science and Engineering, Stanford University, Stanford, CA 94305-2205

ABSTRACT

We have prepared epitaxial Fe(110)/Mo(110) multilayers by sputter deposition. These films exhibit a large uniaxial anisotropy and may be attractive as islanded in-plane recording media. The large uniaxial anisotropy is due to the intrinsic surface anisotropy of the Fe(110)/Mo(110) interface, which is of the same magnitude as the surface anistropy of the Fe(110)/W(110) interface but has a different sign (the surface anisotropy of the Fe(110)/Mo(110) interface prefers the [001] axis of magnetization). The magnetoelastic component of the anisotropy is not large. A novel magneto-optic technique was used to measure the transverse component of the magnetization and deduce information about the anisotropy and domain structure of the multilayers.

INTRODUCTION

Interest in single crystal materials with a large in-plane anisotropy has been spurred recently by attempts to pattern islands in single crystals for in-plane recording media. Each island would represent one bit, and it is therefore essential to have a strong uniaxial anisotropy that will pin the magnetization along a certain axis[1]. An obvious candidate for uniaxial anisotropy is the [110] surface with its two-fold symmetry. However, the [110] surface also has a large biaxial term in the anisotropy[2] which must be overcome with other uniaxial anisotropy mechanisms, such as surface or magnetoelastic anisotropy. Multilayers are a logical choice for enhancing the uniaxial anisotropy because of their propensity to have high surface and magnetoelastic energies[3]. We prepared Fe/Mo multilayers in an attempt to increase the uniaxial component of the anisotropy, studied their mechanical properties, and measured their magnetic anisotropy with torque magnetometry and the Magneto-Optic Kerr Effect (MOKE).

EXPERIMENTAL

Samples were synthesized by DC magnetron sputtering in a custom-built UHV chamber with a base pressure of 3.0×10^{-9} torr. Single crystal ($11\bar{2}0$)-oriented Al_2O_3 substrates were cleaned with solvents and heated above 650°C for deposition of a 600 ml Mo (110) underlayer. Previous work has shown that Mo grows epitaxially on the Al_2O_3 with the [$\bar{1}11$] direction parallel to the [0001] axis of the Al_2O_3[4,5]. The Fe/Mo multilayer was deposited at room temperature by alternately opening and closing shutters in front of the sputtering guns. The first layer deposited on the Mo underlayer was Fe and the top layer was Mo. All samples were capped by at least 50Å of Mo. Thicknesses were determined by calibration runs and rate monitors.

A total of eight multilayers were prepared. All multilayers had approximately 500 monolayers ('ml') each of Fe(110) and Mo(110) with bilayer periods ranging from 15 ml Fe plus 15 ml Mo (51Å) to 100 ml Fe plus 100 ml Mo (426Å). The epitaxies of both the Mo underlayer and the multilayer were determined by x-ray diffraction using laboratory sources and verified at the Stanford Synchrotron Radiation Laboratory (SSRL).

The magnetic properties of the samples were measured *ex-situ* with the Magneto-Optic Kerr Effect and a Digital Measurement Systems torque magnetometer/vibrating sample magnetometer (VSM). In the MOKE experiment, the magnetic field H was applied close to parallel to the in-plane $[1\bar{1}0]$ direction of the sample, which was the sample's hard axis. A He-Ne laser beam (633 nm) polarized 45° to the plane of incidence was incident on the sample at an 18° angle of incidence (angle from the sample normal). A lock-in measurement technique utilizing a photoelastic modulator (PEM) also oriented 45° to the plane of incidence of the light was used to measure the component of the signal oscillating at twice the reference frequency[6]. The signal was detected at a photodiode, whose output voltage was measured by a DC multimeter and read by a lock-in detector. Rotation of the optics by 45° allowed us to measure the transverse component of the magnetization (the component of the magnetization perpendicular to the plane of incidence of the laser beam and, in our experiment, also perpendicular to H) because it effectively rotated the sample by -45°. This technique measures the real part of the quantity $(r_{pp} + r_{ss})/(r_{pp} - r_{ss} - 2r_{ps})$, which is first order in the transverse component of the magnetization (r_{pp}, r_{ss} and r_{ps} are reflection coefficients of the sample that give the amplitudes of respectively reflecting 'p' polarized light into 'p' polarized light, 's' polarized light into 's' polarized light, and 'p' polarized light into 's' polarized light)[7]. The real part of $(r_{pp} + r_{ss})/(r_{pp} - r_{ss} - 2r_{ps})$ will henceforth be designated '$\theta_k(45°)$'. There is also a small term proportional to the longitudinal component of the magnetization present in $\theta_k(45°)$ which gives a non-zero signal at high fields, even after the film has been saturated[7].

RESULTS

Mechanical Properties

Rocking curves were typically 2° wide or less, and there was no twinning about the in-plane $[\bar{1}11]$ direction, as can occur if the deposition temperature is not high enough[7]. Grazing incidence and asymmetric x-ray diffraction were used to observe three $\{121\}$ type reflections for these BCC structures. These three reflections are due to planes which are crystallographically identical, but oriented at different angles to the sample surface and in-plane axes, and therefore experience different strains.

In order to determine the strains (from which the magnetoelastic component of the magnetic anisotropy could be calculated), standard elasticity analysis[8] was used to calculate the d-spacing from an assumed stress state and unstrained lattice parameter a_0. The stress was assumed to be biaxial, with principal stresses applied along the in-plane [001] and $[1\bar{1}0]$ directions. This yielded expressions for each of the observed d-spacings as a function of the three parameters σ_{11}, σ_{22}, and a_0. Inverting these expressions determined these parameters from the x-ray data [5].

The strain (calculated from the stress) and the unstrained lattice parameter of the Fe films are plotted in Figs. 1a - b. The stress in the Fe film was roughly equal in both in-plane directions, but because the $[1\bar{1}0]$ direction is stiffer than the [001] direction, the resulting strains were higher in the [001] direction. This was consistent with our previous *in-situ* observation of anisotropic strain relaxation in an Fe(110)/Mo(110) bilayer except that in the multilayers the strain is *higher* in the thicker Fe films. Our previous study of Fe(110)/Mo(110) multilayers also reported a larger stress (and therefore strain) at smaller bilayer periods, although in that case, the substrate exerted a larger stress in the $[1\bar{1}0]$ direction than in the [001] direction. This was not observed in the current study.

The lower strain at smaller bilayer periods observed in the multilayers could be explainable by interdiffusion between the Fe and Mo multilayer constituents, which was observed in our earlier study[5]. If the thickness of the Fe film is less than the critical thickness for dislocation formation[9], the strain in the Fe film will equal the misfit strain, which decreases as the amount of alloying increases. The increase in unstrained lattice parameter of the Fe film with decreasing bilayer thickness suggests that there is interdiffusion which would produce a smaller strain in the thinner Fe films *if* the thickness of the Fe films is less than the critical thickness for dislocation formation.

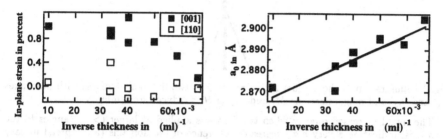

Figs. 1a and 1b. Strain and unstrained lattice parameter of the Fe film as a function of its inverse thickness. The line in Fig. 1b is a guide to the eye.

Magnetic Properties

The magnetic anisotropy was determined with torque magnetometry in an applied field (H) of 12.5 kOe. All the Fe films showed more than a 10:1 ratio of second to fourth order anisotropies. We attempted to use the Miyajima method[10] of plotting the ratio of torque (L) at 45% to H over H squared against L to find the moment and the uniaxial anisotropy component of the film, but found that our extrapolated moment disagreed with the moment obtained from VSM by up to 20%. Our plot of $(L/H)^2$ vs. L indicated that the samples were very close to saturation at $H = 12.5$ kOe. We therefore assumed that the anisotropy energy density was given by the maximum torque divided by the volume of the Fe film. Given that our film volumes are known within only 15%, it is especially useful to calculate the 'anisotropy field' ($\frac{2K_u}{M_s} = \frac{2\times\text{maximum torque}}{\text{moment}}$, where M_s is the saturation magnetization) because the volume cancels out. The error in the film volume was the largest error in the analysis of the anisotropies and is responsible for the error bars in Fig. 4.

The $\theta_k(45°)$ signal was measured in all samples with H close to parallel to the hard axis. The amplitude of the $\theta_k(45°)$ signal reaches a minimum when H is applied parallel to the *average* hard axis of the film; i.e., when one-half of the domains have their hard axes at a positive angle to H and the other half have their hard axes at a negative angle to H. The best way to illustrate this is with an example. Consider the case of two domains with hard axes lying a few degrees to either side of H, which is large enough to saturate the film (see Fig. 2). As the applied field is reversed, the magnetizations of the domains rotate in different directions toward their respective easy axes (indicated by the dashed lines in the figure). Since H is applied exactly between the two hard axes, for every value of the field H, the angles between M_1 and H and between M_2 and H will sum to zero, so that the net transverse magnetization is zero (we assume that M_1 and M_2 have the same absolute value). Although the angular resolution of our sample holder (0.5°) was not high enough to find the orientation at which the transverse component of the magnetization was exactly zero, the amplitude of the $\theta_k(45°)$ signal did reach a minimum and was nearly circular (thus indicating continuous rotation of the magnetization) when H was applied along the hard axis of the sample.

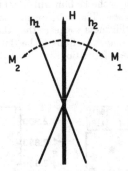

Fig. 2. Illustration of the magnetization rotation in two different domains with hard axes 'h_1' and 'h_2' and magnetizations M_1 and M_2 as H is reduced from infinity.

The signal increased quickly when the field was applied at larger ($> 1°$) angles to the hard axis, indicating a very high degree of coherence of the domains (a $2°$ spread in easy axes). This agreed with the mosaic spread of the crystallites determined by x-ray rocking curves, which indicated that the structural coherence of the films resulted in a coherent domain structure.

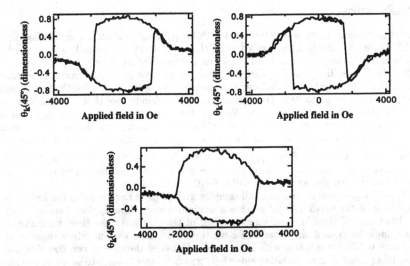

Fig. 3. $\theta_k(45°)$ measured at three different angles to the hard axis of a [30 Fe/30 Mo]×17 multilayer. The upper left-hand transverse magnetization loop was measured with H applied at an angle of $1°$ from the hard axis; the upper right-hand loop with H applied at an angle of $-1°$ from the hard axis, and the lower figure with H within $0.5°$ of the hard axis.

Typical measurements of the transverse magnetization component in a Fe(110)/Mo(110) multilayer are displayed in Fig. 3. Note that when H is applied only $1°$ or so away from the hard axis, there are sharp transitions in the transverse magnetization curve, indicating sudden changes in the magnetization orientation and/or domain wall motion. These transitions correspond to transitions in the signal from the longitudinal magnetization, and are absent when H is closest to parallel to the hard axis (the lower figure).

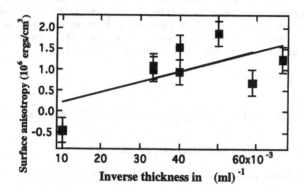

Fig. 4. Resulting surface anisotropy of the Fe film. Error bars are primarily from estimation of the Fe film volume. The straight line is a fit to the data with a slope of 0.5 ergs/cm².

The measurement of $\theta_k(45°)$ was also used to determine the total magnetic anisotropy. We found the orientation of the sample at which the $\theta_k(45°)$ signal had a minimum amplitude and was most circular. The anisotropy field was given by the field at which the circular $\theta_k(45°)$ signal closed on itself; for example, the sample displayed in Fig. 3 had an anisotropy field of approximately 2.2 kOe. The anisotropy field found in this manner agreed with the results of torque magnetometry to within only 30%. This discrepancy might be due to: (1) variation in the anisotropy throughout the multilayer (the MOKE signal is sensitive to only the top few hundred Å of the film[11]) (2) antiferromagnetic coupling in the films with small bilayer spacings (the 12.5 kOe field applied during the torque experiment should saturate the antiferromagnetically coupled layers[12]). Since both of these effects will affect the MOKE hysteresis loop more than the torque results, we used the torque results to determine the total magnetic anisotropy.

Using standard elasticity theory, the strains were calculated from the stresses in the Fe film and the resulting strain anisotropy was deduced (see Osgood et al.[5] for details). For the uniaxial term of the [110] surface, the procedure was straightforward[2]: subtract the bulk crystalline and strain anisotropies from the total measured anisotropy to give the surface anisotropy. Because the strains were small, the strain anisotropy was small and the resulting $1/t$ dependence of the total anisotropy on thickness was attributed primarily to a surface anisotropy (see Fig. 4).

It is useful to compare our results with those of Elmers et al. for a single Fe (110) film evaporated onto a W (110) single crystal at room temperature[2]. The surface anisotropy of the Fe constituent of a Fe(110)/Mo(110) multilayer and a single Fe (110) film on W (110) had the same magnitude (0.5 ergs/cm²) but different signs, indicating a preference for [001] magnetization in the multilayers and [1$\bar{1}$0] magnetization in the single Fe layer.

CONCLUSIONS

We have shown that the surface anisotropy of a Fe(110)/Mo(110) multilayer prefers the [001] direction of magnetization, in agreement with earlier data[5] and in contrast with

Fe(110)/W(110), where the surface anisotropy prefers the $[1\bar{1}0]$ direction of magnetization[2]. This conclusion is based on the fact that the magnetic anisotropy is large while the magnetoelastic component derived from our experimentally measured strains is small. The surface anisotropy is therefore equal in magnitude to the surface anisotropy of the Fe(110)/W(110) interface, but opposite in sign. Whether this is due to the different electronic structures of W and Mo, different morphologies at the Fe(110)/W(110) and Fe(110)/Mo(110) interfaces, or both, still needs to be determined. In addition, we have demonstrated the use of measuring the transverse component of the magnetization with MOKE to determine useful information about the anisotropy and domain structure of the multilayers.

REFERENCES

1. R. M. H. New, F. Pease and R. L. White, Journal of Vac. Sci. and Tech., **12**, 3196 (1994).

2. H. J. Elmers and U. Gradmann, App. Phys. A., **51**, 252 (1990).

3. B. N. Engel, C. D. England, R. A. Van Leeuwen, M. H. Wiedmann, and C. M. Falco, **67**, 1910 (1991).

4. B. M. Clemens, R. M. Osgood, A. P. Payne, B. M. Lairson, S. Brennan, R. L. White, and W. D. Nix, J. Mag. Magnetic Mats., **121**, 37 (1993).

5. R. M. Osgood III, B. M. Clemens, and R. L. White. *Mechanisms of Thin Film Evolution.* (Mats. Res. Soc. Proc. **317**, Pittsburgh, PA, 1994), edited by S. M. Yalisove, C. V. Thompson, and D. J. Eaglesham.

6. J. Badoz, M. Billardon, J. C. Canit, and M. F. Russel, J. Optics (Paris), **8**, 373 (1977).

7. R.M. Osgood III. PhD thesis, Stanford University, 1995.

8. J.A. Bain. PhD thesis, Stanford University, 1993.

9. W. D. Nix, Metall. Trans. A, 20, 2217 (1989).

10. H. Miyajima, K. Sato, and T. Mizoguchi, Journal of App. Phys., **47**, 4669 (1976).

11. E. R. Moog, J. Zak, M. L. Huberman, and S. D. Bader, Phys. Rev. B, **39**, 9496 (1989).

12. M. E. Brubaker, J. E. Mattson, C. H. Sowers, and S. D. Bader, Appl. Phys. Letts., **58**, 2306 (1991).

DIFFERENT TEMPERATURE DEPENDENCIES OF MAGNETIC INTERFACE AND VOLUME ANISOTROPIES IN Gd / W(110)

M. Farle, B. Schulz, A. Aspelmeier, G. Andre, and K. Baberschke

Institut für Experimentalphysik, FU Berlin, Arnimallee 14, D-14195 Berlin, Germany

Abstract

The magnetic anisotropy of epitaxial Gd(0001) films on W(110) is determined as a function of temperature (150 to 350 K) and film thickness (9 to 30 monolayers) by in situ ferromagnetic resonance. It is found that the usual analysis in terms of a thickness independent part K^V and a thickness dependent contribution $2K^S/d$ must be performed at the same reduced temperature $t = T/T_c(d)$. K^V shows qualitatively the same temperature dependence as the magnetocrystalline anisotropy of bulk Gd. It changes in sign near $0.7\ T_c$ and does not vanish at T_c. K^S on the other hand decreases linearly from 1.2 meV/atom at $0.6 \cdot T_c$ to zero at T_c. It appears that the intrinsic origin for K^V and K^S is fundamentally different. The vanishing of K^S at T_c indicates that two-ion anisotropy (spin-spin interaction) is dominating the interface anisotropy. The non- zero $K^V(T \geq T_c)$ is likely due to a single ion magnetic anisotropy which is known for bulk Gd.

Introduction

The orientation of the magnetization in magnetic ultrathin films and multilayers is determined by the magnetic anisotropy. Phenomenologically, the total anisotropy K_u has been found to show a 1/d dependence on magnetic layer thickness d [1-5].:

$$K_u = K^V + 2K^S/d \qquad (1)$$

Here, K^S is considered as an interface contribution, and K^V is a thickness independent volume coefficient composed of bulk magnetocrystalline anisotropy and a thickness independent magnetoelastic contribution arising from residual strain in the film [6]. The effective 1/d dependence of K_u has been associated with Néel's surface anisotropy [1-3] due to the broken symmetry at the interfaces and a thickness dependent relaxation of misfit strain [7,8]. Except for $2\pi M^2$, which favors in-plane magnetization, $2K^S/d$ and K^V may favor either in-plane or out-of-plane orientation of the magnetization.

K^V and K^S are temperature dependent [3,9]. Interestingly, this fact has often been ignored in discussions of thin film anisotropies [10]. A measurement of $K^V(T)$ may be a good identification of the existence of an undisturbed magnetocrystalline volume anisotropy. Also the temperature dependence of the Néel surface anisotropy is unknown. Aside from the phenomenological approach of Eq.(1) the temperature dependence may give new insights on the intrinsic origin of the different coefficients of magnetic anisotropy. Two mechanisms based on spin-orbit interaction have been discussed in the bulk literature [11] to account for the different anisotropic behavior of ferromagnets: the single ion anisotropy and the two-ion model. In the two-ion or pair model the exchange interaction between neighbouring local magnetic moments causes differences in the free energy if the paired moments are aligned along different crystallographic directions. In the single ion model the anisotropy arises from the interaction of the local spin moment with its own non-spherical orbital momentum distribution. Consequently one expects, that a magnetic anisotropy caused by a single ion mechanism may persist above the Curie temperature while a two ion anisotropy should vanish at T_c because exchange is compensated by thermal energy.

The magnetic anisotropy of bulk Gadolinium has been explained in terms of the single ion model in the vicinity of the Curie temperature [12]. It is known to remain finite at T_c and to vanish only far above T_c. One may ask if the interface anisotropy behaves in the same way or if due to the broken symmetry a different mechanism is more important. In the latter case a different temperature dependence of K^S and K^V is expected. In addition, the magnetic behavior of Gd(0001) thin films has shown many surprising features which have been discussed in relation to anisotropies in the films. [13-15].. The only anisotropy values available for Gd films have been obtained near T_c for films grown at 450°C (3 -80 Å) [16], which is known to yield rough films with island formation. Only very recently, magnetic anisotropy data for flat Gd(0001) films on W(110) have been reported by us [9]. Here, we extend our analysis and propose that the thickness independent coefficient K^V is caused by single-ion anisotropy and the thickness dependent term K^S/d is dominated by two-ion anisotropy.

9 to 30 layers of Gd(0001) were grown on W(110) at room temperature as described in detail earlier [9,17]. After deposition our films were annealed in order to make them magnetically

homogeneous [18]. All films were subsequently measured in situ by ferromagnetic resonance at 9.3 GHz. The Curie temperature was independently determined by in situ ac susceptibility measurements. Details of the experiment have been reported previously [9,19,20].

Analysis and Results

The general FMR resonance condition for our system with H applied in the film plane is [9]:

$$\left(\frac{\omega}{\gamma}\right)^2 = H_{R||}\left(H_{R||} + 4\pi(N_\perp - N_{||})M(H_{R||}, T) - 2K_u(T)/M(H_{R||}, T)\right) \tag{2}$$

We neglect a small threefold in-plane anisotropic contribution [21]. The magnetization $M(H_R, T)$ as a function of temperature is determined from the resonance intensity [9], which is known to be proportional to the total magnetic moment of the sample.

Our analysis which in detail is given in [9] yields a strongly enhanced uniaxial anisotropy K_u in comparison to bulk Gd over the full temperature range. It is positive and favours an out-of-plane orientation of the magnetization. But the shape anisotropy still dominates in the 9 to 30 ML range investigated here and forces the easy axis of the magnetization to lie in the film plane. Near Tc K_u becomes very small ($\cong 0.01$meV/atom), but remains finite for all layers even above T_c. This behaviour is known also in bulk Gd. One should note that $M(H_R)$ is not zero at T_c and $2K_u/M(H_R)$ in Eq.(2) does not diverge.

It has been shown in the case of Ni(111) on Re(0001) [3] and of Gd(0001) on W(110) [9] that $K_u(d)$ has to be analyzed at constant reduced temperature $t=T/T_c(d)$. This is expected if the anisotropies predominantly scale with the magnetization and the magnetostriction. One might haved argued for a scaling with the absolute temperature T, if the anisotropy is assumed to be due to elastic strain which is independent on magnetic ordering. Experimentally, however, it is shown that for the above systems the anisotropies scale with $t=T/T_c(d)$ [3,9].

In Fig. 1 K_u is plotted as a function of $1/d$. A linear dependence is observed for all T/T_c and K^V and K^S are determined according to Eq.(1). The temperature dependence of $K^V(t)$ and $K^S(t)$ and the magnetocrystalline anisotropy $K_b^V = k_2 + 2k_4$ of bulk Gd [12] is shown in Fig.2. $K^V(t)$ has the same order of magnitude and changes sign at the same temperature as in the bulk. This result implies that K^V in Eq.(1) does represent the magnetocrystalline anisotropy of bulk Gd. It is modified by a small **thickness independent** magnetoelastic contribution

217

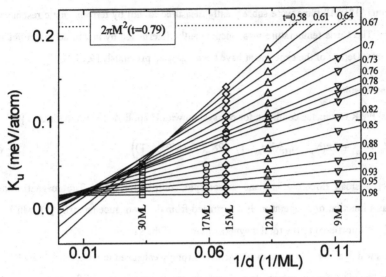

Fig. 1: Total anisotropy K_u as a function of reciprocal film thickness for several reduced temperatures $t=T/T_c(d)$. The demagnetization energy at $t=0.79$ is indicated.

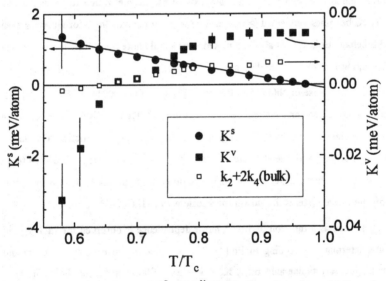

Fig. 2: Temperature dependence of K^s and K^v (Eq.1). Error bars for $t < 0.7$ are larger since not all thicknesses were measured at these temperatures. The magneto-crystalline anisotropy k_2+2k_4 of bulk Gd [12] is plotted for comparison.

$K_{ME}^V = K^V - K_b^V \approx 1.5 \ K_b^V \approx 0.01 \text{meV/atom}$ for t>0.8. The latter also remains finite at T_c indicating the single-ion origin as well. As discussed in [9] this corresponds to an average thickness independent strain of 0.8% for 9 to 30 ML. Note that in the case of Ni(001)/Cu(001) K_{ME}^V is by two to three orders of magnitude larger than K_b^V [4]. The growth of Gd/W(110) is *not* pseudomorphic above 1 ML. A thickness dependent release of strain which leads to an effective 1/d dependence is expected to be small (0.01 meV/atom) for films thicker than 9 ML, because the accomodation of misfit dislocation should be completed at that thickness. K^S is identified in good approximation with the Néel surface anisotropy [9]. $K^S(t)$ in Fig. 2 decreases linearly with increasing temperature and vanishes at T_c. Presently, there exists no theory on the temperature dependence of K^S. In the experimental result $K^S(T/T_c) = k_0 + t_0 T/T_c$ with $k_0 = 2.79 \pm 0.01$ meV/atom and $t_0 = -0.66$ meV/atom k_0 represents the Néel surface anisotropy at T=0 K. It is interesting to note that Néel's model implies an exchange between nearest neighbours or in other words a pair model of anisotropy. This is in agreement with the observation that K^S vanishes at T_c and consequently is caused by a magnetic pair anisotropy between surface atoms. The underlying bulk atoms however predominantly react to a single ion anisotropy.

Summary

A full temperature dependent analysis of bulk K^V and interface K^S anisotropies has been presented for flat Gd(0001)/W(110) films. For this system a temperature dependent analysis must be performed at constant t= $T/T_c(d)$ [3,9] and not at constant T. The temperature dependence of K^S and K^V is different. K^S vanishes at T_c while K^V remains finite. We propose that this indicates that a pair anisotropy is dominating the interface anisotropy and that the thickness independent coefficient K^V for 9 to 30 ML Gd is determined by the same single ion anisotropy which is also observed in the bulk.

This work was supported in part by the DFG Sfb 290, TPA2.

References

[1]   U. Gradmann and P. Müller, Phys. Stat. Sol. **27**, 313 (1968), U. Gradmann; J. Magn. Magn. Mater. **54-57**, 733 (1986).

[2]   P. F. Garcia, A. D. Meinhaldt, and A. Suna; Appl. Phys. Lett. **47**, 178 (1985)

[3]   R. Bergholz and U. Gradmann; J. Magn. Magn. Mat. **45**, 389 (1984)

[4]   B. Schulz, and K. Baberschke; Phys. Rev. B**50**, 13467 (1994)

[5]   R. Jungblut, M. T. Johnson, J. aan de Stegge, A. Reinders, and F. J. A. den Broeder; J. Appl. Phys. **75**, 6424 (1994)

[6]   In our definition K^V does not contain the shape anisotropy $2\pi M^2$. In the literature K^V sometimes includes $2\pi M^2$ and care must be taken when comparing K^V values.

[7]   C. Chappert, and P. Bruno; J. Appl. Phys. **64**, 5736 (1988)

[8]   B. M. Clemens, R. L. White, W. D. Nix, and J. A. Bain; Mat. Res. Soc. Symp. Proc. Vol. **231**, 459 (1991)

[9]   G. Andre, A. Aspelmeier, B. Schulz, M. Farle, and K. Baberschke; Surface Science **326**, 275 (1995)

[10]   See for example Symposium C on Magnetic Thin Films, Multilayers and Surfaces, edited by A. Fert, G. Güntherodt, B. Heinrich, E. E. Marinero, and M. Maurer, Proceedings of the E-MRS Spring 1990 Meeting, Strasbourg [J. Magn. Magn. Mat. **93** (1991)].

[11]   S. Chikazumi, Physics of Magnetism, (Robert E. Krieger Publishing Co., Malabar, 1964) p.147.

[12]   B. Coqblin, The Electronic Structure of Rare-Earth Metals and Alloys: the Magnetic Heavy Rare-Earths (Academic, London, 1977)

[13]   H. Tang, D. Weller, T. G. Walker, J. C. Scott, C. Chappert, H. Hopster, A. W. Pang, D. S. Dessau, and D. P. Pappas; Phys. Rev. Lett. **71**, 444 (1993)

[14]   A. Berger, A. W. Pang, and H. Hopster; J. Magn. Magn. Mat. **137**, L1 (1994)

[15]   R. R. Erickson, and D. L. Mills; Phys. Rev. **B 43**, 11527 (1991)

[16]   M. Farle, A. Berghaus and K. Baberschke; Phys. Rev. B**39**, 4838 (1989)

[17]   A. Aspelmeier, F. Gerhardter, and K. Baberschke; J. Magn. Magn. Mat. **132**, 22 (1994)

[18]   U. Stetter, M. Farle, K. Baberschke, and W. G. Clark; Phys. Rev. B**45**, 503 (1992)

[19]   M. Farle, K. Baberschke, U. Stetter, A. Aspelmeier, and F. Gerhardter; Phys. Rev. **47**, 11571 (1993)

[20]   U. Stetter, A. Aspelmeier, and K. Baberschke; J. Magn. Magn. Mat. **116**, 183 (1992)

[21]   W. A. Lewis, and M. Farle; J. Appl. Phys. **75**, 5604 (1994)

[22]   L. Néel; J. Phys. Rad. **15**, 225 and 376 (1954)

[23]   P. Bruno and J.P. Renard; Appl. Phys. **A49**, 499 (1989)

[24]   P. Bruno; Phys. Rev. B **39**, 865 (1989)

UNUSUAL BEHAVIOR IN THE MAGNETIC ANISOTROPY
OF ULTRA-THIN Co SANDWICHES : THE ROLE OF Au UNDERLAYERS.

CHRISTIAN MARLIÈRE†, BRAD N. ENGEL AND CHARLES M. FALCO
The Optical Sciences Center and the Department of Physics, University of Arizona, Tucson, AZ 85721.
† Permanent address : IOTA, URA 14 du CNRS, B.P. 147, 91403 Orsay Cedex, France.

ABSTRACT

We have used *in situ* polar Kerr effect measurements to study the magnetic anisotropy of X/Co/Y sandwich structures grown by MBE on Cu(111) buffers, where X and Y are variable thicknesses of Au. For fixed values of Y and in the case of an underlayer wedge, e.g. variable X value, we have found a sharp minimum in both coercive field and perpendicular anisotropy at ≈1 atomic layer of the Au underlayer. This anisotropy behavior is opposite to that of an Au overlayer deposited on a Co film, *i.e.* variable Y and fixed X.

INTRODUCTION AND BACKGROUND

In the past few years a great amount of research has been devoted to the study of the magnetic surface and interface anisotropies in ferromagnetic ultra-thin films. However the underlying fundamental mechanism remains a puzzling problem in modern magnetism.

Besides the magnetocrystalline interface anisotropy [1], possible explanations include the strain-induced magnetoelastic anisotropy [2-3] and altered electronic structure at the surfaces and interfaces [4].

The recent advent of sensitive *in situ* magnetic measurement techniques gives the researcher ways of investigating the evolving behavior of anisotropy during the growth of thin films. Thus, it has been discovered that the perpendicular anisotropy of cobalt thin films on different buffer layers such as Pd(111) [5], Au(111) [6], Ag(111) [7] or Cu(111) [8] displays a drastic increase during the deposition of a non-magnetic metallic overlayer. A pronounced peak for both coercivity and anisotropy was observed for an overlayer of about one monoatomic layer (ML). Similar non-monotonic behavior has also been observed for Cu(001)/Co/Cu films with an in-plane anisotropy [9]. As this effect was observed with a large variety of metals used for the buffer layer as well as for overlayer, this phenomenon can be more likely explained by a change in the electronic structure of the overlayer due to its restricted dimensionality, or to hybridization with cobalt, than by a strain-induced magneto-elastic anisotropy. Indeed, there have been recent reports of quantum-well-type confinement effects with Cu on Co(0001) [10] , Co on Cu(111) [11] or Au on W(111) [12]. Thus, by changing the layering sequences of different metallic layers over or under a cobalt thin film, we aim to probe the effect of confinement on the magnetic anisotropy.

We report here a series of experiments using *in situ* polar Magneto-Optical Kerr effect measurements (pMOKE) with cobalt films embedded between gold layers on a Cu(111) buffer layer, showing a pronounced decrease in the anisotropy for a Au underlayer of about 1 ML.

EXPERIMENTAL

Film Growth

The thin films were deposited at room temperature in the growth chamber of our Molecular Beam Epitaxy (MBE) machine on single-crystalline Cu(111) buffer layers epitaxially grown on Si

Mat. Res. Soc. Symp. Proc. Vol. 384 ● 1995 Materials Research Society

(111) substrates. The background pressure during deposition was $\leq 5 \times 10^{-10}$ torr. Optical–feedback–controlled e–beam evaporators were used to deposit the Au (≈ 0.1Å/s), Co (≈ 0.1Å/s) and Cu (≈ 0.15Å/s). All deposition rates were determined from Rutherford Backscattering Spectrometry (RBS) analysis of calibration films and were reproducible to within ± 10 %. Film quality and crystal structure were monitored during and after growth with Reflection High Energy Electron Diffraction (RHEED) and Low Energy Electron Diffraction (LEED), respectively. The samples were made by first evaporating a stepped-wedge gold underlayer on the Cu buffer layer, then a cobalt thin film and finally an Au overlayer. Sample rotation during the deposition of the Cu, Co and Au overlayer assured thickness variations of less than 1% across the full substrate diameter as determined by RBS. The step-wedge Au underlayer was formed by moving a shutter, located very close to the substrate, during deposition using a computer-controlled stepper-motor. For each sample, up to eight 4mm-wide steps were made with varying Au underlayer thicknesses.

The sample was then transferred from the growth chamber to the pMOKE chamber (base pressure $\leq 2 \times 10^{-10}$ torr) where it was aligned between the poles of an external electro-magnet for *in situ* Kerr effect measurements. The magnetic field was applied along the sample normal. The sample could be moved repeatedly between the measurement and the deposition chambers without need for optical realignment.

In situ pMOKE measurements.

For films below a critical Co thickness (dependent on interface material), all of our samples revealed very square polar hysteresis curves indicating a perpendicular magnetic easy-axis. It was shown previously [5] that the coercivity of these square loops is directly related to the total anisotropy energy of the ultra-thin film. Furthermore, the coercivity is very sensitive to changes at the interface. Therefore, we have tracked the coercive field during interface formation in order to get information about the evolution of the interface anisotropy. For all of these measurements, the cobalt thickness was fixed at 11Å. We have also made direct measurements of the total anisotropy energy (K_1). To accomplish this, we chose the characteristics of the sandwich structure to maintain an in-plane easy-axis of moderate anisotropy strength and thus allow the saturation of the sample's magnetization by our maximum field of ± 2.2 kOe. Then K_1 is calculated (see Fig. 1) from the relation :

$$K_1 = -H_k M_s / 2 \qquad (1)$$

where $M_s = 1422$ emu/cm^3 is the Co bulk saturation magnetization. Here we have adopted the convention used by many researchers, where a positive K_1 indicates perpendicular anisotropy.

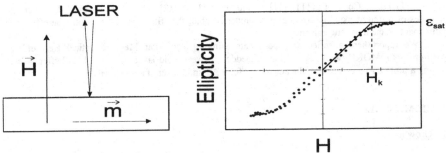

<u>Fig. 1</u> : *Schematic of the in situ anisotropy energy measurement. The magnetization **m** lies in the plane of the sample while the applied field **H** is perpendicular to the film plane.*

RESULTS AND DISCUSSION.

Using stepped-wedge growth, we varied the thickness of the Au underlayer, each step increasing by ≈ 0.5ML and deposited between the Cu(111) buffer and the cobalt film. The Co was then covered by a series of increasing Au overlayer thicknesses, with a complete set of pMOKE measurements taken along the wedge after each coverage.

Figure 2 is a plot of coercive field versus Au underlayer thickness. In contrast to our earlier finding for overlayers, the coercive field displays a pronounced minimum for 1ML Au underlayer thickness. This decrease in coercivity is surprising in view of the strong Co/Au perpendicular interface anisotropy. Indeed, the variation of coercive field versus the Au overlayer thickness (Fig. 3) reveals the previously observed maximum peaked at around 1ML and is independent of the Au underlayer thickness t_{under}.

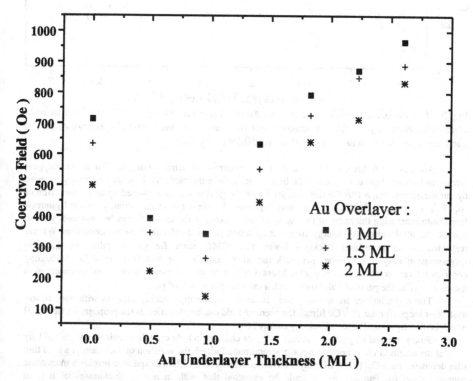

Fig. 2 : *Perpendicular coercive field versus Au underlayer thickness for a Cu(111)/Au(x)/Co/Au sandwich structure for different values of the Au overlayer thickness. Uncertainties in H_c are the size of the data points.*

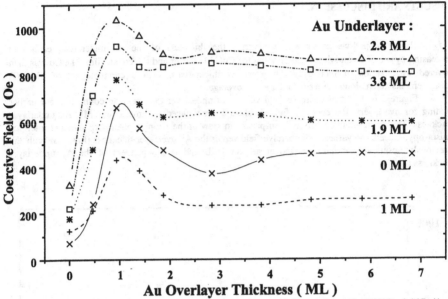

Fig. 3 : *Perpendicular coercive field versus Au overlayer thickness for a Cu(111)/Au(x)/Co/Au sandwich structure for different values of the Au underlayer thickness. For reason of clarity each curve has been successively shifted by +50 Oe along y-axis.*

We have also directly measured the total anisotropy on similar samples with an Au stepped-wedge underlayer deposited on Cu(111). In order to achieve the moderate in-plane anisotropy required by our measurements, a 15Å Co film was grown on the gold wedge and covered by an uniform film of Au (7 Å) on which a 2Å Cu top layer was deposited. Figure 4 shows the variation of the hysteresis loops with increasing thickness of the Au underlayer. Because the moments can be saturated, we can deduce the total anisotropy energy from extrapolation of the hard-axis curve to saturation. We are restricted to Au underlayer thickness lower than ≈2ML, since for greater values of t_{under}, the magnetization easy-axis becomes perpendicular to the sample. The plot (Fig. 5) of the anisotropy constant K1 versus the Au underlayer thickness displays the same non-monotonic behavior observed in the coercivity of the perpendicular films, with a minimum at ≈1ML of gold.

These similarities are not surprising because, in such high quality samples with very square hysteresis loops (for the 11Å Co films), the coercive field can be identified as the propagation field [13] which is directly related to the anisotropy energy [14].

Strain-induced magnetoelastic anisotropy or changes in bulk crystalline anisotropy are unlikely to be at the origin of our observed anisotropy minimum. Indeed, the variation of the strain in a gold thin film deposited on a Cu(111) buffer layer has been monitored by RHEED and also reveals a monotonic behavior (Fig. 6). Furthermore, it could be expected that with increasing thicknesses of the Au underlayer, the crystallographic structure of the cobalt layer changes from f.c.c. when deposited on Cu(111) to h.c.p. when on Au(111). This would cause a significant increase of the magnetocrystalline anisotropy [15] which is contrary to our observations.

More probably this sharp minimum observed in our experiments can be attributed to a variation of the confined underlayer band-structure leading to a change in the hybridization of electronic states at the cobalt/underlayer interface and causing a significant alteration of the total anisotropy energy.

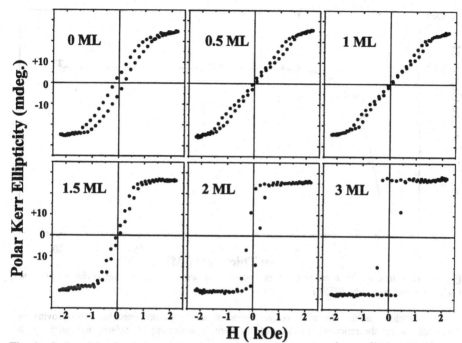

Fig. 4 : *Series of hard-axis (perpendicular) hysteresis loops of a 15Å Co film deposited on a stepped-wedge underlayer of Au(111). The thickness of the Au underlayer is written on the plot.*

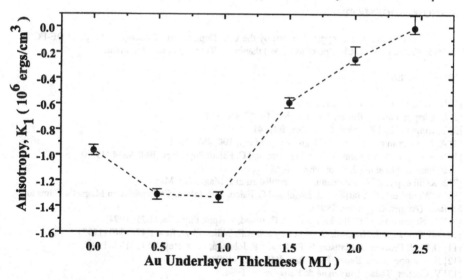

Fig. 5 : *Total anisotropy energy, K_1, versus Au underlayer thickness for a Cu(111)/Au(x)/Co/Au sandwich.*

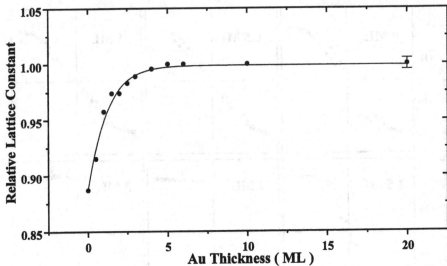

<u>Fig. 6</u> : *Variation of the average in-plane lattice parameter during the deposition of gold on Cu(111) as measured by RHEED.*

We think that the unusual coverage dependent anisotropy presented here will provide an additional test for theoretical explanations of the magnetic anisotropy at surfaces and interfaces in layered metallic systems.

ACKNOWLEDGEMENT

This research was supported in part by the U.S. Department of Energy under grant DE-FG03-93ER45488 and by C.N.R.S. One of us (C.M.) thanks NATO for an exchange grant.

REFERENCES

[1] L. Néel, J. Phys. Rad. **15**, 225 (1954).
[2] C. Chappert and P. Bruno, J. Appl. Phys. **64**, 5736 (1988).
[3] T. Kingetsu and K. Sakai, Phys. Rev. **B48**, 4140 (1993).
[4] A. J. Freeman and R. Wu, J. Magn. Magn. Mater. **100**, 497 (1991).
[5] B. Engel, M. Wiedmann, R. Van Leeuwen and C. Falco, Phys. Rev. **B48**, 9894 (1993).
[6] P. Beauvillain *et al.*, J. Appl. Phys. **76**, 6078 (1994).
[7] J. Kohlhepp and U. Gradmann, to be published in J. Mag. Mag. Mat.
[8] M. Wiedmann, C. Marlière, B. Engel and C. Falco, 14th Intern. Colloquium on Magnetic Films and Surfaces (August-September 1994).
[9] F.O. Schumann, M.E. Buckley and J.A.C. Bland, J. Appl. Phys. **76**, 6075 (1994).
[10] J.E. Ortega, F.J. Himpsel, G.J. Mankey and R.F. Willis, Phys. Rev. **B47**, 1540 (1993).
[11] D. Li, J. Pearson, J. Mattson, S. Bader and P. Johnson, Phys. Rev. **B51**, 7195 (1995).
[12] H. Knoppe and E. Bauer, Phys. Rev. **B48**, 5621 (1993).
[13] V. Grolier, Thèse, Université de Paris-Orsay, 1994.
[14] P. Bruno *et al.*, J. Appl. Phys. **68**, 5759 (1990).
[15] M. Sakurai and T. Shinjo, J. Phys. Soc. Japan **62**, 1853 (1993).

MAGNETIC STRUCTURES IN NONMAG-/MAG-/NONMAG-NETIC SANDWICHES

YOSHIYUKI KAWAZOE AND XIAO HU
Institute for Materials Research, Tohoku University, Sendai 980-77, Japan

ABSTRACT

Spin-reorientation transition in magnetic thin film sandwiched by nonmagnetic materials is clarified by means of the variational formalism. Phase diagrams of the magnetization configuration are presented. Scaling relations among the film thickness, exchange coupling and magnetic anisotropies are revealed. A formula for the interface anisotropy is presented.

INTRODUCTION

Much attention has been paid to metallic thin films in recent years, since sophisticated epitaxial techniques enable us to control the thickness, the surface condition and the interface smoothness between the film under concern and substrate, and many new phenomena have been observed. Among them it is found experimentally in nonmag-/mag-/nonmag-netic sandwich structures that, the easy-axis direction for magnetization changes according to the thickness of the transition metallic film[1]: At lower thickness, it is normal to the film plane, while at higher thickness, it lies in the film plane. A similar phenomenon has also been discussed in magnetic bilayer systems used for magneto-optical recording[2]. From the application point of view, the presence of perpendicular magnetization in thin film of transition metal is of great potential for high density recording. In sandwich and bilayer structures, a surface anisotropy normal to the film is produced by the breaking of translation invariance at the interface, as considered first by Néel[3]. In contrast to 3D systems, the magnetization in film also produces shape anisotropy, which favors in-plane ordering. In the present paper, we will assume the presence of the long-range order in the thin film under a surface anisotropy, and proceed to discuss the spin-reorientation phase transitions with the variance of the thickness.

Although the role played by the anisotropies has been explored by both experiments and theoretical arguments, the exchange stiffness has not been paid enough attention. In most of the literature, the inverse thickness dependence of the effective volume anisotropy is taken as a satisfied fitting of experimental data or as an ansatz a priori in theories. In the present study we explore the role of the exchange stiffness in the competition of the shape anisotropy and the surface anisotropy, both around the transition point and at large thickness limit.

DISCRETE MODEL AND SCALING

The discrete model of magnetic thin film sandwiched by nonmagnetic materials is shown in Fig.1 and the energy per unit area is

$$\gamma = -Jm_s^2 \sum_{i=1}^{N-1} \cos(\varphi_i - \varphi_{i+1}) - K_v' \sum_{i=2}^{N-1} \sin^2 \varphi_i + K_s'(\sin^2 \varphi_1 + \sin^2 \varphi_N). \tag{1}$$

The first term on the right-hand side of the above expression covers the exchange coupling energy between the classical spin-vectors on the nearest-neighboring layers. The remaining terms are for the in-plane shape anisotropy K_v' and the vertical interface anisotropy K_s', respectively, where the dashes denote the difference between the present quantities and those in standard experimental notations.

227

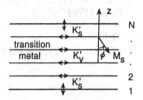

Figure 1: Discrete model[4].

The stable spin configuration is determined by minimizing the energy functional (1). In the present discrete model, no analytic results can be expected. For numerical calculation in this section, we take $\frac{1}{2}Jm_s^2 = 1$ and $\hat{a} = 1$(lattice constant).

Fixing the magnetic constants, we have found a spin-reorientation transition from the perpendicular uniform configuration to a nonuniform one, as the number of layers is increased from $N = 2$. A further transition is observed, where the nonuniform configuration is switched into the in-plane uniform one. The phase diagram of the spin configuration, shown as dashed arrows for $K_v' = 0.10$, with the number of layers and the surface anisotopy as variables is depicted in Fig.2, for two different values of volume anisotropy. The phase boundaries consist of steps, as the result of the discrete variance of the thickness, namely the number of layers. The locations of the phase boundaries depend sensitively on the value of volume anisotropy.

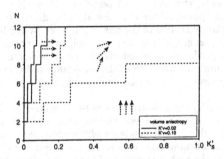

Figure 2: Phase diagram of the spin configuration with fixed volume anisotropies[4].

We have calculated the magnetic configuration in systems of $N = 4 \sim 20$ and various values of K_s' and K_v', and tried to rearrange the data into a single diagram. We then arrive at the conclusion that if one takes the variables as $(N - \Delta N)\sqrt{K_v'}$ and $K_s'/\sqrt{K_v'}$ as in Fig.3 where $\Delta N = 2.2$ is selected to obtain the best plotting, all the phase boundaries, such as those shown in Fig.2, fall into two smooth curves[4]. This fact implies the presence of the following scaling relations among the film thickness, the anisotropies and the exchange coupling in the spin-reorientation transitions: $(N - \Delta N)\sqrt{K_v'/(Jm_s^2)}$ and $K_s'/\sqrt{K_v'Jm_s^2}$.

CONTINUOUS MODEL

In this section, we introduce the continuum approach and present the essential results.

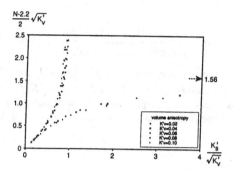

Figure 3: Scaled phase diagrams[4].

Consider a ferromagnetic layer with thickness $2a$: On the surfaces there exist perpendicular anisotropies K_s, within the film the anisotropy K_v is in the film plane, and the exchange stiffness A is ferromagnetic and finite. We assume that the magnetization is uniform in the in-plane directions. Then, half of the total energy stored in the film per unit area is expressed by

$$\gamma = \int_0^a [A(\frac{d\varphi}{dz})^2 - K_v \sin^2 \varphi] dz + K_s \sin^2 \varphi(0), \tag{2}$$

where the z-axis is taken to be normal to the film plane and the origin at the bottom surface[5,6]. The relations among the magnetic quantities in (1), the discrete model, and (2), the continuous model, are given as: $\frac{1}{2} J m_s^2 \hat{a} = A$, $K_v'/\hat{a} = K_v$ and $K_s' = K_s$.

Spin-reorientation transition

The stable spin configuration is determined by solving a variational problem to the above energy functional. It is then revealed that there exists the first critical thickness

$$a_{c1} = \sqrt{\frac{A}{K_v}} \tan^{-1} \frac{K_s}{\sqrt{AK_v}}, \tag{3}$$

so that for $a \leq a_{c1}$ the stable configuration is uniform and magnetization stands normally to the film plane. For materials satisfying $K_s < \sqrt{AK_v}$, there exists the second critical thickness

$$a_{c2} = \sqrt{\frac{A}{K_v}} \tanh^{-1} \frac{K_s}{\sqrt{AK_v}}, \tag{4}$$

so that for $a \geq a_{c2}$, the magnetization is aligned uniformly within the film plane. For $a_{c1} < a < a_{c2}$, the spin configuration is not uniform in the vertical direction. The spin direction at $z = a$ is determined by

$$\frac{K_s}{\sqrt{AK_v}} = \frac{\text{sn}[a\sqrt{K_v/A}, \sin \varphi_a] \text{dn}[a\sqrt{K_v/A}, \sin \varphi_a]}{\text{cn}[a\sqrt{K_v/A}, \sin \varphi_a]}, \tag{5}$$

where $\varphi_a \equiv \varphi(a)$, and the spin configuration is expressed by φ_a as

$$\varphi(z) = \sin^{-1}\{\sin\varphi_a \frac{\mathrm{cn}[(a-z)\sqrt{K_\mathrm{v}/A},\sin\varphi_a]}{\mathrm{dn}[(a-z)\sqrt{K_\mathrm{v}/A},\sin\varphi_a]}\}, \qquad \text{for} \quad 0 \le z \le a. \tag{6}$$

The phase diagram of magnetization configuration obtained from the above formalism is given in Figs.4 and 5. The dashed arrow in Fig.4 denotes $a\sqrt{K_\mathrm{v}/A} = \pi/2$, where the phase boundary (3) saturates. Good coincidence between Fig.3 and Fig.4 must be found, which implies the sufficiency of the continuous approach in the present problem.

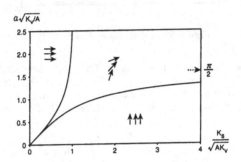

Figure 4: Phase diagram derived by continuous model.

Formula for interface anisotropy

In the large thickness limit, we have found in the case of $K_s < \sqrt{AK_\mathrm{v}}$ that the energy γ behaves asymptotically as[5]

$$\gamma = -aK_\mathrm{v} + E_s, \tag{7}$$

where $E_s = K_s$. In the case of $K_s > \sqrt{AK_\mathrm{v}}$, the above relation is established exactly for $a \ge a_{c2}$. In other words, the inverse thickness dependence of the effective anisotropy is satisfied outside the region of the spin-reorientation transitions. This fact implies insufficiency of the phenomenological argument for the spin-reorientation transition based on the inverse thickness dependence of anisotropy. It is found in the case of $K_s > \sqrt{AK_\mathrm{v}}$ that the constant E_s, which is usually measured by experiments, is different from the surface anisotropy K_s introduced in (2) and is given by[5]

$$E_s = 2\sqrt{AK_\mathrm{v}} - \frac{AK_\mathrm{v}}{K_s}. \tag{8}$$

The difference appears because of the existence of the nonuniform structure near the surface for $a > a_{c1}$ in the case of $K_s > \sqrt{AK_\mathrm{v}}$. The extrapolation formulas (7) and (8) should be important in the interpretation of experimental results and thus for the discovery of the mechanism the interface anisotropy.

As the exchange stiffness A approaches to infinity, a_{c1} and a_{c2} are reduced to a single critical value $a_c = K_s/K_\mathrm{v}$, and the phenomenological theory is deduced. The strong exchange coupling makes the spin rotation a discrete transition and the stable states uniform, denoted by $\varphi = 0$ and $\varphi = \pi/2$ for $a < a_c$ and $a > a_c$, respectively.

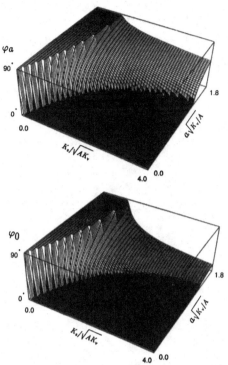

Figure 5: Spin-direction at the central part(top) and at the interface(bottom).

SUMMARY

Spin-reorientation transitions are observed as the film thickness and/or the magnetic constants are varied. Scaling relations among the relevant quantities in these transitions are derived. From the comparison of the results derived from the continuum model and the discrete one, we have found that the continuum approximation is sufficient in the study of the spin-reorientation transitions in magnetic ultrathin films. It is highly expected to perform systematic experimental investigations to verify the present theoretical predictions.

References

[1]See the review article by S.D.Bader, D.Q. Li and Z.Qui, J.Appl.Phys. **76**, 6419(1994).
[2] X.Hu and Y.Kawazoe, Phys.Rev. B **49**, 3294(1994); Y.Kawazoe, X.Hu and S.Honma, Mat.Res.Soc.Symp.Proc. **131**, 513(1993); X.Hu and Y.Kawazoe, J.Appl.Phys. **75**, 6486(1994): X.Hu et al., IEEE Trans. Magn. **29**, 3790(1993).
[3]L.Néel, J.Phys.Rad. **15**, 376(1954).
[4]X.Hu and Y.Kawazoe, submitted to Phys.Rev. B.
[5]X.Hu and Y.Kawazoe, Phys.Rev. B **51**, 311(1995).
[6]A.Thiaville and A.Fert, J.Magn.Magn.Mater. **113**, 161(1992).

INDUCED MAGNETIC ANISOTROPY OF SPUTTER NiFe THIN FILMS ON THIN TANTALUM NITRIDE UNDERLAYER

T. YEH, L. BERG, J. FALENSCHEK, J. YUE
Solid State Electronics Center, Honeywell Inc.
12001 State Highway 55, Plymouth, MN 55441, U.S.A.

ABSTRACT

The structure and properties of sputter NiFe thin film deposited on both thermal oxide and thin tantalum nitride have been studied. The magnetic anisotropy field H_K increases to 8.2 Oe when the NiFe film was deposited on a thin tantalum nitride underlayer. Anisotropic stress was found on the sample film with tantalum nitride underlayer. Results of X-ray diffraction show that a thin tantalum nitride underlayer appears to promote a preferred crystalline orientation formation of the NiFe film. The induced magnetic anisotropy is attributed to the formation of the preferred crystalline orientation and the induced anisotropic magnetoelastic energy which is associated with the anisotropic stress of the sample film.

INTRODUCTION

There is considerable interest in NiFe-based thin films because of their applicability for both anisotropic magnetoresistive (AMR) and giant magnetoresistive (GMR) magnetic sensor devices. Because of their low magnetic anisotropy field and "soft" magnetic characteristics, NiFe-based thin films are gaining wide use for both AMR and GMR sensor devices. The success of utilizing both AMR and GMR as magnetic sensors largely depends upon a better understanding and control of the magnetic anisotropy field.

Higher magnetic anisotropy field (H_K) was found when the NiFe thin films were sputter deposited on a thin tantalum nitride layer. The purpose of this paper is to investigate the effect of thin tantalum nitride underlayer on the structure and magnetic properties of the sputter thin NiFe films. The manner in which the structure and magnetic properties of sputter NiFe thin films are affected by the thin tantalum nitride underlayer is examined. Also, the structure/properties relationship of the sputter NiFe thin films is discussed.

One of the considerations of this study is that the anisotropic magnetoelastic energy contributes to the magnetic anisotropy.[1] The anisotropic magnetoelastic energy in the sample films may arise from the residual stress combined with magnetostriction of the films. Thus, the magnetic anisotropy may be varied by the induced anisotropic magetoelastic energy in the sample films. Another consideration is that it has frequently been reported that body-centered-cubic (bcc) Cr underlayers appear to promote epitaxial formation of hexagonal-closed-packed (hcp) Co-based thin films on a grain-to-grain basis, and result in inducing the hcp c-axes to distribute in the plane of the film.[2-6] The consequence of this is a marked increase in the coercivity of the films. These results demonstrate that the magnetic properties of the sputter films are sensitive to the structure of the films. When the NiFe thin films are sputter deposited on the thin tantalum nitride underlayers, the surface energy of the NiFe film nucleation and growth could be differnet from the NiFe films deposited on thermal oxide. The surface energy could play an important role for sputter thin film nucleation and growth and has significant effect on the structure and properties of the films.[7] Therefore, the structure of the NiFe thin films could be different while the films were deposited on the thin tantalum nitride.

The experiment results obtained from X-ray diffraction and stress measurement show a preferred crystalline orientation of NiFe (111) and anisotropic stress when the film was sputter deposited on a thin tantalum nitride film. The induced magnetic anisotropy may be attributed to the effect of thin tantalum nitride on the formation of the preferred crystalline orientation of the NiFe film and the induced anisotropic magnetoelastic energy.

EXPERIMENT

Samples of NiFe film with and without a thin tantalum nitride underlayer were RF sputter deposited on silicon wafers coated with thermally grow oxide. A 350Å thick NiFe film was sputter deposited on the two different underlayers and then a 75Å thick tantalum nitride layer was deposited on the NiFe film to prevent the surface of the NiFe film from oxidation. After the depositions the magnetic properties of the deposited films were characterized by a B-H loop; the sheet resistance of the films were measured by using a four-point probe, the anisotropic magnetoresistance effect of the sample films was characterized by a four point probe with 100 Oe applied magnetic field.

Higher magnetic anisotropy fields were found when the NiFe film was sputter deposited on a thin tantalum nitride underlayer. In an attempt to study the manner in which the magnetic properties of NiFe films are affected by the thin tantalum nitride underlayers, the stress of the NiFe sample films were determined by a flexus stress gauge along both easy and hard axes. The magnetostriction of the sample films was also measured by optical interferometry method. X-ray diffractometry was used to characterize the crystalline structure of the sample films. X-ray diffration patterns obtained on all the NiFe sample films show a predominantly face-centered-cubic (111) peak. The preferred crystalline orientation of the NiFe films was measured by using $\Delta\theta_{50}$ of X-ray rocking curve of the NiFe (111) diffraction peak.

RESULTS AND DISCUSSION

The averages of the magnetic properties of the NiFe films obtained from B-H looper measurement are summarized in the Table 1. The experiment results obtained show that the underlayers have a great effect on the magnetic properties of the NiFe thin films. Very low skew and dispersion measured on all the sample films is evidence of the unaxial anisotropy of the sample films. Figure 1 shows a typical hysteresis loop of the sample films measured along both easy and hard axes of the NiFe films.

Table 1 Magnetic properties of the NiFe films deposited on both thermal oxide and thin tantalum nitride underlayers

	skew (degree)	Dispersion (degree)	B_S (nW)	H_K (Oe)	$H_{C//}$ (Oe)	R (Ω)	$\Delta R/R$ (%)
thermal oxide	- 0.9	1.1	2.74	5.2	1.70	7.19	2.33
tantalum nitride	0.4	2.3	2.78	8.2	2.29	6.00	2.72

One of the significant effects due to the different underlayers is that the magnetic anisotropy field H_K of the sputter NiFe films increase from 5.2 Oe for the films deposited on thermal oxide to 8.2 Oe for the NiFe deposited on thin tantalum nitride. The sheet resistance was measured on the NiFe films to be 6.0 Ω/Ω, and 7.19 Ω/Ω for the films deposited on thin tantalum nitride and thermal oxide, respectively. The magnetoresistance effect of 2.72% and 2.33% for the NiFe films deposited on thin tantalum nitride and thermal oxide respectively were obtained. The higher magnetoresistance effect was obtained on the NiFe films with lower sheet resistance. The B_S is the nondestructive indirect measurement of the NiFe film thickness; the B_S is linearly dependent on the NiFe film thickness. The results of B_S measurement indicate the variation of thickness on all the sample films is within 2%. This implies that the effect of thickness on the magnetic properties of the NiFe films is negligible.

The coercivity of the NiFe films deposited on both thermal oxide and tantalum nitride were measured to be 1.7 and 2.29 Oe, respectively. The sample film with higher anisotropy field H_K has higher coercivity measured along the easy axis $H_{C//}$. The coercivity is a measure of the magnetic field required to reverse the magnetization of the film, and is related to the domain wall energy gradient.[8] As a result of competition between the exchange and anisotropy energies, the domain wall width will decrease with increasing the anisotropy energy. The higher the H_K, the

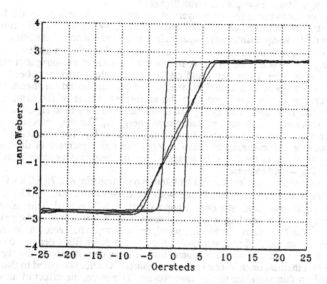

Fig.1 a typical B-H hysteresis loop of the NiFe film measured along easy and hard axes

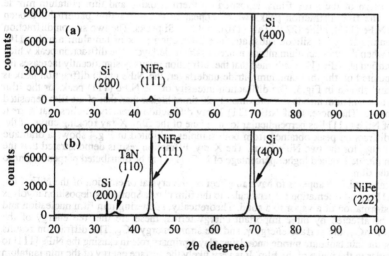

Fig.2 X-ray diffraction patterns of the NiFe films deposited on both thermal oxide (a) and thin tantalum nitride (b)

thinner the domain wall; therefore, the domain wall energy gradient increase with H_K. Thus, the coercivity would be higher for the sample film with higher H_K.

In an attempt to understand how the underlayers affect the magnetic anisotropy field H_K of the NiFe films, the stresses measured along both hard axis and easy axis of the NiFe films and the magnetostriction of the sample films were studied. The magnetostriction of the NiFe sample films obtained from the optical interferometry measurement is 6.1×10^{-7}. The difference of the stress measured along easy axis and the stress measured along hard axis, anisotropic stress, was found on the samples of NiFe films deposited on sputter tantalum nitride to be 76 MPa. Anisotropic stress of 6 MPa was obtained on the NiFe films deposited on thermal oxide underlayer. The magnetostrition combined with the anisotropic stress of the sample films introduce aniostropic magnetoealstic energy.[1] The magnetic anisotropy field H_K of an unaxial anisotropy thin film is determined by the total anisotropy energy K_t. The magnetoelastic energy may contribute to the total anisotropy energy to have an effect on the H_K. The contribution of the anistropy energy from the induced magnetoelastic energy can either increase or decrease the magnetic anisotropy field H_K of the films. In a case where magnetoelastic energy adds to the anistropy energy, the H_K would increase.

When the NiFe films were deposited on thin tantalum nitride underlayers, 76 MPa anistropic stress was found. The anistropic stress combined with the magnetostriction, 6.1×10^{-7}, of the NiFe film gives anistropic magnetoelastic energy approximately 460 erg/cm^3. This anisotropic magnetoelastic energy is corresponding to approximately a 1.1 Oe increase in the magnetic anisotropy filed H_K of the NiFe films. In anothor word, the anisotrpic magnetoelastic enegy is responsible for 30% of the induced magnetic anisotropy of the NiFe films deposited on thin tantalum nitride underlayer. The anisotrpic magnetoelastic energy contribution to the H_K of the NiFe films deposited on thermal oxide is one order of magnitude samller compared to that of the NiFe films deposited on thin tantalum nitride underlayers. Therefore, the effect of anistropic magnetoelastic energy on the H_K of the NiFe film deposited on thermal oxide is approximately 0.1 Oe.

The crystalline structure of the sample films has been studied by X-ray diffreacton. The X-ray diffration patterns obtained from the sample films show that the underlayers have a great effect on the crystalline structure of the sputtered NiFe films. Figure 2 shows the X-ray diffraction pattern of the NiFe films deposited on thermal oxide and thin tantalum nitride underlayers. The five diffraction peaks which appear in the diffraction pattern have been identified as NiFe (111), NiFe (222), TaN (110), and two Si peaks. The two silicon diffraction peaks are coming from the silicon substrate. The diffraction peak of tantalum nitride (110) appears only when the thin tantalum nitride is used as the underlayer. The diffraction peak which has been identified as NiFe (111) indicated that the diffraction intensity significantly increases for the NiFe deposited on the thin tantalum nitride underlayer, the NiFe (111) diffraction peak is highlighted and shown in Fig.3. The diffraction intensity of the NiFe (111) peak for the film deposited on thin tantalum nitride is approximately 20 times higher than that of the film deposited on thermal oxide. The increase of NiFe (111) X-ray diffraction intensity implies that a great percentage of NiFe [111] lies perpendicular to the plane of the film. X-ray rocking curve of the NiFe (111) diffraction peak obtained from both samples exhibited in Fig.4 shows a dramatic difference of $\Delta\theta_{50}$ for the two NiFe films. The X-ray diffraction results demonstrated that the thin tantalum nitride induced higher percentage of NiFe [111] to be distributed perpendicular to the plane of the film.

Thin tantalum nitride appears to have an effect on the crystal orientation of the NiFe film, inducing [111] crystal orientation perpendicular to the film plane. Sputtering deposition involves a phase transformation of a vapor to a solid. Theoretically, sputtering thin film nucleation and growth can be affected by three important energy terms; these are the free energy of the transformation ΔG_V, the surface energy γ, and the strain energy ΔG_ε. The diffraction results indicated that the thin tantalum nitride underlayer play a primary role in causing the NiFe [111] to be perpendicular to the plane of the film. It is very likely the surface energy of the thin tantalum nitride which may be lower than the thermal oxide is one of the important energy terms affecting the nucleation and growth of the NiFe overlayer. The strain energy may also play an important role in affecting the nucleation and growth of the NiFe overlayer.

236

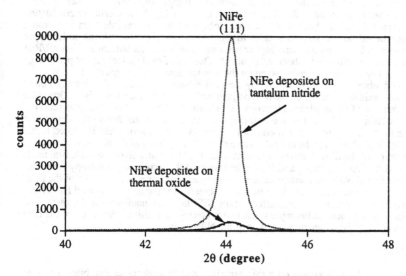

Fig.3 X-ray diffraction peak of the NiFe (111)

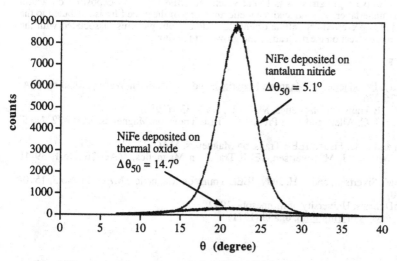

Fig.4 X-ray rocking curve of the NiFe (111) diffraction peak

Now, we move to a discussion of the effect of structure on the properties of the NiFe films. The sheet resistance R measurement of the the sample films is also consistent with the crystaline structure results obtaned from the X-ray diffraction. Higher $\Delta\theta_{50}$ implies more random distribution of the grain structure of the film and result in increasing the scattering of conduction electrons. Therefore, higher sheet resistance was obtained on the sample film with more randomly distributed grain structure and higher $\Delta\theta_{50}$. In the case of the anisotropic magnetoresistance, ΔR remains constant, higher percentage of magnetoresistance effect $\Delta R/R$ is expected on the sample with lower sheet resistance R. This could explain that higher percentage $\Delta R/R$ was obtained on the NiFe film deposited on the thin tantalum nitride underlayer.

As mentioned, observed in-plane anisotropy (unaxial anisotropy) of the NiFe films exhibited easy axis and hard axis in the plane of the film perpendicular to each other. (see Fig.1) One quite possible interpretation of the in-plane anisotropy in the sample films is the induced pair-ordering or directional ordering of like atoms due to the applied magnetic fields during the deposition processes. The physical nature of this interaction is like crystal anisotropy and is related to the spin-orbital coupling. Therefore, the crystal texture of the NiFe film would have an effect on the magnetic anisotropy of the film. A randomly distributed crystal orientation of the NiFe film tends to diminish the directional ordering and results in lowering the magnetic anisotropy field H_K. While a highly oriented NiFe film would induce the directional ordering in the plane of the film. As a consequence, the magnetic anisotropy field H_K increases for the highly oriented NiFe film. The underlayers play an important role in affecting the NiFe thin film nucleation and growth. The consequence of this is induced anisotropic stress and preferred crystalline orientation of the film which results in altering the magnetic properties of the NiFe film.

CONCLUSION

Sputter NiFe thin film deposited on a thin tantalum nitride underlayer exhibited a higher magnetic anisotropy field, in-plane coercivity, magnetoresistance effect and lower sheet resistance compared to the films deposited on thermal oxide. Thin tantalum nitride appears to play an important role in affecting thin film nucleation and growth of the NiFe film and in inducing crystal orientation texture in the film. A very strong preferred crystalline orientation of NiFe [111] and anisotropic stress was found when the NiFe film was deposited on a thin tantalum nitride underlayer. The induced magnetic anistropy of the NiFe film is attributed to the induced NiFe [111] preferred crystalline orientation and anisotropic stress associated with the thin tantalum nitride effect on the nucleation and growth of the film.

REFERENCES

1. B. D. Cullity, Introduction to Magnetic Materials, 2nd ed. (Addison-Wesley, Reading, MA, 1972) p. 226-275
2. G. Chen, IEEE Trans. on Magnetics, MAG-22, No.5, 334 (1986)
3. R. D. Fisher, J. C. Allan, and J. L. Pressesky, IEEE Trans. on Magnetics, MAG-22, No.5, 352 (1986)
4. J. C. Allan and R. D. Fisher, IEEE Trans. on Magnetics, MAG-23, 112 (1986)
5. J. Lin, C. Wu, and J. M. Sivertsen, IEEE Trans. on Magnetics, MAG-26, No.1, 39-41 (1990)
6. T. Yeh, J. M. Sivertsen, and J. H. Judy, IEEE Trans. on Magnetics, MAG-26, No.5, 1590-1592 (1990)
7. T. Yeh, PhD thesis, University of Minnesota, 1992
8. A. Yelon, Physics of Thin Film, 6, 238 (1971)

PAIR ORDERING ANISOTROPY IN AMORPHOUS Tb-Fe THIN FILMS

T.C. Hufnagel,* S. Brennan,** and B.M. Clemens*
*Department of Materials Science and Engineering, Stanford University, Stanford, CA 94305-2205
**Stanford Synchrotron Radiation Laboratory, Stanford, CA 94309-0210

ABSTRACT

We have studied the structural origins of perpendicular magnetic anisotropy in amorphous Tb-Fe thin films by employing high energy x-ray scattering. The as-deposited films show a clear structural anisotropy, with a preference for Fe-Tb near-neighbors to align in the out-of-plane direction. Upon annealing, the magnetic anisotropy energy drops significantly, and we see a corresponding reduction in the structural anisotropy. The radial distribution functions indicate that the number of Fe-Tb near-neighbors increases in the in-plane direction, but does not change in the out-of-plane direction. Therefore, the distribution of Fe-Tb near-neighbors becomes more uniform upon annealing. We conclude that the observed reduction in perpendicular magnetic anisotropy energy is a result of this change in structure.

INTRODUCTION

Amorphous RE-TM alloy thin films are in widespread use as magnetooptic recording media. One important property that these materials have is a strong perpendicular magnetic anisotropy; that is, an easy axis of magnetization perpendicular to the plane of the film. While the physical origins of magnetic anisotropy (such as dipolar interactions and single-ion anisotropy) are reasonably well understood, each of these mechanisms requires an underlying structural anisotropy to produce a macroscopic magnetic anisotropy. The nature of this structural anisotropy in amorphous RE-TM films has been the subject of considerable debate. A variety of different theories have been proposed, including pair ordering anisotropy[1] and bond orientation anisotropy[2].

Recently, two independent observations of atomic-scale structural anisotropy in amorphous RE-TM thin films have been reported. The first compared x-ray scattering from an amorphous $Tb_{.26}Fe_{.62}Co_{.12}$ in the symmetric reflection geometry (which gives in-plane structural information) with grazing-incidence scattering (which gives out-of-plane structural information)[2]. While no real-space structural information was presented, the difference in scattering between the two geometries lead the authors to conclude that bond orientation anisotropy was present in their sample. Bond orientation anisotropy is characterized by a different near-neighbor spacing and coordination number in the in-plane and out-of-plane directions, but not by any difference in chemical ordering between the two directions.

The presence of pair-ordering anisotropy in amorphous $Tb_{.26}Fe_{.74}$ has been reported by Harris and coworkers[3]. These authors measured the polarization-dependent EXAFS

239

from their samples and then fit calculated EXAFS spectra based on a structural model to the experimental data. Their results showed that there was a slight preference for Fe–Tb near-neighbors to align in the out-of-plane direction. They did not see any difference in overall coordination numbers or near-neighbor spacings between the two directions.

The present experiment represents an effort to clarify the nature of the structural anisotropy by making a detailed study of the near-neighbor environment in amorphous Tb-Fe thin films. We have employed x-ray scattering at high energies (20-30 keV) in two geometries: reflection (which is sensitive to atomic correlations in the out-of-plane direction) and transmission (which is sensitive to in-plane atomic correlations). The high x-ray energy allows us to examine a wide range of reciprocal space ($q = 1 - 20$ Å^{-1}), providing a detailed picture of the local atomic environments.

EXPERIMENTAL

Amorphous Tb$_{.25}$Fe$_{.75}$ thin films were deposited by dc magnetron sputter codeposition from elemental targets in a chamber with a base pressure of $< 2 \times 10^{-9}$ torr. The sputtering gas was 1.5 mtorr Ar purified by a Ti gettering furnace. A quartz crystal rate monitor was used to monitor the deposition rates of the elements to ensure that the chemical composition of the film (which has a dramatic effect on the magnetic properties) did not vary during the deposition. The thickness of the film used for this study was 8900 Å; the film was capped with a reactively sputtered 300 Å thick layer of SiN to prevent oxidation. The substrate was a free-standing 3000 Å thick SiN membrane.

The x-ray scattering experiments were performed on beamline 7-2 at the Stanford Synchrotron Radiation Laboratory. The beamline was operated in an unfocused mode with dual Si (220) monochromator crystals. The transmission scattering measurements were conducted at an x-ray energy of 30 keV, and the reflection experiments were done at 21.5 keV, primarily to avoid experimental difficulties associated with making measurements at very low incident angles. We did perform some reflection experiments at the higher energy as a check; the results were consistent with the measurements at the lower energy. The measured scattering was corrected for the effects of detector nonlinearity, substrate scattering, absorption, multiple scattering, and polarization of the incident beam [4]. The corrected intensity data were then placed on an absolute scale using the method of Norman [5], and finally Fourier transformed to real space to obtain radial distribution functions.

The magnetic properties of the sample were measured with a vibrating-sample magnetometer and a torque magnetometer. The measured perpendicular anisotropy of the as-deposited film was 7×10^6 ergs/cm^3. Upon annealing under vacuum for one hour at 250 °C, the magnetic anisotropy was reduced to 2×10^6 ergs/cm^3. X-ray scattering measurements were made before and after the anneal in an attempt to correlate structural changes with the observed reduction in magnetic anisotropy.

RESULTS

The reduced radial distribution functions for the in-plane and out-of-plane direction from the as-deposited sample are shown in Figure 1. There is a clear structural anisotropy in the near-neighbor shell (r=2–4 Å). Also indicated on the figure are the approximate near-neighbor distances for Fe–Fe, Fe–Tb, and Tb–Tb near-neighbor pairs. By comparing the amplitude of the Fe–Tb correlation with that of the Tb–Tb correlation between the in-plane and out-of-plane directions, we conclude that there are relatively fewer Fe–Tb near-neighbor pairs in the in-plane direction. We should note that the amplitude of the correlations in the reduced radial distribution function depends on both the coordination number and the sharpness of the distribution; we will address this point later.

The scattering experiments were repeated after annealing the sample. The reduced radial distribution functions after annealing are shown in Figure 2. One significant effect of the annealing is that the atomic distributions become sharper, but there is no overall change in coordination number. There is still an apparent structural anisotropy between the in-plane and out-of-plane directions, but the magnitude of the anisotropy (relative to the as-deposited sample) is reduced. In particular, the amplitude of the in-plane Fe–Tb correlation relative to the Tb–Tb correlation has increased.

To examine the structure of the near-neighbor shell in more detail, we fit each of the observed RDFs with a sum of three Gaussian peaks in the near-neighbor region. To obtain rational results, it was necessary to constrain the three peaks in each fit to have the same width (although the width was allowed to vary between fits). The partial coordination number for type of correlation was then calculated from the peak areas. Table I contains the partial coordination numbers for Fe atoms around Tb atoms (Z_{TbFe}) and for Tb around Fe (Z_{FeTb}); note that Z_{TbFe} and Z_{FeTb} are not independent but are related by the ratio of concentrations of Fe and Tb. The confidence intervals given in the table are based on the estimated experimental error as well as a 3σ confidence interval in the parameters of the RDF fitting function. Note that the reported coordination numbers are spherically averaged; that is, the coordination number for each direction is reported as though it were the coordination number for a spherically symmetric atomic environment.

	As deposited		After annealing	
	Z_{FeTb}	Z_{TbFe}	Z_{FeTb}	Z_{TbFe}
Out-of-plane	3.9 ± 0.2	11.8 ± 0.6	3.9 ± 0.2	11.8 ± 0.7
In-plane	3.1 ± 0.2	9.4 ± 0.8	3.5 ± 0.3	10.6 ± 0.8

Table I: Fe–Tb and Tb–Fe partial coordination numbers.

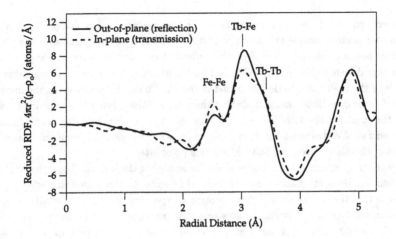

Figure 1: Reduced radial distribution functions from the as-deposited sample in the in-plane and out-of-plane directions. Approximate distances of different coordinations (Fe–Fe, Fe–Tb, and Tb–Tb) are shown.

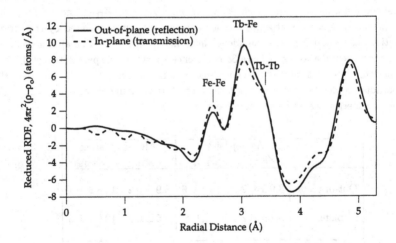

Figure 2: Reduced radial distribution functions from the same sample as Figure 1 after annealing.

DISCUSSION

The partial coordination numbers shown in Table I indicate that in the as-deposited sample there are more Fe–Tb correlations in the out-of-plane direction than in the in-plane direction. We propose that it is this chemical anisotropy which is responsible for the observed magnetic anisotropy. While we have presented these results based on a single sample, we have observed the same anisotropy in other samples measured under different x-ray scattering conditions. In each case, we observe a preference for unlike near-neighbors to align in the out-of-plane direction.

Upon annealing, the number of Fe–Tb correlations out-of-plane does not change. (The increase in amplitude of the reduced RDF is due the distribution of atomic spacings becoming sharper, not a change in coordination number.) In-plane, however, the number of Fe–Tb correlations increases; on average, each Tb atom is coordinated by one additional Fe atom. Therefore, the effect of the annealing is to make the distribution of Fe–Tb near-neighbors more uniform. The remaining magnetic anisotropy after annealing is due to the residual anisotropy in near-neighbor distribution, as well as to inverse magnetostriction.

These results are in general agreement with those of Harris and coworkers, who reported more Fe–Tb correlations in the out-of-plane direction than in the in-plane direction for an as-deposited sample[3]. They reported that this anisotropy disappeared upon annealing, but did not report whether the pair correlations changed in both directions, or only in the in-plane direction, as we have observed.

CONCLUSIONS

We have observed an anisotropy in the local atomic environments in amorphous Tb-Fe thin films. The as-deposited films show a preference for unlike near-neighbor (Fe–Tb pairs) to align perpendicular to the plane of the film. Upon annealing, the magnetic anisotropy is reduced. The the number of Fe–Tb pairs perpendicular to the plane of the film does not change, but the number of Fe–Tb pairs in-plane increases. This makes the overall distribution of Fe–Tb pairs more uniform. We conclude that the observed magnetic anisotropy is a result of the anisotropy in the distribution of unlike pair correlations.

ACKNOWLEDGEMENT

This work was carried out with financial support from the Kobe Steel U.S.A. Applied Electronics Center; SSRL is supported by the U.S. Department of Energy, Office of Basic Energy Sciences under contract DE-AC03-76SF00515. T.C.H. acknowledges support from the U.S. Department of Defense through the National Defense Science and Engineering Graduate Fellowship program. The authors gratefully acknowledge helpful discussions with A. Bienenstock, E.E. Marinero, and V.G. Harris. We are indebted to R. Sermiaa for the use of her computer code for analyzing x-ray scattering data.

REFERENCES

1. T. Mizoguchi and G.S. Cargill III, J. of Appl. Phys. **50**, 3750 (1979).

2. X. Yan, M. Hirscher, T. Egami, and E.E. Marinero, Phys. Rev. B **43**, 9300 (1991).

3. V.G. Harris, K.D. Aylesworth, B.N. Das, W.T. Elam, and N.C. Koon, Phys. Rev. Lett. **69**, 1939 (1992).

4. C.N.J. Wagner, J. Non-Crys. Solids **31**, 1 (1978).

5. N. Norman, Acta. Cryst. **10**, 370 (1957).

Part V
Ultrathin Films, Magnetic Domains

DEFECTS AND MAGNETIC PROPERTIES: THE Cr/Fe(001) INTERFACES

D. Stoeffler, A. Vega*, H. Dreyssé and C. Demangeat
Institut de Physique et de Chimie des Matériaux de Strasbourg (U.M.R. 46 du C.N.R.S.),
Groupe d'Etude des Matériaux Métalliques, 23 rue du Loess, 67037 Strasbourg, France

ABSTRACT

We report self-consistent band-structure calculations for the magnetism of Cr overlayers adsorbed on Fe(001) in order to study the role played by imperfect Fe-Cr interfaces, at the microscopic and macroscopic scales, on the total magnetisation. The surprisingly large reduction of the total magnetisation ($\approx$ - 5 μ_B/interfacial atom) recently observed in Cr/Fe(001) through *in situ* magnetometer measurements is shown to be reproduced only when an interchange of one Cr and Fe monolayer at the interface after deposition of the second Cr monolayer occurs.

The magnetic behaviour in artificial structures has been extensively studied both theoretically and experimentally during the last years. One of the most astonishing cases is the Fe-Cr system which has been investigated in overlayer systems as well as in multilayers. The discovery in Fe/Cr multilayers of the giant magneto resistance [1] and the oscillating interlayer coupling [2] effects has stimulated a large number of studies motivated by the desire to progress in the fundamental understanding of the mechanism and by the promising industrial applications like magnetic sensors components. It is now well established [3] that the layered antiferromagnetic (LAF) order, consisting of parallel magnetic planes with antiferromagnetic coupling, in thick Cr layers is stable up to high temperatures (550 K) and is at the origin of the short period in the oscillations of the interlayer couplings [4, 5].

Although the recent experimental results for Cr ultra thin films deposited on a Fe (001) substrate obtained with different techniques [3, 6, 7, 8] are in good qualitative agreement with each other, the magnetic behaviour during the deposition of the first few Cr monolayer on Fe is found to be completely different from that obtained for thicker coverage. Indeed, when the LAF order in Cr is obtained, the magnetisation of the topmost atomic layer changes its sign with the parity of the number of deposited Cr monolayers, but in some cases such a behaviour is not observed for Cr thicknesses smaller than 5 monolayers [6]. Moreover, a change of approximately -5 μ_B/interfacial atom in the total magnetisation has been observed by Turtur and Bayreuther through *in situ* magnetometer measurements when a few Cr monolayers are deposited on a Fe(001) substrate [7]. This large reduction has been tentatively ascribed to a strong deviation with respect to the LAF order. These authors have suggested that the magnetic moments of the first two Cr monolayers are both opposite to the magnetisation of the Fe(001) substrate, the LAF order being recovered only with the deposition of the third Cr monolayer. Using this hypothetical magnetic moments profile with a Monte Carlo simulation of a non-layer-by-layer growth process, the observed variation of the total magnetisation with the Cr thickness was reproduced [7]. However, the model-profile used by these authors does not result from band-structure calculations. It does not take into account the large reduction of the magnetic moment of Cr at the interface when more than one monolayer is deposited, this reduction being particularly important when the magnetic moments of two adjacent Cr monolayers point in the same direction (frustration effect). More recently, a similar quantitative measurement of the magnetisation change due to the Cr deposition has been made [8] using Spin Polarized Secondary Electron Emission (SPSEE). Even if a similar decreasing magnetisation has been obtained, its value is approximately 3 times smaller as compared to the one of Turtur and Bayreuther.

In this paper we show that the large reduction of the total magnetisation experimentally observed cannot be traced back to the deviation from the LAF order in Cr. We propose different growth processes which lead to a more realistic magnetic moments profile obtained through self-consistent band-structure calculations. We show that a satisfactory agreement between our results and the experimental ones can been obtained only by assuming a partial interchange between interfacial Fe and Cr atoms. However, we point out that, in order to account for the most spectacular result of Turtur and Bayreuther we have to assume a complete interchange of a Cr and an Fe monolayer at the interface which is certainly highly improbable.

The spin-polarised electronic-charge distribution was determined self-consistently by solving a tight-binding model Hamiltonian in the unrestricted Hartree-Fock approximation for the valence "3d" electrons. This model contains an accurate description of the itinerant character of the "3d" elements and the essence of the directional bonding and electronic correlations [9]. The calculations have been done with the real space recursion technique [10] which allows the accurate determination of the magnetic moments distribution of systems involving a large number of inequivalent sites [9, 11], which are beyond the possibilities of the *ab initio* methods.

In order to investigate the possibility to have frustration of the LAF order, we have first determined the magnetic moments distributions for n Cr monolayers deposited on a perfect Fe(001) substrate with and without considering a LAF order in the first two Cr monolayers (figures 1-a and 2-a). After having determined the magnetic moments distributions for the LAF order in the Cr overlayer, we have started the second calculation changing *a priori* the sign of the magnetic moments in the Cr layers so that the LAF order in Cr was frustrated at this starting point. A different self-consistent solution is obtained in the frustrated situation only for n > 5 monolayers; for n ≤ 5, all the magnetic moments are reversed during the self-consistent procedure giving finally a unique solution, that with the LAF order. For example, for n = 6 (figure 1-a)

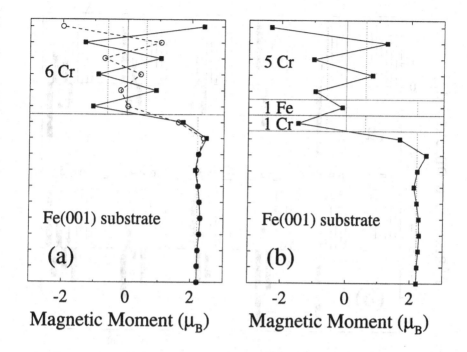

Figure 1 Magnetic moments distributions for (a): $Cr_6/Fe(001)$ with (open circles) and without (black squares) having considered frustration of the layered antiferromagnetic order in Cr, and (b): $Cr_5/Fe_1/Cr_1/Fe(001)$. The horizontal dotted lines separate the layers of different chemical nature and the vertical ones give the values of the magnetic moment in the bulk.

when the LAF order is frustrated, the interfacial Cr magnetic moment is nearly equal to zero, and the other Cr moments are strongly reduced as compared to the values obtained for the LAF order situation. Similarly, due to the strong Fe-Cr coupling, the magnetic moments of the first Fe monolayers are slightly reduced. These results indicate that the Fe-Cr as well as the Cr-Cr antiparallel couplings are strong and cannot be frustrated without strong reductions of the Cr local magnetic moments. If we choose the magnetisation of the free Fe substrate as a reference, we obtain a magnetisation variation per surface atom ΔM of -3.23 μ_B, 0.85 μ_B, -2.31 μ_B, 1.18 μ_B, -2.34 μ_B, 1.04 μ_B, -2.39 μ_B, 1.06 μ_B, -2.33 μ_B and 1.08 μ_B for n = 1 to 10 in the non-frustrated situation. These results show that for layer-by-layer growth we obtain a total magnetisation oscillating with the parity of the Cr overlayer thickness around the average value $<M> = -1.25$ μ_B with an amplitude of $\cong 3.4$ μ_B. The average value obtained here is smaller than the one found by Turtur and Bayreuther with their LAF magnetic profile [7], due to the fact that

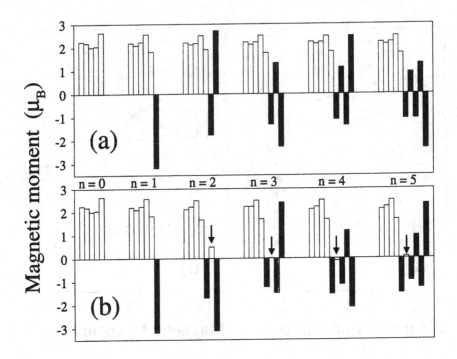

Figure 2 Magnetic moments profiles for 1 to 5 Cr monolayers on Fe(001), (a): $Cr_n/Fe(001)$, (b) $Cr_1/Fe(001)$ and $Cr_{n-1}/Fe_1/Cr_1/Fe(001)$ for $n \geq 2$. The Fe (Cr) moments are given in white (grey) and the arrow indicates the exchanged Fe monolayer.

their model profile does not take into account the large reduction of the interfacial Cr magnetic moment when more than one monolayer is adsorbed on the substrate. Thus, our band-structure calculations indicate that the two model profiles used by these authors [7] overestimate the reduction in the total magnetization due to the Cr layer and therefore, the experimental behavior cannot be traced back to a deviation from the LAF order in the first two Cr monolayers. However, their main qualitative result, i.e., the fact that the essential contribution to the change of the total magnetization is an interfacial property, is certainly the only way to reproduce their experimental values. The mechanism we propose, in order to obtain the observed reduction of the total magnetization, consists in the interchange of a complete Cr and Fe monolayer at the interface after deposition of the second Cr monolayer (figures 1-b and 2-b). In this way, for $n > 1$, the moments of the two Cr layers at the interface remain large and negative whereas the Fe layer in between displays a moment nearly equal to zero. Consequently, if we assume that for $n = 1$ we have the

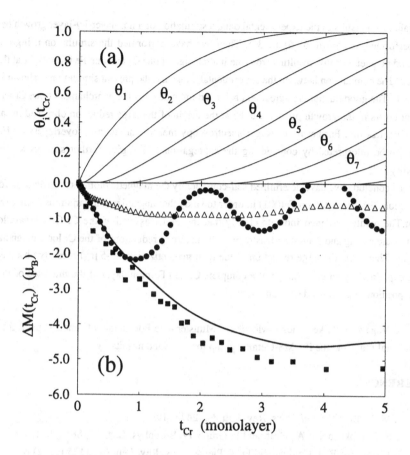

Figure 3 (a) Non layer by layer growth. θ_i represents the percentage of the i-th layer present at the considered coverage. (b) Total change of the magnetisation as a function of the deposited Cr thickness. Filled circles correspond to the layered antiferromagnetic order magnetic moments profile $\Delta M(n)$ assuming a nearly layer by layer growth. Open triangles correspond to the same profile but with the non layer by layer growth. The full line corresponds to our proposed magnetic moments profile $\Delta M'(n)$ assuming the same non layer by layer growth. The experimental values (filled squares) are taken from Turtur and Bayreuther [7].

$Cr_1/Fe(001)$ situation, and for $n > 1$ we have $Cr_{n-1}/Fe_1/Cr_1/Fe(001)$, we obtain a change in the total magnetization $\Delta M'$ of -3.23 μ_B, -6.73 μ_B, -2.70 μ_B, -6.14 μ_B, -2.80 , -6.24 μ_B, -2.95 μ_B, -6.30 μ_B, -2.94 μ_B, -6.31 μ_B and -2.93 for $n = 1$ to 11.This time, the average value is $\langle M' \rangle$ = -4.6 μ_B, in good agreement with the asymptotic experimental value. In order to compare directly

251

these calculated results to the experimental data, a simulation of a non-layer-by-layer growth has been performed as shown in figure.3-a. Then, we have performed the simulation using our calculated magnetic profile resulting from the interchange of one Cr and one Fe monolayer at the interface. The comparison between the experimental value and the present simulation is shown in figure 3-b. The good qualitative agreement achieved indicates that the interchange of one Cr and Fe monolayers in the growth process can be at the origin of the large reduction observed in the total magnetization. Furthermore, Auger spectroscopy measurements in Cr overlayers on Fe, which can be interpreted by considering the segregation of Fe, give further support to this possibility [12].

To conclude, we have determined self-consistently the magnetic moments distribution for n Cr monolayers deposited on Fe(001) in order to study the magnetization variations during the growth.The results obtained indicate clearly that the Fe-Cr as well as the Cr-Cr antiparallel couplings are strong and cannot be frustrated without strong reductions of the Cr local magnetic moments. Therefore, the large reduction of the total magnetization ($\cong$ -5 μ_B) recently observed can be explained by an interchange of a complete Cr and Fe monolayer at the interface occurs after deposition of the second Cr monolayer.

A. Vega would like to acknowledge the Ministerio de Educacion y Ciencia (Spain) for a post-doctoral grant and the IPCMS-Gemme group for their kind hospitality.

REFERENCES

1. M. N. Baibish et al, Phys. Rev. Lett. **61**, 2472 (1988)

2. S. Demokritov, J. A. Wolf, and P. Grünberg, Europhys. Lett. **15**, 881 (1991)

3. J. Unguris, R. J. Celotta, and D. T. Pierce, Phys. Rev. Lett. **69**, 1125 (1992)

4. D. Stoeffler and F. Gautier, Progress of Theoretical Physics Suppl. **101**, 139 (1990)

5. M. van Schilfgaarde and F. Herman, Phys. Rev. Lett. **71**, 1923 (1993)

6. T. G. Walker et al, Phys. Rev. Lett. **69**, 1121 (1992)

7. C. Turtur and G. Bayreuther, Phys. Rev. Lett. **72**, 1557 (1994)

8. P. Fuchs, K. Totland, and M. Landolt, Proceedings of ICMFS-EMRS (1994)

9. A. Vega et al, Phys. Rev. B **49**, 12797 (1994)

10. R. Haydock, in *Solid State Physics*, edited by H. Ehrenreich, F. Seitz, and D. Turnbull (Academic, New York, 1980), Vol. **35**, p. 215

11. D. Stoeffler and F. Gautier, J. Magn. Magn. Mater. 1995 (in press)

12. C. Carbone and S. F. Alvarado, Phys. Rev. B **36**, 2433 (1987)

MAGNETISM OF BCT IRON GROWN IN (001) FeIr SUPERLATTICES

Stéphane ANDRIEU *, Philippe BAUER *, Fils LAHATRA-RAZAFINDRAMISA **,
Louis HENNET *, Etienne SNOECK ***, Michel BRUNEL **, Michel PIECUCH *
* Laboratoire de Métallurgie Physique, CNRS, 54506 Vandoeuvre, France
** Laboratoire de Cristallographie, CNRS, 38042, Grenoble, France
*** CEMES/LOE, CNRS, 31055, Toulouse, France

Abstract

In this paper, we show that iron can be grown by MBE in a body centered tetragonal structure in (001) FeIr superlattices. The growth, structure and morphology of these superlattices are briefly resumed. A variation of the BCT Fe magnetic moment depending on the Ir thickness is observed. This variation is demonstrated to come from a variation of the BCT Fe atomic volume, due to the competition of the Fe and Ir stresses. A magnetic transition from a non-magnetic to a low spin ferromagnetic state depending on the atomic volume is thus observed.

The heteroepitaxy of metals with unusual cristallographic structure has became possible using the Molecular Beam Epitaxy technique (MBE). This is of great interrest for the understanding of the relationship between the cristallographic structure and magnetic properties of metals of the first transition series. Indeed, if striking structural similarities can be noticed between the metals belonging to the second and first series, it does not extend to the first transition series. These discreepancies are due to the magnetic contribution to the total energy of the system. For instance, ruthenium and osmium exhibit a non-magnetic HCP structure, but a ferromagnetic BCC structure is observed for iron. This paper reports a work done to improve our understanding of the interrelation between the crystalline structure and magnetism of iron.

Several bands calculation methods have been used in order to study the magnetic behavior of BCC and FCC iron phases [1-5]. As the BCC iron phase was always found to be ferromagnetic whatever the value of the lattice parameter is, the situation is more complex for the FCC phase : by taking into account an isotropic variation of the atomic volume, several authors found that FCC iron can be non-magnetic (NM), low-spin ferromagnetic (LS), high spin ferromagnetic (HS), or antiferromagnetic (AF). Peng and Hansen [5] also predicted these magnetic behaviour for a tetragonalisation of the FCC phase.

Mat. Res. Soc. Symp. Proc. Vol. 384 ° 1995 Materials Research Society

In order to check these theoretical predictions, iron with unusual structure must be prepared. For this purpose, the MBE technique is used here. It is wellknown that BCC iron can be grown on (001) Ag [6] ((100) BCC Fe // (110) FCC Ag) and FCC iron on (001) Cu [7] ((100) FCC Fe // (100) FCC Cu), because of the very small mismatch in both cases. The main idea of this work is to grow iron on a non-magnetic metal with a first neighbour distance in the (001) plane intermediate between those of Ag and Cu, i.e. intermediate between the (110) FCC Fe distance (2.54Å at room temperature) and the (100) BCC Fe distance (2.8664Å). FCC Ir is a good candidate since its (110) distance is equal to 2.715Å. A strained BCC or FCC Fe structure was thus expected to grow on (001) Ir.

The growth and structure of iron on this (100) Ir surface was ever reported in several papers [8,9]. The main results are briefly summarized. Two-dimensional (2D) growth was observed during the epitaxy of Fe on (001) Ir up to a minimum of 4 atomic planes (fig.1). The growth of Ir on this resulting Fe surface was also observed to follow a layer by layer growth mechanism [9]. This is the first time that a "complete" 2D growth was observed on a metallic system with a so large mismatch (7%). As a consequence, superlattices with very flat interfaces are obtained as shown on figure 2. On the contrary, some periodic roughness was observed when the Fe thickness exceeded 5 atomic planes [9]. This behaviour can be understood by analysing the structure of the Fe layers with respect to their thickness. Figure 1 shows the variation of the in plane parameter with the number of Fe deposited planes. Up to 4 atomic planes (i.e. during the 2D regime as shown by RHEED oscillations), the in plane parameter determined by RHEED is equal to the Ir one. In that case, no misfit dislocations are observed on cross sectionnal Transmission Electron Microscopy (TEM) images. However, above 5 atomic planes, the in plane parameter begins to significantly vary (fig.1), and a large amount of misfit dislocations are thus observed for

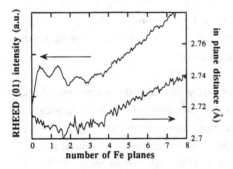

Figure 1 : RHEED oscillations on the (01) streak and variation of the in plane distance with the number of Fe planes.

Figure 2 : TEM image in cross section on a superlattice with 5 atomic planes. The thickness of the superlattice is 900Å.

superlattices with more than 10 planes in each Fe layers [9]. Moreover, the structural analysis performed by X-Ray diffraction [8] and EXAFS [10] demonstrate that iron is in a BCT structure up to 5 atomic planes, and partly relaxes to the BCC structure for larger thicknesses. To conclude, these results show that iron is totally strained by Ir up to 5 deposited Fe atomic planes. We are only interrested here by this strained structure where Fe is pseudomorphic to Ir in the plane of growth.

A simple elastic calculation shows that this BCT structure can be the consequence of an elastic deformation of the FCC phase, and not of the BCC phase [8]. The magnetic properties of this phase were investigated by SQUID measurements. First, no magnetic moment was detected up to 2 Fe atomic planes. There is consequently 2 dead layers. We have also observed that, for a series of superlatttices with 4 Fe atomic planes, the magnetic moment of iron depends on the Ir thickness, as shown on figure 3. Moreover, No magnetic moment was detected by SQUID on superlattices grown with one plane of iridium. For larger Ir thicknesses, an increasing average moment was detected. This behaviour can be explained by a Fe atomic volume variation due to the increasing strain imposed by Ir when the Ir thickness is increased [8]. Elastic calculations can thus be performed and such variations are actually predicted : the atomic volume was found to vary from 11.9 to 12.2 $Å^3$/at, and the c/a ratio from 1.23 to 1.29 [8].

The variation of the magnetic moment of Fe in the BCT phase can thus be related to the variation of the atomic volume variation. This is the keypoint of this paper. However, a number of points should be verified in order to ensure the validity of this approach. Indeed, we first assume that the superlattices are uniformly strained. Secondly, it implies that Ir in the Ir layers of the superlattice is also strained. Thirdly, it implies that the Fe and Ir in plane distance are equal in the superlattice.

Samples	e_{Fe}(Å)	e_{Ir}(Å)	Λ (Å)	N	calculation $a_{//}$ (±0.02Å)	grazing X-Ray $<a_{//}>$ (±.001Å)	DAFS $a_{//}^{Ir}$ (±.02Å)
1	6.8	5.8	12.6	50	2.653	2.669	2.65
2	6.2	14.8	21	35	2.687	2.694	
3	6.8	22.5	29.3	25	2.694	2.695	
4	6.8	9.4	16.2	40	2.671	2.687	
5	6.8	2.5	9.3	50	2.614	2.63	2.62
6	8	31.5	39.5	35	2.698	2.701	
7	8	22	30	30	2.690	2.682	
8	8	62.9	70.9	10	2.705	2.701	

Table I : comparison of the theoretical in plane parameter with the average in plane parameter measured by grazing X-Ray and with the Ir in plane distance determined by DAFS.

The first point is verified using the TEM technique. As the mismatch between the unstrained FCC Fe and Ir phases is around 7%, misfit dislocations should be observed every ≈40Å according to the Frank-Van der Merwe criterion. On TEM images with a scale of several hundred angström, no misfit dislocations are observed, which demonstrates that the superlattices are actually uniformly strained. Concerning the second point, DAFS experiments were performed [11] and the in plane distance in the Ir layers was determined as shown in Table I. This distance is actually found to be smaller than its bulk value, which demonstrates that Ir is actually strained by Fe. Finally, the third point is verified since the in plane distance calculated using the elastic modeling is actually in good agreement with the distance measured by grazing X-Ray diffraction (Table I). These values are also in agreement with the DAFS determination. We can thus conclude that the hypothesis of a uniform strain in the superlattices is correct.

We can now calculate the atomic volume variation and relate it to the magnetic moment. The first method consists to calculate the in plane parameter $a_{//}$ of the superlattice by minimizing the total elastic energy [8], which gives :

$$a_{//} = \frac{1+C}{\frac{1}{a_{Ir}^0} + \frac{C}{a_{Fe}^0}} \quad \text{with} \quad C = \frac{B_{Fe} a_{Ir}^0 e_{Fe}}{B_{Ir} a_{Fe}^0 e_{Ir}} \tag{1}$$

where e is the thickness, a^0 the parameter of the unstrained FCC structure, and $B=E/(1-v)$ where E is the Young modulus and v the Poisson's ratio. As the stress is uniform in the plane of growth, the out of plane parameter c_{Fe} is thus deduced using the relation :

$$\varepsilon_{zz}^{Fe} = -\left(\frac{2v}{1-v}\right)_{Fe} \varepsilon_{xx}^{Fe} \quad \text{with} \quad \varepsilon_{zz}^{Fe} = \frac{c_{Fe} - c_{Fe}^0}{c_{Fe}^0} \quad \text{and} \quad \varepsilon_{xx}^{Fe} = \frac{a_{//} - a_{Fe}^0}{a_{Fe}^0} \tag{2}$$

where $c^0 = \sqrt{2} a^0$ in the FCC structure. Note that the same relations exist for Ir. $c^0 = 3.59$Å for the Fe FCC structure at room temperature [12]. Moreover, the FCC Fe Young modulus can be estimated at room temperature around 170 GPa [13]. The Poisson's ratio is not known but can be estimated : indeed, the average peak of the superlattices determined by X-Ray diffraction can be compared to the calculated one using the equation (2) [8]. The best fit is obtained for $v=0.35$. The Fe atomic volume can now be calculated but the accuracy is not very good because of the uncertainty on the Fe and Ir thicknesses. However, this volume is accurately determined since the in plane parameter is accurately determined by grazing X-Ray diffraction (Table I). In order to verify that this estimated atomic volume variation is correct, we have performed Mössbauer experiments at room temperature. As this BCT phase is paramagnetic at room temperature, the Mössbauer spectra can

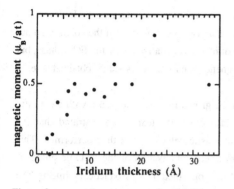

Figure 3 : magnetic moment vs the Ir thikness for superlattices with 4 Fe atomic planes.

Figure 4 : variation of the isomer shift with the atomic volume calculated by the elastic theory.

be fitted using two contributions : a disymetrical quadrupole doublet (which confirms the tetragonalisation of the FCC Fe phase) and a additionnal contibution attributed to the interface layer. The isomer shift of the former contribution is significative of an atomic volume variation. On figure 4 is plotted the variation of the measured isomer shift with the atomic volume variation compared to the BCC phase. A linear variation is obtained which definitely confirmes this atomic volume evolution and demonstrates the validity of the elastic calculation.

The magnetic moment is thus plotted with the atomic volume variation as shown on figure 5. A transition from a non magnetic to a low spin ferromagnetic phase is clearly observed. A sharp transition occurs at a volume equal to 12.2 ± 0.3 $Å^3$/at, which corresponds to a Wigner-Seitz radius of 2.700 ± 0.025 a.u., the incertaincy coming from the incertaincy on v and c^0. This result is in

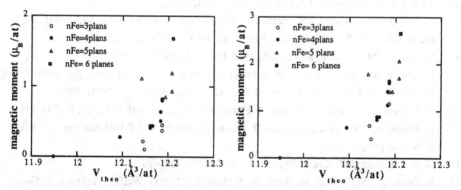

Figure 5 : on left, variation of the average magnetic moment with the atomic volume. On right, same variations but the magnetic moment is calculated assuming the occurence of two dead layers.

good agreement with the theoretical prediction of Moruzzi et al [2] and Grasco [3]. However, this agreement is surprising since these authors considered an isotropic variation of the volume. To our knowledge, only Peng and Hansen [5] took into account a tetragonalisation of the FCC phase, but they predicted the occurence of a high spin ferromagnetic state for a the c/a values obtained here.

To conclude, we have shown that iron can be grown in a BCT structure with a c/a ratio ranging from 1.24 to 1.29 in (001) FeIr superlattices. It is clearly demonstrated that the superlattices are thus uniformly strained. Regular elastic calculation is thus performed. This approach shows that this BCT phase results from a tetragonalisation of the FCC Fe phase. A variation of the BCT Fe atomic volume is predicted and confirmed by Mössbauer experiments. The macroscopic magnetic moment variation is thus related to this atomic volume variation. A transition from a non-magnetic state to a low spin ferromagnetic state is thus observed. This transition is actually predicted by the theory but for isotropically strain films. However, a number of questions still remain about the microscopic magnetic behaviour. In particular, we do not know up to now if the non-magnetic state is really non-magnetic at very low temperatures. Mössbauer experiments performed at low temperatures on superlattices grown with ^{57}Fe are in progress.

References

1- D.Bagayoko, J.Callaway, Phys. Rev. B, 28, (1983), 5419

2- V.L.Moruzzi, P.M. Marcus, J. Kübler, Phys. Rev. B, 39, (1989), 6957

3- G.L. Grasco, Phys. Rev. B, 36, (1987), 8565

4- J. Häglund, Phys. Rev. B, 47, (1993), 566

5- S.Peng et H.J.F.Hansen, J.Appl.Phys., 67, (1990), 4567

6- N.C. Koon, B.T. Jonker, F.A. Volkening, G.A. Prinz, Phys. Rev. Lett., 59, (1987), 2463

7- see M.T.Kief, W.F.Egelhoff, Phys. Rev. B, 47, (1993), 10785 and ref. therein

8- S. Andrieu, M. Hennion, M. Piecuch, MRS Proc., 313, (1993), 135,
S.Andrieu, M. Piecuch, L. Hennet, J. Hubsch, E. Snoeck, Europhys. Lett., 26, (1994),189

9- S. Andrieu, E. Snoeck, Ph. Arcade, M. Piecuch, J. App. Phys., 77, (1994), 1308

10- A. Traverse, S. Pizzini, S. Andrieu, A. Fontaine, M.Piecuch, Surf. Sci., 319, (1994), 131

11- H. Renevier, J. Weigelt, S. Andrieu, R. Frahm, D. Raoux, XAFS VIII conference, Berlin, (1994), in press

12- W.A.A. Macedo, W. Keune, Phys. Rev. Lett., 61, (1988), 475

13- R. Boehler, J.M. Besson, M. Nicol, M. Nielsen, J.P. Itie, G. Weill, S. Johnson, F. Grey, J. Appl. Phys., 65, (1989), 1795

TWO DIMENSIONAL MAGNETIC PROPERTIES OF PdFe LAYERS

F. PETROFF, V. CROS, A. FERT, S. LAMOLLE, M. WIEDMANN, AND A. SCHUHL
Unité Mixte CNRS - Thomson-CSF, Domaine de Corbeville, 91405 Orsay, France

ABSTRACT

We have studied the magnetic properties of very thin PdFe films grown by molecular beam epitaxy. The behavior expected for a 2D Heisenberg system - that is a variation of the susceptibility as $\exp(B/T)$ and a logarithmic dependence of the magnetization on the applied field - is observed in a certain range of temperature and field. This is in agreement with recent results of Webb et al [1].

INTRODUCTION

Two-dimensional (2D) magnetism has been a subject of investigation for many years. The famous Mermin-Wagner [2] theorem shows that long-range magnetic order cannot exist at finite temperature for a 2D isotropic system of Heisenberg spins with short ranged interactions. Unfortunately, real 2D magnetic systems are never perfect and always exhibit a certain degree of anisotropy. The addition of anisotropic terms to the basic Heisenberg Hamiltonian allows finite temperature ordering to occur. The influence of finite size effects, magnetocristalline anisotropy and long-range dipolar interactions on the onset of long-range magnetic order has been studied theoretically by many authors [3,4,5,6]. On the experimental side, 2D magnetism has been widely investigated recently in ultra-thin films of 3d transition metals ferromagnets [7] but there are few clear reports of 2D Heisenberg magnetism [8]. Evidence for a 2D Heisenberg behavior has been reported recently by Webb et al [1] in Pd with dilute Fe impurities (PdFe) thin films. As pointed out by these authors, PdFe is a very interesting system to investigate 2D magnetism. Bulk PdFe is a prototype "giant moment" magnet extensively studied in the litterature [9]. In contrast with PdCo and PdNi [10], PdFe is a 3D Heisenberg magnet with anisotropy and dipolar energies typically 2 orders of magnitude smaller than the exchange energy. From an experimental point of view it has also two additional advantages compared to 3d ferromagnets like Co or Fe. The distance between neighboring magnetic atoms (a critical length scale for 2D magnetism) is typically one order of magnitude larger which permits the use of non-monolayer films. Secondly, the (bulk) Curie temperature is strongly dependent on the Fe concentration and can be adjusted to be in a suitable temperature range for experimental investigation.

In this paper, we report on the magnetic behavior of thin PdFe films grown by molecular beam epitaxy and confirm that PdFe behaves like a 2D Heisenberg magnet down to a given temperature where a cross-over to another behavior is observed.

Mat. Res. Soc. Symp. Proc. Vol. 384 ©1995 Materials Research Society

EXPERIMENTAL DETAILS

The samples were grown by molecular beam epitaxy on chemically etched and in-situ annealed Si(001) and Si(111) single-crystal substrates. The PdFe dilute-alloy layers were grown at room temperature using separate effusion cells for Fe and Pd. As the magnetization of a single PdFe layer is expected to be weak, we have used a multilayer geometry : the samples were made of three to six layers of 17Å thick PdFe layers spaced by thick enough (80Å) Pd spacer layers to avoid magnetic coupling between the PdFe films and reduce the strain related magnetic anisotropy. Prior to the multilayer deposition, buffer layers of Pd 80Å and Ag 500Å/Pd80Å were used on Si(001) and Si(111) respectively. During deposition, the sample holder was rotated in order to avoid anisotropic distribution. The alloy concentration was determined by separate calibrations of the two cells and by controlling the cell temperatures.

Structural characterization based on in-situ RHEED and ex-situ X-Ray Diffraction (XRD), Scanning electron microscopy (SEM) and Atomic force microscopy (AFM) give the following results. On Si(001) the films are polycristalline and the closely packed grains have in-plane sizes of 350-400Å. For Si(111), the films are epitaxially (111)-oriented with in-plane grain sizes of 1500-2000Å. For both orientations, the typical rms surface roughness deduced from AFM is around 10Å. The magnetization studies were performed with a Quantum Design SQUID magnetometer. For the measurements in low applied magnetic field, a Pd reference sample was used to determine the remnant field of the superconducting magnet with a typical accuracy of 0.2 Oe.

RESULTS AND DISCUSSION

We will focus here on two samples with a nominal concentration of 2% and the following characteristics:
sample I is Si(001)//Pd80Å//(17ÅPdFe/Pd80Å)x3
sample II is Si(111)//Ag500Å//Pd80Å//(17ÅPdFe/Pd80Å)x6.
The temperature dependence of the magnetization in a small in-plane magnetic field is shown Fig. 1 for the two samples. The general shape of M(T) and the splitting of the field-cooled (FC) and zero field cooled (ZFC) curves (below 10K in sample I, for example) suggest the existence of magnetic ordering at low temperature. In the low temperature ordered phase, the magnetization versus field curves (not shown) are characteristic of a soft ferromagnet with a coercive field of 10 Oe at T=5K. This ferromagnetic-like ordered phase could be due to some amount of anisotropy, dipolar interactions or finite size effects (grains). In contrast with the results of Webb et al [1], we see no relaxation of the remnant magnetization with time. One can also notice that the ordering temperature is different for samples I and II. We attribute this difference mainly to a higher than expected Fe concentration in sample II.

Our main goal is to investigate the magnetic properties of our samples in the paramagnetic phase far enough above the ordering temperature in the temperature range where one can expect 2D Heisenberg behavior. Before giving our

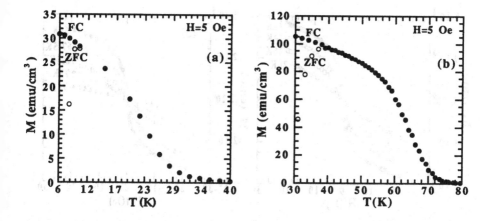

Fig. 1 : Zero field-cooled (ZFC) and field-cooled (FC) magnetization for samples I (a) and II (b) (defined in the text) in a small in-plane field.

experimental results we recall briefly the predictions for 2D Heisenberg ferromagnets.

The general theory of 2D ferromagnets was reviewed by Pokrovsky [11]. The 2D Heisenberg magnet is described in terms of "spin blocks" which act like a giant spin. These blocks can move freely and their size is strongly dependent on temperature. This follows from the temperature dependence of the correlation length and the magnetic susceptibility χ. There is no ordering transition at finite temperature but a divergence of the susceptibility when T tends toward zero. Recent calculations of the T dependence of χ for an isotropic 2D Heisenberg magnet give [12] :

$$\chi \propto \exp(4\pi JS^2/T) \qquad (1)$$

where J is the nearest-neighbor exchange energy and S the spin per atom. The magnetic field H dependence of the magnetization M(H) is also expected to follow a peculiar behavior [13] :

$$M(H)=A Ln(H/H^*) \qquad (2)$$

where A is proportional to T and $H^* \propto \chi^{-1}$ This logarithmic dependence is predicted to be valid at intermediate fields (far from saturation) and preceded at very low H by a linear dependence.

We show in Fig. 2 typical low field M(H) curves at several T for samples I and II. In each case it is clear that features expected for a 2D Heisenberg system are

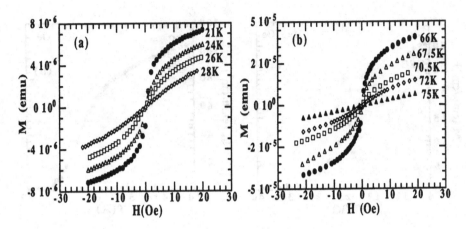

Fig. 2 : Low-field magnetization curves at several temperatures for samples I (a) and II (b)

observed: (i) a sizeable magnetization is generated by very small fields well above the ordering temperature and (ii) a strong T dependence of the magnetic susceptibility is observed. We plot in Fig.3 χ versus 1/T on a logarithmic scale. χ was extracted for each T from the M(H) curves using linear or polynomial fit depending on the linearity of the data and then corrected from the diamagnetic background of the Si substrate.

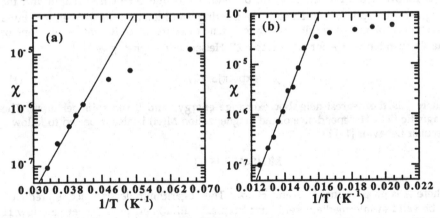

Fig. 3 : Magnetic susceptibility on a logarithmic scale as a function of 1/T for samples I (a) and II (b). The lines are exponential fits using Eq. (1).

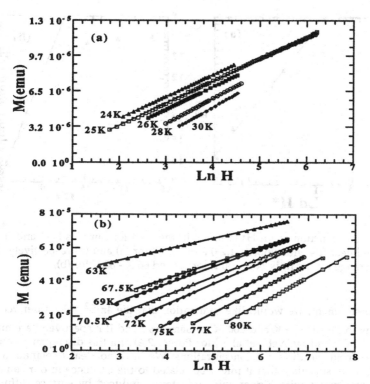

Fig. 4 : Magnetization as a function of Ln H at intermediate fields for several temperatures in samples I (a) and II (b). The lines are linear fits using Eq (2).

The range of T for the so-derived χ is limited at high T by the diamagnetism of Si and at low T by the high non-linearity of the data. We see that equation (1) is obeyed in a given range of T for each sample. This exponential divergence is characteristic of the transition at T=0 of the 2D Heisenberg system but, in our samples, cannot be followed down to zero. A significant departure from the exponential law occurs when approaching the ordering transition due to anisotropy or dipolar fields. The M(H) curves for higher fields in the high-T range are displayed in Fig.4. The logarithmic dependence expected from (2) is observed and can be followed at larger H in the case of sample II (larger magnetization). Furthermore, the slope is found to increase with T. In Fig. 5 we plot Lnχ as a function of LnH* deduced from fitting the M(H) curves (shown in Fig. 4) with Eq. (2) . In the T range where the exponential law is valid, the slopes are close to predicted value (-1) . We note that, although at lower T there is (as for χ) a clear deviation, the M logarithmic dependence on H is observed in a given range of H up to the ordering T as already reported by Webb et al [1]

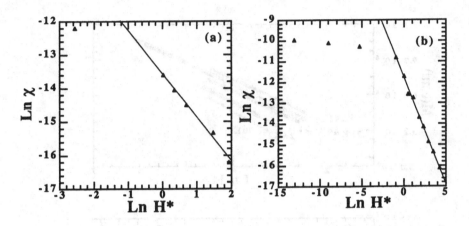

Fig. 5 : Ln χ as a function of Ln H* (defined in the text) for samples I (a) and II (b). The lines are linear fits with slopes (-1.29) and (-1.02) for (a) and (b) respectively. The temperature range for the fit is 24 K - 30 K for (a) and 66 K - 80 K for (b).

As a final comment, we would like to mention that the $4\pi JS^2$ values derived by fitting $\chi(T)$ with Eq. (1) (300K and 2000K for samples I and II respectively) are much higher than 53K found by Webb et al [1] for PdFe(1.2%) and this discrepancy cannot be explained from the different concentrations. We have no clear explanation for this result but we speculate that it might be related to the existence in our samples of Fe pairs or clusters with higher spins (clustering induced by surface diffusion processes in our MBE grown samples).

REFERENCES

[1] D.J. Webb and J. D. McKinley, Phys. Rev. Lett. **70**, 509 (1993).
[2] N.D. Mermin and H. Wagner, Phys. Rev. Lett. **17**, 1133 (1966).
[3] V.L. Berezinskii and A. Blank, Sov. Phys. JETP **37**, 369 (1973)
[4] Y. Yafet, J. Kwo, and E.M. Gyorgy, Phys. Rev. B **33**, 6519 (1986)
[5] M. Bander and D.L. Mills, Phys. Rev. B **38**, 12015 (1988)
[6] P. Bruno, Phys. Rev. B **43**, 6015 (1991)
[7] see for example R. Allenspach, J. Magn. Magn. Mater. **129**, 160-185 (1994) and references therein
[8] D. Kerkmann, D. Pescia, R. Allenspach, Phys. Rev. Lett. **68**, 686 (1992)
[9] G.J. Nieuwenhuys, Adv. Phys. **24**, 515 (1975)
[10] S. Senoussi, I.A. Campbell, A. Fert, Solid State Commun. **21**, 269 (1977)
[11] V.L. Pokrovsky, Adv. Phys. **28**, 595 (1979)
[12] M. Takahashi, Phys. Rev. Lett. **58**, 168 (1987)
[13] S.B. Khokhlachev, Sov. Phys. JETP **44**, 427 (1976)

STUDIES ON MAGNETIC CONFIGURATIONS IN MULTILAYERS BY A QUANTUM SPIN MODEL

YOSHIYUKI KAWAZOE*, MANABU TAKAHASHI* XIAO HU*, AND RUIBAO TAO **
* Institute for Materials Research, Tohoku University, Sendai 980-77, Japan
** Department of Physics, Fudan University, Shanghai 200433, China

ABSTRACT

A method based on the variation of the magnetization direction of each layer and Holstein-Primakoff transformation is presented to estimate the magnetization configuration in a superlattice of quantum ferromagnetic system with an interface between perpendicular and in-plane easy-axis layers. Numerical results on the magnetization configurations under different applied magnetic fields, critical field, and critical anisotropic parameter are given.

INTRODUCTION

Applying a magnetic bilayer disk of a magnetic thin film with in-plane anisotropy (capping layer) coupled to the bulk medium for memory (recording layer) with perpendicular easy axis, one can significantly improve the recording properties[1]. Some magnetic multilayer systems have already shown to have a good potential for higher recording density and to reduce the recording time in magneto-optical recording[2]. Theoretical studies have also been done for double-layer systems [3-6]. Among them a theory for magnetic bilayer system has been developed by two of the present authors using classical continuum model[4]. The theory addressed the competition between the vertical anisotropy of the recording layer and the in-plane one of the capping layer. It clarified the mechanism of transition between two different magnetization configurations; namely a uniformly perpendicular and a bent structures, with the variances of magnetic constants, the thickness of the capping layer, and the temperature. The critical thickness of the above transition gives the minimal thickness of capping layer that shows the capping effect in multilayer structures[1]. The theory explained successfully the main experimental observations. Nevertheless, it still deserves to consider the effect of quantum fluctuation and that of the discreteness of lattice structure in the spin-reorientation transition. The first effect is particularly interesting from the theoretical point of view, and the second one is important for the quantitative estimation of critical values. In the present work, we would extend the former theory to quantum case. As a first step, we consider a multilayer lattice model with an interface between perpendicular and in-plane easy axis layers. To express it explicitly, an anisotropic quantum Heisenberg ferromagnetic model is studied.

HAMILTONIAN

$$\mathbf{H} = \sum_{m,m'} \sum_{\mathbf{R},\mathbf{R}'} \mathbf{H}_{m,m'}(\mathbf{R}, \mathbf{R}') + \sum_{m,\mathbf{R}} [D_m(S_m^z(\mathbf{R}))^2 - hS_m^z(\mathbf{R})], \qquad (1)$$

where

$$\mathbf{H}_{m,m'}(\mathbf{R}, \mathbf{R}') = -\frac{1}{2} I_{m,m'}(\mathbf{R}, \mathbf{R}') \mathbf{S}_m(\mathbf{R}) \cdot \mathbf{S}_{m'}(\mathbf{R}'), \qquad (2)$$

and the subscripts $\{m, m'\}$ denote the layer numbers, $\mathbf{R}$ and $\mathbf{R}'$ are the vectors of a lattice site on the layer, h is related to the applied magnetic field. We only discuss the simple cubic case, in which the layers are arranged along with the [001] direction. For a ferromagnetic system, the coupling constant $I_{m,m'}$ in the Hamiltonian $\mathbf{H}$ is positive. Figure 1 shows the present geometry of the layer structure where z direction is perpendicular to the layer planes. The parameter D_m describes the anisotropy of magnetization.

LOCAL COORDINATES AND BOSE TRANSFORMATION

The local coordinates (LC) are introduced for each layer as shown in Fig.1.

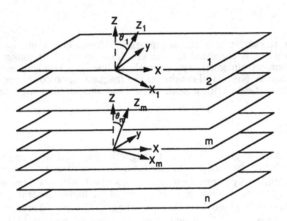

Figure 1: Lattice model of layer structure and local coordinate systems.

The $\mathbf{y}$ axis is always kept along the original direction, and the $\mathbf{x}, \mathbf{z}$ axes are rotated by an angle θ for each layer which may be different from layer to layer. The spins $\mathbf{S}_m(\mathbf{R})$ are expressed in LC systems:

$$
\begin{aligned}
\mathbf{S}_m(\mathbf{R}) &= [\cos(\theta_m)S_m^{z_m}(\mathbf{R}) - \sin(\theta_m)S_m^{x_m}(\mathbf{R})]\mathbf{z} \\
&+ [\cos(\theta_m)S_m^{x_m}(\mathbf{R}) + \sin(\theta_m)S_m^{z_m}(\mathbf{R})]\mathbf{x} \\
&+ S_m^y(\mathbf{R})\mathbf{y}.
\end{aligned}
\tag{3}
$$

The Hamiltonian can be expressed by a function of the components of the spins in LC and the angles $\{\theta_m\}$:

$$
\mathbf{H} = \mathbf{H}(\{S_m^{x_m}(\mathbf{R}), S_m^{y_m}(\mathbf{R}), S_m^{z_m}(\mathbf{R})\}, \{\theta_m\}).
\tag{4}
$$

Then, we apply the Holstein-Primakoff transformation for each spin operator in transformed Hamiltonian (4) and obtain

$$
\tilde{\mathbf{H}} = U_0 + \mathbf{H}_1 + \mathbf{H}_2 + \cdots.
\tag{5}
$$

It is easy to have the expressions of U_0, $\mathbf{H}_1$ and $\mathbf{H}_2$; for example

$$
\begin{aligned}
U_0 &= N_s S^2 \sum_m D_m - \frac{N_s S^2}{2} \sum_m \sum_{m' \neq m} I_{mm'} \cos(\theta_m - \theta'_m) \\
&- h N_s S \sum_m \cos(\theta_m) + \frac{N_s S}{2}(1 - 2S) \sum_m D_m \sin^2(\theta_m).
\end{aligned}
\tag{6}
$$

GROUND STATE IN THE FIRST ORDER APPROXIMATION

In the first approximation, only three terms of $U_0, \mathbf{H_1}$ and $\mathbf{H_2}$ are included in the Hamiltonian $\tilde{\mathbf{H}}$. The ground state energy E_0 can be obtained from the minimum of U_0 by means of the variations of the parameters $\{\theta_m\}$. The necessary conditions are

$$\frac{\delta U_0}{\delta \theta_m} = 0, \quad m = 1, 2, \cdots. \tag{7}$$

They yield the following non-linear equations:

$$S \sum_{m'} I_{m,m'} \sin(\theta_m - \theta'_m) + \frac{1}{2}(1 - 2S)D_m \sin(2\theta_m) + h \sin(\theta_m) = 0. \tag{8}$$

The above equations are the same as the condition of $\mathbf{H_1} = 0$.

In the classical case, the spin is a vector $\mathbf{S}$ with the value S and its direction can be changed continuously so that the energy of the system can be obtained easily from the Hamiltonian (1) where the spins $\{\mathbf{S_m(R)}\}$ are vectors but not like the operators in the quantum case. The classical energy of the system is

$$U_0 = N_s S^2 \sum_m D_m - \frac{N_s S^2}{2} \sum_m \sum_{m' \neq m} I_{m,m'} \cos(\theta_m - \theta'_m).$$

$$- hN_s S \sum_m \cos(\theta_m) - N_s S^2 \sum_m D_m \sin^2(\theta_m). \tag{9}$$

Comparing eq.(9) with eq.(6), there is an additional term $N_s S/2 \sum_m D_m \sin^2(\theta_m)$ in quantum case. If S becomes large, the additional term gives a negligible contribution so that it reduces to the classical results. The classical energy (9) in our discrete model is converted easily to the equation (1) in the previous paper [4] if we take the continuous limit. However, the difference is significant for small S, particularly for $S = \frac{1}{2}$ since the anisotropic term of single ions in the Hamiltonian become the constants for the case of $S = \frac{1}{2}$. Therefore, in this case there exists no spin-reorientation induced by the anisotropic term of single ions. This special case will be treated in the forthcoming paper.

Equation (8) always has trivial solutions $\{\theta_m = 0, \text{ or } 180^0\}$. However, we can find a non-trivial solution if the applied magnetic field is not too high. In that case, the system will have the spin reorientation and presents a non-trivial configuration of the magnetization. We have solved eq.(9) numerically, and the results are shown in Figs.2-5. The first example has ten layers of perpendicular easy axis ferromagnet ($D_2 < 0$) covered by two layers of ferromagnet ($D_1 > 0$) with in-plane easy-axis. The coupling between the layers is denoted by $I_1 = I_{m,m\pm1}$ and on layers by $I_2 = I_{m,m}(\mathbf{R}, \mathbf{R'})$. I_{12} denotes the coupling of the interface between easy axis and easy plane ferromagnets. In Figs.2-3 and Fig.5, as an example to show the basic physical features of the system, without loosing the generalization, we have fixed the parameters as:

$$I_1/I_2 = 1.0, \qquad I_{12}/I_2 = 1.0,$$
$$\tilde{D}_2 = (2 - \tfrac{1}{S})D_2/I_2 = -2.5$$

and

$$\tilde{D}_1 = (2 - \tfrac{1}{S})D_1/I_2 = 0.5,$$

where S can be any value which corresponds to the different values of D_2/I_2, but not be $S = \frac{1}{2}$. Figure 2 gives the obtained spin configurations for the cases of $h'(= h/SI_2) = 0, 0.05$ and 0.1.

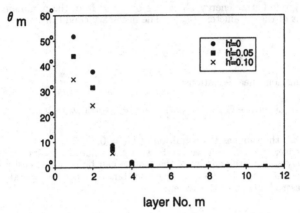

Figure 2: Spin configurations: the angle distribution as a function of number of layers for three different applied magnetic fields: $h' = 0.0, 0.05$ and 0.1.

The changes of spin directions will be significant only for few layers near the interface. Increasing the applied field h results in decreasing angles $\{\theta_m\}$ of the layers.

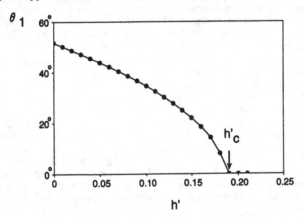

Figure 3: Resulting angle of the magnetization in the first layer related to field h'.

Figure 3 shows the relation of the angle of the magnetization in the 1st layer with applied field h and a critical value $h'_c(\approx 0.185)$ can be found. The critical applied field h_c is proportional to h'_c, $h_c = SI_2h'_c$. The larger S or larger ferromagnetic exchange coupling I_2 offers the larger critical field. If $h \geq h_c$, all of the spins in the system will point to the direction of the field and the deviations of the spin directions can appear only in the case of $h < h_c$. Meanwhile, the value of h_c is changed if we change $\tilde{D}_1$ or $\tilde{D}_2$. The anisotropy $\tilde{D}_1$

dependence of the critical field h_c is shown in Fig.4.

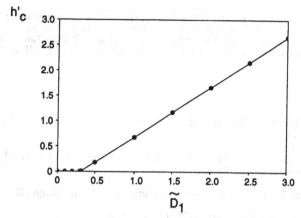

Figure 4: Obtained relation of anisotropic parameter $\tilde{D}_1^c$ and critical field h_c.

We can find another critical value $\tilde{D}_1^c$. The value of h_c is zero when the $\tilde{D}_1 \leq \tilde{D}_1^c$. It means that the spin deviations can appear only when the effective anisotropic parameter $\tilde{D}_1$ is larger than a critical value. We also calculated 21-layer model where the first 10 layers are the in-plane easy axis ferromagnet with $\tilde{D}_1 = 0.5$ and others with $\tilde{D}_2 = -2.5$ and the result is presented in Fig.5. We can see a big jump of angle in crossing the interface. The magnetizations in the most of the regions which are far from interface approach to the bulk value.

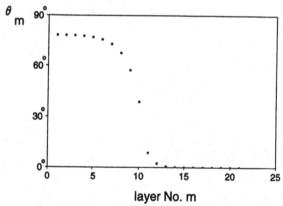

Figure 5: Spin configuration of the model with 21 layers. The interface is located at the boundary between 10th and 11th layers.

We fix D_1 and decrease D_2, and perform some calculations to show that the deviation angle θ_m near the interface and the critical applied field h_c will be increased. Finally, we must mention that all calculations are only for $S \neq \frac{1}{2}$.

ACKNOWLEDGEMENTS

The present numerical calculation has been performed at the Supercomputing Center of the Institute for Materials Research, Tohoku University. The authors are grateful to the crew for the continuous support of the system. One of authors, Ruibao Tao , would express his sincere thanks to the Institute to invite him as the chair donated by the Hitachi Company.

References

[1] S.Ohnuki, K.Shimazaki, N.Ohta and H.Fujiwara J. Magn. Soc. Jpn., **15** Supplement No.S1 (1991), 399.

[2] M.Kaneko, K.Aratani, Y.Mutoh, A.Nakaoki, K.Watanabe and H.Makino, Jpn. J. Appl. Phys. Supplement **28-3** (1989), 927.

[3] X.Hu, T.Yorozu, Y.Kawazoe, S.Ohnuki and Ohta: IEEE Trans. Magn., **29** (1993), 3790.

[4] X.Hu and Y.Kawazoe: Phys. Rev. **B49** (1994), 3294.

[5] Y.Kawazoe, X.Hu and S.Honma: MRS Symp. Proc., **313** (1993), 513.

[6] X.Hu and Y.Kawazoe: J.Appl. Phys., **75** (1994), 6486.

[7] T. Holstein and H.Primakoff: Phys. Rev. **59**,(1940), 1098.

MAGNETIC PROPERTIES OF FE$_X$MN$_{1-X}$/IR(100) SUPERLATTICES.

H. FISCHER, S. ANDRIEU, PH. BAUER AND M. PIECUCH
Laboratoire de métallurgie physique, Sciences des matériaux,URA CNRS N°155,
Université Henri Poincaré, BP 239, 54506 Vandoeuvre-les-Nancy Cedex, France

ABSTRACT

We have produced pseudomorphic Fe$_x$Mn$_{1-x}$/Ir(100) superlattices having different stoichiometry ($0.8 \geq x \geq 0.3$). The alloy crystalline structure is body centered tetragonal with a c/a ratio between 1.18 and 1.26. Iron rich alloys are ferromagnetic when the corresponding bulk alloys are antiferromagnetic. Manganese rich alloys are certainly antiferromagnetic according to bulk magnetization measurements and Mössbauer effect results. The transition from a ferromagnet with a vanishing moment when $x = 0.5$ to an antiferromagnet is associated with a volume expansion.

INTRODUCTION:

The elaboration of epitaxial new phases of transition metals offers unique opportunities to study the relationship between crystalline structure and magnetism [1]. During the recent years, we have studied the properties of body centered tetragonal iron on (001) Iridium. The main result of this study [2] is the appearance of a magnetic moment at a critical volume of about 0.012 nm^3. This volume is very close to the prediction of appearance of magnetism in fcc iron [3]. Manganese is also an attractive transition metal since some theoretical work [4,5] have predicted the possibility of having a ferromagnetic phase with a large magnetic moment in the body centered cubic phase of manganese. However, despite a lot of work (see eg[6]), no one has succeded in producing bcc manganese. Moreover, the body centered tetragonal phases realised by several workers [6] seems to be antiferromagnetic as explained by Oguchi and Freeman [7]. We try another route to have ferromagnetic manganese. Since we have grown ferromagnetic iron on Ir (001) and non magnetic (or antiferromagnetic) manganese on the same Ir (001)[8], we believe that in the Fe$_x$Mn$_{1-x}$ alloys on Ir (001) we must have somewhere a magnetic transition between ferromagnetism and antiferromagnetism. The goal of this paper is to investigate this possibility.

GROWTH AND STRUCTURE

Using MBE technique, we have grown Fe$_x$Mn$_{1-x}$ with $0.8 \geq x \geq 0.3$ on Ir(100). A thick buffer layer of Ir(100) was first grown on (100) MgO. The growth of the alloys on Ir is pseudomorphic, layer by layer with RHEED oscillations up to about 20 atomic planes at room temperature and up to 10 atomic planes at 100°C. The atomic planes present then a square symetry and a parameter equal to 0.2715 nm. We have studied pseudomorphic superlattices of about 9 atomic planes of Fe$_x$Mn$_{1-x}$ alloys and 2 nm of Ir. The exact thicknesses of the superlattices components was determined by Small Angle X-Ray Reflectivity. Results of the simulation of the reflectivity spectra are reported in table I, t is the thickness and σ the roughness of the interfaces. The results of reflectivity were confirmed by cross section transmission electron microscopy.

The crystalline structure of these superlattices was determined by X- Ray Scattering. We have fitted the diffraction results with a model taking into account a plane of intermediate concentration at the interfaces. We have deduced from the fits the structural parameters reported in table II. One

Mat. Res. Soc. Symp. Proc. Vol. 384 © 1995 Materials Research Society

can see that the interplane distances in FeMn alloys, d_{FeMn}, are between 0.161nm and 0.171nm. The in-plane parameter is equal that of iridium because of psudomorphy : we are then able to conclude that the alloys are body centered tetragonal with a c/a ratio between 1.19 and 1.26. The atomic volume, between $0.01187nm^3$ et $0.0126nm^3$, are large compared to fcc Mn and fcc Fe. We have explained these results with the elastic theory : we have shown that the crystalline structure in the alloy layer can be viewed as a tetragonal deformation of the bulk fcc alloys. Moreover a large volume expansion is observed for a concentration of iron smaller than 50%, while the volume remains about constant for large concentration of iron (see table II).

%Fe	t_{buffer}(nm)	σ_{buffer}(nm)	t_{FeMn}(nm)	σ_{FeMn}(nm)	t_{Ir}(nm)	σ_{Ir}(nm)
74	84	0.2	1.28	0.22 to 0.85	1.86	0.22 to 0.85
59	49.5	0.2	1.27	0.35	1.86	0.35
57	60	0.25	1.25	0.25 to 0.9	2	0.25 to 0.9
48	49.5	0.2	1.15	0.3	1.87	0.3
39	50	0.1	1.08	0.18 to 0.3	1.95	0.18 to 0.3
32	48	0.22	1.37	0.22 to 0.62	1.54	0.22 to 0.62

<u>TableI.</u> : Structural parameters of the Fe_xMn_{1-x}/Ir(100) superlattices

%Fe	$\bar{d}$(nm)	d_{FeMn}(nm)	V_{atFeMn} (nm^3)	c/a
74	0.1774	0.161	0.0119	1.19
59	0.1784	0.162	0.0119	1.20
57	0.1795	0.163	0.0120	1.20
48	0.1806	0.165	0.0122	1.21
39	0.1823	0.167	0.0123	1.23
32	0.1817	0.171	0.0126	1.26

<u>Table II.</u> : Crystalline parameters of Fe_xMn_{1-x}/Ir(100) superlattices. $\bar{d}$ is the average d-spacing in the growth direction of the superlattice. d_{FeMn} is the d-spacing in the growth direction of the alloy. The c/a ratio and the atomic volume V_{atFeMn} in the alloy have been calculated with reference to the centered tetragonal structure.

MAGNETIC PROPERTIES :

<u>Bulk magnetization results:</u>

The Fe_xMn_{1-x}/Ir superlattice have been studied by a SQUID magnetometer from 10 to 400K. The magnetic properties depend dramatically on the alloy compositions : iron alloys with iron contents greater than 50% are ferromagnetic at room temperature while the alloys with 32% and 39% iron show only a weak magnetic response at 10K. The alloy of 47% of iron is intermediate between these two groups. Figure 1 shows the magnetization curves of all the samples at 10 K.The three alloys with a large iron content are clearly ferromagnetic. Their Curie temperatures are larger than 400 K : we determine a saturation magnetization of 1200 Gauss at 10 K, and of 1000 Gauss at 300 K for an iron content of 74%. For an iron content of 57%, we measure a saturation magnetization of 500 Gauss at 10 K, and 400 Gauss at 300 K.

For the other alloys, the situation is very different. For the manganese rich alloys, the magnetic signal is weak and close to the detectability limit of our SQUID. These magnetic properties are independant of the temperature. We can conclude that these alloys are either Pauli paramagnets or antiferromagnets with a Neel temperature larger than 400K. The magnetic susceptibility is however relatively large : 5.10^{-5} cm^3/g. The sample with 48% of iron has a M(T) curve which seems to be close to that of a ferrimagnet. (See figure 2.).

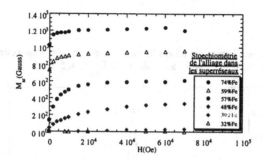

Figure 1. : Magnetization curves of $Fe_xMn_{1-x}/Ir(100)$ superlattices.

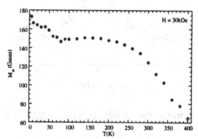

Figure 2 : Magnetization versus temperature for the $Fe_xMn_{1-x}/Ir(100)$ superlattice with 48% iron.

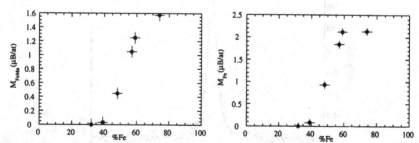

Figure 3 : Magnetic moments per atoms, right if only iron bears the moment, left with an equal moment on iron and manganese.

For the calculation of the magnetic moment of the atoms, we have made two different hypothesis : first, we have supposed that only the iron atoms bear a moment ; secondly, we have supposed an equal magnetic moment on the iron and the manganese atoms with a ferromagnetic coupling between iron and manganese. The results of the two calculations are shown in Figure 3. Figure 3 shows clearly a magnetic transition for the same concentration of about 50% of iron. The left curve of figure 3 is nearly parallel to the Slater Pauling curves of Co-Mn and Ni-Mn bulk alloys.

Mössbauer results :

We have studied three representative alloys by conversion electron Mössbauer spectroscopy at room temperature. The Mössbauer spectra are shown in figure 4, 5 and 6.

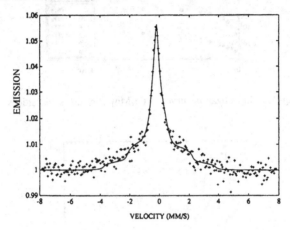

Figure 4. : Mössbauer spectrum of the Fe_xMn_{1-x}/Ir(100) superlattice with 59% of iron.

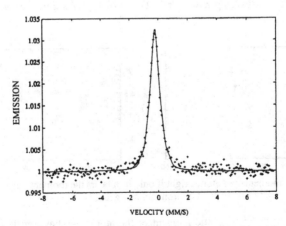

Figure 5: Mössbauer spectrum of the Fe_xMn_{1-x}/Ir(100) superlattice with 48% of iron.

We have simulated these spectra with a distribution of hyperfine fields. In the ferromagnetic region, x > 50% iron, the main results of these simulation is the existence of two parts in the hyperfine field distribution, a part with a very low hyperfine field (16 kOe) and a part with a moderate one (160kOe). The weigth of these two contibutions is about the same. As in these alloys the magnetic moment per atom is rather high (about 1μB per atom) this result seems to be strange. But, we can recall that in bulk alloys [9] which are however antiferromagnetic whatever their composition, the hyperfine field on iron is weak (30KOe) while the iron magnetic moment is also about one Bohr magneton per atom (determined by neutron diffraction). A possible explanation of our Mösbauer spectra is that the iron atoms close to the interface have a larger hyperfine field as in the iron iridium alloys in the iridium rich side (the hyperfine field is weak in iron rich fcc Fe-Ir alloys but is of the order of 150kOe at 4K for a 30% of iron Fe-Ir alloy [10]). Then the weak hyperfine field corresponds to the central iron atoms which are in a crystalline structure close to fcc iron-manganese alloys and have a small hyperfine field like in the bulk alloys, but, may be, a magnetic moment of about 1μB per atom.

The Mössbauer spectrum of the alloy with 48% of iron has only one central part which corresponds to a very weak hyperfine field of about 10kOe. The magnetic moment of this alloys is 0.33μB per atom at room temperature. We can compare this hyperfine field with the bulk alloys of Endoh et Ishikawa [9]. If we suppose that the magnetic moment and the hyperfine field have the same ratio in bulk alloys and in our superlattice, we found 11KOe for iron in the superlattice, a value very close to our experimental determination.

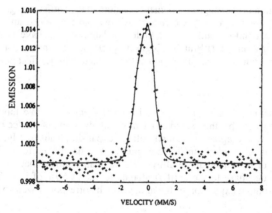

Figure 6. Mössbauer spectrum of $Fe_xMn_{1-x}/Ir(100)$ superlattice with 39% of iron

The spectrum of the alloy with 39% of iron is broader than the former one and corresponds to an hyperfine field of 40kOe. This spectrum is very close to the spectra obtained [9] on bulk alloys. Moreover the magnetization of this superlattice is very weak. By comparison with the bulk alloys, we conclude that this alloy is certainly an antiferromagnet like the bulk fcc alloys.

DISCUSSION

When the iron content is larger than 50%, the alloys are ferromagnetic with a magnetic moment decreasing sharply at 50% of iron. Conversely, the atomic volume increases sharply at the same concentration of iron (see table 2). We can interpret this magnetovolumic transition in the framework of the theory of Moriya [11] : Moriya gives a relationship between the variation of volume and the variation of the square of the magnetic moment as:

$$\frac{\Delta V}{V} \approx \frac{D_0}{B} \Delta S_L^2 = \frac{D_0}{B}\left(S_L^2(AF) - S_L^2(F)\right) \tag{1}$$

where B is the bulk modulus, D_0 the magnetovolumic coupling constant and ΔS_L^2 the variation of the square of the magnetic moment during the transition. Then the observed increase of the atomic volume around 50% of iron can be related to an abrupt increase of the local magnetic moment. As we observe a decrease of the magnetic moment in the ferromagnetic phase when we approach 50% of iron, we must have an antiferromagnetic phase with a large magnetic moment for the concentration smaller than 50% of iron. According to Moriya [11], one has:

$$\frac{D_0}{B} \approx 2.10^{-6}\left(emu/g\right)^2. \tag{2}$$

Our experimental results are:

$$S_L^2(F)-> 0 \quad \text{and} \quad \frac{\Delta V}{V} \approx 2.7\%. \tag{3}$$

Then, the calculation of the magnetic moment per atom in the antiferromagnetic phase leads to a value of 1.2µB/at. For the manganese rich alloys, we have obtained an hyperfine field of 40kOe. For the corresponding bulk alloys, Endoh et Ishikawa [9] have found 33kOe and a magnetic moment per atom of 1µB/at. A single proportionnality shows that the value of the magnetic moment determined by our calculation of the magnetovolume effect is correct.

CONCLUSION:

We have prepared new epitaxial alloys of iron and manganese. We have shown that the structure of these alloys can be understand as a tetragonal distortion of the fcc bulk alloys. In the iron rich side of the phase diagram, the small volume expansion leads to a ferromagnetic phase even though the bulk alloys are antiferromagnetic. We have therefore obtained the first ferromagnetic phase of iron-manganese alloys. We have also shown that we have a transition from the ferromagnetic phase to the antiferromagnetic one at 50% of iron. This transition is accompagnied by a volume expansion of the alloys which corresponds well to the prediction by the theory of Moriya.

References:
- [1] G. A. Prinz, Phys. Rev. Let., 54, 1051 (1985)
- [2] S. Andrieu, M. Piecuch, L. Hennet, J. Hubsch and E. Snoeck, Europhys. Lett., 26, 189 (1994)
- [3] V. L. Moruzzi, P. M. Marcus and J. Kübler, Phys. Rev. B, 39, 6957 (1989).
- [4] J. Kübler, J. Magn. Magn. Mater., 20, 107 (1980).
- [5] P. M. Marcus and V. L. Moruzzi, J. Appl. Phys., 63, 4045 (1988).
- [6] B. T. Jonker, J. J. Krebs and G. A. Prinz, Phys. Rev. B, 39, 1399 (1989)
- [7] T. Oguchi and A. J. Freeman, J. Magn. Magn. Mat., 46, L1 (1984).
- [8] H. Fischer, Thesis, Université Henri Poincaré Nancy, France, (1995).
- [9] Y. Endoh and Y. Ishikawa, J. Phys. Soc. Jpn, 30, 1614 (1971).
- [10] M. Shiga, Phys. Stat. Sol. (b) 43, K37 (1971).
- [11] Tôru Moriya, Spin Fluctuations in Itinerant Electron Magnetism, Springer Verlag, Berlin, Heidelberg, (1985).

DIRECT EXPERIMENTAL STUDY OF DOMAIN STRUCTURE IN MAGNETIC MULTILAYERS

V.I. Nikitenko[*], V.S. Gornakov[*], L.M. Dedukh[*], L.H. Bennett[**], R.D. McMichael[**], L.J. Swartzendruber[**], S. Hua[***], D.L. Lashmore[***], and A.J. Shapiro[**]
[*]Institute of Solid State Physics, Russian Academy of Sciences, Chernogolovka, Moscow Dist., 142432, Russia
[**]National Institute of Standards and Technology, Gaithersburg, Maryland 20899, USA
[***]Materials Innovation, West Lebanon, NH 03784, USA

ABSTRACT

We have applied a polarized light optical microscope in reflective mode with a Bi-substituted yttrium-iron-garnet indicator film with in-plane anisotropy for visualization of the magnetostatic fields produced by nanostructured magnetic CoNiCu/Cu electrodeposited multilayers. By analysis of the magneto-optical stray field image, detailed information is obtained not only on the as-grown multilayer magnetic structure but on its change during the magnetization reversal processes. The influences of crystal lattice defects and nonuniformity of the nonmagnetic spacers thicknesses on the domain wall nucleation and motion are studied. Peculiarities of the re-magnetization of antiferromagnetically exchange coupled multilayers are discussed, including real-time observations of domain wall creep in a constant applied field.

INTRODUCTION

Multilayer systems composed of ferromagnetic layers separated by nonmagnetic metallic spacers (with nanoscale range thicknesses) have become a subject of great interest as a new type of material in the last few years [1,2]. It has been discovered that such artificially modulated structures can possess unique properties. The exchange interaction between the layers can oscillate from ferromagnetic to antiferromagnetic with the thickness of the spacers. Inversion of the magnetization direction in adjacent layers leads to the giant magnetoresistance effect (GMR). Utilization of this effect opens up perspectives in development of new devices based on noninductive reading of magnetically recorded information for future generations of computers.

The character of the spin distribution in antiferromagnetically coupled multilayers can be disturbed by nonuniformities in the nonmagnetic spacer thicknesses, crystal structure defects, and by magnetization reversal processes. As a result, peculiar magnetic domain structures with unusual walls can be formed. Some examples were discussed in reference [3]. If domain walls are nucleated in only one or in only a few layers, their motion is accompanied by changes in the area of "pseudo-domain boundaries" (parallel to the interfaces) and therefore the value of the exchange coupling energy between layers. Changes in "pseudo-domain boundary" area also give rise to changes in resistance through the GMR effect. All

277

Mat. Res. Soc. Symp. Proc. Vol. 384 ® 1995 Materials Research Society of this emphasizes the necessity of the development of nondestructive methods for characterization of the multilayer microstructure and real time investigation of the magnetization processes. Recently we have shown the capabilities of the magneto-optical indicator film (MOIF) method in solving such problems [4]. The MOIF method is much simpler and no less sensitive than well known methods such as magnetic force, optical Kerr, and Lorentz microscopes. The results of investigation of the fundamental processes of magnetization in electrodeposited multilayers are described in the present paper.

EXPERIMENTAL

The CoNiCu/Cu multilayer composed of 200 bilayers was electrodeposited on a (100) oriented copper single crystal substrate from a sulfamate electrolyte containing Co^{2+}, Ni^{2+}, and Cu^{2+} ions in a single cell [5]. Depending on the cathode potential, the deposition of either CoNiCu or Cu layers takes place. The magnetic layer composition is estimated to be $Co_{64}Ni_{31}Cu_5$. The thickness of the magnetic layer is 2 nm, and the Cu layer is 1 nm thick. As previously reported [6], similar superlattices exhibit the giant magnetoresistance effect.

The domain structure was investigated using an advanced high-resolution magneto-optical technique that was first used for the investigation of magnetic flux penetration in high temperature superconductors [7]. A Bi-doped iron-garnet film was placed on the multilayer sample surface. Observation of the magneto-optical patterns is conducted in reflected polarized light through the double Faraday rotation of the light polarization in the indicator film. The distribution of the normal magnetic field component above a magnetic sample surface is revealed as appropriate spatial variation of light intensity across the image.

Fig.1a schematically shows a magneto-optical image of an as-deposited multilayer. Black and white rings near the perimeter of the disk are the result of an alternating polarity of the specimen stray magnetic field (Fig. 1b) which is believed to arise from oscillation of the ferromagnetic - antiferromagnetic coupling

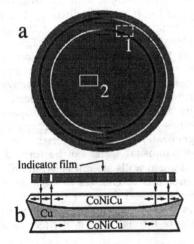

Fig.1. Schematic drawing (a) of the image of magnetostatic stray fields above the 20 mm diameter multilayer disk. (b) The image was obtained using a magnetooptical indicator film, with an aluminum layer on the bottom as a reflector.

between layers which, in turn, is caused by nonuniform spacer thickness. Similar magnetic structures were observed earlier by scanning electron microscopy with polarization analysis [8] and Kerr effect imaging [9] in other layered structures. The ring pattern shown in Fig.1a is destroyed by magnetization of the sample, and does not return upon demagnetization at room temperature. The domains that appear upon reversal of the field are nucleated on defects and sample edges and are described below.

RESULTS AND DISCUSSION

In Fig.2a is shown a micrograph taken near the perimeter of the disk magnetized in an in-plane magnetic field close to an easy direction. The region investigated is shown schematically on Fig.1a by rectangle 1. Only small stray fields near defects can be seen. As the field is reduced and reversed, nucleation of domain walls typically begins near sample defects. In Figs 2b and 2c such nuclei are seen in the lower part of the picture. In order to enhance the image contrast, these and the following figures were adjusted by subtracting the image shown in Fig.2a. The new domains are magnetized in a direction opposite to that of the main volume of the sample. Dark and light colors of the wall images are determined by opposite components of the indicator film magnetization parallel to the light beam and, as a result, by opposite signs of the Faraday effect. The majority of the boundaries observed are charged head-to-head and tail-to-tail domain walls.

At higher fields, when nucleated domains have coalesced, the domain walls take on a sawtooth shape (Fig.2d). Segments of the sawtooth are imaged as broadened dark bands with a visible fine structure of darker sub-bands. The width of the wall image and its fine structure are attributed to nonuniform progress of domain walls in the various layers of the sample. Analysis of image changes induced by increasing external magnetic field, showed that the regions of the crystal swept by these walls were completely remagnetized. When the walls were displaced towards the disk center their images became more and more "diffuse" (Fig.2f) up to complete disappearance.

The thin and less bright magnetooptic stray field images (Figs 2d and 2e) are connected with magnetization reversal in individual magnetic layers. Separation of such thin walls which lead the main wall can be seen in Figs 2d and 2e. Nucleation of the thin walls also occurred at sample defects (Fig.2e).

Fig.3 shows micrographs of the sample cut from the central region of the multilayer. This region is indicated on Fig.1a by rectangle 2. The stray fields at the sample edge, which is almost perpendicular to the magnetization direction (Fig.3a), are imaged as a black narrow band. As the field is reduced and reversed, the nucleation of reversed domains with head-to-head and tail-to-tail walls is observed at $\mu_0 H = -6$ mT. In Figs 3b - 3d domain walls are seen to nucleate near defects and at the sample edge.

One important characteristic property of the domain wall behavior in the material studied is the jump-wise motion of the domain walls in a time-period after the field is changed as shown by the time series in Figs 3b - 3d. The jump-wise, jerky motion of the walls can continue for several minutes after a step wise change of the magnetic field. Distances covered by such domain wall jumps were as large as

279

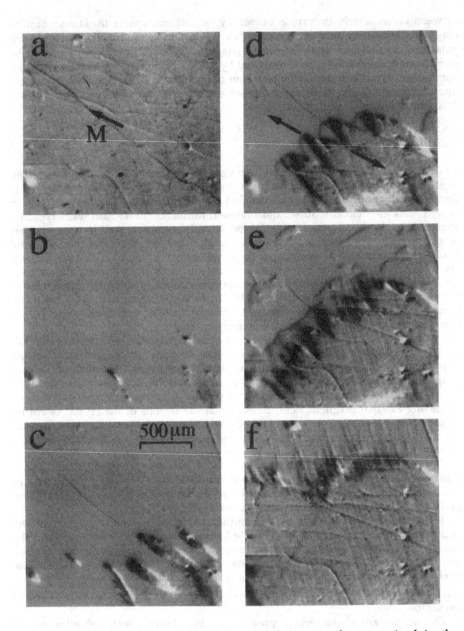

Fig.2. Magneto-optical micrograph of the multilayer sample magnetized in the sample plane in the easy direction (a - μ_0H = -20 mT). Images b - f were obtained at μ_0H = 5 mT (b), 6 mT (c), 7 mT (d), 8 mT (e), and 10 mT (f), and are shown after subtraction of the image in (a).

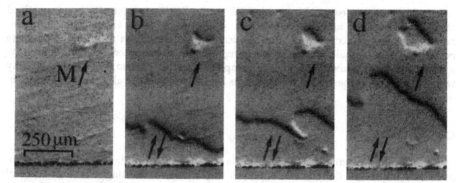

Fig.3. a - Micrograph of the sample with an applied field of +20 mT. The field is in the direction shown by the arrow. Micrographs b - d are obtained during remagnetization at $\mu_0 H = - 6$ mT (b). They are obtained at different times after DW nucleation: b - 1sec, c - 15 sec, d - 60sec.

hundreds of micrometers, depending on the real structure of the crystal.

With increasing field, these walls move towards each other, sweeping over the entire sample. When the processes of domain wall displacement are complete, the stray fields at the bottom sample edge are very weak. However, with increased field, this edge of the sample becomes gradually bright, and changing intensity of color over the area of the sample was observed. Magnetization reversal was almost complete at $\mu_0 H = 25$ mT.

The image of the bottom sample edge in Fig.3 is distorted as a result of over contrasting the images in order to enhance the quality of the wall images. In Fig.4, raw micrographs taken near the edge are shown without using the image over contrasting technique when the sample had been magnetized in opposite directions, (Figs 4a and 4c) and just after displacing the walls (Fig.4b).

These data can be explained by taking into account an antiferromagnetic exchange coupling between layers. In a sufficently strong field, the initial relative antiparallel alignment of magnetization in adjacent magnetic layers changes to a parallel alignment. As the result, high stray fields are observed at the sample edges

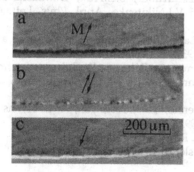

Fig. 4. Image of stray fields at edge of the sample in the multilayer sample with opposite magnetization directions (a and c) and (b) just after displacing domain walls similar to those shown in Fig. 3.

where flux enter or leaves the sample. When the field is turned off, the magnetized state is preserved, and the nucleation of domains with antiparallel spins in neighboring layers occurs only after application of a certain critical field of opposite polarity. These domains form by reversing spins in every other magnetic layer. As a result, the stray fields at edges of the sample disappear. The edge fields appear only when the field reaches a high magnitude leading to parallel alignment of spins in neighboring layers.

CONCLUSION

The results presented show that in the CoNiCu/Cu multilayer system, head-to-head and tail-to-tail domain walls play a major role in the magnetization reversal process. In cases where Cu spacer thicknesses give rise to predominantly ferromagnetic coupling between layers, the magnetizations in adjacent layers are aligned, and the domain walls separate regions with oppositely directed magnetization. The area swept out by these domain walls is completely remagnetized. The fine structure of the domain wall images reveals nonuniform progress of domain walls in individual layers or groups of layers. In cases where the coupling between layers is predominantly antiferromagnetic, parallel alignment induced by a large field is maintained up to a critical oppositely directed field value necessary for nucleation of domain walls. These walls appear to separate regions of parallel alignment of magnetization in adjacent layers from regions of antiparallel alignment. Observed in real time, these domain walls are observed to "creep" in a constant field. The area swept out by these domain walls appears to be demagnetized. Remagnetization occurs continuously without visible domain wall formation.

References

1. A.Fert, P.Grünberg, A.Barthelemy, F.Petroff, and W.Zinn, J. Magn. Magn. Mater., 1-8, 140 (1995),
2. Magnetic Multilayers, edited by L.H.Bennett and R.E.Watson (World Scientific, River Edge, New York, 1994).
3. H.Fujiwara, T.Ishikawa, and W.D.Doyle, J. Appl. Phys. 75, 6446 (1994).
4. L.H.Bennett, R.D.McMichael, L.J.Swartzendruber, S.Z.Hua, D.S.Lashmore, A.J.Shapiro, V.S.Gornakov, L.M.Dedukh, and V.I.Nikitenko, Appl. Phys. Lett. 66, 888 (1995).
5. M.Alper, K.Attenborough, R.Hart, S.J.Lane, D.S.Lashmore, C.Younes, and W.Schwarzacher, Appl. Phys. Lett. 63, 2144 (1993).
6. S.Z.Hua, D.S.Lashmore, L.Salamanca-Riba, W.Schwarzacher, L.J.Shwarzendruber, R.D. McMichael, L.H. Bennett, and R. Hart, J. Appl. Phys. 76, 6519 (1994).
7. L.A.Dorosinskii, M.V.Indenbom, V.I.Nikitenko, Yu.A.Ossip'yan, A.A.Polyanskii, and V.K.Vlasko-Vlasov, Physica C, 203, 149, 1992,
8. J.Unguris, R.J.Celotta, and D.T.Pierce, Phys. Rev. Lett. 67, 140, (1991),
9. M. Ruhrig, R. Schafer, A. Hubert, R. Mosler, J.A. Wolf, S. Demokritov, P. Grunberg, Phys. Stat Sol. (a) 125, 635 (1991).

MAGNETIC PROPERTIES OF GADOLINIUM SILICIDE THIN FILMS FOR DIFFERENT HEAT TREATMENTS

C. PESCHER*, J. PIERRE**, A. ERMOLIEFF* [a], C. VANNUFFEL*
* LETI, CEA, Technologies Avancées DOPT/CPM, 17 Avenue des Martyrs, 38054 Grenoble, cedex 9, France
** CNRS, Laboratoire de Magnétisme Louis Néel, 25 avenue des Martyrs, BP166, 38042, Grenoble cedex 9, France

ABSTRACT

The magnetic properties of heavy rare earth silicide GdSi2-x thin films are investigated as a function of the annealing temperature of the films. Resistivity measurements reveal in the two films annealed at high and low temperatures, but for a short time, the existence of two transition temperatures corresponding to the presence of an ordered and a disordered structure. In the film annealed at high temperature for a long time, only one transition temperature occurs. It corresponds to a magnetic structure transformation.

I INTRODUCTION

Rare earth silicides have retained some attraction due to their potential applications in silicon technology, as heavy Rare Earth (RE) silicides may be obtained as epitaxial layers on (111) Si surface, with stoechiometry close to RE_3Si_5.

Resistivity and magnetic properties of gadolinium silicide films are studied depending on the different heat treatments and on their structure.

II EXPERIMENTAL

Gadolinium silicide films were co-evaporated on (111) Si with a Si/Gd ratio close to 1.7 [1]. Heat treatments were then performed on three different layers. Sample A was annealed at 650°C for 40 min then at 720°C for 15 min, sample B at 740°C for 15 min and sample C at 450°C for 15 min.

The films, approximately 10 to 20 nm thick, were analysed in-situ by Low Energy Electron Diffraction (LEED) and ex-situ by Transmission Electron Microscopy (TEM).

Resistivity measurements were performed below 100K using a 4-probe current method. I(V) measurements were also performed on sample A. The curve obtained (fig 1) is characteristic of a Schottky diode with a low barrier height equal to 0.3 eV.

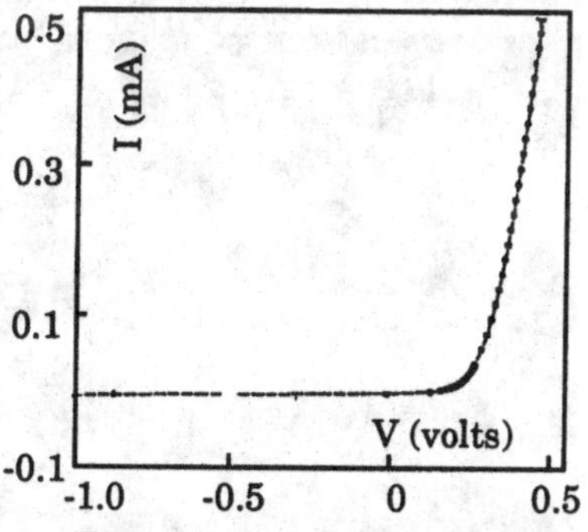

Figure 1: Intensity variation versus the voltage for sample A at 20°C

II a Compound structures

LEED analysis results in two different pattern types depending on the annealing temperatures [1]. Samples A and B present a sharp $\sqrt{3}.\sqrt{3}R30°$ pattern, characteristic of a silicide having the AlB_2 hexagonal structure, with vacancies arranged in a

a: corresponding author

Mat. Res. Soc. Symp. Proc. Vol. 384 © 1995 Materials Research Society

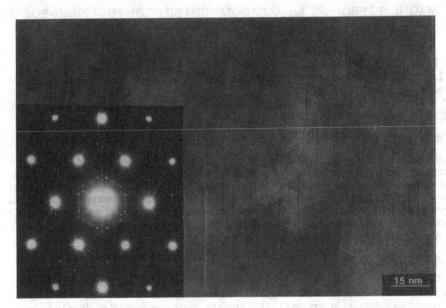

Figure 2: A-film TEM plan view and its associated electron diffraction

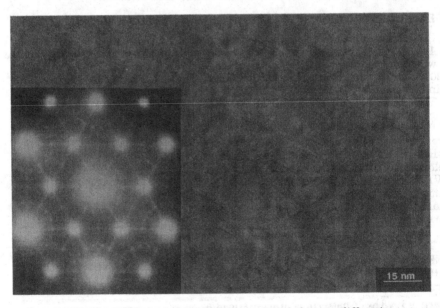

Figure 3: C-film TEM plan view and its associated electron diffraction

hexagonal superstructure.

The C diffraction pattern is of the 1*1 type. It is characteristic [1] of the AlB_2 diagram, the silicide having no Si vacancy.

A TEM (Transmission Electron Microscopy) study on A and C layers was performed (fig 2 and 3). They clearly are single crystals. They contain a rather high density of planar defects. The annealing process dramatically decreases this defect density. Diffraction superlattice reflections similar to those found for other rare-earth silicides (eg T.L. Lee [2] and F.H. Kaatz [3]) strongly suggest the ordering of vacancies in the Si sublattice of $GdSi_2$. A more detailed TEM study is reported elsewhere [3].

II b Resistivity measurements

From resistivity measurements it is possible to deduce the residual resistivity ρ_0 due to the collision of conduction electrons with impurity atoms or mechanical stress in the lattice, and also the magnetic resistivity ρ_m due to crystal field and to spin disorder.

The values obtained are summarised in Table I. The residual resistivity is lower for films elaborated at higher temperature. The high temperature decreases the crystalline defect density, as seen on previous micrographs.

Magnetic-ordering temperatures are better determined from the first derivative of the resistivity curves (fig. 4, 5, 6). Two well separated anomalies are observed for samples B and C, whereas one anomaly or two closely located anomalies occur for sample A. Characteristic temperatures are given in Table II, the highest temperature is always close to 50 K and is obviously a Néel temperature. Note that the knee on the curves around 15 K is not related to any magnetic transition, but to the rate of the thermal population of excited magnetic levels.

TABLE I: Thin film residual and magnetic resistivity ρ_0 and ρ_m

resistivity ($\mu\Omega..cm$)	ρ_0	ρ_m
film A	15.25	8.75
film B	59	14
film C	108.5	15.5

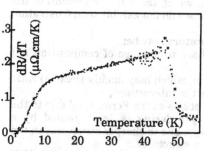

Figure 4: First derivative of the A-film resistivity versus the temperature

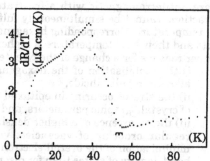

Figure 5: First derivative of the B-film resistivity versus the temperature

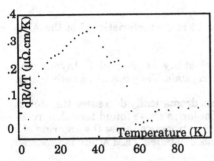

Figure 6: First derivative of the
C-film resistivity versus the
temperature

TABLE II: Thin film critical temperatures

critical temperature (K)	T_N	T_x
film A	50	48
film B	49	40
film C	55	41

III. Discussion

Magnetic and transport measurements have been previously performed in bulk silicides. Orthorhombic GdSi (FeB type) is reported to order at 50K. The hexagonal phase $GdSi_{1.65}$ has a Néel point near 33 K, whereas the orthorhombic phase $GdSi_{1.8}$ orders at 25 K [5, 6] and presents a modification of its magnetic structure near 23 K.

The high transition temperature is close to that of bulk GdSi, whereas the low one is closer to that of hexagonal $GdSi_{1.65}$, thus a first explanation would be that these two phases are present in B and C samples. Indeed X Ray Electron Spectroscopy (XPS) experiments [1,7] show that some concentration of GdSi compound is present in several gadolinium silicide thin films. However, if this hypothesis were true, no $GdSi_{1.67}$ phase would be present in sample A, whereas this sample is the most homogeneous one with the best hexagonal $GdSi_{1.67}$ characteritics.

Conversely, it appears that the Néel temperature of $GdSi_x$ phases (1.6 <x< 1.85) varies rapidly with the silicon content [5]. One reason is related to the hexagonal to orthorhombic transformation: orthorhombic silicides with rare earths from Tb to Ho are antiferromagnetic with a frustrated magnetic structure where magnetic interactions cannot be simultaneously fulfilled, which leads to a reduction of the Néel temperature. Corresponding hexagonal phases do not exhibit such frustration effects and their Néel temperature is higher.

Other reasons for a change of the Néel temperature may be:
   i) the stabilisation of the hexagonal phase in a range of composition not allowed in bulk silicide,
   ii) the strains occuring in epitaxial layers, which may modify the (c/a) ratio of crystallographic parameters, and thus the interactions,
   iii) the occurrence of a higher density of states at the Fermi level due to the regular ordering of vacancies: vacancy ordering is accompanied by a minimisation of the overall electron energy, due to sharper structures in the band density of states than for disordered structures.

Regarding now the occurrence of a second transition temperature around 40 K for samples B and C, it can be attributed to two origins: either the existence of two different ordering temperatures T_{N1} and T_{N2} corresponding to two cristallographic phases (with different compositions or different types of vacancy ordering), or a magnetic structure tranformation at temperature T_x in the case of a unique phase.

The ordering of vacancies may have a strong effect on the ordering temperature. In the tetragonal phases of $CeSi_{1.86}$ [8], $CeGe_{2-x}$ [9] and $PrGe_{1.6}$ [10], ordered and disordered phases coexist. It has been shown that these phases in germanides have rather different ordering temperatures.

In the case of a single crystallographic phase, the second anomaly at T_x may correspond to a transition temperature from a non commensurate magnetic structure to a commensurate one. Such a transformation has already been observed in orthorhombic $GdSi_{1.8}$ [5] and in hexagonal $TbSi_{1.67}$ [11].

We now compare our resistivity data to the theoretical predictions for a simple colinear antiferromagnet. H.Yamada and S.Takada [12] proposed a theory which relies on a mean field calculation taking into account longitudinal and transverse spin fluctuations. The reduced magnetic resistivity $R(T)/R(T_N)$ versus T/T_N was computed by substracting the phonon contribution corresponding to a Debye temperature of 350 K. Its temperature dependence is given in figure 7 for samples A, B and C, and compared to the variation computed within the frame of the Yamada-Takada theory. For each sample, T_N is the temperature of the highest anomaly. It appears that the resistivity of the three samples is higher than the theoretical prediction.

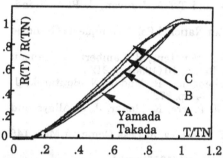

Figure 7: Reduced magnetic resistivity R(T)/R(TN) of the A, B and C films versus T/TN compared to the Yamada and Takada theoretical curve.

Sample A, which was annealed at higher temperature and for a longer time, has an homogeneous structure with ordered vacancies, thus it may be supposed that only one crystallographic phase and one critical temperature exists. The experimental curve is close to the theoretical curve, although slightly above it. The same result was also found for orthorhombic $GdSi_{1.8}$ and hexagonal $GdSi_{1.65}$ polycrystalline samples. This can be attributed to non-colinear or modulated structure occuring below the Néel point. Another reason for the discrepancy may be the fact that the present model is a mean field model: Yamada and Takada found a higher resistivity at low temperature when spin waves are taken into account.

For the other two samples, annealed at two different temperatures for a rather short time, most probably there are two different structures, ordered and disordered, with two different Néel temperatures T_{N1} and T_{N2}.

IV. CONCLUSION

Transport properties of gadolinium silicide thin films strongly depend on the annealing process. Resistivities for samples B and C, annealed at two different temperatures for a short time, seem similar in spite of differences in their LEED diagrams. They are characterised by two Néel temperatures corresponding to an ordered and a disordered phases. Sample A, annealed at higher temperature, has the same LEED diagram as sample B. However, its residual resistivity is lower and the thermal dependence is close to that of a colinear antiferromagnet, which leads to the conclusion that this film contains very few disordered domains. However the Néel temperature encountered in this film is higher than for bulk silicides, which is not completely understood at present.

Acknowledgment

The authors are very grateful to J.Y Veuillen and Nguyen Tan for the fabrication of the GdSi films. One of us, C. P., is indebted to Fisons Instruments for her PhD financial support.

References

1. C. Pescher, A. Ermolieff, J.Y. Veuillen, T. Nguyen Tan, submitted to Solid State Communication
2. T.L. Lee, L.J. Chen, F.R. Chen, J. Appl. Phys. **33**,(1992), p. 2089
3. F.H. Kaatz, W.R. Graham, J. Van der Spiegel, Appl. Phys. Lett. **62** (15) (1993), p. 1748
4. C. Vannuffel, to be published in the Proceedings of the Inst. Phys. Conf., Micros. Semic. Mat., Oxford, 20-23 March 1995
5. S. Auffret, J. Pierre, B. Lambert-Andron, R. Madar, E. Houssay, D. Schmitt and E. Siaud, Physica B **173** (1991), p. 265, and references therein
6. J. Pierre, S. Auffret, J.A. Chroboczek and T.T.A. Nguyen, J. Phys. Cond. Matter **6** (1994), p. 79
7. C. Pescher: thèse de Doctorat de l'Institut National Polytechnique de Grenoble 1995
8. R. Madar, E. Houssay, A. Rouault, J.P. Senateur, B. Lambert, C. Meneau d'Anterroches, J. Pierre, J. Pelissier, J. Mat. Res. 5, **10** (1990), p. 2126
9. B. Lambert- Andron, J. Pierre, B. Chenevier, R. Madar, N. Boutarek, J. Rodriguez- Carvajal, J. Phys. Cond. Mat. **6** (1994), p. 8725
10. B. Lambert- Andron, N. Boutarek, J. Pierre, R. Madar, J. of Alloys and compounds **203** (1994), p. 1
11. P. Schobinger-Papamantellos, K.H.J. Buschow, J. Less Common Met. **146** (1989), p. 279
12. H. Yamada, S. Takada, J. Phys Soc Japan, **34** (1975), p. 51

Part VI

Giant Magnetoresistance I

First Principles Calculation of Electrical Conductivity and Giant Magnetoresistance of Co|Cu Multilayers

W. H. Butler * , X. -G. Zhang * , D. M. C. Nicholson ** and J. M. MacLaren †
*Metals and Ceramics Division, Oak Ridge National Laboratory, Oak Ridge, Tennessee 37831-6114
**Computational Physics and Engineering Division, Oak Ridge National Laboratory, Oak Ridge, Tennessee 37831-6114
†Department of Physics Tulane University New Orleans, LA 70118

ABSTRACT

We show that the Kubo formula can be used to calculate the non-local electrical conductivity of layered systems from first principles. We use the Layer Korringa Kohn Rostoker method to calculate the electronic structure and the Green function of Co|Cu|Co trilayers within the local density approximation to density functional theory. This Green function is used to calculate the conductivity through the Kubo formula for both majority and minority spins and for alignment and anti-alignment of the Co moments on either side of the Cu spacer layer. This allows us to determine the giant magnetoresistance from first principles. We investigate three possibilities for the scattering in Co|Cu|Co: (1) equal electron lifetimes for Cu, majority spin Co, and minority spin Co, (2) equal electron lifetimes for majority and minority Co, weaker scattering in Cu and spin dependent interfacial scattering, (3) electron lifetimes for majority and minority spin cobalt proportional to their Fermi energy densities of states and spin dependent interfacial scattering.

Introduction

Recently there has been great interest in the transport properties of layered magnetic materials because of the discovery of a new form of magnetoresistance[1, 2] called the giant magnetoresistance (GMR). GMR is a change (generally a pronounced decrease) in the electrical resistance of an inhomogeneous system that is observed when an applied magnetic field causes an alignment of the magnetic moments in different parts of the material. GMR has been observed in several geometries, but the most promising and interesting GMR systems are composed of thin layers of ferromagnetic material separated by non-magnetic or very weakly magnetic spacer layers.

The transport properties of layered materials have been the subject of several theoretical investigations based on the model of free electrons with random point scatterers (FERPS). Using this model, Fuchs[3] and later Sondheimer[4] obtained a solution to the semi-classical Boltzmann equation with boundary conditions appropriate to free electrons in a thin film. Barnas and coworkers[5] extended this approach to the case in which the film has several layers with differing scattering rates. Levy and coworkers[6, 7, 8, 9, 10] applied the more rigorous Kubo-Greenwood[11, 12] formula to the FERPS model and developed two different approximations for transport in magnetic multilayers. Zhang and Butler[13] have recently evaluated the Kubo-Greenwood formula exactly for the FERPS model applied to multilayers. Their results allow a comparison of the relative success of the various approximations in representing the conductivity of the free electron model. They found that the semi-classical

Mat. Res. Soc. Symp. Proc. Vol. 384 ©1995 Materials Research Society

approximation works surprisingly well for the FERPS model applied to multilayers.

In addition to theoretical treatments of GMR based on the FERPS model there have been a few previous applications of first principles techniques. Butler, *et al.*[14] calculated the GMR for periodic multilayers of copper and cobalt and of copper and permalloy ($Ni_{.8}Fe_{.2}$). They calculated the complex energy bands using the coherent potential approximation and showed that the imaginary part of the crystal momentum can be interpreted as the inverse of twice the electron mean free path. Their calculations showed that there is the potential for a very large GMR due to spin dependent interfacial scattering because the Fermi energy scattering amplitudes for majority spin cobalt, majority spin nickel and majority spin iron (as an impurity in nickel) are all very similar. Nesbet[15] reached a similar conclusion in studies of periodic Cu_2Co multilayers. Oguchi[16] found that there could be a signifcant band structure effect on the GMR because of differences in the Fermi velocities between the parallel and anti-parallel moment configurations. Schep *et al.*[17] have investigated a very different form of GMR from that seen experimentally by assuming that electron transport is *ballistic* rather than diffusive.

In this paper we report on first-principles calculations of the electronic structure of cobalt-copper multilayers. Using this electronic structure we calculate the conductivity by evaluating the Kubo-Greenwood linear response formula. We do not assume that the scattering is weak or that the electron wave functions are those of free electrons nor do we make the semi-classical approximations necessary to apply Boltzmann theory. It should also be noted that our approach does not require periodicity perpendicular to the layers so that it can be applied to spin valves and trilayers.

Conductivity of Inhomogeneous Systems

We define the nonlocal conductivity $\sigma_{\mu\nu}^{s}(\mathbf{r}, \mathbf{r}')$ as the linear response of the current of electrons of spin s at point $\mathbf{r}$ in direction μ to the local applied field at point $\mathbf{r}'$ in direction ν,

$$J_\mu^s(\mathbf{r}) = \int d\mathbf{r}' \sum_\nu \sigma_{\mu\nu}^s(\mathbf{r}, \mathbf{r}') E_\nu^s(\mathbf{r}'). \tag{1}$$

Here "local applied field" means the change in the local electrostatic field that arises due to the application of a potential difference across the sample. For an inhomogeneous system this may differ from the average applied field and it may be different for different spins[8].

For a homogeneous system, the current and applied field can be assumed to be uniform so that one can define a single conductivity which is also uniform, $J_\mu^s = \sum_\nu \sigma_{\mu\nu}^s E_\nu$. This is the conductivity which is given by the Kubo-Greenwood formula[11, 12],

$$\sigma_{\mu\nu}^s = \frac{\pi\hbar}{N\Omega} \left\langle \sum_{\alpha,\alpha'} \langle\alpha|j_\mu|\alpha'\rangle\langle\alpha'|j_\nu|\alpha\rangle \delta(\epsilon_F - \epsilon_\alpha)\delta(\epsilon_F - \epsilon_{\alpha'}) \right\rangle \tag{2}$$

where j_μ is the current operator, $j_\mu \equiv (-i\hbar e/m_e)\partial/\partial r_\mu$, Ω is the volume per atom and N is the number of atoms. The quantum states $|\alpha\rangle$ in Eq. (2) represent the exact eigenfunctions of a particular configuration of the random potential, and the large angle brackets indicate an average over configurations.

In order to define a non-local site dependent conductivity, $\sigma_{\mu\nu}^{ij,s}$, we define the current density at site i for spin s as the average of the current density over the atomic cell at that site, $J_\mu^{i,s} = \Omega_i^{-1}\int_{\Omega_i} d\mathbf{r} J_\mu^s(\mathbf{r})$. We also assume that the local field, $E_\nu^s(\mathbf{r})$, is constant over each atomic cell. Thus we write Ohm's law in a discrete form in which the current at site i is related to the local electric field at site j through the two point conductivity function, σ^{ij},

$$J_\mu^{i,s} = \sum_{j\nu} \sigma_{\mu\nu}^{ij,s} E_\nu^{j,s}. \tag{3}$$

The superscript s on the local field indicates that it can be spin dependent. The local field will be determined *after* the non-local conductivity is determined by the requirement of current continuity in the steady state, $\sum_\mu \partial J_\mu^s(\mathbf{r})/\partial r_\mu = 0$.

The intersite conductivity, $\sigma_{\mu\nu}^{ij,s}$, is given by Eq.(2) with the matrix element integrals $\langle \alpha | j_\mu | \alpha' \rangle$ and $\langle \alpha' | j_\nu | \alpha \rangle$ restricted to sites i and j respectively and can be seen to depend on the *imaginary part* of the Green function, $\sum_\alpha |\alpha\rangle\langle\alpha| \delta(\epsilon_F - \epsilon_\alpha)$. It can be written in terms of the Green function, $G(\mathbf{r}, \mathbf{r}'; \epsilon_F)$, by writing,

$$\sigma_{\mu\nu}^{ij,s} = \frac{1}{4} \sum_{p,p'=\pm 1} (pp') \tilde{\sigma}_{\mu\nu}^{ij,s}(\epsilon_F + i\eta p, \epsilon_F + i\eta p'), \tag{4}$$

where η is infinitesimal and where

$$\tilde{\sigma}_{\mu\nu}^{ij,s}(z_1, z_2) = \frac{-\hbar}{\pi\Omega_i} \int_{\Omega_i} d\mathbf{r} \int_{\Omega_j} d\mathbf{r}' \, \langle j_\mu(\mathbf{r}) G^s(\mathbf{r}, \mathbf{r}'; z_1) j_\nu(\mathbf{r}') G^s(\mathbf{r}', \mathbf{r}; z_2) \rangle. \tag{5}$$

Following[18] we can write the Green function in terms of the scattering path operator of multiple scattering theory, and the local solutions to the Schrödinger equation. These are determined by the atomic potentials which are obtained self-consistently by using the local spin density approximation to density functional theory. For the case in which the only scattering is due to impurities or to alloying one can use the Coherent Potential Approximation to average the two particle Green function[19]. In this paper we shall take a simpler and more general approach. In realistic GMR systems the scattering usually comes from several sources: impurities, grain boundaries, vacancies, voids, static displacements, phonons, static moment misalignment and magnons. The proper first principles treatment of any one of these scattering mechanisms is quite tedious and the simultaneous treatment of all of them would be difficult and probably pointless since we do not have a sufficiently detailed characterization of experimental GMR systems to know the strengths, concentrations and other relevant parameters of these defects. In this paper we approximate the scattering processes by a phenomenological local scattering rate. Thus we average the two Green functions independently and assume that the effect of this averaging is that each atomic potential acquires an imaginary term which describes the scattering rate in its vicinity.

Application to Layered Systems

We now consider the special case of layered systems. We assume that the system has a two dimensional periodicity, but that its properties may vary in the third dimension. Thus different atomic layers may consist of different types of atoms and have different concentrations of impurities, but there is a common periodicity to all of the layers after averaging over impurity configurations. We use a notation in which a site labeled by i in the preceding section and representing any lattice site in the three dimensional crystal acquires two labels $i \to Ii$, where the upper case I distinguishes different *atomic* layers and the lower case i

labels a site within layer I. The interlayer conductivity can then be written in the form $\bar{\sigma}^{IJ} = N_I^{-1} \sum_{ij} \tilde{\sigma}^{IiJj}$, where N_I is the number of atoms per layer.

Because of the two dimensional periodicity we can relate the Green function which connects any two sites G^{IiJj} to a Green function which connects layers through an integral over the two dimensional Brillouin zone, of area Ω_x.

$$G^{IiJj} = \Omega_x^{-1} \int_{\Omega_x} d^2\mathbf{q} \ G^{IJ}(\mathbf{q}) \ e^{i\mathbf{q}\cdot(\mathbf{R}_i - \mathbf{R}_j)}. \tag{6}$$

These layer Green functions G^{IJ} can be calculated using the layer KKR formalism[20]. The final expression for the conductivity is expressed in terms of matrices indexed by the layer numbers,

$$\tilde{\sigma}_{\mu\nu}^{IJ} = \Omega_x^{-1} \int_{\Omega_x} d^2\mathbf{q} \ M_\mu^I G^{IJ}(\mathbf{q}) M_\nu^J G^{JI}(\mathbf{q}) \tag{7}$$

where M_μ^I represents a dipole matrix element in direction μ evaluated for a site in layer I. Details of the conductivity formalism will be given elsewhere[21].

The local fields can be determined after σ^{IJs} is obtained by using $J^{Is} = \sum_K \sigma^{IKs} E^{Ks}$ and the condition that the current for each spin must be continuous in the steady state. Two geometries are commonly discussed. If the field is applied parallel to the layers, a geometry sometimes referred to as "CIP" for "current in the plane", the local fields will be uniform by symmetry and equal to the average applied field. Thus the overall conductivity will be given by $\sigma = d^{-1} \sum_{IKs} d_I \sigma^{IKs}$ where d_I is the thickness of layer I, and d is the total film thickness. If the field is applied perpendicular to the layers, a geometry referred to as "CPP" for current perpendicular to the planes, then J^{Is}, will be independent of I for each spin. Thus $J^s = \sum_K \sigma^{IKs} E^{Ks}$ and the local fields can be obtained (at least in principle) by inverting σ^{IKs},

$$E^{Is} = \sum_K [(\sigma^s)^{-1}]^{IK} J^s = \sum_K \rho_s^{IK} J^s. \tag{8}$$

Non-Local Conductivity of Free Electrons, Copper, and Cobalt

It is important to understand the non-local conductivity if one wants to understand GMR because it is the non-local nature of the conductivity that leads to GMR. We shall see that the form of the non-local conductivity is a fairly sensitive function of the electronic structure.

We used Eq.(7) to calculate the non-local electrical conductivity for free electrons, for copper and for cobalt using various values for the scattering rate, $\Delta = \hbar/\tau$. Figure 1 shows the non-local layer dependent conductivities for free electrons calculated using our first principles codes compared with exact results from the analytic formulas obtained by Zhang and Butler[13]. The atomic layers were taken to be perpendicular to the (111) direction. These calculations assumed a scattering rate, $\Delta = \hbar/\tau$ of 0.01 Hartree (0.272 eV), one electron per atom, and a lattice constant appropriate to copper (6.8165 Bohr). They were performed as a check of the first-principles code, the validity of approximating the atomic cells by spheres, and the degree of convergence of the integration over the two dimensional Fermi surface.

The agreement is quite satisfactory. We believe that most of the small discrepancy between the analytic and first principles results actually arises from a small difference in the way the spatial averages over layers I and J are performed in the two cases. The first-principles σ^{IJ} involves volume averages of the microscopic non-local conductivity $\sigma(\mathbf{r}, \mathbf{r}')$

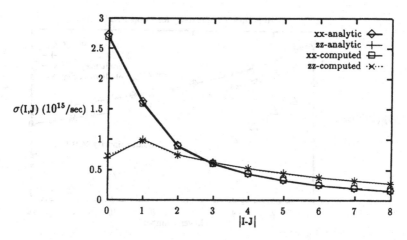

Figure 1: Non-local layer dependent conductivity for free electrons. Diamonds (◇) and squares (□) represent exact analytic results and results calculated using the first-principles code, respectively for CIP. Pluses (+) and crosses (×) represent exact analytic and first-principles results respectively for CPP. The analytic results are averaged over a slab with the thickness of an atomic layer. The first principles results are averaged over the atomic spheres in a plane. The Fermi Energy is 0.2595 Hartrees.

over the atomic cells (here approximated by spheres) in layers I and J. For the analytic free electron results, however, the averages are over slabs with a thickness equal to the interlayer spacing and bounded by planes perpendicular to the z axis.

Figure 2 shows the calculated values of the non-local layer dependent conductivity, $\sigma_{\mu\nu}^{IJ}$, for copper and for cobalt at their respective Fermi energies using a scattering rate of 0.005 Hartree (0.136 eV). The atomic planes were again taken to be perpendicular to the (111) direction. In addition to the calculated non-local conductivities we show attempts to fit these results with the free electron model. For copper, one can obtain a reasonable fit to the non-local conductivity both parallel to the planes, σ_{xx}^{IJ}, and perpendicular to them, σ_{zz}^{IJ}. The fit shown assumes that the Fermi energy is appropriate to one electron per atom (0.26 Hartree) and the effective mass is 1.52 times the free electron mass.

Figure 2 also shows the non-local layer dependent conductivities for majority and minority spin cobalt. The majority spin conductivity was fit to the free electron results using an effective Fermi energy of 0.111 Hartree which agrees qualitatively with a model for majority carriers in cobalt which assumes that the Fermi surface for the majority spins contains less than 0.5 electrons. The scattering rate used in the fit was 0.0046 Hartree. The fit works well for large values of $|I - J|$ but significantly underestimates the conductivity for small values. This can be interpreted as indicating the presence of two types of majority spin cobalt electrons. One type has a relatively short mean free path. The other type has a longer mean free path and fits reasonably well to the free electron model. Also shown is the non-local conductivity of minority spin cobalt. Note that these data points have been multiplied by 0.1 to shift them downwards on the plot. We were unable to obtain a good fit to the free electron model for this data. The free electron model will need to be extended, e.g. by having at least two kinds of carriers, in order to represent the calculated non-local

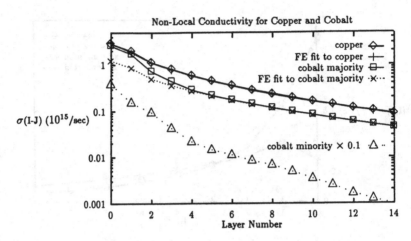

Figure 2: Non-local layer dependent conductivity for copper and cobalt. Diamonds (◇) represent the non-local conductivity for copper, plusses (+) a free-electron fit. Squares (□)represent the non-local conductivity for majority cobalt, crosses (×) a free-electron fit. Triangles (△) represent cobalt minority.

conductivity of minority spin cobalt. It should be noted that the current carried by the minority electrons is not negligible.

The results for cobalt illustrate the difficulty associated with applying free electron models to transition metals. The assumption of the same lifetime for both the majority and minority spins yields, according to our calculations, very nearly the same conductivities for the two channels, $e.g.$ for a scattering rate $\hbar/\tau$ of 0.005 Hartrees we calculate a single channel majority spin resistivity of 58.5 $\mu\Omega$cm and a minority spin resistivity of 60.8 $\mu\Omega$cm. It is clear, however, that the mean free paths are very different for the two channels and that for minority spin cobalt one needs at least two mean free paths to represent the non-local conductivity. This is consistent with our knowledge of d-band metals. The Fermi velocity can vary by large factors over the Fermi surface. Typically the flat portions of the bands contribute strongly to the density of states and they can also contribute moderately to the conductivity but the contribution will be relatively local in nature. The more dispersive portions of the Fermi surface contribute weakly to the density of states but contribute significantly to the conductivity and especially to the non-local conductivity.

Electronic Structure of Copper layers embedded in Cobalt

As a model of the electronic structure of a Co|Cu|Co trilayer we calculated the self-consistent electronic structure of cobalt at its experimental lattice constant. Then we inserted differing numbers of interface cobalt and copper layers into the bulk cobalt, and again solved the electronic structure self-consistently holding the Fermi energy fixed at that of bulk cobalt. We used the Green function technique so that we could treat an infinite system without the need of assuming artificial periodicities. The largest system that we treated had 24 (111) atomic layers that were calculated self-consistently: 7 cobalt followed by 10 copper

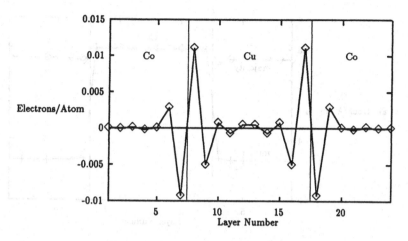

Figure 3: Calculated charge on each layer.

followed by 7 cobalt. These 24 layers were embedded in an infinite matrix of self-consistently determined cobalt (111) atomic layers.

The calculated charge on each of the layers is shown in Fig. 3. The net charge transfer between cobalt and copper is quite small. We calculate that approximately .01 electrons are transferred to the copper, but this number might change slightly if the lattice were relaxed. In these calculations the copper has the same lattice spacing as cobalt. We neglected the small (2%) difference between the lattice constants of bulk cobalt and bulk copper. We also calculated the self-consistent moments and charges for the anti-parallel arrangement of the cobalt moments. The change in the charges and in the magnitude of the moments between the parallel and anti-parallel alignments was less than .001 electrons for every layer.

Figure 4 shows how the valence electrons are divided between the majority and minority spin channels. Note that there is a reasonably close match between the majority Co and the Cu in terms of the number of electrons per atom. The number of valence electrons on the Cu and Co sites differ by less than 0.2 electrons. For the minority spin electrons on the other hand the difference is much larger, more than 1.8 electrons. To a good approximation the electronic structure of ideal Co|Cu interfaces can be understood in terms of a very simple picture. First, there is very little charge transfer between the Co and the Cu. Second, the moment changes are relatively small near the interfaces so that Co moments are all around 1.7 Bohr magnetons. The consequence of this is that the number of valence electrons per atom per spin channel is 5.5 for Cu and approximately 5.35 for majority spin Co and 3.65 for minority spin cobalt.

This approximate "matching" of the number of valence electrons per atom in the majority spin channel means that the atomic cobalt and copper potentials appear very similar to majority spin electrons. This can be verified by considering the scattering phase shifts for electrons at the Fermi energy. These are very similar for copper majority cobalt, but differ greatly for copper and cobalt minority, particularly for the d-phase shifts which because of the large d Fermi energy density of states and the large magnitude of the phase shifts are the primary determinants of the scattering. Another important qualitative difference between the majority and minority spin channels is a large difference in the Fermi energy

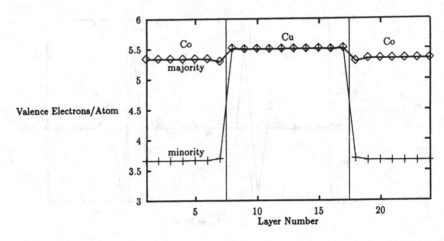

Figure 4: Calculated numbers of majority and minority valence electrons per atom for each layer.

density of states on the cobalt layers. The Fermi Energy DOS for the minority spin channel is approximately 7.34 times as large as for the majority channel for the cobalt layers.

Non-Local Conductivities Near Interfaces

Figure 5 shows calculated non-local layer dependent conductivities for 10 layers of copper embedded in cobalt. This figure shows the conductivity for currents in the plane of the layers, the usual experimental geometry. For this calculation, we assumed the same lifetime, $\hbar/\tau = 0.005$ Hartrees, for the copper layers as for the cobalt layers. Because the majority spin cobalt potential "matches" that of the copper, the non-local layer dependent conductivity for the Co|Cu|Co trilayer in the majority spin channel (Fig. 5a) is very similar to that of pure cobalt (majority spin) or pure copper. The major difference being that the local conductivity is reduced for the copper layer at the interface.

For the minority spin electrons, however, the interfaces greatly modify the conductivities as is shown in Fig. 5b. The conductivities of the copper layers near the interface are greatly reduced. Those on the cobalt layers near the interface are also affected. The local conductivity (peak at $I = J$) is enhanced but the non-local contributions drop off much faster as a function of distance.

The calculated conductivity for anti-parallel alignment for the majority spin channel (relative to the left hand side of the film) is shown in Fig. 5c. These calculations were based on electronic structures calculated self-consistently for the anti-parallel alignment. As might be expected the conductivities on the left hand side appear similar to those of the majority channel for parallel alignment and those on the right hand side appear similar to those of the minority spin for parallel alignment. The conductivity for the other spin channel is identical except reversed left to right.

The difference between the total conductivities for the two alignments is the GMR or more precisely the giant magnetoconductance and is shown in Fig. 5d. The contributions to

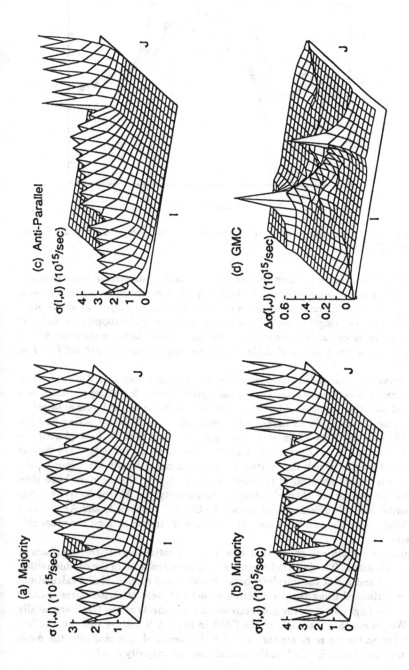

(a) Majority

$\sigma(I,J)$ (10^{15}/sec)

(b) Minority

$\sigma(I,J)$ (10^{15}/sec)

(c) Anti-Parallel

$\sigma(I,J)$ (10^{15}/sec)

(d) GMC

$\Delta\sigma(I,J)$ (10^{15}/sec)

Figure 5: Non-local layer dependent conductivities for 10 layers of copper embedded in cobalt. A scattering rate of 0.005 Hartree was assumed for all layers.

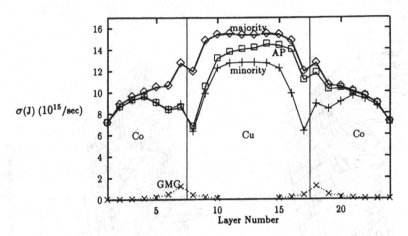

Figure 6: Layer dependent conductivities corresponding to the non-local layer dependent conductivities of Figure 5.

the giant magnetoconductance are seen to arise from completely different regions than the major contributions to the conductivity. The highest peaks correspond to currents flowing in cobalt layers (I) at one interface which sense the field in the cobalt layer (J) at the other interface. There is also a "ridge" of contributions running through the copper, i.e. currents flowing in one copper layer due to fields sensed its mirror image layer on the other side of the interface. There is also a region of slightly negative magnetoconductance for $I \approx J$ in the cobalt layers.

Figure 6 shows layer dependent conductivities, i.e. the sum over I or J of $\sigma(I, J)$ or $\Delta\sigma(I, J)$. It can be seen that the assumption of equal lifetimes for all layers leads to a small GMR and that the magnetoconductance flows mainly in the cobalt layers adjacent to the interface. One can also gain an insight into the origin of the GMR by noting how the anti-parallel (AP) conductivity varies with layer number. Note that we have plotted the conductivity of only one of AP channels because the other is its mirror image. On the left hand side of the figure for which the plotted AP spin channel is locally the minority, the AP conductivity is almost identical to the minority conductivity. On the right hand side, however, where the plotted AP channel is locally the majority, the AP conductivity is less than the majority. It is this difference that causes the GMR. The majority and AP currents on the right hand side of the plot can sense the regions of mean free path on the other side of the interface.

Our calculations contain any effects that arise from potential steps at the interfaces or from quantum well states. It is clear that there are discontinuities in all of the conductivities; majority, minority, and AP at the interfaces between cobalt and copper. Model calculations which we have performed using the free electron model and steps of various sizes indicate that they may have large effects on the conductivities but the effects on the GMR are usually quite small. We believe that the origin of the GMR in figure 6 is not the step but the fact that although the scattering rates are the same for all layers and spin channels, the mean free path for minority cobalt is significantly smaller than for majority cobalt.

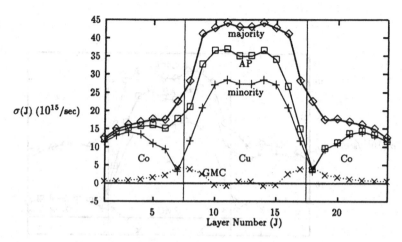

Figure 7: Layer dependent conductivities for strong interfacial scattering.

Interfacial Scattering

In order to evaluate the effect of strong interfacial scattering we calculated the non-local layer dependent conductivities for 10 copper (111) planes embedded in cobalt. We attempted to model a system in which the copper resistivity is 2.8 $\mu\Omega$ cm and the cobalt resistivity is 14.8 $\mu\Omega$ cm. These values seem to be typical of sputtered films[22]. We assumed that the electron lifetimes in the majority and minority spin channels were the same in cobalt (.0025 Hartrees). This would lead to very nearly the same conductivities in the two channels for a system that is entirely cobalt. In addition we assumed that due to intermixing at the interface, the scattering rate for majority spin cobalt at the interfacial layer is twice that in the bulk and for the minority spin it is 24 times that of the bulk. This factor of 12 between the scattering rates of minority and majority spin electrons is based on coherent potential approximation calculations that we performed of the resistivity due to copper impurities in cobalt and (spin aligned) cobalt impurities in copper. The scattering rates for the copper interfacial layer were chosen to be 3.4 and 6.8 times that in bulk copper (.0006 Hartree) respectively for the majority and minority spins. The calculated GMR ($\Delta R/R_p$) for the assumed geometry and scattering rates is 0.035.

Figure 7 shows the layer dependent conductivities. The effect of the interfacial scattering is to strongly depress the minority conductivity in the vicinity of the interface where the strong scattering was assumed. The GMR is seen to be greater for the case with interfacial scattering than for the case in which it was ignored.

Bulk and Interfacial Scattering

The calculations presented in the previous section assumed that the majority and minority lifetimes are the same in the cobalt layers. The scattering rates that occur in practice will depend on the scattering mechanism. For each spin channel, the probability of a scattering event is proportional to the number of final states. For most scattering mechanisms such as nonmagnetic impurities or phonon scattering it means that the scattering rate is

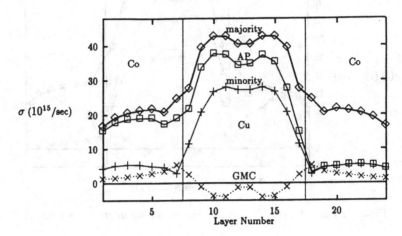

Figure 8: Layer dependent conductivity in the presence of both interfacial and bulk scattering.

proportional to the density of states of the given spin channel at the Fermi energy. Because the density of states is usually much higher at the Fermi energy for the minority spin, the lifetime of minority electrons is usually much shorter than that of majority electrons. Figure 8 shows the calculated layer dependent conductivity assuming that the electron lifetime for majority carriers in the cobalt layers is seven times that in the minority layers due to the difference in Fermi Energy density of states, $\hbar/\tau = .0014, .01008$ for majority and minority, respectively. The GMR in this case is $\Delta R/R_p = .024$.

It is interesting that the introduction of an asymmetry in the bulk scattering rates actually decreased the GMR compared to the result of figure 7. The GMR actually increases substantially in the cobalt but this is more than offset by the decrease in the copper. The decrease in the copper can be traced to the fact that decreasing the scattering rate for the majority electrons in the cobalt *lowers* the conductivity of these electrons in the copper. The overall conductivity is however increased as expected and the GMR would probably have increased if the bulk cobalt layers had been thicker.

Conclusions

We have shown that the Kubo formalism can be evaluated with the Layer-KKR formalism to calculate the non-local layer dependent conductivities and GMR from first principles. We have shown that GMR in the CIP geometry is an inherently non-local phenomenon and that the largest contributions to the GMR come from currents carried near one interface arising from fields sensed near the other interface. Our results also demonstrate that the effects of electronic structure and scattering rates on the conductivity and GMR can be quite subtle.

Acknowledgements

Work at Oak Ridge was sponsored by DOE Assistant Secretary of Defense Programs, Technology Management Group, Technology Transfer Initiative under contract DEAC05-84OR21400 with Martin Marietta Energy Systems and by the High Performance Computing and Communication Initiative. Work at Tulane University was partially supported by the Louisiana Quality Education Support Fund under grant number LEQSF (1991-1994)-RD-A-30.

REFERENCES

1. M. N. Baibich, J. M. Broto, A. Fert, F. Nguyen Van Dau, F. Petroff, P. Etienne, G. Creuzet, A. Friederich, and J. Chaezelas, Phys. Rev. Lett. **61**, 2472, (1988).

2. G. Binasch, P. Grünberg, F. Sauerbach, and W. Zinn, Phys. Rev. B **39**, 4828 (1989).

3. K. Fuchs, Proc. Camb. Phil. Soc. **34**, 100 (1938).

4. E. H. Sondheimer, Adv. Phys. **1**, 1 (1952).

5. J. Barnas, A. Fuss, R. E. Cameley, P. Grünberg, and W. Zinn, Phys. Rev. B **42**, 8110 (1990).

6. P. M. Levy, S. Zhang, and A. Fert, Phys. Rev. Lett. **65**, 1643 (1990).

7. S. Zhang, P. M. Levy, and A. Fert, Phys. Rev. B **45**, 8689 (1992).

8. H. E. Camblong, S. Zhang, and P. M. Levy, Phys. Rev. B **47**, 4735 (1993); H. E. Camblong, Phys. Rev. B **51**, 1855 (1995).

9. H. E. Camblong, P. M. Levy, Phys. Rev. Lett, **69**, 2835 (1992).

10. H. E. Camblong and P. M. Levy, J. Appl. Phys. **73**, 5533 (1993).

11. R. Kubo, J. Phys. Soc. Jpn. **12**, 570 (1957).

12. D. A. Greenwood, Proc. Phys. Soc. London **71**, 585 (1958).

13. X.-G. Zhang and W. H. Butler, Phys. Rev. B **51**, (in press) (1995).

14. W. H. Butler, J. M. MacLaren, and X.-G. Zhang, *Materials Research Society Symposium Proceedings* **313**, 59 (1993).

15. R. K. Nesbet, J. Phys. Condens. Matter **6**, L449 (1994).

16. T. Oguchi, J. Mag. and Mag. Mater., **126**, 519 (1993).

17. K. M. Schep, P. J. Kelly, and G. E. Bauer, Phys. Rev. Lett. **74** 586 (1995).

18. W. H. Butler, Phys. Rev. **31**, 3260 (1985).

19. W. H. Butler, X.-G. Zhang, D. M. C. Nicholson, and J. M. MacLaren, J. Appl. Phys. **76** 6808 (1994).

20. J. M. MacLaren, S. Crampin, D. D. Vvednsky, R. C. Albers, and J. B. Pendry, Computer Physics Communications, **60** 365 (1990).

21. W. H. Butler, X.-G. Zhang, D. M. C. Nicholson, and J. M. MacLaren, *to be published.*

22. Bruce A. Gurney, *private communication.*

GIANT MAGNETORESISTANCE AND ELECTRONIC STRUCTURE

KEES M. SCHEP*,**, PAUL J. KELLY*, AND GERRIT E.W. BAUER**
* Philips Research Laboratories, Prof. Holstlaan 4, 5656 AA Eindhoven, The Netherlands
** Faculty of Applied Physics and Delft Institute of Microelectronics and Submicrontechnology, Delft University of Technology, Lorentzweg 1, 2628 CJ Delft, The Netherlands

ABSTRACT

The electronic structure of magnetic multilayers is expected to play an important role in determining their transport properties. We explain how the conductance through a ballistic point contact is related to simple geometrical projections of the Fermi surface. The essential physics is first discussed for simple model systems and then realistic results for magnetic metallic multilayers based on first principles band structure calculations are presented. The electronic structure is shown to make an important contribution to the perpendicular giant magnetoresistance.

INTRODUCTION

Electrical transport in metallic multilayers has been subject to extensive experimental and theoretical investigation. Most attention has been paid to the giant magnetoresistance (GMR) effect that arises in antiferromagnetically coupled magnetic multilayers when the anti-parallel (AP) magnetizations of adjacent magnetic layers are forced to become parallel (P) by an external magnetic field [1].

All of the experiments which have been performed so far have been in the *diffusive* transport regime, in which the sample dimensions are much larger than the mean free path. In this regime the conductivity is determined both by the electronic structure of the material and by the scattering at defects. This can be illustrated within the free electron model with two commonly used expressions for the Drude conductivity,

$$\sigma_{Drude} = e^2 \left\{ \frac{n}{m} \right\} \tau = \frac{2e^2}{h} \left\{ \frac{k_F^2}{3\pi} \right\} \ell. \qquad (1)$$

σ_{Drude} depends both on electronic structure parameters (in curly brackets) such as the Fermi wave vector k_F or the ratio between the density n and the mass m of the electrons, and on scattering parameters such as the mean free path ℓ or the relaxation time τ. The GMR is usually ascribed to a spin-dependence of the scattering properties which, within the free electron model, corresponds to assuming a spin-dependent ℓ or τ. The electronic structure parameters are on the other hand often taken to be spin-independent and constant throughout the multilayer. Recently, it was pointed out that the difference in the band structures for the AP and the P configurations can also make a large contribution to the GMR [2,3]. We identify the determination of the relative importance of electronic structure and scattering effects as a central issue in any study of the microscopic origin of GMR. Because both of these effects contribute to the diffusive conductivity it is difficult to distinguish between them on the basis of present transport measurements. In the *ballistic* transport regime it *is* possible to evaluate the effect of electronic structure on the GMR unambiguously, both experimentally and theoretically.

Mat. Res. Soc. Symp. Proc. Vol. 384 ©1995 Materials Research Society

BALLISTIC TRANSPORT

Consider two semi-infinite electrodes separated by an insulating barrier and only connected via a small opening in the barrier. When the diameter of the opening is much smaller than the mean free path and much larger than the electron wavelength, such a structure is referred to as a classical ballistic point contact. The resistance of such a point contact is determined by the ballistic motion of the electrons through the opening [4]. Even though the electrons passing through the constriction are not scattered out of their Bloch states, the conductance of the point contact is finite due to its finite cross section A. The net current is given by the difference in the number of electrons incident upon the opening from each side per unit time. Thus for a small voltage difference V between the electrodes (and at low temperatures) the current I in the transport direction $\hat{n}$ is [4]:

$$I = A \, eV \, e \, \frac{1}{2} \sum_{\nu\sigma} \sum_{q} |\hat{n} \cdot \vec{v}_{\nu\sigma}(\vec{q})| \delta(\varepsilon_{\nu\sigma}(\vec{q}) - E_F), \tag{2}$$

where $\vec{v}_{\nu\sigma}(\vec{q})$ and $\varepsilon_{\nu\sigma}(\vec{q})$ are the velocity and the energy, respectively, for a state with Bloch vector $\vec{q}$, band index ν and spin index σ. The factor $1/2$ appears because only electrons moving towards the opening contribute to the current. The summation over $\vec{q}$ can be replaced by an integral over the corresponding sheet of the Fermi surface $FS(\nu\sigma)$. The (Sharvin) conductance $G(\hat{n}) = I/V$ can then be written as

$$G(\hat{n}) = \frac{e^2}{h} \frac{A}{4\pi^2} \frac{1}{2} \sum_{\nu\sigma} \int_{FS(\nu\sigma)} \frac{dS}{|\vec{v}_{\nu\sigma}(\vec{q})|} |\hat{n} \cdot \vec{v}_{\nu\sigma}(\vec{q})| = \frac{e^2}{h} \frac{A}{4\pi^2} \frac{1}{2} \sum_{\nu\sigma} S_{\nu\sigma}(\hat{n}) = \frac{e^2}{h} \sum_{\sigma} N_\sigma(\hat{n}), \tag{3}$$

where $S_{\nu\sigma}(\hat{n})$ is the projection of $FS(\nu\sigma)$ in the direction $\hat{n}$. To express $G(\hat{n})$ in terms of $S_{\nu\sigma}(\hat{n})$ recall that $\vec{v}_{\nu\sigma}(\vec{q})$ at E_F is always normal to the Fermi surface. In the language of the Landauer-Büttiker formalism $G(\hat{n})$ is simply the conductance quantum e^2/h times the number of conduction channels $N(\hat{n})$ [5].

It follows from Eq. (3) that the conductance of a ballistic point contact is completely determined by the electronic structure. Given the Fermi surface, its projection in direction $\hat{n}$, and thus $G(\hat{n})$, can be calculated. This allows for a rigorous theoretical evaluation of the electronic structure effects on the transport properties and, in the case of magnetic multilayers, on the GMR. The Fermi surface of a multilayer structure and its dependence on the magnetic configuration can be calculated using first principles band structure calculations based on the local-spin-density approximation. Ballistic metallic point contacts have already been fabricated (see e.g. [6]), so that the ballistic regime can also be studied experimentally.

MODEL CALCULATIONS

Before evaluating Eq. (3) from first principles, it is instructive to first consider the results for several simple models. The simplest model for the electronic structure is the free electron model with a single parabolic band. The Fermi surface of a free electron gas is a sphere, the projections of the two semi-spheres are circles with radius k_F, thus $S_\sigma(\hat{n}) = 2\pi k_F^2$, independent of $\hat{n}$. By substitution of $S_\sigma(\hat{n})$ in Eq. (3) and using spin-degeneracy the well known free electron expression for the Sharvin conductance G_{FE} [4,7] is obtained:

$$G_{FE} = \frac{2e^2}{h} \frac{Ak_F^2}{4\pi}. \tag{4}$$

Note the resemblance of the electronic structure part of σ_{Drude} to G_{FE}.

Several studies of diffusive transport in multilayers use a single spin-independent para-bolic band to describe the electronic structure throughout the multilayer [1]. For this band model the Sharvin conductances in the AP (G_{AP}) and in the P (G_P) configuration are both equal to G_{FE} and the GMR [GMR $= (G_P - G_{AP})/G_{AP}$] will vanish of course.

In general the Sharvin conductance depends on the transport direction. This can be illustrated using a nearest neighbour tight binding model for a square lattice in two dimen-sions. The dispersion relation depends on the on-site potential ε_0 and the hopping matrix element t:

$$\varepsilon(\vec{q}) = \varepsilon_0 - 2t \left(\cos q_x a + \cos q_y a\right), \tag{5}$$

where a is the lattice parameter. From Eq. (5) the shape of the Fermi surface as a function of the band filling can be obtained. If the Fermi energy is close to the bottom (top) of the band, the Fermi surface resembles the free electron (hole) Fermi circle, as shown schematically in Fig. 1a (Fig. 1c). As E_F gets closer to the middle of the band the Fermi surface starts to de-viate from the free electron behaviour and exactly at half filling it becomes square (Fig. 1b). The Fermi surface always has fourfold symmetry because of the symmetry of the underlying square lattice. From Fig. 1b it follows that the projection and therefore the conductance, depends on $\hat{n}$, e.g. $G(01)$ is a factor $\sqrt{2}$ larger than $G(11)$ at half filling. Thus $G(\hat{n})$ is anisotropic even for a square lattice. For a cubic tight binding model in three dimensions and also for bulk fcc copper [5] the anisotropy in $G(\hat{n})$ is less pronounced. In multilayers, which have uniaxial symmetry, the main anisotropy is between the perpendicular (CPP geo-metry) and the in-plane (CIP geometry) directions, the in-plane anisotropy being very small.

The Kronig-Penney model has been used by several authors to study the effect of a periodic potential in the growth direction z on the GMR [1,7]. The effect of such a potential on the conductance can be significant in the CPP geometry and it is instructive to study this model in the ballistic limit. Consider the potential as a perturbation to the free electron result. In Fig. 2a a cross section of the unperturbed Fermi sphere is plotted in an extended zone scheme. When the multilayer period is larger than the Fermi wavelength λ_F, several Bragg planes cut the Fermi sphere. The perturbed Fermi surface will (almost always) intersect these Bragg planes perpendicularly, as is well known from nearly free electron theory. Gaps open in the projection of the Fermi surface in the z- or CPP-direction, as

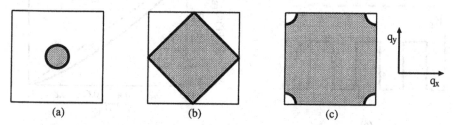

Figure 1 The shape of the Fermi surface (thick line) for a two dimensional tight binding model as a function of the band filling. The large squares are the boundaries of the first Brillouin zone (BZ), the shaded areas represents the filled states. (a) $\varepsilon_0 - 4t \le E_F \ll \varepsilon_0$; (b) $E_F = \varepsilon_0$; (c) $\varepsilon_0 \ll E_F \le \varepsilon_0 + 4t$.

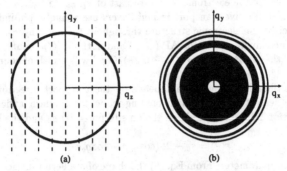

Figure 2 (a) The cross section of the unperturbed Fermi sphere (thick line) at $q_x = 0$. The dashed lines represent the Bragg planes that correspond to the multilayer period in the z-direction. (b) The projection of the perturbed Fermi surface in the z- or CPP-direction. The positions of the gaps (white rings) in the perturbed Fermi surface correspond to the positions where the Bragg planes cut the unperturbed Fermi sphere.

shown in Fig. 2b, reducing the projected area and thus $G(CPP)$. The position of the gaps is determined by the multilayer period, their size by the strength of the potential.

Consider the potential landscape shown in Fig. 3a. When the layer thicknesses are much larger than λ_F, $G(CPP)$ is independent of layer thickness. Fig. 3b shows the dependence of $G(CPP)$ on the height of the potential step U calculated numerically. The main reduction of $G(CPP)$ comes from states with kinetic energy normal to the layers smaller then U, as identified by Bauer [7] (dashed line in Fig. 3b). An additional reduction of $G(CPP)$ is caused by perturbation of the electrons above the barrier by the potential. The dependence of the ballistic conductance $G(CPP)$ and the diffusive conductivity [1,7] on U is similar but not identical.

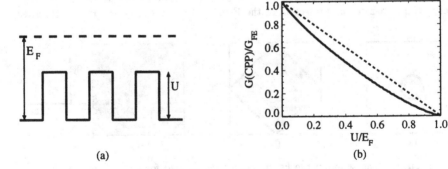

Figure 3 (a) Potential landscape for the Kronig-Penney model. (b) The dependence of $G(CPP)$ (solid line) on U for large layer thicknesses ($\gg \lambda_F$). Dashed line represents the approximate result of Bauer [7].

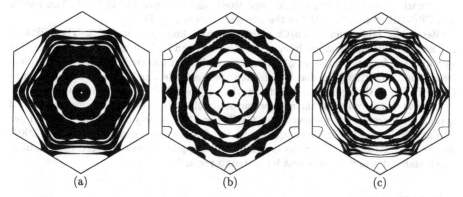

Figure 4 Projections of the Fermi surfaces of a Co_3/Cu_3 (111) oriented multilayer on a plane parallel to the interfaces. The Γ-point is in the middle of each of the figures. The hexagon indicates the boundary of the first Brillouin zone. (a) majority spin in the parallel configuration; (b) minority spin in the parallel configuration; (c) anti-parallel configuration (both spins are the same).

FIRST PRINCIPLES CALCULATIONS

To evaluate Eq. (3) from first principles we calculated band structures in the local-spin-density approximation using the linear muffin-tin orbital method in the atomic spheres approximation. The projections of the Fermi surfaces were calculated using a suitable adaptation of the tetrahedron method. Convergence as a function of the number of k points was obtained using different meshes containing up to 55000 k points in the full Brillouin zone.

Fig. 4 shows the projections in the CPP direction of the Fermi surfaces of a (111) oriented Co_3/Cu_3 multilayer, i.e., each layer is 3 monolayers thick. The projection of the majority spin resembles a free electron projection but the circle is distorted is a way which reflects the sixfold symmetry of the underlying (111) plane. This is similar to the effect found in the tight binding model (Fig. 1b). For (100) oriented multilayers the distortion has fourfold symmetry [3]. As in the Kronig-Penney model (Fig. 2b) gaps - the 'white' rings - have opened due to the multilayer potential. For the minority spin the deviation from free electron behaviour is more pronounced. The total projected area is smaller, which in the language of the Kronig-Penney model corresponds to a higher potential step. Some states (the thin lines) have no dispersion in the direction normal to the multilayer planes and can be identified as quantum well states. Because their velocity is normal to $\hat{n}$, their projected area and thus their contribution to the CPP transport is negligible. Fig. 4c shows the projection of the Fermi surface of the same multilayer in the AP configuration. Because the projections for spin-up and spin-down electrons are identical only one is shown. Fig. 4c is similar to Fig. 4b but the number of gaps is twice as large which is due to the doubling of the unit cell. The qualitative features of Fig. 4 can be understood in terms of the simple models presented in the previous section. To obtain quantitative results, the more realistic band structures have to be used. The calculated Sharvin conductances in the CPP geometry for the majority and the minority spin in the P configuration and for each spin in the AP configuration are 0.41, 0.32 and 0.26 respectively, all in units of 10^{15} $\Omega^{-1}m^{-2}$. This results in a CPP-MR of 40%. The MR in the CIP geometry is only 4%.

Results for (100) oriented Co/Cu multilayers have been given in Ref. 3. The CPP-MR is found to increase for larger layer thicknesses and can be high as 120%. This result is very different from the vanishing of the ballistic GMR that was predicted from the free electron model, which indicates that free electron theories neglect an important contribution to GMR. In Ref. 3 the hybridization between the free electron-like s electrons and the heavy d electrons was shown to be the microscopic origin of the ballistic GMR; neglecting the s-d hybridization results in a collapse of the GMR from 120% to only 3%. We have carried out similar calculations for Fe/Cr multilayers [8] and found magnetoresistances of up to 200% for the CPP geometry. The method can easily be applied to new material combinations which makes it a useful instrument for materials research.

CONCLUSIONS

We have shown that in the ballistic limit transport properties can be evaluated rigorously using parameter-free calculations. Qualitatively, the results can be understood in terms of simple models. Quantitatively, the calculated ballistic CPP-MR is comparable to experimental values in the diffusive regime. Electronic structure effects make an important contribution to the GMR, most probably also in the diffusive regime[†]. We hope that our detailed predictions will stimulate experimental studies of transport in multilayers in the ballistic regime.

REFERENCES

1. P.M. Levy, Solid State Phys. **47**, 367 (1994), and references therein.

2. T. Oguchi, J. Magn. Magn. Mater. **126**, 519 (1993). In calculations which treated the electronic structure realistically and included spin-dependent scattering, large values for the magnetoresistance were found. See W.H. Butler, J.M. MacLaren, and X.-G. Zhang, Mater. Res. Soc. Symp. Proc. **313**, 59 (1993); R.K. Nesbet, J. Phys.: Condens. Matter 6, L449 (1994); Ref. 1.

3. K.M. Schep, P.J. Kelly, and G.E.W. Bauer, Phys. Rev. Lett. **74**, 586 (1995); J. Magn. Magn. Mater. **140-144**, 503 (1995).

4. Yu.V. Sharvin, Zh. Eksp. Teor. Fiz. **48**, 984 (1965) [Sov. Phys. JETP **21**, 655 (1965)]; G. Wexler, Proc. Phys. Soc. London **89**, 927 (1966); A.G.M. Jansen, A.P. van Gelder, and P. Wyder, J. Phys. C: Solid St. Phys. **13**, 6073 (1980).

5. G.E.W. Bauer, A. Brataas, K.M. Schep, and P.J. Kelly, J. Appl. Phys. **75**, 6704 (1994).

6. P.A.M. Holweg et al., Phys. Rev. Lett. **67**, 2549 (1991).

7. G.E.W. Bauer, Phys. Rev. Lett. **69**, 1676 (1992).

8. K.M. Schep, P.J. Kelly, and G.E.W. Bauer (unpublished).

[†]It is not possible to compare the absolute values of the calculated *ballistic* point contact conductance (in units $\Omega^{-1}m^{-2}$) with the *diffusive* conductivity (in units $\Omega^{-1}m^{-1}$) measured in published experiments. The present theory can be tested only by future experiments in the ballistic regime.

UNIFIED SEMI-CLASSICAL THEORY OF PARALLEL AND PERPENDICULAR GIANT MAGNETORESISTANCE IN SUPERLATTICES

V.V. USTINOV AND E.A. KRAVTSOV
Institute of Metal Physics, GSP-170, Ekaterinburg, 620219, Russia

ABSTRACT

The giant magnetoresistance in magnetic superlattices for the current perpendicular to and in the layer planes is studied within a unified semi-classical approach that is based on the Boltzman equation with exact boundary conditions for the spin-dependent distribution functions of conduction electrons. We show that the main differences between the in-plane and perpendicular-to-plane magnetoresistance result from the fact that they originate from different interface processes responsible for spin-dependent scattering. A correlation between the giant magnetoresistance and the superlattice magnetization is also discussed and it is shown that its study has much potential for yielding information about properties of spin-dependent scattering in magnetic superlattices.

INTRODUCTION

A great deal of attention has been devoted recently to the giant magnetoresistance (GMR) that is observed in magnetic superlattices for the current flowing in the layer planes [1] (CIP case) and perpendicular to the layer planes [2] (CPP case). When experimental results for the CIP case are compared with those for the CPP one, it is apparent that, although both the CIP and CPP-GMR are accounted for by spin-dependent scattering, they differ essentially in general behavior (GMR magnitude, magnetic field, thickness and temperature dependences). In order to appreciate physical mechanisms for this difference, the CIP and CPP-GMR should both be considered within a unified theory.

MODEL

Consider an infinite superlattice composed of single-domain ferromagnetic layers with magnetic moments in the layer plane, each layer being L in thickness. The nonmagnetic-spacer thickness is assumed negligible compared with L. Neighboring magnetic moments are considered to be rotated through an angle Θ relative to each other, Θ being equal to Θ_0 in the initial state. An external magnetic field H applied in the layer plane rotates the magnetizations to the parallel arrangement and changes the angle $\Theta = \Theta(H)$. When the magnetic field is strong enough ($H \geq H_s$), the magnetizations are forced to lie in the same direction (Θ=0). The relative superlattice magnetization μ is given as $\mu(H)=M(H)/M_s=\cos(\Theta/2)$ where M is the superlattice magnetization and M_s is the saturation magnetization.

We define the maximum magnetoresistance ratio as $\Delta^G = [\rho^G(H_s) - \rho^G(0)] / \rho^G(0)$ and the relative magnetoresistance as $\delta^G(H) = [\rho^G(H) - \rho^G(0)] / [\rho^G(H_s) - \rho^G(0)]$. Here G defines the geometry under discussion (CIP or CPP). Having obtained $\delta^G(H)$ and $\mu(H)$ from experimental $\rho^G(H)$ and $M(H)$ dependences, one can find a correlation between δ^G and μ and eliminate the common variable H. When experimental data are represented in the form $\delta(\mu^2)$, they should be compared with results of the present theory to estimate microscopic parameters of spin-dependent scattering.

Semi-classical formalism

In our approach the electron energy spectrum $\varepsilon_\sigma(\mathbf{k})$ is assumed to have the form corresponding to the octahedral model of the Fermi surface:

$$\varepsilon_\pm(k) = v^\pm \left(|k_x| + |k_y| + |k_z| \right) \mp \varepsilon_a \tag{1}$$

where ε_a is the spin-splitting energy. Transport properties of the ferromagnet depend on the electron velocity $v^\pm = \partial \varepsilon_\sigma / \partial k$ and the intralayer relaxation time of momentum τ_σ.

The rigorous semi-classical treatment of the superlattice response to an applied electric field E requires solving a system of equations for non-equilibrium parts of the distribution function in each layer $\phi_\sigma(r,k)$ together with a set of boundary conditions that establish a link between the distribution function of electrons moving away from the boundary and that of electrons incident on the boundary from both layers. If x is the axis perpendicular to the layer plane, the Boltzman equation for $\phi_\sigma(\mathbf{r},\mathbf{k})$ in the relaxation-time approximation takes the form

$$v_x \partial \phi_\sigma / \partial x + evE(x)\delta(\varepsilon_\sigma - \varsigma) = -\left(\phi_\sigma - \langle \phi_\sigma \rangle\right)/\tau_\sigma \tag{2}$$

where ς is chemical potential and $\langle \phi_\sigma \rangle$ is the local-equilibrium part of $\phi_\sigma(\mathbf{r},\mathbf{k})$ defined by

$$\langle \phi_\sigma \rangle = \delta(\varepsilon_\sigma - \varsigma) \int dk \phi_\sigma / \int dk \delta(\varepsilon_\sigma - \varsigma). \tag{3}$$

The integrals in (6) are taken over the Fermi surface.

To write the boundary conditions, we introduce quantities characterizing the interaction of electrons with the interface. Let R_σ (P_σ) be the specular (diffusive) reflection probability for an electron of spin σ and $T_{\sigma\sigma'}$ ($Q_{\sigma\sigma'}$) be the probability for an electron of spin σ (with respect to the magnetization in layer i-1) to pass coherently (diffusively) through the interface into spin state σ' (with respect to the magnetization in layer i). Then the boundary condition at the interfaces $x = x_i$ can be written as

$$\phi_\sigma^{(i)}(k_x) = R_\sigma \phi_\sigma^{(i)}(-k_x) + \sum_{\sigma'=\pm} T_{\sigma\sigma'} \phi_{\sigma'}^{(i-1)}(k_x) + P_\sigma \langle \phi_\sigma^{(i)}(-k_x) \rangle$$
$$+ \sum_{\sigma'=\pm} Q_{\sigma\sigma'} \langle \phi_{\sigma'}^{(i-1)}(k_x) \rangle \tag{4}$$

Let us specify the angular dependence of the probabilities introduced that is due to rotating the spin quantization direction of an electron in traveling from one layer to its neighbor. It can be shown [3] that the probabilities take the form:

$$T_{\sigma\sigma}(\Theta) = t_\sigma \cos^2(\Theta/2); \quad T_{-\sigma\sigma}(\Theta) = t \sin^2(\Theta/2); \quad Q_{\sigma\sigma}(\Theta) = q_\sigma \cos^2(\Theta/2);$$
$$Q_{-\sigma\sigma}(\Theta) = q \sin^2(\Theta/2); \quad P_\sigma(\Theta) = p_\sigma^f \cos^2(\Theta/2) + p_\sigma^{af} \sin^2(\Theta/2). \tag{5}$$

Here the various t, q and p are parameters of the theory.

Having found $\phi_\sigma(x,\mathbf{k})$ from (3-4), we calculate the non-equilibrium charge density $n(x)$ and the current density $j(x)$ by using

$$n(x) = \frac{e}{(2\pi\hbar)^3} \sum_{\sigma=\pm} \int dk \phi_\sigma, \qquad j(x) = \frac{e}{(2\pi\hbar)^3} \sum_{\sigma=\pm} \int dk v \phi_\sigma. \tag{6}$$

The calculational techniques to be used are different for different geometries. In the CIP case the electric field is uniform ($n(x)=0$) but there is non-uniform current density. The in-plane magnetoresistance ρ^{CIP} is obtained by averaging $j(x)$ over the layer thickness. As to the CPP case, there is uniform current density but non-uniform electric field $E(x)$. The charge density $n(x)$ given by (6) is a functional of $E(x)$. By using the Maxwell equation $dE / dx = 4\pi n(x)$, one comes to a integro-differential equation for $E(x)$. Having averaged $E(x)$ over the layer thickness, one obtains the perpendicular-to-plane magnetoresistance ρ^{CPP} .

Results

The analytical results obtained are too complex because many independent parameters enter the final formulae. To simplify the problem, we take additional assumptions and ignore any difference between the Fermi velocities ($v^+=v^-= v_F$) and the Fermi-surface areas ($A^+=A^-=A$) for electrons of opposite spin. Then the magnetoresistance ρ^G has the same form in both geometries:

$$\rho^G = \frac{\rho^G_+ \rho^G_- + (\rho^G_+ + \rho^G_-) \rho^G_{mix}}{\rho^G_+ + \rho^G_- + 4\rho^G_{mix}}. \tag{7}$$

Here the "partial resistance" ρ_σ^G is given by the sum $\rho_\sigma^G = \rho_\sigma + r_\sigma^G$ where $\rho_\sigma = (2\pi\hbar)^3/e^2 A v_F \tau_\sigma$ is the resistivity of the spin-subzone σ in the bulk ferromagnet and r_σ^G is an interface contribution. The resistivity ρ^G_{mix} has been introduced to take into account "mixing" processes which are due to the transmission of electrons between layers with different magnetizations. Being considered in different geometries, the interface contributions to ρ_σ^G and ρ^G_{mix} are determined by different combinations of probabilities P, R, Q, T. In case of the CPP geometry the total penetration probabilities $W_\sigma = T_{\sigma\sigma} + Q_{\sigma\sigma}$ and $W = T_{+-} + Q_{+-}$ descriptive of the electron transmission from spin-subzone σ in one layer to spin-subzone σ and $(-\sigma)$ respectively in the neighboring layer are of importance:

$$r_\sigma^{CPP} = \frac{1-[W_\sigma + \dfrac{W}{2}(1+\dfrac{W_\sigma}{W_{-\sigma}})]}{W_\sigma + \dfrac{W}{2}(1+\dfrac{W_\sigma}{W_{-\sigma}})} r_L, \qquad \rho^{CPP}_{mix} = \frac{Wr_L}{W_+ W_- + \dfrac{W}{2}(W_+ + W_-)} \tag{8}$$

where $r_L = \rho_\sigma l_\sigma/L$ ($l_\sigma = v_F \tau_\sigma$ is the mean free path). With the CIP problem, the total probability of diffusive scattering $S^\sigma = P_\sigma + Q_{\sigma\sigma} + Q_{\sigma(-\sigma)}$ and the factor T_{+-} are crucial:

$$r_\sigma^{CIP} = S_\sigma r_L, \qquad \rho^{CIP}_{mix} = T_{+-} r_L. \tag{9}$$

The correlation between the relative magnetoresistance δ^G and the relative superlattice magnetization μ takes the form

$$\delta^G(\mu) = 1 - \frac{1-[\alpha^G \mu^2 + \beta^G \mu^4]/[1-(1-\alpha^G-\beta^G-\gamma^G)\mu^2 - \gamma^G \mu^4]}{1-[\alpha^G \mu_0^2 + \beta^G \mu_0^4]/[1-(1-\alpha^G-\beta^G-\gamma^G)\mu_0^2 - \gamma^G \mu_0^4]} \tag{10}$$

where $\mu_0 = \cos(\Theta_0/2)$ and parameters α^G, β^G, γ^G depend on properties of bulk and interface spin-dependent scattering, with γ^{CIP} being equal to zero.

A comprehensive analysis of our results will be given elsewhere [3] . Here we touch on some interesting physical consequences of the results obtained.

1. The GMR magnitude in the CIP case and the same in the CPP case are defined by different sets of microscopic parameters characterising the interface properties. Consequently, there is no definite relationship between the CIP-GMR and the CPP-GMR. As a rule $\Delta^{CPP} > \Delta^{CIP}$, but

it is not inconceivable that one can discover layered systems where the CIP-GMR exceeds the CPP-GMR.

2. In superlattices with thin magnetic layers ($L \ll l_\sigma$) the GMR may be observed, even if the magnetizations are not ordered antiferromagnetically in zero magnetic field. For this to happen, the angle θ_0 between the neiboring magnetizations at $H = 0$ must exceed a critical value $\Delta\theta$ which can be estimated $\Delta\theta \approx \max\{L/l_\sigma; S_\sigma\}$ for the CIP geometry and $\Delta\theta \approx \max\{L/l_\sigma; 1 - W_\sigma\}$ for the CPP one.

3. With the proviso that $L \gg l_\sigma$, it is possible to expect that the CPP magnetoresistance peaks at a magnetic field H_0 less than the saturation field H_s; rough estimates give $H_0/H_s \propto \sqrt{l_\sigma/L} \ll 1$.

4. There are essential differences between Δ^{CIP} and Δ^{CPP} in their thickness dependence. The first may have either one or two damping length, depending on the ratio of diffusive scattering to coherent transmission at the interfaces. In the limit $L \gg l_\sigma$ we always find $\Delta^{CIP} \propto L^{-1}$. Generally, there are three scaling lengths in the latter case. These are the Debye screening length, the spin diffusion length l_{sf} and a characteristic length at which the interface contributions to the resistance vanish. In the limit $l_\sigma \ll L \ll l_{sf}$ one finds $\Delta^{CPP}(L) \to$ const.

COMPARISON WITH EXPERIMENTS

We take as our first example the experimental data of Pratt et al.[2] who observe Δ^{CIP}=-0.127 and Δ^{CPP}=-0.415 in Ag/Co. The experimental results and corresponding theoretical curves are depicted in Fig 1.

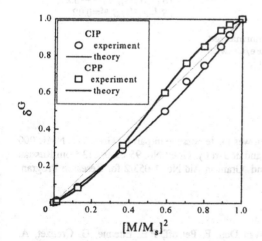

The fitting parameters are found to be α^{CPP}=0.46, β^{CPP}=0.39, γ^{CPP}=-1.04, α^{CIP}=0.76, β^{CIP}=0. Having analyzed the results of the fit, we can give some conclusions about the nature of the GMR effect in Ag/Co. It can be shown in general that the value β^{CPP} is defined by the asymmetry in diffusive scattering. The fact that β^{CPP}=0 allows us to say with certainty that there is no asymmetry in diffusive scattering, i. e., interfacial diffusive scattering is spin-independent and does not depend on what kind of the magnetic ordering appears in the superlattice.

Figure 1: Theoretical and experimental [2] normalized magnetoresistance in Ag/Co as a function of $(M/M_s)^2$

So the CIP-GMR originates from bulk spin-dependent scattering. By assuming that diffusive scattering is insignificant, we obtain the following estimates of microscopic parameters: $l_-/l_+ = 4.2$; $tL/l_+ = 0.09$; $tL/l_- = 0.41$. With the CPP-GMR, it is significant that both interface (from specular reflection) and intralayer contributions to the resistivity play important parts in the magnetoresistance behaviour.

We consider as the second example our data on the in-plane magnetoresistance in superlattices with non-collinear magnetic ordering. The measurements were taken at room temperature on $[Fe(23A)/Cr(8A)]_{30}$ multilayer grown by MBE on MgO substrate and characterized by X-ray diffraction. The magnetic moments of neighboring layers are non-collinear and in zero magnetic field $\Theta_0 = 134°$ The experimental results and the corresponding theoretical curve are depicted in Fig. 2.

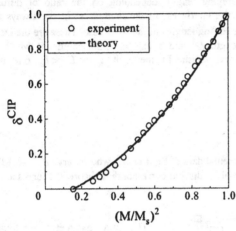

The fitting parameters are found to be $\alpha^{CIP}=0.34$, $\beta^{CIP}=0.41$. If the assumption is made that interface spin-dependent scattering is dominant in the system , we obtain the following values of interface diffusive scattering parameters S_σ at antiferromagnetic (af) and ferromagnetic (f) arrangement of the magnetic moments of the multilayer:

$S_-^{af}/S_+^{af}= 13.8$;
$[S_+^f-S_+^{af}]/S_+^{af}=-0.27$;
$[S_-^f-S_-^{af}]/S_-^{af}=0.02$.

Figure 2: Theoretical and experimental normalized in-plain magnetoresistance in Fe/Cr with non-collinear magnetic ordering.

ACKNOWLEDGMENT

The research described in this publication was made possible in part by Grant No. NMK 000 from the International Science Foundation and in part by Grant No. 95-02-04813 from Russian Foundation of Fundamental Researches and Grant-in-Aid No. 1-053/2 for Research Program from the Ministry of Science of Russia.

REFERENCES

1. M.N. Baibich, J.M. Broto, F. Nguyen van Dau, F. Petroff, P.E. Eitenne, G. Creuzet, A. Friederich, and J. Chazelas Phys. Rev. Lett. 61, 2472 (1988) .
2. W.P. Pratt, Jr., S.F. Lee, J.M. Slaughter, R. Loloee, P.A. Schroeder, and J. Bass Phys. Rev. Lett. 66, 3060 (1991).
3. V.V. Ustinov and E.A. Kravtsov J. Phys.: Condens. Matter 7 (1995), in press.

EFFECT OF THE ORIENTATION OF THE MAGNETIC FIELD ON THE GIANT MAGNETORESISTANCE OF Fe/Cr SUPERLATTICES

V.V. USTINOV, V.I. MININ, L.N. ROMASHEV, A.B. SEMERIKOV AND A.R. DEL
Institute of Metal Physics, GSP-170, Ekaterinburg, 620219, Russia

ABSTRACT

We study the magnetoresistance of $[Fe/Cr]_{30}/MgO$ superlattices grown by molecular beam epitaxy at a various magnetic field directions. The theory of the orientation dependence of the effect is developed. It is shown that the magnetic field strength dependence of magnetoresistance can be calculated for arbitrary orientation of magnetic field if this dependence is known for in-plane and perpendicular-to-plane magnetic fields. It is noted that the magnetization curve can be obtained by making use of the results of the magnetoresistance measurements.

INTRODUCTION

Since discovering the giant magnetoresistance (MR) effect in metallic superlattice [1,2], the MR measurement is a standard procedure for every new multilayer. The magnetoresistance $r(H)$ is usually defined by the relation

$$r(H) = (R_H - R_0) / R_0 \qquad (1)$$

where R_H is a resistance of a sample placed in a magnetic field H, R_0 is a resistance at $H=0$. As a rule the electric current j flows in the sample plane. Three types of MR can be distinguished depending on the direction of H with respect to j: longitudinal MR ($r_{\parallel}$) with in-plane H and $H \parallel j$, transverse MR (r_T) with in-plane H and $H \perp j$, and perpendicular MR ($r_{\perp}$) with H perpendicular to the film plane. The magnetic field dependence of $r_{\parallel}$ and $r_{\perp}$ in Fe/Cr superlattice turns out to be different [1-3]. It has been supposed that this difference is caused by the magnetic anisotropy, but a detailed analysis has not been made.

To our opinion, two problems have to be solved. First it is not clear how the magnetic field dependence of MR changes if the angle Φ between H and the film plane changes from $0°$ to $90°$. Further it is interesting to know whether one can calculate $r(H,\Phi)$ for arbitrary angle provided that the functions $r_{\parallel}(H)$ and $r_{\perp}(H)$ are known. If the orientation dependence of MR really correlates with that of magnetization, one could obtain the magnetization curves by making use of the MR measurements.

EXPERIMENTAL RESULTS

We studied Fe/Cr multilayers grown by molecular beam epitaxy method on MgO (100) substrates. The resistance was measured by standard four-probe method. The temperature was 290K and 77K.

The longitudinal and transverse MR are practically identical. Shown in Fig.1 is $r_\parallel(H)$ and $r_\perp(H)$ curves for the epitaxial superlattice [Fe(23Å)/Cr(8Å)]$_{30}$.

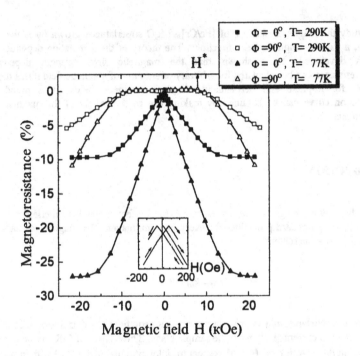

Figure 1

One can see that the longitudinal MR decreases linearly with growing magnetic field at low H and is practically constant if H exceeds the saturation field H_s . The perpendicular MR practically does not change if the field is less than $H^*=8$ kOe (we define H^* as non-zero value of magnetic field vanishing $r_\perp$, i.e. $r_\perp(H^*)=0$, see Fig 1); moreover, at the range from 0 to 6 kOe the resistance slightly grows. The saturation value of longitudinal MR is equal to that of $r_\perp$. The insert in the figure shows how $r_\parallel$ changes in weak fields. The measurement of a hysteresis loop gave the coercivity H_c to be about 50 Oe.

In Fig.2 we present $r (H, \Phi)$ for different Φ at the room temperature; the results for $T=77K$ are similar. If the magnetic field is directed at a small angle with respect to the layers plane, MR changes insignificantly but even a slight deviation of Φ from 90° results in a drastic decrease of the MR value.

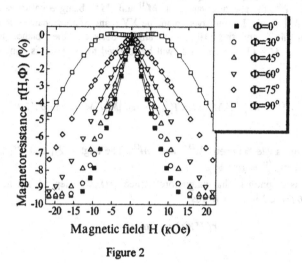

Figure 2

This fact is illustrated by Fig.3 in which the angular dependence of MR is given for $H=1-5$ kOe

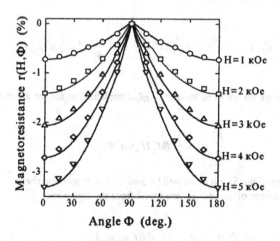

Figure 3

THEORY

Let us assume that the superlattice can be considered as a sum of two magnetic sublattices. The magnetization of a layer belonging to the first sublattice is $M^{(1)}$, the magnetization of the second one is $M^{(2)}$, the absolute values of $M^{(1)}$ and $M^{(2)}$ being equal to each other, namely $|M^{(1)}| = = |M^{(2)}| = M_0$. Let the layers plane be XY plane of coordinate system and Ψ_i be the angle between $M^{(i)}$ and this plane; the angle between the plane and the mean magnetization $M = (M^{(1)} + M^{(2)})/2$ will be referred to as Ψ. It is convenient to use the variables

$$\mu_z^{(i)} = M_z^{(i)} / M_0 = \sin \Psi_i, \qquad \mu = |M| / M_0 = \cos(\theta / 2) \qquad (2)$$

where θ is the angle between $M^{(1)}$ and $M^{(2)}$. The angles Ψ_1 and Ψ_2 are equal due to symmetry, so that $\mu_z^{(1)} = \mu_z^{(2)} \equiv \mu_z$.

Our basic assumption is that magnetoresistance $r(H, \Phi)$ depends on its arguments only through $\theta = \theta(H, \Phi)$, in other words

$$r(H, \Phi) = r(\mu(H, \Phi)) \qquad (3)$$

The energy of the superlattice per unit area and per one magnetic layer can be written as a sum of the Zeeman energy $E_H = -dH \cdot M$, demagnetizing energy $E_M = 2\pi dM_0^2 \mu_z^2$, exchange energy $E_{12} = E_{12}(\mu)$, and anisotropy energy E_A that depends in general on both μ_z and μ. The anisotropy in XY plane is weak and can be neglected. One can easily write a formal expansion of the energy in a power series:

$$E = E_H + E_M + \frac{a_1}{2} \mu^2 + \frac{a_2}{4} \mu^4 + \frac{b_1}{2} \mu_z^2 + \frac{b_2}{4} \mu_z^4 + \frac{c}{2} \mu^2 \mu_z^2 + \dots \qquad (4)$$

For our purposes we do not need to know $E_{12}(\mu)$ explicitly; as for the anisotropy energy, we keep only

$$\frac{b_1}{2} \mu_z^2 = \frac{1}{2} dM_0 H_A \sin^2 \Psi. \qquad (5)$$

The role of the terms of the fourth order will be analyzed in details elsewhere.

The equilibrium values of μ and μ_z are found from the equations:

$$\frac{\partial E(\mu, \mu_z)}{\partial \mu} = 0, \quad \frac{\partial E(\mu, \mu_z)}{\partial \mu_z} = 0. \qquad (6)$$

320

After some manipulations we obtain following equation for μ

$$\frac{\sin^2\Phi}{\left[\left(4\pi M_0 + H_A\right)\mu + H_{12}(\mu)\right]^2} + \frac{\cos^2\Phi}{H_{12}^2(\mu)} = \frac{1}{H^2} \tag{7}$$

where

$$H_{12}(\mu) = \frac{1}{dM_0}\frac{\partial E_{12}(\mu)}{\partial\mu}. \tag{8}$$

Let $H_r(\Phi)$ be a magnetic field strength at which magnetoresistance is equal to a certain value r, i.e.

$$r\left(\mu\left(H_r(\Phi),\Phi\right)\right) = r. \tag{9}$$

It follows

$$\frac{\partial\mu}{\partial H_r}\frac{\partial H_r(\Phi)}{\partial\Phi} + \frac{\partial\mu}{\partial\Phi} = 0. \tag{10}$$

Finding $\partial\mu/\partial H$ and $\partial\mu/\partial\Phi$ from (7) and inserting them in (10), we obtain the differential equation for $H_r(\Phi)$:

$$\frac{\partial}{\partial\Phi}\left(\frac{1}{\sin(2\Phi)}\frac{\partial}{\partial\Phi}\frac{1}{H_r^2(\Phi)}\right) = 0. \tag{11}$$

The general solution of (11) can be written as

$$H_r(\Phi) = \left(\frac{\cos^2\Phi}{H_r^2(0°)} + \frac{\sin^2\Phi}{H_r^2(90°)}\right)^{-\frac{1}{2}}. \tag{12}$$

If we suppose that $r(H, 0°) \propto H$ at $H \ll H^*$ and $r(H, 90°) \propto (H^* - H)$ at $0 < H - H^* \ll H^*$ then on can show from (12) that the magnetoresistance $r(H,\Phi)$ as a function of Φ near $\Phi = 90°$ has the following asymptotic behaviour in the range $0 < H \leq H^*$:

$$r(H,\Phi) \propto \frac{H/H^*}{\sqrt{1-(H/H^*)^2}}\left|\Phi - \frac{\pi}{2}\right|, \quad \text{if } H < H^*;$$

$$r(H,\Phi) \propto (\Phi - \frac{\pi}{2})^{2/3}, \qquad\qquad \text{if } H = H^*. \tag{13}$$

It must be noted that derivative $\dfrac{\partial\, r(H,\Phi)}{\partial\Phi} \to \infty$ at $H \to H^*$.

The relation (12) can be immediately applied to interpretation of the experimental data. Upon finding $H_r(0°)$ and $H_r(90°)$ by making use of results of measuring MR at $\Phi=0°$ and $\Phi=90°$, one can calculate $r(H, \Phi)$ at arbitrary H and Φ. In Figs.2-3 the calculated values of MR are presented as solid lines. One can see that our theoretical curves are in a good agreement with experiment.

It follows from (7) and (12) that $H_r(90°) - H_r(0°) \equiv \Delta H_r$ for every given r is nothing but $(4\pi M_0 + H_A)\mu$ corresponding to this r. Since the greatest possible r takes place at $H=H_s$, when $\mu=1$, we have $\Delta H_{r_{max}} = 4\pi M_0 + H_A$. Hence the relation $\Delta H_r / \Delta H_{r_{max}}$ is nothing but the relative magnetization μ of the superlattice in the in-plane H corresponding to given r:

$$\mu_r = \Delta H_r / \Delta H_{r_{max}}.$$

CONCLUSION

Thus both the field and the angle dependencies of magnetoresistance for our epitaxial Fe/Cr superlattices can be described satisfactorily on the base of assumption about existing two subsystems of magnetic moments with energy of coupling depending on the angle between these moments. We have found that the difference between magnetoresistances $r(H, \Phi=90°)$ and $r(H, \Phi=0°)$ is due to the anisotropy and demagnetizing field making it more difficult to align the layers' magnetic moments in the out-of-plane direction. As a result, one can find the magnetization curve by making use of the data on magnetoresistance.

ACKNOWLEDGMENT

The research described in this publication was made possible in part by Grant No. NMK 000 from the International Science Foundation and in part by Grant No. 95-02-04813 from the Russian Foundation of Fundamental Researches and Grant-in-Aid No. 1-053/2 for Research Program from the Ministry of Science of Russia.

REFERENCES

1. M.N.Baibich, J.M.Broto, F.Nguyen van Dau, F.Petroff, P.E.Eitenne, G.Creuzet, A.Friederich, and J.Chazelas. *Phys. Rev. Lett.* **61**, 2472 (1988) .

2. F.Nguyen van Dau., A.Fert, P.Etienne, M.N.Babich, J.M.Broto, S.Chazelas, G.Creuzet, A.Fredrich., S.Hadjoudj, H.Hurdequint, J.P.Redoules and J. Massies. *J. de Phys.* **49**, C8-1633 (1988)

3. T.Miyazaki, H.Kubota, S.Ishio. *J. Magn. Magn. Mater.* **103**, 131 (1992)

Calculation of Electrical Conductivity and Giant Magnetoresistance within the Free Electron Model

X.-G. Zhang* and W. H. Butler *

*Metals and Ceramics Division, Oak Ridge National Laboratory, Oak Ridge, Tennessee 37831-6114

ABSTRACT

We use the model of free electrons with random point scatterers (FERPS) to calculate the electrical conductivity and giant magnetoresistance (GMR) for FeCr multilayer systems and compare our results with the experimental values. Our analysis suggests that the primary cause of the GMR in FeCr systems is regions of interdiffusion near the interfaces. We find that in the samples analyzed, these regions of interdiffusion occupy about 8.5Å of the magnetic layer near each interface.

Introduction

Previous calculations of the conductivity and Giant Magnetoresistance in magnetic multilayers have generally employed the model of Free Electrons with Random Point Scatterers (FERPS) and have approximated the conductivity within this model by using either a semi-classical approximation[1, 2, 3] or an approximate solution[4] to the Kubo formula[5, 6]. In a previous study[7] we evaluated the Kubo formula exactly within the FERPS model with a local self-energy, and compared it with the other methods. We investigated the relationships among the various approaches and found that under most circumstances the semi-classical approach agrees surprisingly well with the numerical solution, while the solution of Zhang, Levy and Fert[4] (ZLF) generally yields a conductivity which is lower than the numerical solution when the mean free path is comparable to the layer thicknesses[8].

In light of these results, the question arises as to whether the past analyses of experimental data using ZLF and the conclusions based on them should be re-examined. Specifically, since ZLF theory tends to give results that are closer to the thin limit, it usually over-emphasizes the effects of regions with strong scattering, e.g., interface regions. Therefore, one needs to reconsider the conclusion drawn from these studies that the dominant effect in these GMR systems is the interfacial scattering.

In this paper we calculate the conductivity and GMR exactly within the FERPS model for FeCr multilayer systems, and compare the results with previous studies[9]. Our study suggests that although interface roughness can be important, there may also be a region of interdiffusion that is larger than the rough regions near the interfaces, and this interdiffusion may be an important contributor to GMR. We further speculate that GMR may be significantly increased if this interdiffusion region can be increased while maintaining spin alignment.

Conductivity in the Free Electron Model

Our model of the multilayer is described by a complex local self-energy $\Sigma(z)$ that is constant in the xy directions (parallel to the layers), and is assumed to be constant in the

323

Mat. Res. Soc. Symp. Proc. Vol. 384 °1995 Materials Research Society

z direction within each layer. It can be shown that variations of the real part of the self-energy contribute very little to the GMR. Therefore we assume that the real part of $\Sigma(z)$ is a constant throughout space. The imaginary part of $\Sigma(z)$ is determined by the mean free path within each layer,

$$\text{Im}\Sigma_I = -\frac{k_F}{2\ell_I},\tag{1}$$

where k_F is the Fermi momentum, and the subscript I denotes the layer I.

The quantum solution to such a multilayer cannot be obtained analytically. To obtain a numerical solution we embed a finite number of multilayers (about 1000Å thick) into an infinite square well, and calculate the Green function using[3]

$$G(k_\parallel; z, z') = \frac{\psi_L(z_<)\psi_R(z_>)}{W},\tag{2}$$

where $k_\parallel$ is the parallel component of the wave vector, and ψ_L and ψ_R are solutions to the differential equation,

$$[E + \frac{\hbar^2}{2m}(\frac{\partial^2}{\partial z^2} - k_\parallel^2) - \Sigma(z)]\psi(z) = 0,\tag{3}$$

and satisfy the boundary conditions on the left and right sides of the system, respectively. For a multilayer system of total thickness d we used the boundary conditions, $\psi_L(0) = 0$ and $\psi_R(d) = 0$. W is the Wronskian of ψ_L and ψ_R.

The conductivity for current-in-plane (CIP) can be calculated from the Kubo formula which gives,

$$\sigma = -\frac{1}{8\pi^3}\frac{e^2\hbar^3}{m^2 d}\int_0^d dz \int_0^d dz' \int d^2k_\parallel k_\parallel^2 \text{Im}G(k_\parallel; z, z')\text{Im}G(k_\parallel; z', z).\tag{4}$$

We first use this to calculate the conductivity of a simple multilayer system and compare with the semi-classical results and those obtained using the theory of ZLF. The comparison is shown in Fig. 1, as a function of the thickness of one period. In these calculations it is assumed that the scattering rates for the two layers correspond to bulk mean free paths of 36.0555 and 360.555 atomic units (1 a.u.=0.529Å) and that the thickness of the dirty layers is twice that of the clean layers. No additional scattering at the interfaces is included. In all of the multilayer calculations we used a sufficient number of periods of the multilayer to avoid the physical quantum size effects for the exact results and the large unphysical size effects that occur for the semi-classical and ZLF theories.

There are two limits that all theories approach correctly: The thin limit in which the layer thicknesses are small compared to the mean free path, and the thick limit in which the layer thicknesses are much larger than the mean free path. In the thin limit the conductivity is determined by the average of the scattering rate, which gives,

$$\frac{1}{\ell_{\text{thin}}} = \sum_I \frac{d_I}{d}\frac{1}{\ell_I}.\tag{5}$$

In the thick limit, the mean free paths are averaged,

$$\ell_{\text{thick}} = \sum_I \frac{d_I}{d}\ell_I.\tag{6}$$

There is a surprisingly good agreement between the semi-classical theory and the FERPS model. On the other hand, the ZLF theory seems to approach the thin limit too fast.

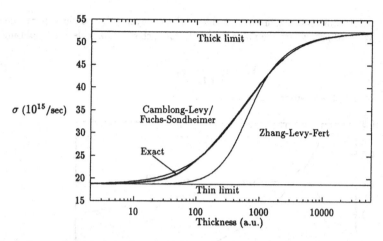

Figure 1: Conductivity as a function of the total thickness of a period of a multilayer system. The period contains two layers, with thicknesses 2/3 and 1/3 of the period, and mean free paths 36.0555a.u. and 360.555a.u., respectively.

Because in most multilayer systems the layer thicknesses are less than 100Å(about 189a.u.), this deviation would cause the ZLF theory to over-emphasize the effects of strong scattering regions, e.g., the interfacial regions.

Application to Multilayer Systems

We have applied our FERPS model to the FeCr multilayer system, and compared the results with the experiment of Baibich *et al*[10, 11] results of which are shown in Fig. 2. Like most of the data on GMR as a function of the layer thicknesses it has three qualitative features:(1)For fixed thickness of the magnetic layer the GMR ($\Delta R/R$) decreases with spacer layer thickness. (2)For sufficiently large magnetic layer thickness GMR decreases with magnetic layer thickness. (3) As the magnetic layer thickness decreases the GMR reaches a maximum and then falls rapidly to zero for zero magnetic layer thickness. In relating the experimental data to the FERPS model we find that feature (1) gives information concerning the mean free path in the spacer layer, feature (2) is related to the relative strength of the bulk scattering (asymmetry between the majority and minority scattering rates in the ferromagnetic layer) and the interfacial scattering. The thickness at which feature (3) occurs indicates the thickness of the region which contributes most strongly to the GMR.

In order to fit the data using the FERPS model we chose (somewhat arbitrarily) $E_F = 0.1$Ha. With this value of E_F a value of the spacer layer mean free path of approximately $\lambda^{Cr} = 79$Åwas necessary to fit the decrease in GMR with spacer layer thickness. In order to fit the dependence of the GMR on the ferromagnetic layer thickness we found it necessary to assume that there is a region in the Fe layers where there is a significant amount of Cr impurities. From our experience with the bandstructure of FeCr alloys, we made an assumption that the mean free path for the minority spin channel in the interdiffusion region is the same as that of the Fe layer. Therefore there are four free parameters to fit, the thickness of the interdiffusion region on each side of the Fe layer, t^{int}, and three parameters from λ_ℓ^{Fe}, p_ℓ^{Fe}, λ_ℓ^{int}, and p_ℓ^{int}, where the mean free paths for each spin channels can be obtained from,

$$\lambda_\sigma = \frac{\lambda_\ell}{(1 + p_\ell\sigma)^2},$$ (7)

with $\sigma = \pm 1$ for up (down) spins.

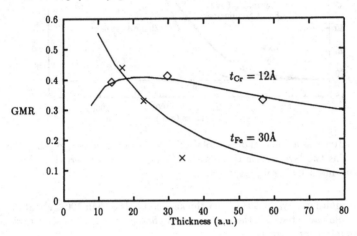

Figure 2: The GMR ratio of FeCr multilayers as a function of Fe layer thicknesses with $t_{Cr} = 12\text{Å}$ ($\Diamond$), and as a function of Cr layer thicknesses with $t_{Fe} = 30\text{Å}$ ($\times$). The solid lines are calculations with the FERPS model.

We were able to fit both curves with $t^{int} = 8.5\text{Å}$, $\lambda_\ell^{Fe} = 33\text{Å}$, $p_\ell^{Fe} = 0.44$, $\lambda_\ell^{int} = 11\text{Å}$, and $p_\ell^{int} = 0.68$. The zero field resistivity calculated from these parameters is about $45\mu\Omega$cm, in good agreement with the measured residual resistivity $37\mu\Omega$cm. The existence of the rather thick interdiffusion regions has important consequences. If the scattering rates in these regions are made the same as those in the Fe layers, the GMR ratio is reduced significantly. This is consistent with the experimental result that samples with sharper interfaces yielded smaller GMR ratios.

Although it seems that we have many parameters to fit only a few data points, we believe the results are significant for the following reasons. First, several parameters, such as the Fermi energy, the mean free path in the Cr layers, and the mean free path of the majority channel in the Fe layers, have little effect on the general trend of the GMR, although they may affect the resistivities significantly. Second, the decrease of the GMR as a function of the Fe layer thickness for thick Fe layers determines quite unambiguously the ratio of bulk scattering versus interfacial scattering in the Fe layers. Lastly, the reduction of the GMR ratio for very thin Fe layers gives a good estimate of the thickness of the interfacial regions where there is a strong spin dependent scattering. Most experimental data on multilayers show a sharp downturn of the GMR ratio for a magnetic layer thickness near 10Å to 20Å. This thickness is much greater than a typical interfacial roughness of about 2Å to 4Å which suggests the existence of relatively thick magnetic regions where there is a significant concentration of nonmagnetic impurities due to diffusion in these multilayer systems. This is illustrated by

Figure 3 which shows the GMR ratio for a typical CuCo multilayer system[12]. A careful study of this system will be presented in a future publication.

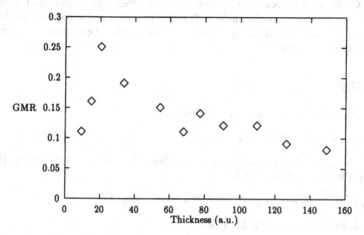

Figure 3: The GMR ratio of CuCo multilayers as a function of Co layer thicknesses with $t_{Cu} = 19\text{Å}$ (◇).

Our results are differ from those obtained by Zhang, Levy and Fert from the same data[4] in several aspects. First, they approximated the interfacial regions with a δ function type scattering potential. This leads to the unphysical consequence that the GMR extrapolates to a finite constant as the thickness of the magnetic layers goes to zero. Our analysis indicates that interfacial interdiffusion region is actually rather thick, about 8.5Å on each side of the Fe layer (17Å for every Fe layer). Secondly, their calculated zero field resistivity is too high, ($83\mu\Omega$cm). Although the resistivities tend to fluctuate from sample to sample, it is important that a theory give the correct resistivity as well as the GMR ratio for the same sample. This difference in the resistivity is due to the fact that the ZLF theory over-emphasizes the strong scattering regions compared to the exact solution of the FERPS model.

Conclusions

We have analyzed the experimental data on FeCr multilayers using the FERPS model. We found that there may exist rather thick interdiffusion regions in the magnetic layers, and these regions with very strong spin dependent scatterings are the primary contributors to the GMR. Petroff et al[9] found that FeCr multilayers with sharp interfaces and small residual resistivities exhibit significantly reduced GMR. Annealing these samples usually increasase the GMR ratio. These observations are consistent with our results. Other experiments on FeCr or CuCo multilayers showed similar trends.

Our results suggest that the GMR may be increased significantly in these multilayer systems by increasing the interdiffusion regions, either by annealing or impurity doping. It is important, however, that these regions should maintain their magnetic moment, which is the source of spin dependence in the scattering rates according to our first-principles studies[13, 14].

Acknowledgements

Work at Oak Ridge was sponsored by DOE Assistant Secretary of Defense Programs, Technology Management Group, Technology Transfer Initiative under contract DEAC05-84OR21400 with Martin Marietta Energy Systems and by the High Performance Computing and Communication Initiative.

REFERENCES

1. K. Fuchs, Proc. Camb. Phil. Soc. **34**, 100 (1938).

2. E. H. Sondheimer, Adv. Phys. **1**, 1 (1952).

3. H. E. Camblong, S. Zhang, and P. M. Levy, Phys. Rev. B **47**, 4735 (1993); H. E. Camblong, Phys. Rev. B 51, 1855 (1995).

4. S. Zhang, P. M. Levy, and A. Fert, Phys. Rev. B **45**, 8689 (1992).

5. R. Kubo, J. Phys. Soc. Jpn. **12**, 570 (1957).

6. D. A. Greenwood, Proc. Phys. Soc. London **71**, 585 (1958).

7. X.-G. Zhang and W. H. Butler, Phys. Rev. B **51**, *in press, April 15* (1995).

8. Professor Levy and Dr. Zhang have suggested that the Zhang-Levy-Fert theory may contain some additional terms in the self-energy that are omitted by the FERPS with local self-energy model.

9. F. Petroff, A. Barthelemy, A. Hamzic, A. Fert, P. Etienne, S. Lequien and G. Creuzet, J. Magn. Magn. Mater. 93, 95 (1991).

10. M. N. Baibich, J. M. Broto, A. Fert, F. Nguyen Van Dau, F. Petroff, P. Etienne, G. Creuzet, A. Friederich, and J. Chaezelas, Phys. Rev. Lett. **61**, 2472, (1988).

11. A. Barthelemy, A. Fert, M.N. Baibich, S. Hadjoudj, F. Petroff, P. Etienne, R. Cabanel, S. Lequien and G. Creuzet, J. Appl. Phys. 67, 5908 (1990).

12. A.M. Shukh, D.H. Shin and H. Hoffmann, J. Appl. Phys. 76, 6507 (1994).

13. W. H. Butler, J. M. MacLaren, and X.-G. Zhang, *Materials Research Society Symposium Proceedings* **313**, 59, (1993).

14. W. H. Butler, X.-G. Zhang, D. M. C. Nicholson, and J. M. MacLaren, *in this volume.*

PERPENDICULAR RESISTANCE OF Co/Cu MULTILAYERS PREPARED BY MOLECULAR BEAM EPITAXY.

N.J. LIST, W.P. PRATT JR.*, M.A. HOWSON, J. XU, M.J. WALKER, B.J. HICKEY and D. GREIG

Department of Physics, University of Leeds, Leeds LS2 9JT, UK,
*Department of Physics and Astronomy, Michigan State University, East Lansing, MI 48824, USA

ABSTRACT

Results are presented of the magnetoresistance of MBE-grown (111) Co/Cu multilayers measured with the current *perpendicular* to the plane of the layers (CPP). Although for measurements made with the more common geometry of current in the plane of the layers (CIP) there are large differences between the results on samples made by sputtering and those prepared by MBE, for these new CPP data the results on samples made by the two techniques are very much alike. For copper layers with thicknesses between 0.9nm to 6nm the magnetoresistance shows oscillations with copper thickness that were almost non-existent in the earlier CIP data. At the second peak the magnetoresistance in the CPP geometry is an order of magnitude greater than that in the CIP configuration. Although the interfaces in these samples have been shown to be very sharp, they appear to form a mosaic structure with the antiferromagnetic regions embedded in a ferromagnetic structure. It is argued that for CIP measurements the GMR is greatly reduced by these ferromagnetic correlations over length scales long compared to the electron mean free path. For CPP measurements, on the other hand, it is the spin diffusion length that is the determining factor with the mean free path no longer a key parameter and with values of the GMR virtually independent of the growth process.

INTRODUCTION

In nearly all studies of the so-called 'giant' magnetoresistance (GMR) observed in magnetic multilayers the measurements have been made with the current directed *along* the layers in the so-called Current-In-Plane (CIP) mode. Experimentally the resistance is of the order of ohms and easily measured, but on the other hand the detailed analysis of the magnetoresistance in the CIP configuration is complicated by the non-uniform distribution of current between the layers. A GMR is observed only when the mean free path of the electrons, λ, is considerably greater than L_M, the length scale of the magnetic inhomogeneity of the specimen. The minimum value of L_M is clearly the bilayer spacing. This condition is essential to ensure that over the length of a mean free path conduction electrons of a given spin can sample regions of opposing magnetisation that can be changed on application of an external field. This is the criterion for the existence of the GMR. On the other hand when the mean free path is very short ($\lambda < L_M$) the above condition is no longer true and the magnetoresistance from this source is effectively zero.

More recently an alternative method of studying the resistance of the multilayers by measuring the resistance *through* the layers in the Current-Perpendicular-to-Plane (CPP) geometry has gained in prominence. With the CPP arrangement current density is uniform across the sample, and, as the electrons pass sequentially through the layers and interfaces, modelling is more straightforward and the above restriction on the value of λ relaxed. In addition conduction through a definite series of magnetic and non-magnetic layers has focussed attention on a number of new phenomena such as spin-dependent interfacial scattering [1,2], the importance of the spin-diffusion length [3,4], spatial oscillations in charge density and electric field [5] the possibility of ballistic transport and the role of s-d scattering [6]. However from an experimental point of view the actual measurements are appreciably more difficult than in CIP both in defining the geometry of the probes and in measuring resistances across thicknesses of much less than one micron. Experiments of this sort are therefore at a comparatively early stage of development and in this paper we present a comprehensive set of CPP measurements on samples grown by MBE.

CPP MEASUREMENTS

At present three experimental methods are being developed to deal with this problem. They are (i) to form samples of micron-sized pillars by techniques of photolithography [7,8], (ii) to grow multilayers in the form of nanowires by electrolysis [9,10], and (iii) to sandwich multilayers between superconducting niobium films and measure the emfs of less than $10^{-9}V$ with a low temperature potentiometric circuit incorporating a SQUID as null detector and a precision current comparator [11,12]. The measurements reported here have been made by the third of these methods with the Co/Cu multilayers formed between niobium strips in one growth cycle [13]. The advantage of this technique is that the superconducting layers provide equipotential contacts injecting the current in a clearly defined geometry perpendicular to the layers, but the disadvantage, compared to the other two methods, is that the measurements must be made below about 4.2K.

The resistance through a metallic film of less than 1 micron thick is smaller than $100n\Omega$ which, in a current of 10mA involves the measurement of voltages of order $10^{-9}V$. The *change* of such voltages in a magnetic field is naturally smaller still so that a SQUID-based current comparator operating at 0.1% to 0.01% relative precision is required. (See reference [12] for further details). The magnetoresistance measurements were made using a small hand-wound NbTi superconducting magnet designed to fit closely round the sample and low enough in field (<1T) not to influence the operation of the SQUID or to drive the superconducting contact pads normal.

SPECIMEN GROWTH

The epitaxial films were grown in a VG80M facility operating in UHV conditions at a base pressure of about 3×10^{-11} mbar. The choice of sapphire (1120) substrates was determined by the fact that niobium has been grown on sapphire substrates by a number of groups (see for example references [14,15]) although the temperature of the substrate holder must be held at about 950°C -- well above the recommended maximum for the facility. However, this high temperature is a key feature of successful Nb growth, and has given us RHEED patterns that are markedly superior to those seen on any previous substrate. (And, incidentally, lead to GMRs in the CIP geometry that are almost twice as great as any that we had measured previously).

The CPP samples, sandwiched between Nb superconducting strips 1mm in width, are of the form sapphire/Nb/Cu[Co/Cu]$_n$/Nb/Au, where the Nb strips are 100nm thick, the cobalt thickness is fixed at 1.15nm while the copper thickness in different multilayers takes various values between 0.9nm and 6nm. The first copper layer, deposited ahead of the multilayer, is essentially a further buffer layer of order 3nm thick to seed the growth in the (111) direction, and this is formed at 375°C. However, the actual Co/Cu multilayers, the top Nb strips and the Au protective caps are all grown 'cold' at about 30°C, with the Nb and Au no longer epitaxial. A schematic diagram of the specimen and Nb contacts is shown in figure 1. The crossed-strip geometry and the cross-sectional area of the multilayer stack were all determined by evaporating through a series of masks. It was found that

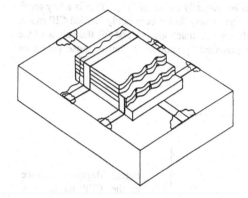

Fig.1: Schematic diagram of the specimen and Nb contacts.

scrupulous care in cleaning the sapphire substrates was absolutely essential otherwise the multilayers were short-circuited by contacts between the top and bottom Nb strips.

RESULTS AND DISCUSSION

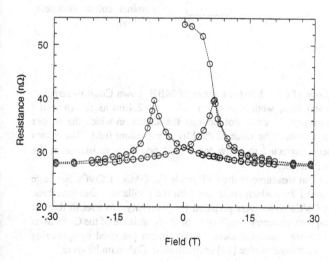

Fig. 2: Field dependence of resistance in a Co/Cu specimen with 0.96nm copper layers.

An example of the measurements is shown in figure 2 where we see that for 3nm copper layers the magnetoresistance is as great as -93%. There are two points that are immediately apparent.

(a) There is a very marked hysteresis effect with the resistance never returning to its high initial value. We attribute this to a change in the domain structure following saturation with an increase in the ferromagnetic component in interlayer coupling. The whole question of the detailed magnetic structure of specimens is something that has been largely neglected in this work on the GMR, and its relevance to magnetoresistance has only begun to be addressed in a recent paper by Zhang and Levy[16].

(b) In comparing figure 2 with CPP measurements made on similar multilayers grown by *sputtering* we find that the two sets of results are virtually *identical* [17]. This is a key result as one of the major issues of the whole GMR saga is why, in the commonly studied CIP mode, (i) the magnitudes of the GMR have invariably been so much greater and (ii) the values of the saturation fields so much smaller in samples prepared by sputtering than in those grown by MBE.

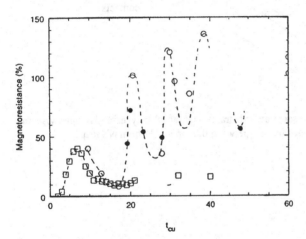

Fig.3: Magnetoresistance in the CPP mode as a function of copper thickness. Cobalt layers are 2.4nm (open circles) and 1.7nm (closed circles). The squares represent CIP data for which the nominal cobalt thickness is 1.5nm.

In figure 3 we show the magnitude of the GMR for a series of MBE grown Co_x/Cu_t samples. Here t is varied between 0.8 and 4nm, while x is either 1.7nm or 2.4nm as shown in the diagram. With the small NbTi magnet we could not saturate the sample in which the copper layers are 0.96nm so the datum shown is the value of the MR at maximum field. The values shown in figure 3 have all been corrected to allow for a $6n\Omega$ boundary resistance at the superconducting/ferromagnetic interfaces.

It is again immediately apparent that measured in the CPP mode the GMR of Co/Cu has large oscillations with a period of about 1.2nm which correlates with the oscillations that have been observed in the CIP mode for Co/Cu multilayers prepared by *sputtering*. We recall that for measurements in the simpler in-plane geometry oscillations in the magnitude of the GMR as a function of spacer layer thickness are seen quite clearly in specimens prepared by sputtering but, apart from a set of results on a trilayer wedge [18] oscillations in Co/Cu multilayers prepared by MBE have never been observed. An example of a set of CIP measurements made on specimens grown on sapphire substrates by MBE is shown in figure 3. In an effort to clarify this problem we have previously reported oscillations in the magnitude of magnetic coupling in MBE grown samples by *magnetisation* measurements [19], although they were extremely difficult to observe. We have attributed the illusive nature of these oscillations to the large ferromagnetic contribution to the magnetisation of these MBE grown samples. By way of complete contrast we see that for CPP data the oscillations are extremely clear.

A second point of importance is that for these MBE grown specimens the extremal values of the CPP-MR -- i.e. those in the regions where we expect antiferromagnetic coupling -- are very similar in magnitude to those in Co/Cu multilayers prepared by sputtering.

CIP AND CPP

From a purely experimental point of view the important question to address is therefore why, on the one hand, are the CPP values of the GMR for specimens prepared by either MBE or sputtering virtually identical, while, on the other hand, for the CIP mode it has proved almost impossible to obtain either large values or oscillations in the GMR for Co/Cu specimens grown by MBE?

It is certainly true that our experience has been that it is possible to increase the value of the GMR in material grown by MBE from only a few percent to almost 50% by very careful control of the growth conditions as a means of improving the quality of the layers. However although we estimated from magnetisation data that in specimens for which the GMR was 26% only about 20% of the sample was antiferromagnetically coupled, the improved growth conditions has not lead to any increase in this fraction.

A very elementary explanation of the much larger GMR in the CPP mode is to assume that with the better defined geometry although the antiferromagnetic regions are still 'embedded' in a ferromagnetic matrix they are clearly separated in well defined collumnar paths. If these paths have a relatively low resistance compared to the rest of the multilayer then it is the change of *their* conductivity in a magnetic field that dominates the conduction process in CPP. In fact a closer analysis shows that this is not a possible explanation as it would predict the zero field resistivity for CIP to be larger than for the CPP configuration whereas experimentally we find exactly the opposite.

As mentioned in the Introduction the magnitude of the mean free path is a key parameter in any transport measurement, and, for the CIP mode, values of $\lambda < L_M$ can nullify the GMR. NMR and XRD [20] studies on samples prepared in this way suggest very sharp interfaces but relatively short distances, suggesting that the layers are a mosaic structure. This would clearly lead to a complex magnetic structure with the sample ordered ferromagnetically on a length scale longer than the expected L_M and a reduction in the volume of the sample that is coupled antiferromagnetically. As the mean free paths in our samples are of the order of 5nm it is not difficult to envisage ferromagnetic correlations on a length scale considerably greater than that. For the CPP mode the mean free path is no longer a key parameter, and the spin diffusion length that determines the scale of spin accumulation at the interfaces -- and is believed to be the determining factor in CPP conduction -- is certainly considerably greater than the thicknesses of these very thin layers.

ACKNOWLEDGEMENTS
We are most grateful for the continuing financial support of the UK SERC, the EPSRC, the University of Leeds and NATO.

REFERENCES

1. J.Barnas and A. Fert, J. Magn. Magn. Mater. **136**, 260 (1994).

2. L. Sheng and D.Y. Xing, J. Phys: Cond. Matter **50**, 6089 (1994).

3. S.F. Zhang and P.M. Levy, Phys. Rev. B. **50**, 6089 (1994).

4. A. Fert, T. Valet and J. Barnas, J. Applied Phys. **75**, 6693 (1994).

5. A. Vedyaev, C. Cowache, A. Ryzhanova and B. Dieny, Phys. Letters A, **198**, 267 (1995).

6. K.M. Schep, P.J. Kelly and G.E.W. Bauer, Phys. Rev. Letts. **74**, 586 (1995)

7. M.A.M. Gijs, S.K.J. Lenczowski and J.B. Giesbers, Phys. Rev. Lett. **70**, 3343 (1993).

8. M.A.M.Gijs, J.B. Giesbers, M.T. Johnson, J.B.F.A. Destegge, S.K.J. Lenczowski, R.J.M. Vandeveerdonk and W.J.M. Dejonge, J. Applied Phys. **75**, 6709 (1994)

9. L. Piraux, J.M. George, J.F Despres, C Leroy, E Ferain, R Legras, K. Ounadjela and A. Fert, Applied Phys. Letts. **65**, 2484 (1994).

10. A. Blondel, J.P. Meier, B Doudin and J.P Ansermet, Applied Phys. Letts, **65**, 3019 (1994)

11. W.P. Pratt Jr, S-F. Lee, J.M. Slaughter, R. Loloee, P.A. Schroeder and J. Bass, Phys. Rev. Letts, **66**, 3060 (1991).

12. J.M. Slaughter, W.P. Pratt Jr and P.A. Schroeder, Rev. Sci. Instrum. **60**, 127 (1988).

13. A preliminary description of this experiment is given in N.J. List, W.P. Pratt Jr, M.A. Howson, J. Xu, M.J. Walker and D. Greig, J. Magn.and Magn. Materials (in press).

14. A. Schreyer, K. Brohl, J.F. Ankner, C.F. Majkrzak, T. Zeidler, P. Bodeker, N. Metoki and H. Zabel, Phys. Rev. B **47**, 15334 (1993).

15. For earlier references on Nb growth on sapphire see F.J. Lamelas, Hui He and R. Clarke, Phys. Rev. B **38**, 6334 (1988).

16. S/ Zhang and P.M.Levy, Phys.Rev. B**50**, 6089 (1994)

17. W.P. Pratt Jr, S-F. Lee, P. Holody, Q. Yang, R. Loloee, J. Bass and P.A. Schroeder, J. Magn. Magn. Materials **126**, 406 (1993).

18. V. Grolier, D. Renard, B. Bartenlian, P. Beauvillain, C. Chappert, C. Dupas, J. Ferre, M. Galtier, E. Kolb, M. Mulloy, J.P Renard and P. Veillet, Phys. Rev. Letts. **71**, 3023 (1993).

19. M.A. Howson, B.J. Hickey, J. Xu, D. Greig and N. Wiser, Phys. Rev. B **48**, 1322 (1993)

20. T. Thomson, P.C. Riedi and D. Greig, Phys. Rev. B **50**, 10319 (1994).

GMR MULTILAYER PATTERNED STRUCTURES

L.V. MELO [1,2], L.M. RODRIGUES[1], A.T. SOUSA[1,2] and P.P. FREITAS[1,2]

[1]INESC, Rua Alves Redol 9, Apartado 13069, 1000 Lisboa, Portugal
[2]IST, Av. Rovisco Pais, 1000 Lisboa, Portugal

ABSTRACT

Co-Cu-NiFe-Cu weakly coupled multilayers, were patterned into stripes with dimensions from 1x6μm² to 2x20μm². After proper biasing using a DC field created by an adjacent bias conductor, these structures show large linear field spans (±150Oe), with MR values up to 6%. These can be used for current or position sensor applications. Sensor response to AC fields was studied up to 1Mhz, using the bias conductors as sources of the AC field. We find that the signal output is reduced above 10 to 100kHz. For comparison, sensors were also fabricated out Co-Cu AF-coupled multilayer (3x100μm² structures) needed for field detection up to several kOe.

INTRODUCTION

GMR (giant magnetoresistance) multilayers have been develloped in the recent years in view of applications in the magnetic digital recording(Hs<100 Oe) and field sensor(Hs<1000 Oe) industries. In the first case, spin-valve heads seem the more promising candidate, having been fabricated and tested with good recording characteristiscs[1,2,3,4]. Alternative solutions for low field, high sensitivity GMR materials have been demonstrated, as NiFe/Ag multilayers deposited at 77K[5], or heat treated[6,7], uncoupled Co/Cu/NiFe/Cu multilayers[8],and NiFeCo/Cu weakly AF-coupled multilayers[8,9,10,11]. For field sensor industrial applications where requirements are for low thermal drifts, small voltage offset, and highly linear response with field ranges up to 1 or 2 kOe, AF coupled Co/Cu multilayers or granular systems become of interest together with mechanisms to control the linearity range and saturation field value. In these field ranges, these sensors have to compete with Hall effect devices. First GMR sensor applications were already demonstrated[12]. In this paper we compare the behavior of two types of sensors, one of the first category made of uncoupled Co/Cu/NiFe/Cu multilayers, and the second made of AF coupled Co-Cu multilayers. Sensor performance, biasing mechanisms, Barkhausen noise, and AC response are described and compared with simpler spin-valve structures when appropriate.

EXPERIMENT

The multilayers were deposited by magnetron sputtering at Ar pressures of 2 mtorr in a system with base pressures of $2x10^{-8}$ torr. Sputtering rates were typically of 1 to 2Å/s. The Cr buffer (50Å), when used, was deposited in the same system by e-beam evaporation at rates of the order of 2Å/s. The Co-Cu-NiFe-Cu multilayers used were of the form: Substrate/$Cr_{50Å}/(Co_{15Å}/Cu_{35Å}/NiFe_{30Å}/Cu_{35Å})_{x10}$ or Substrate/$Fe_{50Å}/(Co_{15Å}/Cu_{55Å}/NiFe_{30Å}/Cu_{55Å})_{x10}$. Different Cu layer thicknesses were used as the maximum saturation MR is

335

obtained with thicker Cu layers for the samples deposited without a Cr buffer. Here NiFe stands for Permalloy, $Ni_{81}Fe_{19}$. The Co-Cu multilayers were of the form: Substrate/$Fe_{50Å}$/($Co_{11Å}$/$Cu_{8Å}$)$_{x16}$. This Cu thickness is near the first maximum of the MR vs. t_{Cu} curve. The samples were prepared on Corning 7059 glass or (100) Si substrates. For the sensors with a biasing conductor, the Al conductor line 0.3 micron thick is first defined on the substrate by conventional sputtering and dry-etch techniques. This process is followed by the deposition of a 2500Å thick SiO_2 insulating layer by RF sputtering. The next step, and first step for the sensors without biasing conductor, is the deposition of the magnetic multilayer, capped by a 1500Å $TiW(N_2)$ film. The sensor pattern is now defined by laser-direct write on 1.4 micron thick positive photoresist. The $TiW(N_2)$ is now etched in a SF_6 plasma. This mask is then used for defining the sensor element by sputter etching. A $TiW(N_2)$ capping layer of about 300-400Å remains on top of the multilayer stripe. Al contact pads were deposited and defined by lift-off. The layouts of both types of sensors are shown in Fig.1. No magnetic field was applied during the deposition of the magnetic multilayer. For magnetoresistance measurements on patterned devices, a 2-probe set-up was utilized with the magnetic field applied transverse to the current direction. The AC measurements were performed using the biasing conductor to generate the magnetic field at different frequencies.

RESULTS AND DISCUSSION

Fig.2a shows the static magnetoresistance versus applied magnetic field for a glass/$Cr_{50Å}$/($Co_{15Å}$/$Cu_{35Å}$/$NiFe_{30Å}$/$Cu_{35Å}$)$_{x10}$ coupon sample. In these uncoupled samples the antiparallel alignment of the magnetizations is due to the different coercive fields of the Co and NiFe layers. The magnetizations of the Co and NiFe layers are parallel at saturation, and then the magnetizations of the NiFe layers rotate between 0 and -40 Oe. At this point the antiparallel alignment of the magnetizations is maximum. At -40 Oe the Co layers magnetizations also start rotating, and the sample resistance starts decreasing. Typical MR values for these structures range from 6 to 10%.

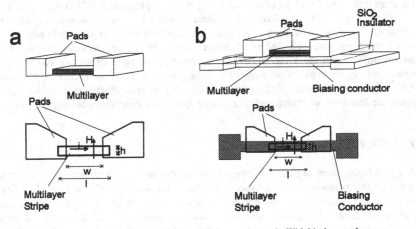

Fig. 1: Layout of the structures fabricated. a: Without biasing conductor; b: With biasing conductor.

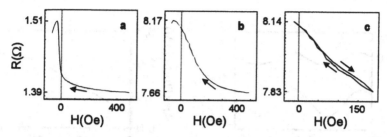

Fig. 2: DC R vs. applied field H for A:Glass/Cr$_{50Å}$/(Co$_{15Å}$/Cu$_{35Å}$/NiFe$_{30Å}$/Cu$_{35Å}$)$_{x10}$ multilayer in the field range corresponding to the NiFe magnetization reversal. a: Coupon sample; b: Patterned into a 1x20 μm^2 stripe; c: Minor loop showing linear behavior. The sense current in the sensor measurements is 2.5mA.

Fig.2b shows the signal output for a sensor made from this sample with dimensions (l,w,h)=(20,16,1)μm where l is the length of the multilayer stripe, w the distance between the contact pads (active region) and h is the sensor height. The sensor is defined as a long and narrow stripe in order to profit from the demagnetizing field to align the NiFe and Co layers in the longitudinal direction. The curve shown in Fig.2b for the patterned sensor has the field applied transverse to the NiFe and Co magnetizations, in order to produce an hard axis response as the NiFe layers switch. The NiFe is seen to start to rotate at positive fields (H>100Oe), and the MR rises till -40Oe. Measurements with the magnetic field applied parallel to the current direction showed a behavior similar to the the the one in Fig.2a. The MR changes linearly from 170Oe to -40Oe. Fig.2c shows a complete minor histeresis loop for the same sensor in this field range. The curve is linear over a 200 Oe span centered at 65 Oe and shows some Barkhausen noise. In order to achieve linear behavior, this 65 Oe field offset has to be removed. The necessary biasing field of 50 to 100 Oe will be provided by the biasing conductor previously patterned below the sensing element. The fields btained in this way are of the order of 1Oe/mA for a 5μm-wide conductor. The multilayers used for this purpose were of the form: Si/(Co$_{15Å}$/Cu$_{55Å}$/NiFe$_{30Å}$/Cu$_{55Å}$)$_{x10}$.

Fig.3 shows such a linearized loop. This curve shows relatively low Barkhausen noise. A similar biasing effect is obtained with the bias field created by the sense current in spin-valve heads[3,4].

Fig.4a shows the AC signal from a 1x20μm^2 structure above at 1kHz with a 115 Oe DC bias field. The bottom curves correspond to the AC magnetic field. This was the highest offset

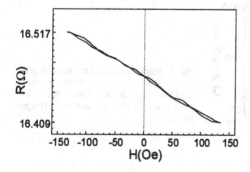

Fig. 3: DC minor loop for (l,w,h)=(20,16,1) μm Co-Cu-NiFe-Cu structures with -60Oe bias field.

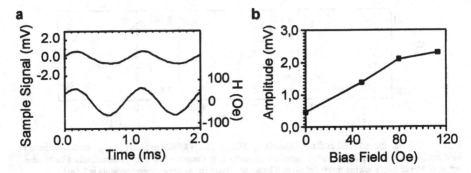

Fig. 4: **a:** AC signals from a (l,w,h)=(20,16,1)μm structure patterned from the same multilayer at 1kHz. The excitation (field) signal is also shown. The plot was obtained for a 110 Oe field amplitudes (pp) and 115Oe offset. **b:** Sensor signal amplitude vs. bias field generated by the wire.

available in our system. A higher field offset should improve the linearity, as at 115 Oe bias the bottom part of the curve is still distorted. The signal amplitude (pp) is 2.3mV for a 110Oe field amplitude (pp).

These results are consistent with the ones obtained by Noguchi *et al.*[8]. Notice also a small phase (time) shift between the applied field and the MR. Fig. 4b shows the dependence on the DC bias field of the signal amplitude (pp). At zero bias field the amplitude is very low, increasing regularly up to a 80 Oe bias field, where it starts to stabilize. The signal amplitude should reduce if the field bias was still increased. Fig.5 shows the amplitude of the sample signal and phase shift vs. frequency plots for a 2x20μm² stripe. The amplitude decreases and the phase shift increases at high frequency. The phase shift indicates that the sample magnetization is changing slowly and is not able to cope with the the field changing at higher rates. The reduction of the amplitude at higher frequency could be due to the increase in phase shift. This behavior is very different from the the one shown by pinned spin-valve sensors[4] or by coupled multilayers. These systems show no amplitude losses for frequencies even higher. Exchange coupling-dominated systems seem to be more appropriated for high frequency use. Studies on NiFeCo-Cu AF-coupled multilayer coupon samples showed that the samples permeability was kept almost unchanged up to 50MHz frequencies[9].

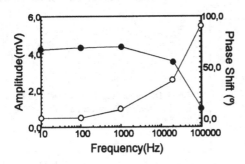

Fig. 5: Amplitude (closed circles) and phase shift (open circles) vs. frequency for a structures patterned from the same multilayer. with dimensions (l,w,h)= (20,16,2)μm.

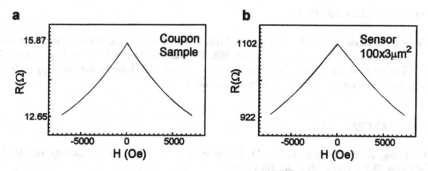

Fig. 6: R vs. applied field for a glass/Fe$_{50Å}$/(Co$_{11Å}$/Cu$_{8Å}$)$_{x16}$ multilayer. a: Coupon sample; b: Patterned into a sensor with dimensions (l,w,h)=(100,60,3)μm.

Figs.6a and 6b show static MR vs. applied field data for a glass/Fe$_{50Å}$/(Co$_{11Å}$/Cu$_{8Å}$)$_{x16}$ sample and for a sensor with dimensions (l,w,h)=(100,60,3)μm patterned from the same multilayer. Both curves show strong AF behavior with high MR values of the order of 20% and large saturation fields (none of the samples is saturated). The patterning does not seem to noticeably change the behavior of the MR and the patterned sample does not show Barkhausen noise, in contrast with the Co-Cu-NiFe-Cu samples. This should be due to the large AF exchange coupling, which dominates the behavior of the magnetic domains. This, together with the large MR values, indicates that these systems may be suitable for high frequency field sensing applications when low saturation fields are not required.

CONCLUSIONS

In the Co-Cu-NiFe-Cu stripe structures fabricated, the magnetic behavior of the stripes seems to be demagnetizing field-dominated. These uncoupled patterned structures show Barkhausen noise in DC measurements. The linear field span reaches 280 Oe for a 1x20μm² Co-Cu-NiFe-Cu stripe. This linear span is centered at fields between 50 Oe and 100Oe. In order to remove this field offset samples were fabricated with a conductor underneath the multilayer stripe. The fields generated by the current in this conductor are of the order of 1Oe/mA for currents up to 200mA The sensitivity depends on the dimensions and reaches 4.5mV for a 170Oe field amplitude for a 2x20μm² stripe. The samples magnetization is not able to follow the field when it changes at high rate. This translates into loss of sensitivity when the field frequency increases. In every case, the signal is reduced by a factor of at least four when the frequency goes from 10 to 100kHz. Co-Cu AF-coupled multilayer sensors were also fabricated. The magnetic behavior shows no modifications after patterning, and the sensor shows no Barkhausen noise. Studies will continue on on these systems.

ACKNOWLEDGEMENTS

We acknowledge Prof. C. B. Almeida and Prof. J. Serralheiro for the use of their laboratories for AC measurements. We also aknowledge MR. Faustino for patient wire bonding. We thank Dr. T. S. Plaskett for reviewing this paper. L.V.Melo also acknowledges travelling support from the Fundação Luso Americana para o Desenvolvimento.

REFERENCES

1. C. Tsang, R. E. Fontana, T. Lin, D. E. Heim, V. S. Speriosu, B. A. Gurney and M. L. Williams, IEEE Trans. Mag. 30, 3801 (1994).
2. W. Folkerts, J. Kools, T. Rijks, R. Coehoorn, M. de Nooijer, G. Somers, J. Ruigrok and L. Pastma, IEEE Trans. Mag. 30, 3813 (1994).
3. P. P. Freitas, J. L. Leal, L. V. Melo, N. J. Oliveira, L. Rodrigues and A. T. Sousa, Appl. Phys. Lett. 45, 493 (1994); J. Leal, N. Oliveira, L. Rodrigues, A. Sousa and P.P.Freitas, IEEE Trans. Mag. 30, 3831 (1994).
4. P. ten Berge, N. J. Oliveira, T. S. Plaskett, J. L. Leal, H. J. Boeve, G. Albuquerque, J. Ferreira, A. R. Morais, A. T. Sousa, L. Rodrigues and P. P. Freitas, presented at the 1995 Intermag Conference, San Antonio, Texas (submitted for publication).
5. J. Mouchot, P. Gerard and B. Rodmacq, IEEE Trans. Mag. 29, 2732 (1993).
6. T. L. Hylton, K. R. Coffey, M. A. Parker and J. K. Howard, J. Appl. Phys. 75, 7058 (1994).
7. M. A. Parker, T. L. Hylton, K. R. Coffey and J. K. Howard, J. Appl. Phys. 75, 6832 (1994).
8. K. Noguchi, S. Araki, T. Chou, D. Miyauchi, Y. Honda, A. Kamijima, O. Shinoura and Y. Narumiya, J. Appl. Phys. 75, 6379 (1994).
9. W. D. Doyle, H. Fujiwara, S. Hossain, A. Matsuzono and M. R. Parker, IEEE Trans. Mag. 30, 3828 (1994).
10. S. Tsunashima, M. Jimbo, T. Kanda, S. Goto and S. Udima, Mater. Res. Soc. Symp. Proc. 313, 271 (1993).
11. R. William Cross, S. E. Russek, S. C. Sanders, M. R. Parker, J. A. Barnard ans S. A. Hossein, IEEE Trans. Mag. 30, 3825 (1994).
12. E. Y. Chen, A. V. Pohm, J. M. Daughton, J. Brown and W. C. Black, IEEE Trans. Mag. 30, 3816 (1994).

DETERMINATION OF SPIN-DEPENDENT SCATTERING PARAMETERS OF NiFe/Cu AND Co/Cu MULTILAYERS

S.K.J. LENCZOWSKI,*‡ M.A.M. GIJS,‡ R.J.M. van de VEERDONK,*‡ J.B. GIESBERS,‡ AND W.J.M. de JONGE*

* Physics Dept., Eindhoven University of Technology, 5600 MB Eindhoven, The Netherlands
‡ Philips Research Laboratories, Professor Holstlaan 4, 5656 AA Eindhoven, The Netherlands

ABSTRACT

We present magnetoresistance (MR) data of high-vacuum magnetron sputtered NiFe/Cu multilayers (NiFe=$Ni_{80}Fe_{20}$) grown on Si(100) substrates with a Cu buffer layer and compare these with earlier results on Co/Cu(100) multilayers [1]. Measured MR values are directly proportional to the antiferromagnetically coupled fraction in the multilayers. Extrapolating to full antiparallel alignment, we can make a reliable comparison of the MR with the magnetoresistance model of Levy, Zhang, and Fert [2,3]. For the NiFe/Cu multilayers the deduced spin-asymmetry parameters are $\alpha_i^{NiFe/Cu} = 5.0 \pm 0.4$ and $\alpha_b^{NiFe} = 2.1 \pm 0.3$ for interface and bulk scattering, respectively. Although much smaller than in our Co/Cu multilayers, where $\alpha_i^{Co/Cu} = 21 \pm 3$ and $\alpha_b^{Co} = 2.6 \pm 0.3$, it is still the spin dependence of the interface scattering that is the main cause for the large MR values.

INTRODUCTION

The giant magnetoresistance (MR) effect in magnetic multilayers has been the subject of numerous studies in recent years (see, for instance, Refs. 4–7 and references therein). A spin dependence of electron scattering processes is at the basis of the giant MR effect, but whether this spin-dependent scattering takes place within the interior of the magnetic layers or at the interfaces between magnetic and nonmagnetic materials still remains a matter of dispute. Since Camley and Barnaś originally introduced their semiclassical MR model based on the spin-dependent Boltzmann equation [8], this way of modeling has been widely applied in modified forms to several multilayer and spin-valve systems [9–15]. The semiclassical approach gives a quantitative agreement with the experimental data in terms of model parameters [16]. An essential drawback, however, is the different treatment of the bulk and interface contributions to the resistivity. Levy, Zhang, and Fert on the other hand developed a model of the MR [2,3], which describes both bulk and interface scattering on equal footing. This model was used to explain the giant MR of the Fe/Cr system [17,18]. This is a good model system, since due to the large antiferromagnetic- (AF-) coupling strength, full antiparallel alignments at zero magnetic field can be realized relatively easily up to Cr spacer layer thicknesses of 2.8 nm, corresponding to the second AF-coupling maximum. For the Fe/Cr case it was concluded that the spin-dependent scattering processes are primarily of interfacial nature. In the case of NiFe/Cu and Co/Cu multilayers such a comparison is more difficult, because the much weaker AF-coupling strengths easily cause an imperfect antiparallel alignment already in the first AF-coupling peak. This reduces the measured MR value leading to data inappropriate for comparison with the model.

In this paper we present MR measurements on NiFe/Cu(100) multilayers (NiFe=$Ni_{80}Fe_{20}$) grown by sputtering techniques and compare these with earlier results on Co/Cu(100) multilayers [1]. These systems are known to exhibit large MR values and can be grown coherently in rather an easy way. Measured MR values are directly proportional to the AF-coupled

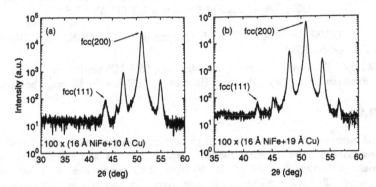

FIG. 1: High-angle X-ray diffraction patterns of (a) a $100\times(16\,\text{Å NiFe}+10\,\text{Å Cu})$ multilayer and (b) a $100\times(16\,\text{Å NiFe}+19\,\text{Å Cu})$ multilayer. Both samples were grown on Cu buffer layers of $200\,\text{Å}$.

fraction in the multilayers. Extrapolating our MR data to full antiparallel alignment, we can make a reliable comparison of the MR of NiFe/Cu(100) and Co/Cu(100) at $4\,\text{K}$ with the magnetoresistance model of Levy, Zhang, and Fert [2,3].

SAMPLE FABRICATION AND CHARACTERIZATION

The multilayer samples were grown by high-vacuum magnetron sputtering. The base pressure of the system prior to the deposition was $4 \times 10^{-7}\,\text{Torr}$ and the Ar pressure during sputtering was $7 \times 10^{-3}\,\text{Torr}$. The samples were deposited at a rate of $2\text{-}4\,\text{Å/s}$ onto $4 \times 12\,\text{mm}^2$ Si(100) substrates held at room temparature. Before sputtering, the samples were *ex-situ* cleaned by a HF-dip and *in-situ* by a $30\,\text{min}$ glow-discharge treatment. In order to obtain a highly face centered cubic (fcc) (100)-oriented texture, we used a Cu buffer layer with a thickness, t_b, of $200\,\text{Å}$. For values of t_b below $200\,\text{Å}$ we get a mixed (111)-(100) growth. After deposition of the multilayers, they were covered with a $50\,\text{Å}$ Au protection layer. The (100) orientation was checked by X-ray diffraction (XRD) measurements using Cu-$K\alpha$ radiation. Figures 1(a) and 1(b) show typical high-angle XRD patterns for the superlattices $100\times(16\,\text{Å NiFe}+10\,\text{Å Cu})$ and $100\times(16\,\text{Å NiFe}+19\,\text{Å Cu})$, respectively. In both cases we observe the main NiFe/Cu(200) Bragg reflection with several multilayer satellite peaks. There is only little intensity of NiFe/Cu(111) reflections present. From the full width at half maximum (FWHM) of the NiFe/Cu(200) reflection, which is $0.40°$, we deduce a perpendicular crystal coherence length of about $230\,\text{Å}$. This value is nearly the same for all our (100)-oriented NiFe multilayers and comparable to previously grown Co/Cu(100) multilayers [1]. The FWHM of the rocking curves about the (200) maximum (not shown) varied between $1.5°$ and $2.0°$, indicating the good texture of the crystallites.

RESULTS AND DISCUSSION

In Figs. 2(a) and 2(b) we show typical MR curves both at room temperature (RT) and at $4\,\text{K}$ for two NiFe/Cu multilayers with Cu spacer layer thicknesses, t_{Cu}, of 10 and $19\,\text{Å}$, respectively. We define MR as $(R_{max} - R_{sat})/R_{sat}$, with R_{max} the maximum resistance around zero

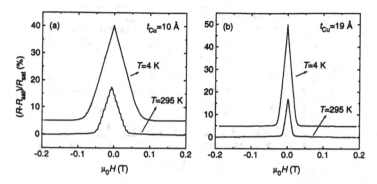

FIG. 2: Magnetoresistance curves at room temperature and at 4 K for (100)-oriented multilayers (a) 100×(16 Å NiFe+10 Å Cu) and (b) 100×(16 Å NiFe+19 Å Cu) grown on Cu buffer layers of 200 Å. The magnetic field was in the plane of the layers and perpendicular to the current direction. The curves at 4 K have been shifted upwards by 5% for clarity.

magnetic field and R_{sat} the value at saturation. We observe high MR values at RT of 18% for the multilayer with $t_{Cu} = 10$ Å (35% at 4 K) and 17% for the multilayer with $t_{Cu} = 19$ Å (45% at 4 K). These two copper thicknesses correspond to the first two antiferromagnetic (AF) maxima of the interlayer exchange coupling in the NiFe system. From the magnetic field at which the MR saturates, H_s, one can deduce the magnitude of this AF-coupling strength, J_{AF}, via the relation $J_{AF} = H_s M_s t_F/4$, with M_s and t_F the saturation magnetization and the thickness of the magnetic layers, respectively. For $t_{Cu} = 10$ Å a coupling strength $J_{AF} = 0.013 \, \text{mJ/m}^2$ at RT is found and $J_{AF} = 0.023 \, \text{mJ/m}^2$ at 4 K. These values are about a factor of twelve smaller than in our sputtered Co/Cu samples [1]. Similarly, for $t_{Cu} = 19$ Å we find $J_{AF} = 4.3 \times 10^{-3} \, \text{mJ/m}^2$ at RT and $J_{AF} = 7.4 \times 10^{-3} \, \text{mJ/m}^2$ at 4 K, approximately 16 times smaller than in comparable Co/Cu samples [1]. This relatively small coupling strength, and hence H_s, leads to a sizable sensitivity of $\Delta R/(R \, \Delta H) = 0.19\%/\text{Oe}$ at 4 K for $t_{Cu} = 19$ Å.

The temperature dependence of J_{AF} in both AF maxima appears to be the same. This is in marked contrast to the results of Parkin [19], who found that J_{AF} in the second AF peak (around $t_{Cu} = 20$ Å) falls off dramatically with increasing temperature, resulting in no measurable AF coupling or MR at RT. This different behavior was attributed to the dissolution of NiFe and Cu at the NiFe/Cu interfaces being more important for thicker t_{Cu}. Figure 3 depicts the MR behavior of NiFe(16 Å)/Cu multilayers as a function of t_{Cu}. Both at RT and at 4 K an oscillatory behavior can be observed with a period of approximately 10 Å. Apparently, in our case, there is no substantial increase of the mixed region at the NiFe/Cu interfaces for the multilayers with thicker t_{Cu} (say 20 Å). This conclusion is supported by measurements of the saturation magnetic moment of NiFe/Cu(19 Å) multilayers as a function of t_{NiFe}, which is shown in the inset of Fig. 3. The slope of this curve reflects the bulk value of M_s of NiFe of 1.0 T. The offset from the origin can be attributed to a nonmagnetic layer at the NiFe/Cu interfaces with a thickness of about only 1.5 Å as was also found by Speriosu and coworkers [20].

To compare the MR data of the first and second AF maximum properly a correction for

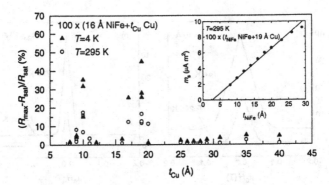

FIG. 3: Dependence of the magnetoresistance ratio on the Cu spacer layer thickness, t_{Cu}, for (100)-oriented $100\times(16\,\text{Å NiFe}+t_{Cu}\,\text{Å Cu})$ multilayers grown on Cu buffer layers of 200 Å. *Inset:* Saturation magnetic moment, m_s, versus the NiFe layer thickness, t_{NiFe}, for a series of multilayers with $t_{Cu} = 19$ Å.

the fraction of AF coupling, F_{AF}, in the multilayers has to be made. For example, the sample of Fig. 2(a) exhibits only a F_{AF} of 63% as determined from the remanent magnetization measured with a vibrating sample magnetometer. In Figs. 4(a) and 4(b) plots of MR versus F_{AF} for our NiFe/Cu multilayers are given for $t_{Cu} \approx 10$ Å (1st AF peak) and $t_{Cu} \approx 19$ Å (2nd AF peak), respectively. From these plots we can deduce the MR values for full antiparallel alignment by extrapolation. At 4 K we find MR values of 55% for $t_{Cu} \approx 10$ Å and 45% for $t_{Cu} \approx 19$ Å. We only compare these low-temperature data with the magnetoresistance model of Levy et al. [2,3], since it is assumed that electric conduction takes place in two separate spin channels. At elevated temperatures spin-flip scattering, which effectively mixes both spin channels, may disturb a proper analysis. Also another set of MR data was measured at $T = 4$ K in which the magnetic layer thickness, t_{NiFe}, has been varied. Again, these MR

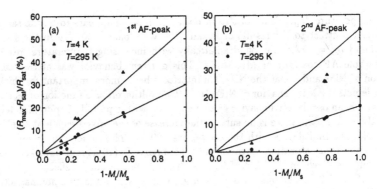

FIG. 4: Magnetoresistance ratio versus $(1 - M_r/M_s)$ for (100)-oriented $100\times(16\,\text{Å NiFe}+t_{Cu}\,\text{Å Cu})$ multilayers with (a) t_{Cu} around the first antiferromagnetic- (AF-) coupling peak and (b) t_{Cu} around the second AF-coupling peak.

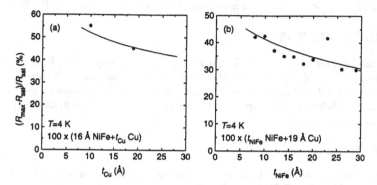

FIG. 5: Dependence of the magnetoresistance ratio on the Cu spacer layer thickness, t_{Cu}, at $T = 4\,K$. The solid line is calculated using the magnetoresistance model of Levy et al. (Refs. 2,3). (b) Dependence of the magnetoresistance ratio on the magnetic layer thickness, t_{NiFe}, at $T = 4\,K$. The solid line represents the model calculation. Fitting parameters are indicated in the text.

values are corrected for incomplete antiparallel alignment. A detailed description of the fitting procedure is given elsewhere [1]. Important fitting parameters of the model are p_i and p_b, representing the ratio of spin-dependent to spin-independent scattering at the interfaces or bulk lattice-planes, respectively. These parameters are related to the more conventional spin-asymmetry parameter $\alpha = \lambda_\uparrow / \lambda_\downarrow$ according to $\alpha = (1 + p)^2/(1 - p)^2$. In Figs. 5(a) and 5(b) we plot the results of the fits with the parameters $p_i = 0.38$ ($\alpha_i^{NiFe/Cu} = 5.0 \pm 0.4$) and $p_b = 0.18$ ($\alpha_b^{NiFe} = 2.1 \pm 0.3$). The values for the mean-free path, λ, for NiFe and Cu used in the calculations are $\lambda^{NiFe} = 60\,Å$ and $\lambda^{Cu} = 200\,Å$. Both data series have been fitted with the same parameter set. These spin-asymmetry parameters of NiFe and NiFe/Cu are significantly smaller than those for our Co/Cu multilayers ($\alpha_i^{Co/Cu} = 21 \pm 3$ and $\alpha_b^{Co} = 2.6 \pm 0.3$) [1], but it is still the spin dependence of the interface scattering that plays the dominant role. A similar conclusion can be drawn from current-perpendicular-to-plane MR measurements by Yang et al. [21,22] on Co/Cu/NiFe/Cu multilayers with rather large individual layer-thicknesses. However, our α-values are somewhat smaller. In general, for various multilayer systems, α_i seems to depend highly on the exact microstructure, orientation, and deposition conditions of the samples. Although still within experimental uncertainty, a possible explanation for the discrepancy between the experimental bulk values α_b^{NiFe} was recently suggested by Rijks et al. [23] who showed that grain-boundary scattering may reduce the effective spin dependence of the scattering.

CONCLUSIONS

We have grown sputtered NiFe/Cu(100) multilayers (NiFe=$Ni_{80}Fe_{20}$) on Cu buffer layers and observed an oscillatory behavior of the magnetoresistance (MR) as a function of the Cu spacer-layer thickness even at room temperature. The measured MR values are directly proportional to the antiferromagnetically coupled fraction in the multilayers. We have interpreted low-temperature MR data, corrected for incomplete antiparallel alignment, with the magnetoresistance model of Levy et al. [2,3] and found that the spin dependence of the

NiFe/Cu-interface scattering is the most important source for the giant MR effect.

ACKNOWLEDGMENTS

Part of this research was supported by the European Community Science Project ESPRIT3: Basic Research, *'Study of Magnetic Multilayers for Magnetoresistive Sensors' (SmMmS)*. The authors would like to thank H. T. Munsters and H. C. Donkersloot for the sample preparation, T. Schudelaro for experimental assistance and R. Coehoorn, and M. T. Johnson for stimulating discussions.

REFERENCES

[1] S. K. J. Lenczowski, M. A. M. Gijs, J. B. Giesbers, R. J. M. van de Veerdonk, and W. J. M. de Jonge, Phys. Rev. B **50**, 9982 (1994).

[2] P. M. Levy, S. Zhang, and A. Fert, Phys. Rev. Lett. **65**, 1643 (1990).

[3] S. Zhang, P. M. Levy, and A. Fert, Phys. Rev. B **45**, 8689 (1992).

[4] *Proceedings of the International Symposium on Magnetic Ultrathin Films, Multilayers and Surfaces*, 7-10 September 1992, Lyon [J. Magn. Magn. Mater. **121**(1-3) (1993)].

[5] *Magnetic Ultrathin Films: Multilayers and Surfaces, Interfaces and Characterization*, MRS Symposium Proceedings Volume 313 edited by B. T. Jonker *et al.* (Materials Research Society, Pittsburgh, 1993).

[6] *Proceedings of the MMM Conference*, 15-19 November 1993, Minneapolis [J. Appl. Phys. **75**(10-II) (1994)].

[7] J. A. C. Bland and B. Heinrich, eds., *Ultrathin Magnetic Structures I and II* (Springer, Berlin Heidelberg, 1994).

[8] R. E. Camley and J. Barnaś, Phys. Rev. Lett. **63**, 664 (1989).

[9] A. Barthélémy and A. Fert, Phys. Rev. B **43**, 13124 (1991).

[10] B. L. Johnson and R. E. Camley, Phys. Rev. B **44**, 9997 (1991).

[11] B. Dieny, Europhys. Lett. **17**, 261 (1992).

[12] B. Dieny, J. Phys. Condens. Matter **4**, 8009 (1992).

[13] T. G. S. M. Rijks, R. Coehoorn, and W. J. M. de Jonge, Mat. Res. Soc. Symp. Proc. Vol. **313**, 283 (1993).

[14] L. M. Falicov and R. Q. Hood, J. Magn. Magn. Mater. **121**, 362 (1993).

[15] J. L. Duvail, A. Fert, L. G. Pereira, and D. K. Lottis, J. Appl. Phys. **75**, 7070 (1994).

[16] B. Dieny, V. S. Speriosu, J. P. Nozières, B. A. Gurney, A. Vedyayev, and N. Ryzhanova, in *Magnetism and Structure in Systems of Reduced Dimension*, Vol. 309 of *NATO Advanced Study Institute, Series B: Physics* edited by R. F. C. Farrow *et al.* (Plenum, New York, 1993), p. 279.

[17] M. A. M. Gijs and M. Okada, Phys. Rev. B **46**, 2908 (1992).

[18] M. A. M. Gijs and M. Okada, J. Magn. Magn. Mater. **113**, 105 (1992).

[19] S. S. P. Parkin, Appl. Phys. Lett. **61**, 1358 (1992).

[20] V. S. Speriosu, J. P. Nozières, B. A. Gurney, B. Dieny, T. C. Huang, and H. Lefakis, Phys. Rev. B **47**, 11579 (1993).

[21] Q. Yang, P. Holody, R. Loloee, L. L. Henry, W. P. Pratt, Jr., P. A. Schroeder, and J. Bass, Phys. Rev. B **51**, 3226 (1995).

[22] P. A. Schroeder, J. Bass, P. Holody, S.-F. Lee, R. Loloee, W. P. Pratt, Jr., and Q. Yang, Mat. Res. Soc. Symp. Proc. Vol. **313**, 47 (1993).

[23] T. G. S. M. Rijks, R. L. H. Sour, D. G. Neerinck, A. E. M. De Veirman, R. Coehoorn, J. C. S. Kools, M. F. Gillies, and W. J. M. de Jonge, IEEE Trans. Magn. (1995), *to be published*.

THE EFFECT OF IMPURITY DOPING OF THE MAGNETIC LAYER ON THE MAGNETORESISTANCE AND SATURATION FIELD OF FeCr/Cr AND CoCu/Cu MULTILAYERS

B.J. DANIELS and B.M. CLEMENS
Department of Materials Science and Engineering, Stanford University, Stanford, CA 94305-2205

ABSTRACT

The effect of doping the magnetic layer on the magnetoresistance and saturation field of epitaxial, sputter-deposited FeCr/Cr(001) and CoCu/Cu(001) multilayers was examined. FeCr/Cr films with a composition of $[14 \text{ Å } Fe_{1-x}Cr_x(001)/8 \text{ Å } Cr(001)]_{50}$ where x was varied from 0 to 0.5 and CoCu/Cu films with a composition of $[8 \text{ Å } Co_{1-x}Cu_x(001)/21 \text{ Å } Cu(001)]_{40}$ where x was varied from 0 to 0.2 were deposited onto single crystal MgO(001). The bilayer period and epitaxial orientation of these films were determined by x-ray diffraction. In the FeCr/Cr system the room temperature magnetoresistance was constant at approximately 31% for $x \leq 0.2$ and decreased for larger Cr concentrations. The spin-dependent scattering $(\Delta\rho)$ is *increased* over the $x=0$ value for $x=0.1$ and $x=0.2$. The magnetic field required to saturate these multilayers decreases linearly with increasing Cr concentration. The net result is that the sensitivity of these films is increased by Cr doping of the Fe layer. In the CoCu/Cu system the room temperature magnetoresistance, saturation field, and saturation magnetization decrease with the addition of Cu to the Co layer. In contrast to the FeCr/Cr system $\Delta\rho$ does *not* increase with doping of the ferromagnetic layer. However, as in the FeCr/Cr films the sensitivity of these multilayers is increased with respect to that of the $x=0$ CoCu/Cu multilayer.

INTRODUCTION

The discovery of the "giant" magnetoresistance (GMR) effect in Fe/Cr superlattices by Baibich, et al.[1] has touched off an explosion of research in this area[2]. Of the materials systems which exhibit GMR, antiferromagnetically-coupled Fe/Cr and Co/Cu multilayer films have large values of magnetoresistance at room temperature and have been studied in great detail. The presence of GMR in these systems is currently believed to be due to spin-dependent scattering of conduction electrons. Although this preferential scattering has been postulated to occur at both interfaces and impurity sites, the relative importance of each mechanism is still under debate.

Using the conventional definition, GMR=$\Delta\rho/\rho_{sat}$, the GMR of a given system can be increased by increasing $\Delta\rho$ (the spin-dependent scattering) or by decreasing ρ_{sat} (the spin-independent scattering). Although the reduction of ρ_{sat} can be accomplished by fabricating films with a greater degree of crystalline perfection[3], methods for increasing $\Delta\rho$ depend upon the relative importance of interface and bulk scattering in these multilayers. In the Fe/Cr system it has been shown that $\Delta\rho$ can be increased by the addition of Cr impurities to the Fe layer in both polycrystalline[4,5] and epitaxial[6,7] multilayers. In addition to increasing $\Delta\rho$, alloying has also been shown to increase the sensitivity, $S=\partial(\text{GMR})/\partial H$ of these films[7]. This technique harbors great potential in the magnetic recording industry as the sensitivity is the figure of merit for magnetoresistive read heads. In this work we report the results of doping the magnetic layer on the magnetic and magnetotransport properties of both FeCr/Cr and CoCu/Cu multilayers.

EXPERIMENTAL

Samples were fabricated in a UHV sputter deposition chamber equipped with 3 confocally-aimed, shuttered dc magnetron sputtering guns. The substrates were mounted vertically at a sample-to-target distance of 6 in. Details of this deposition chamber are described elsewhere[8]. Single crystal MgO(001) substrates were organically cleaned and placed in the chamber which was evacuated to 1×10^{-9} Torr. The substrates were heated to 650°C for 15 min and then cooled to 600°C. Seed layers of 100 Å Cr and 50 Å Pd were deposited for the Fe/Cr and Co/Cu system, respectively. These seed layers allow high quality, epitaxial multilayers to be grown at lower temperatures, reducing the amount of interdiffusion[3,9,10]. The substrates were then cooled to 180°C (FeCr/Cr) or 130°C (CoCu/Cu) whereupon multilayers were sequentially deposited at rates of less than 1 Å/s. Alloy layers were formed by cosputtering from elemental targets. For the FeCr/Cr system 50 layers of 14 Å $Fe_{1-x}Cr_x$/8 Å Cr where $x = 0, 0.1, 0.2, 0.3, 0.4, 0.5$ were deposited. The CoCu/Cu multilayers consisted of 40 layers of 8 Å $Co_{1-x}Cu_x$/21 Å Cu where $x = 0, 0.1, 0.2$. A 30 Å Pd capping layer was deposited at 130°C for the CoCu/Cu multilayers in order to prevent oxidation of the film.

Sample structure was characterized via x-ray diffraction (XRD). Low-angle reflectivity measurements for the FeCr/Cr films were made at the Stanford Synchrotron Radiation Laboratory (SSRL) using an x-ray wavelength of 1.5405 Å. Low-angle scans for the CoCu/Cu multilayers as well as high angle symmetric scans for both the FeCr/Cr and CoCu/Cu multilayers were obtained from a powder diffractometer using CuKα radiation and equipped with an exit beam monochromator. Asymmetric XRD scans for the FeCr/Cr multilayers were obtained from a three-circle diffractometer. A four-circle diffractometer with a Ge detector was used for the CoCu/Cu asymmetric XRD scans.

Magnetotransport measurements for the FeCr/Cr system were made at the Francis Bitter National Magnet Laboratory using a warm bore superconducting magnet. The sample temperature during the measurements was between 15 and 20°C. The applied field was swept from -1 to 30 kOe at a rate of approximately 100 Oe/s while the resistance of the sample was measured using a standard dc four-point probe setup. The magnetoresistance measurements for the CoCu/Cu multilayers were made at room temperature using a standard electromagnet. The magnetic field was continuously swept from -4 to 4 kOe at a rate of approximately 50 Oe/s. The resistance of the multilayer was measured using a four-point probe attached to an LCR meter. In-plane vibrating sample magnetometry (VSM) was used to characterize the magnetization behavior of the CoCu/Cu multilayer films.

RESULTS AND DISCUSSION

High angle symmetric x-ray diffraction scans for both the FeCr/Cr and CoCu/Cu multilayers showed only (002) film peaks. For the Co/Cu system the first and second order satellite peaks could also be resolved. This indicated that these films were strongly oriented out-of-plane, a phenomenon that could result from either strong fiber texture or epitaxy. Epitaxial growth by sputter deposition has been reported in both of these systems under similar growth conditions[3,9,10]. Since the lattice parameters of Fe and Cr differ by only 0.7% and those of FCC Co and Cu by only 2% we expect that there will be no relative

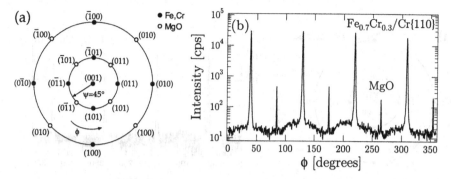

Figure 1: (a) Stereographic projection for FeCr/Cr(001) multilayers in the cube-on-cube epitaxial orientation. The path taken by the asymmetric XRD scan in (b) is shown. (b) ϕ scan of the {110} peaks for a $Fe_{0.7}Cr_{0.3}/Cr$ multilayer. The fourfold symmetry is indicative of epitaxial growth in the cube-on-cube orientation. The four smaller peaks are due to the MgO(001) substrate.

rotation between the surface nets of the components in either system. This epitaxial orientation is often referred to as cube-on-cube epitaxy and is shown on the stereographic projections for the FeCr/Cr and CoCu/Cu systems in Figs 1(a) and 2(a), respectively. In order to investigate for epitaxy the {110} film peaks for each multilayer were studied using asymmetric XRD. The angle between the scattering vector and the sample normal, referred to as ψ, was moved to 45°. The sample was then rotated about its normal by an angle, ϕ. The net result is that the circular trajectories shown on the stereographic projections in Figs. 1(a) and 2(a) were followed. For epitaxial films we expect to observe the distinct peaks shown in the stereograms, while for textured samples we expect a ring of constant intensity, i.e. no variation with ϕ. For both systems ϕ scans show four film peaks separated by 90° (Figs. 1(b) and 2(b)). This fourfold symmetry is indicative of (001) epitaxy with the cube-on-cube epitaxial orientation. The full width at half maximum (FWHM) of these peaks for each system is $\leq 1°$ indicating excellent crystalline quality of the film.

Low-angle XRD scans were used to determine the bilayer period, Λ, of each multilayer. For the FeCr/Cr system two low angle superlattice peaks were obtained. By plotting n^2 vs $\sin^2\theta$, where n is the order of the superlattice peak and θ is half of the scattering angle for the peak, the index of refraction-corrected bilayer period can be obtained. Using this technique Λ was determined to be 20.6 Å for the FeCr/Cr multilayers. This value is approximately 7% lower than the nominal value and was a constant offset for all samples. Two low angle peaks were also obtained for the CoCu/Cu multilayers. Using the same method of analysis, the bilayer period was found to be 27.5 Å. This value is lower than the nominal bilayer period by approximately 5%. The sample-to-sample variation in Λ for the CoCu/Cu films was approximately 0.1 Å.

Room temperature magnetoresistance vs applied field data measured in the "current in plane" (CIP) configuration for the six different $Fe_{1-x}Cr_x/Cr$ multilayers are shown in Fig. 3(a). Although these measurements were made for $H\|I\|[110]$, we also measured the magnetoresistance for the cases where $H\perp I\|[110]$ and $H\|I\|[100]$. There were no noticeable difference in these curves, however. The magnetoresistance is approximately composition independent at 31% for $0\leq x\leq 0.2$. This constant magnetoresistance results from an *in-*

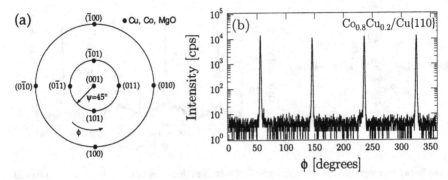

Figure 2: (a) Stereographic projection for CoCu/Cu(001) multilayers in the cube-on-cube epitaxial orientation. The path taken by the asymmetric XRD scan in (b) is shown. (b) ϕ scan of the {110} peaks for a $Co_{0.8}Cu_{0.2}/Cu$ multilayer. The fourfold symmetry is indicative of epitaxial growth in the cube-on-cube orientation.

crease in the spin-dependent scattering ($\Delta\rho$) which is offset by a corresponding increase in the spin-independent scattering (ρ_{sat}). For larger values of x the spin-dependent scattering and the magnetoresistance decrease. The saturation field for these films decreases linearly with increasing Cr content for all values of x. The net effect is that the sensitivity, $S \approx \partial(GMR)/\partial H$, of these multilayers is increased by Cr doping (Fig. 3(b)).

Room temperature CIP magnetoresistance vs applied field data for the $Co_{1-x}Cu_x/Cu$ system are shown in Fig. 4(a). Both the magnetoresistance and the saturation field decrease as Cu is added to the Co layer. This is presumed to be due to a decrease in the antiferromagnetic coupling strength in these multilayers. The cusp-like peak in the GMR vs H curve (Fig. 4(a)) for $x=0.2$ implies incomplete antiferromagnetic alignment at zero field which would be a consequence of decreased coupling. Figure 5 shows the in-plane saturation magnetization, M_{sat}, and the saturation field, H_{sat}, vs Cu concentration in the CoCu layer. From this figure we see that both M_{sat} and H_{sat} decrease linearly with Cu alloying. As a

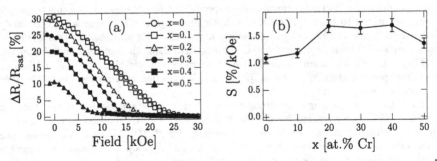

Figure 3: (a) Room temperature magnetoresistance vs field and (b) sensitivity vs x for 100 Å $Cr(001)/[14$ Å $Fe_{1-x}Cr_x(001)/8$ Å $Cr(001)]_{50}$ films.

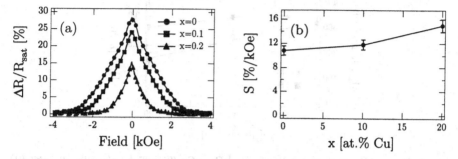

Figure 4: (a) Room temperature magnetoresistance vs field and (b) sensitivity vs x for 50 Å Pd(001)/[8 Å $Co_{1-x}Cu_x$(001)/21 Å Cu(001)]$_{40}$/30 Å Pd(001) films.

rough estimate of the coupling strength, J, we can use the expression $J \propto H_{sat} \cdot M_{sat}$ used by Chen et al.[4]. Since both H_{sat} and M_{sat} decrease linearly, we conclude that the coupling strength decreases roughly parabolically with the addition of Cu to the Co layer in this system. In contrast to the FeCr/Cr system the spin-dependent scattering does *not* increase with the addition of scattering centers to the ferromagnetic layer. The sensitivity of these films, however, is increased by doping the Co layer with Cr (Fig. 4(b)).

At this point it is worth comparing the effects of alloying in the CoCu/Cu system with those observed in the FeCr/Cr system. It is not surprising that the spin-dependent scattering in the CoCu/Cu multilayers is not increased by Cu doping. Since the Cu layer is over 2.5 times thicker than the CoCu layer and is also lower in resistivity, we expect most of the current to be shunted through this lower resistance path. As only a small fraction of the total current would be carried by the higher resistance CoCu layer, it is likely that changes in the spin-dependent scattering in this layer would be difficult to observe. This complication does not occur in FeCr/Cr multilayers because both the resistivities and the thicknesses of the two layers are comparable. In order to circumvent this problem in the CoCu/Cu system, the Cu layer thickness must be decreased and/or the CoCu layer thickness must be increased. To this end we have grown CoCu/Cu multilayers with the structure [8 Å $Co_{1-x}Cu_x$(001)/11 Å Cu(001)]$_{40}$ where x=0, 0.1, and 0.2. This Cu spacer layer thickness is near that where the first peak in the antiferromagnetic coupling is expected. Although magnetoresistance data has not been obtained for these films, preliminary magnetization data indicate that the saturation field is decreased markedly for the films with alloy layers.

CONCLUSIONS

High-quality, epitaxial FeCr/Cr(001) and CoCu/Cu(001) multilayers have been fabricated by sputter deposition. Cr doping of the Fe layer in $Fe_{1-x}Cr_x$/Cr(001) multilayers results in an increase in the spin-dependent scattering for $0.1 \leq x \leq 0.2$. The magnetoresistance remains constant at 31% for $0 \leq x \leq 0.2$ and decreases for larger Cr concentrations while the saturation field decreases linearly with Cr concentration. The net result is that the sensitivity of these multilayers is increased over that of the pure Fe/Cr multilayer for $0.1 \leq x \leq 0.5$. In contrast, Cu doping of the Co layer in $Co_{1-x}Cu_x$/Cu(001) multilayers does not appear

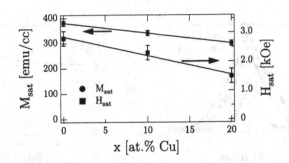

Figure 5: M_{sat} and H_{sat} vs x for 50 Å Pd(001)/[8 Å Co$_{1-x}$Cu$_x$(001)/21 Å Cu(001)]$_{40}$/30 Å Pd(001) films. The lines are least squares fits to the data.

to increase the spin-dependent scattering, although this is likely to be the result of current shunting through the Cu spacer layer. The magnetoresistance for this system is maximum (28%) for the pure Co/Cu multilayer. Both the saturation magnetization and saturation field vary linearly with Cu concentration over the range $0 \leq x \leq 0.2$, implying parabolic dependence of the coupling strength on Cu concentration. As in the FeCr/Cr system the sensitivity of Co$_{1-x}$Cu$_x$/Cu(001) multilayers is also increased by doping the ferromagnetic layer with the spacer layer component. It is hoped that this technique will prove useful for increasing the sensitivity of other multilayer systems which exhibit GMR.

ACKNOWLEDGEMENTS

The authors would like to thank the Francis Bitter National Magnet Laboratory for the use of their facilities for the FeCr/Cr magnetotransport measurements and Kobe Steel, USA for the use of their VSM. This work was supported by NSF Grant Nos. DMR-9022248 and DMR-9408552. SSRL is supported by the U.S. Department of Energy, Office of Basic Energy Sciences.

REFERENCES

1. M.N. Baibich, J.M. Broto, A. Fert, F. Nguyen Van Dau, F. Petroff, P. Etienne, G. Creuzet, A. Friederich, and J. Chazelas, *Phys. Rev. Lett.* **61**, 2472 (1988).
2. B. Dieny, *J. Magn. Magn. Mater.* **136**, 335 (1994) and references therein.
3. E.E. Fullerton, M.J. Conover, J.E. Mattson, C.H. Sowers, and S.D. Bader, *Appl. Phys. Lett.* **63**, 1699 (1993).
4. L.H. Chen, T.H. Tiefel, S. Jin, R.B. Van Dover, E.M. Gyorgy, R.M. Fleming, *Appl. Phys. Lett.* **63**, 1279 (1993).
5. N.M. Rensing, *Ph.D. Dissertation*, Stanford University, 1994.
6. F. Petroff, A. Barthelemy, A. Hamzic, A. Fert, P. Etienne, S. Lequien, and G. Creuzet, *J. Magn. Magn. Mater.* **93**, 95 (1991).
7. B.J. Daniels and B.M. Clemens, *Appl. Phys. Lett.* **66**, 520 (1995).
8. A.P. Payne, S. Brennan, and B.M. Clemens, *Rev. Sci. Instr.* **63**, 1147 (1992).
9. B.J. Daniels, W.D. Nix, and B.M. Clemens, *Thin Solid Films* **253**, 218 (1994).
10. G.R. Harp and S.S.P. Parkin, *Appl. Phys. Lett.* **65**, 3063 (1994).

GIANT MAGNETORESISTANCE AND OSCILLATION IN EPITAXIAL Fe/Cr(111) MULTILAYERS

Wen-C. Chiang, David V. Baxter, and Yang-Tse Cheng*
Dept. of Physics, Indiana University, Bloomington, IN 47405
*General Motors Research and Development Center, Warren, MI 48090

ABSTRACT

We report on the first studies of the giant magnetoresistance and oscillatory coupling in epitaxial Fe/Cr(111) multilayers. A series of samples were grown on hydrogen terminated Si(111) substrates at room temperature by UHV electron beam evaporation; with the thickness of Fe layer fixed at 30 Å, and the thickness of Cr layer varied from 10–47 Å. Giant magnetoresistance (GMR) is observed at 4.2 K in these samples, with a maximum value of 13% for a Cr layer thickness of 13 Å. The associated oscillatory coupling is comparable to that reported in other crystallographic orientations in terms of both its period and phase.

INTRODUCTION

Sandwiches and superlattices of magnetic films separated by a thin layer of nonmagnetic material have recently attracted wide interest because of their unusual magnetic and transport properties. Baibich et al. first discovered that the Fe/Cr multilayers display unexpectedly high values of magnetoresistance (The resistance dropped by almost a factor of 2 in a magnetic field of 2 T for some samples) [1]. This so called giant magnetoresistance (GMR) has great potential for technical applications such as recording heads and magnetic sensors. The GMR has since then inspired a lot of experimental and theoretical investigations in many systems [2, 3, 4, 5, 6, 7, 8, 9, 10, 11, 12], with significant progress being made both in increasing the size of the GMR effect and in reducing the size of the field required to effect the change in R. The large reduction in sample resistance has its origins in the dependence of the resistance on the relative orientation of neighboring ferromagnetic (FM) layers within the sample (this being smallest for a parallel arrangement) [13]. The largest effects are seen in samples where neighboring FM layers are oriented anti-ferromagnetically (AFM; the configuration of maximum resistance) through indirect exchange coupling across the non-magnetic spacer layer. From this configuration the low resistivity parallel arrangement can be induced by an external field of sufficient strength to overcome the exchange coupling.

A phenomenon of interest equal to the GMR effect itself is the observation that for many systems the coupling between FM layers alternates periodically between FM and AFM with changing spacer thickness. This phenomenon is very reminiscent of the RKKY coupling seen between isolated magnetic atoms in a non-magnetic host, but in many cases the observed value for the period is much longer than that expected from a simple free-electron model of the RKKY interaction (π/k_F) [14]. Moreover in some cases two different periods have been observed with one being comparable to π/k_F and the other longer. Many variations on the RKKY theme have been put forward to explain the long periods with many authors pointing out that aliasing associated with the periodic sampling of the spacer thickness imposed by its crystalline structure affords an opportunity for long effective periods. Herman and Schrieffer[11] have pointed out that such a mechanism is very susceptible to disruption through roughness at the interface but that a full description of the coupling through Bloch states (rather than free electron states) can also lead to long periods. Irrespective of their details however, most current theories of the oscillatory coupling relate the period to reciprocal lattice vectors that span the Fermi surface (joining states with antiparallel group velocities along the multilayer growth direction) [10, 12].

353

In light of this current thinking about the origins of the oscillatory coupling, a somewhat remarkable observation is that the observed period is very similar $(10 \pm 1 \text{ Å})$ for the vast majority of spacer layers for which AFM coupling has been observed [15]. Naively one would expect a variety of periods from spacer materials with such different Fermi surfaces. Very recently models invoking resonance structures in quantum wells provided by the multilayers have been put forward as a possible solution to this quandary [16, 17]. There is clearly considerable need for additional experiments investigating the connection between the properties of the spacer material and the period of the coupling it provides. One avenue for such investigations is to look more closely at those spacer materials that appear as exceptions to the above noted trend. Most notable among these materials is Cr which exhibits a long period of roughly 18 Å and, for samples with very flat interfaces, a short period closer to 3 Å. An obvious direction for such investigations is to look for variations in the period with the crystallographic orientation of the film for a given spacer material. Such investigations are of interest not only because different orientations probe different structures in a complicated Fermi surface, but different interface energies can lead to different dislocation densities and interfacial structure. It has been argued that in high quality samples such differences will influence various proposed coupling mechanisms in different ways [10]. One such study has recently compared Fe/Cr multilayers in the traditional (100) orientation with the (211) orientation [5]. There has also been at least one study looking at Fe/Cr multilayers grown with a (110) orientation[18]. All three of these orientations show very similar periods and phase for the oscillatory coupling between the Fe layers. The Fe/Cr system is one of the most extensively investigated of all GMR systems which makes it an obvious candidate for further investigations of the orientation dependence of the GMR effect, particularly in light of the apparent anomalous behavior of Cr. In the present work we investigate a fourth orientation in the Fe/Cr system, (111), and find that this orientation exhibits coupling whose dependence on spacer thickness is very similar to that seen in other orientations.

EXPERIMENT

The Fe/Cr(111) multilayers were prepared by UHV electron beam evaporation onto hydrogen terminated Si(111) (etched in 10 % HF) substrates. The base pressure of the chamber was better than 2×10^{-9} torr, and remained less than 10^{-8} torr during deposition. The substrates were introduced into the chamber via a load lock. The deposition rate for both Fe and Cr was 0.5 Å/s. The multilayered structure was achieved by opening and closing the shutters over each source alternatively, so that the substrate was exposed to one evaporating material after another, and repeated throughout the deposition. A series of samples was made at room temperature, with the thickness of the Fe layer fixed at 30 Å, and the thickness of Cr layer varied from 10 to 47 Å (in steps of 3-4 Å). Each sample was made with a total of 10 bilayers. The growth of all samples was monitored *in situ* by RHEED and the diffraction pattern observed during deposition was consistent with those seen in earlier studies of epitaxial Fe growth on Si(111). Detailed analysis from these earlier studies indicated that the in-plane orientation was $[\bar{1}10]_{Fe} \parallel [1\bar{1}0]_{Si};[19]$. It is reasonable to assume that this same orientation is maintained in the Fe/Cr multilayers. More detailed structural analysis on the multilayers, including a direct check on this assumption will be the subject of a future publication.

The Fe/Cr multilayer samples were characterized by x-ray diffraction. Fig. 1(a) shows the low-angle diffraction patterns for representative samples. The effects of layering are apparent in these scans although the diffraction peaks are not extraordinarily well resolved above the background. This is due primarily to a combination of the low x-ray contrast between Cr and Fe and the small number of periods which leads to significant interference structure associated with the finite thickness of the films, but it may also indicate that layering in these samples is not perfect. The origin of the bump observed $2\theta \sim 6°$ in the diffraction pattern of some samples

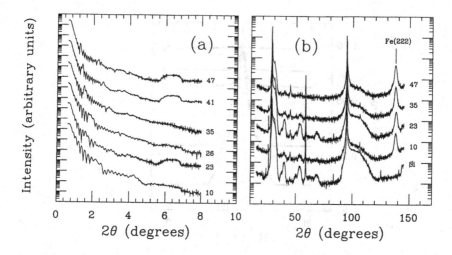

Figure 1: (a) Low-angle x-ray diffraction patterns of selected Fe/Cr(111) multilayers using Cu K_α radiation. (b) High-angle XRD patterns of selected Fe/Cr(111) multilayers and Si(111) substrate. In both graphs, the numbers following the curves denote t_{Cr}'s in Å.

is not clear to us at this time.

The modulation wavelengths of these samples, as determined from the superlattice lines, are within ± 2 Å of the expected values. High-angle θ-2θ scans were also performed on these samples from 10° to 145° in 2θ, and representative scans are shown in Fig. 1(b). The only peak in these scans that cannot be attributed to the substrate is the Fe(222) peak near $2\theta = 138°$. The absence of other allowed Fe peaks, especially the Fe(110) peak (2θ at 44.67°) which is about 17 times more intense than the (222) peak in the powder diffraction pattern, strongly suggests that the (111) planes of the Fe layers are parallel to the (111) plane of the substrate, indicating a highly oriented multilayer. The results of both high-angle XRD and RHEED support the conclusion that these samples are epitaxially grown with the (111) planes of the multilayer perpendicular to the growth direction.

The magnetoresistance was measured at 4.2 K using a cryostat equipped with a superconducting magnet that is capable of providing magnetic fields up to 8 T. The resistance was measured by standard four-probe technique, with the current applied in the plane of the sample and the field also in plane but transverse to the current direction. The data were collected under computer control using a digital nanovoltmeter (Keithley 182) and a programmable constant current source (Keithley 220). The DC current (typical current density roughly 5 kA/cm^2) was reversed for each data point to cancel out any thermal voltages and the field was swept continuously in both positive and negative directions.

RESULTS AND DISCUSSION

The dependence of magnetoresistance *vs.* applied field is shown in Fig. 2 for a few representative samples. The resistance decreases as the field increases and becomes virtually constant beyond a certain field (H_s). We define the MR here as $-[(R-R_o)/R_o] \times 100\%$ (i.e. normalized with respect to the zero field value) with the negative multiplier emphasizing that the GMR is a negative effect. The values of $\Delta R/R_o$ and the saturation field H_s were determined by fitting

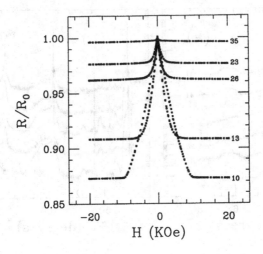

Figure 2: Magnetoresistance of selected Fe/Cr(111) multilayers measured at 4.2 K. The field is applied in the plane of the layers, and transverse to the direction of the current. The number at the end of each curve denotes t_{Cr} of the sample. t_{Fe} is held constant at 30Å for all samples.

straight lines to the high field and low field behavior respectively. The intersection of these two lines then gives the desired MR and H_s values. Alternative definitions for these parameters give slightly different absolute values, but the trends shown in Fig. 3 are maintained.

Fig. 3(a) is a plot of MR changes ($\Delta R/R_o$, in percentage) versus Cr thickness (t_{Cr}) over the range from 10 Å to 47 Å, and the saturation field (H_s) vs. t_{Cr} is shown in Fig. 3(b). The maximum value of $\Delta R/R_o$ (close to 13%) is found in sample with the thinnest Cr thickness (10 Å). As t_{Cr} is increased, the magnitude of $\Delta R/R_o$ becomes smaller and reaches a first minimum at t_{Cr} near 23 Å. A second maximum of $\Delta R/R_o \sim 4\%$ is found at $t_{Cr} \sim 26$ Å. It is difficult to distinguish a third oscillation period due to small number of data points and the noise in the data but the displayed results are consistent with a third maximum appearing near $t_{Cr} = 43$ Å. Similar trends are seen in the values of H_s. Combining the results from fig. 3 (a and b) we find that the positions of the second and third peaks in the oscillatory coupling for Fe/Cr(111) multilayers are at 26 ± 2 Å and 43 ± 2 Å respectively. Given the error bars, these results indicate an for the oscillatory coupling whose period and phase are both consistent with those reported for other orientations [5, 18, 20]. Although the greatest $\Delta R/R_o$ we observe is at $t_{Cr} = 10$ Å, it is possible that the GMR could reach an even higher value for thinner t_{Cr}. We have, unfortunately, experienced difficulty growing samples of sufficient quality with thinner spacer layers using our current shutter and deposition rate control systems, so the precise position of the first peak cannot be determined from the present data.

The results given above, when combined with published work[5, 8, 20], indicate that both the period and the phase of the oscillations are substantially constant for four different orientations of the growth direction in Fe/Cr multilayers. It is important to keep in mind, however, that the size of the error bars on these measurements are such that small residual differences between the investigated directions could escape detection. The present results do show that if any feature of the Fermi surface is responsible for the long period oscillations in Cr then this feature must be quite isotropic. Detailed and precise calculations such as those described by Stiles [12] need to be performed for all three directions to see if these observations are consistent with the Fermi

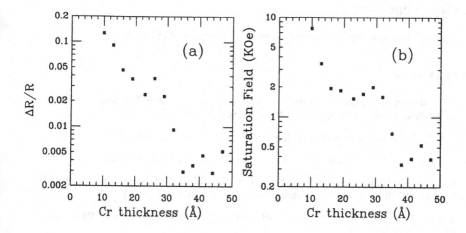

Figure 3: (a) Saturation magnetoresistance *vs.* Cr layer thickness, (b) Saturation field *vs.* Cr layer thickness for Fe/Cr(111) multilayers, deposited on Si(111) substrates at room temperature.

surface of Cr. Of course, it is also possible that the Fermi surface has nothing to do with the long period oscillations across Cr spacer layers, just as has been suggested to be the case for other transition metals, but then we are left with the problem of explaining why the period of Cr is so much different from the majority of other spacer materials. Additional investigations on this question are clearly warranted.

The average size of GMR effect observed near $t_{Cr} = 26$ Å in our samples is comparable to that seen by Gijs *et al.* for samples in the (110) orientation[18], but is about a factor of 5 smaller than the magnitude quoted by Fullerton *et al.* [5] for the (100) orientation. Fullerton *et al.* reported a substantial difference in the size of the MR for the two orientations they investigated (17% for (211) and 22% for (100), normalized to the zero field value at 4.2 K and at t_{Cr} corresponding to the second peak). The magnitude of the discrepancies is too large to be explained by the different t_{Fe} used in the various studies but, given the strong dependence of the GMR effect on the detailed structure of the interface, it is not clear whether the differences reflect intrinsic differences in the electronic interfacial states associated with various orientations or merely different interface structure (dislocation densities etc.) associated with the different surface energies and growth conditions. This too is a problem worthy of further investigation, but such studies must obviously involve samples of the highest possible quality grown with different orientations under equivalent conditions combined with detailed characterization of the interface structure.

CONCLUSION

We have measured the GMR effect in evaporated Fe/Cr multilayers grown in the (111) orientation. We find that the period and phase of the oscillatory coupling between the Fe layers for this orientation is consistent with that seen by other investigators for the (100), (110), and (211) orientations in the same system. The size of the effect we observe is smaller than that seen for the other orientations but it is impossible to tell whether this is an intrinsic property of the interfaces involved or (as is perhaps more likely) reflects differences in growth conditions in different deposition systems.

ACKNOWLEDGEMENT

This work was supported by the NSF under grant number DMR93-14018.

REFERENCES

1. M. N. Baibich, J. M. Broto, A. Fert, F. Nguyen Van Dau, F. Petroff, P. Eitenne, G. Creuzet, A. Friederich, and J. Chazelas, Phys. Rev. Lett. **61**, 2472 (1988).

2. S. S. P. Parkin, N. More, and K. P. Roche, Phys. Rev. Lett. **64**, 2304 (1990).

3. J. Unguris, R. J. Celotta, and D. T. Pierce, Phys. Rev. Lett. **67**, 140 (1991).

4. S. T. Purcell, W. Folkerts, M. T. Johnson, N. W. E. McGee, K. Jager, J. ann de Stegge, W. B. Zeper, W. Hoving, and P. Grunberg, Phys. Rev. Lett. **67**, 903 (1991).

5. E. E. Fullerton, M. J. Conover, J. E. Mattson, C. H. Sowers, and S. D. Bader, Phys. Rev. B **48**, 15755 (1993).

6. J. E. Mattson, M. E. Brubaker, C. H. Sowers, M. Conover, Z. Qiu, and S. D. Bader, Phys. Rev. B**44**, 9378 (1991).

7. W. Folkerts, W. Hoving, and W. Coene, J. Appl. Phys. **71**, 362 (1992).

8. M. A. M. Gijs, S. K. Lenczowski, and J. B. Giesbers, Phys. Rev. Lett. **70**, 3343 (1993).

9. Y. Wang, P. M. Levy, and J. L. Fry, Phys. Rev. Lett. **65**, 2732 (1990).

10. P. Bruno and C. Chappert, Phys. Rev. Lett. **67**, 1602 (1991).

11. F. Herman and R. Schrieffer, Phys. Rev. B**46**, 5806 (1992).

12. M. D. Stiles, Phys. Rev. B**48**, 7238 (1993).

13. A. Fert and P. Bruno, in Ultrathin Magnetic Structures II, ed. by B. Heinrich and J. A. C. Bland, (Springer Verlag, Berlin, 1994), pp. 82-117.

14. C. Kittel, in Solid State Physics, ed. by F. Seitz, D. Turnbull, and H. Ehrenreich, (Academic, New York, 1968) **22**, pp. 1-26.

15. S. S. P. Parkin, Phys. Rev. Lett. **67**, 3598 (1991).

16. M. C. Munoz, and J. L. Perez-Diaz, Phys. Rev. Lett. **72** 2482 (1994).

17. J. E. Ortega, F. J. Himpsel, G. J. Mankey, and R. F. Willis, Phys. Rev. B**47** 1540 (1993).

18. M. A. M. Gijs, and M. Okada, Phys. Rev. B **46** 2908 (1992).

19. Y-T Chen, Y-L Chen, W-J Meng, and Y. Li, Phys. Rev. B **48** 14729 (1993).

20. S. S. P. Parkin, N. More, and K. P. Roche, Phys. Rev. Lett. **64**, 2304 (1990).

STRUCTURAL STUDIES AND MAGNETOTRANSPORT PROPERTIES OF SPUTTERED Ni/Co MULTILAYERS

J.M. Freitag, X. Bian, Z. Altounian and J.O. Ström-Olsen
Centre for the Physics of Materials and Department of Physics, McGill University,
3600 University St., Montréal, Québec, Canada H3A 2T8;
R.W. Cochrane
Département de Physique et Groupe de recherche en physique et technologie des couches
minces, Université de Montréal, C.P. 6128, Succ. Centre-Ville, Montréal, Québec,
Canada H3A 3J7.

Abstract

Ferromagnetic/ferromagnetic Ni/Co multilayers were prepared by DC-magnetron sputtering with component layer thicknesses ranging from 40 Å down to 5 Å. Structural characterizations by x-ray diffractometry show a well-defined compositional modulation along the film growth direction and a preferred (111) crystalline orientation. A longitudinal magnetoresistance $\Delta R/R$ over 2.7% with a sensitivity of $\sim$ 0.11%/Oe was measured at room temperature in small fields less than 20 Oe. The highest room temperature sensitivity obtained in this system was 0.16%/Oe. Magnetoresistive sensitivity was found to vary inversely with the number of bilayers in the multilayers. The magnetic anisotropy of the films as determined by MOKE magnetometry is correlated to the magnetoresistance and indicative of an AMR effect.

I. Introduction

The observation of giant magnetoresistance (GMR) in a wide class of antiferromagnetically coupled ferromagnetic/nonmagnetic multilayers[1,2] and in uncoupled granular alloy thin films[3,4] has spurred much research in this field toward a better understanding of the fundamental principles as well as to develop possible applications. In this context, it has been reported recently that ferromagnetic/ferromagnetic Ni/Co multilayers[5] grown by molecular beam epitaxy (MBE) show large magnetoresistance (MR) effects with small saturation fields, leading to high MR sensitivities which compare well to other current and potential magnetic sensor materials.

In this paper, we report similar results from *sputtered* Ni/Co multilayers, and attribute the effect to anisotropic magnetoresistance (AMR). The AMR effect[6], which depends on the orientation of the magnetization with respect to the direction of electric current and the motion of magnetic domains in an external field, is common in ferromagnetic metals with magnitudes $\Delta\rho/\rho$ greater than 5% for some bulk ferromagnetic alloys such as $Ni_{.70}Co_{.30}$. Thin films exhibiting large AMR are currently used in the fabrication of magnetoresistance devices, and therefore the development of new materials with higher sensitivities is of practical importance.

Since AMR is related to the intrinsic magnetization of the material, it is therefore important to relate the magnitude and sensitivity of the MR effect to the growth conditions and multilayer composition.

II. Experimental

A series of multilayers was prepared by DC-magnetron sputtering from separate targets of Ni and Co onto glass, silicon (100), and oxidized silicon (100) substrates with individual layer thicknesses ranging from 40 Å down to 5 Å and total number of bilayers between 6 and 48. Background pressure was $< 8 \times 10^{-7}$ Torr. With a sputtering pressure of 8.0 mTorr of argon, the deposition rates for Ni and Co, as determined by low-angle x-ray reflectivity measurements on single layer films, were 1.32 Å/s and 1.63 to 2.08 Å/s respectively. A contact mask defined a 16mm × 4mm deposition area.

Structural characterizations of the samples were performed by low- and high-angle x-ray diffraction using Cu-$K\alpha$ radiation with the scattering vector perpendicular to the film surface. Magnetization data at 300K was obtained using a magneto-optic Kerr effect (MOKE) magnetometer with a 6328Å He-Ne laser and the applied field in the plane of the film. The magnetoresistance measurements between 77K and 300K were carried out using a four-terminal geometry and a high-resolution ac bridge[7]. The current was in the plane of the film and the magnetic field was either in plane, parallel to the current (longitudinal MR), in plane, perpendicular to the current (transverse MR), or perpendicular to the film surface (perpendicular MR).

III. Results and Discussion

Figure 1 shows low-angle reflectivity spectra for the $SiO_2/(Ni40Å/Co10Å) \times N$ samples with different bilayer number N. Superlattice peaks up to fourth order are visible, indicating a well-defined composition modulation along the film growth direction. Total thickness oscillations between the superlattice peaks are also clearly visible. The critical angle θ_c for

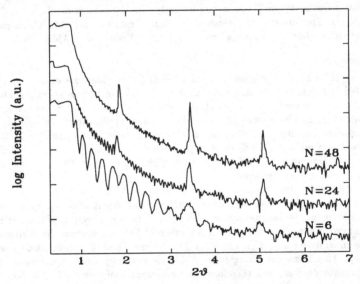

Figure 1: Low-angle x-ray reflectivity spectra for the multilayers $SiO_2/(Ni40Å/Co10Å) \times N$, where the bilayer number $N = 6, 24, 48$.

total reflection in this series of samples is $\approx 0.35°$. The actual bilayer period Λ can be obtained from the modified Bragg law[8]:

$$\sin^2\theta = \sin^2\theta_B + \sin^2\theta_c \tag{1}$$

where θ_B is the position of the Bragg peak $\sin\theta_B = n\lambda/2\Lambda$. With the first order superlattice peak located at $2\theta = 1.85°$ for these samples (nominal bilayer period of 50Å), we get

$\Lambda \approx 51.5\text{Å}$. Measured bilayer periods were found to be within 10% of nominal values for all samples.

High-angle x-ray structural characterizations were also performed on these samples. Figure 2 shows the spectrum for a Si/(Ni40Å/Co30Å)×12 multilayer. Whereas in most samples no superlattice peaks were observed due to the low electronic contrast between Ni and Co and the structural disorder caused by alloying at the interfaces, for the multilayer in Fig. 2 faint secondary peaks are visible. Their position is consistent with the multilayer period Λ obtained by small-angle x-ray spectroscopy. All samples gave two distinct diffraction peaks. The first peak at $\sim 44.5°$ consists of a weighted average of Ni(111) and Co(111). Co grows in its fcc phase as demonstrated by the fact that this first peak moves from the position corresponding to $d_{\text{Ni}}(111)=2.034\text{Å}$ to the position corresponding to $d_{\text{Co}}^{fcc}(111)=2.046\text{Å}$ as the total proportion of Co is increased. If Co grew in its hcp phase, then the diffraction peak would move to the $d_{\text{Co}}^{\text{hcp}}(0002)=2.023\text{Å}$ position. The second peak at $\sim 51.6°$ for Ni(200) and Co(200) also behaves in the same manner, shifting from $d_{\text{Ni}}(200)=1.762\text{Å}$ to $d_{\text{Co}}^{fcc}(200)=1.772\text{Å}$. The ratio of the two main diffraction peaks for the sample of Fig. 2

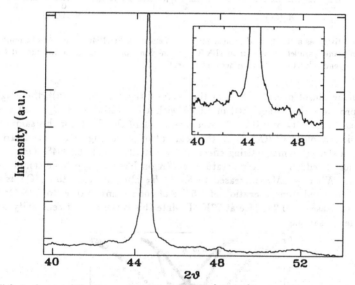

Figure 2: High-Angle x-ray diffraction spectrum for a Si/(Ni40Å/Co30Å)×12 multilayer. Inset, is a close-up showing the superlattice secondary peaks around the main (111) diffraction peak.

indicates an exceptionally good texturing in the (111) orientation. In most other samples, the ratio of the integrated intensities of these diffraction lines $I_{111}/I_{200} \approx 3.0$, as calculated from a fit to the data, indicates that the films have a polycrystalline structure with a preferred (111) orientation normal to the plane of the films and a fraction of (200) structural domains. An ideal polycrystalline structure would have a ratio $I_{111}/I_{200} \approx 2.0$.

Figure 3 shows the MR as a function of applied field for a Si/(Ni40Å/Co20Å)×12 sample at room temperature. In agreement with the usual convention[9], for field parallel to the current (LMR), the resistance increases at low fields, whereas for field perpendicular to the current (TMR, PMR), the resistance decreases with increasing field. The MR exhibited a

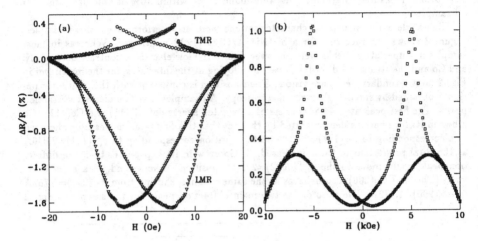

Figure 3: Resistivity as a function of applied magnetic field for a Si/(Ni40Å/Co20Å)×12 multilayer. In (a), field is in-plane, parallel to the current (LMR), and in-plane, perpendicular to the current (TMR). In (b), field is perpendicular to the film surface and current.

strong in-plane anisotropy. For field in the plane of the film [Fig. 3(a)], the longitudinal geometry produces the highest MR of 1.65% with a saturation field $H_S = 20$ Oe, giving rise to a peak sensitivity of 0.16%/Oe over a range of 10 Oe. TMR for this sample is 0.4% with a sensitivity of 0.03%/Oe. With the field out of plane [Fig. 3(b)], the PMR is 1.0% and, due to a strong de-magnetizing effect, the sensitivity is only 0.0004%/Oe.

The effect of reducing the temperature to 77K on LMR for the same sample is shown in Figure 4. Whereas LMR increased to 5.5%, H_S also increased to 40 Oe as the two MR peaks separated from a location of ±5 Oe at room temperature, to ±18 Oe at 77K. Sensitivity increased to 0.25%/Oe at 77K. To date this is the highest sensitivity measured for this series of samples.

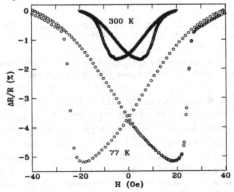

Figure 4: Resistivity as a function of applied magnetic field for a Si/(Ni40Å/Co20Å)×12 multilayer at 300K and at 77K.

MR as a function of number of bilayers for the multilayers $SiO_2/(Ni40\text{Å}/Co10\text{Å}) \times N$ was also measured. Whereas the magnitude of the effect in both the LMR and TMR geometries remained roughly the same, the saturation field increased from 50 Oe for the $N = 6$ multilayer to 500 Oe for the $N = 48$ one, leading to an inverse dependence of sensitivity on number of bilayers.

In-plane MOKE magnetic hysteresis loops for the $Si/(Ni40\text{Å}/Co20\text{Å}) \times 12$ multilayer are presented in Figure 5. An easy axis of magnetization oriented in-plane along the width was evident for all compositions and substrates, except in samples with very thin (5Å) Co layers. In these samples, a previously reported perpendicular anisotropy[10] is perhaps present. The saturation fields obtained by magnetometry closely match the saturation

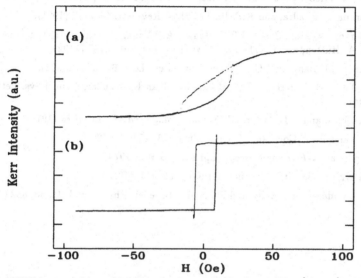

Figure 5: MOKE hysteresis loops with field in the plane of the film for a $Si/(Ni40\text{Å}/Co20\text{Å}) \times 12$ multilayer. In (a) field is along the length of the film (longitudinal), and in (b) field is along the width (transverse).

fields of the MR curves. The anisotropic magnetic domain structure, which is swept away by application of small external fields, together with an anisotropic scattering mechanism such as spin-orbit coupling may therefore be invoked to explain the MR in the present multilayer system.

For comparison, 1000Å thick Ni and Co films were also deposited onto the same substrates. The MOKE curves for these samples showed a magnetic structure similar to that of the multilayer system. Room temperature MR measurements for these single-layer films gave the following results for LMR and TMR (sensitivities): 0.22% (0.013%/Oe), 0.45% (0.013%/Oe) for Ni; and 0.34% (0.004%/Oe), 4.3% (0.11%/Oe) for Co. These values are consistent with the results of AMR studies, e.g. in Ref. 6 where LMR and TMR (sensitivities) are given as: 1.6% (0.04%/Oe), 1.75% (0.05%/Oe) for Ni; and 0.5% (0.02%/Oe), 2.1% (0.07%/Oe) for Co.

In summary, well-defined, textured ferromagnetic/ferromagnetic Ni/Co multilayers can be produced by sputter deposition. These multilayers give rise to a large MR with high sensitivities comparable to present alloys used in the fabrication of magnetoresistive devices. The magnitude of the MR was found to depend on the composition of the multilayers, whereas the number of bilayers controlled the sensitivity. The overall shape, magnitude, and anisotropy of the MR, as well as the magnetic structure of the multilayers is indicative of an AMR effect.

References

[1] M.N. Baibich, J.M. Broto, A. Fert. F. Nguyen van Dau, F. Petroff, P.E. Etienne, G. Creuzet, A. Friederich, and J. Chazelas, Phys. Rev. Lett. **61**, 2472 (1988).

[2] S.S. Parkin, P. Bhadra, and K.P. Roche, Phys. Rev. lett. **66** 2152 (1991).

[3] A.E. Berkowitz, M.J. Carey, J.R. Mitchell, A.P.Young, S. Zhang, F.E. Spada, F.T. Parker, A. Hutten, and G. Thomas, Phys. Rev. Lett. **68**, 3745 (1992).

[4] J.Q. Xiao, J.S. Jiang, and C.L. Chien, Phys. Rev. Lett. **68**, 3749 (1992).

[5] J.M. Gallego, D. Lederman, T.J. Moran, and Ivan K. Schuller, Appl. Phys. Lett., **64** 19 (1994).

[6] T.R. McGuire and R.I. Potter, IEEE Trans. Magn. **MAG-11**, 1018 (1975).

[7] Y. Huai, and R.W. Cochrane, J. Appl. Phys. **72**, 2523 (1992).

[8] B.K. Agarwal, *X-Ray Spectroscopy* (Springer, Berlin, 1979).

[9] S.S.P. Parkin, Solid State Physics, RJ 9311 (82241), (1993).

[10] G.H.O. Daalderop, P.J. Kelly, and F.J.A. den Broeder, Phys. Rev. Lett. **68**, 682 (1992).

MICROSTRUCTURE AND GMR IN (111) SPUTTER-DEPOSITED Co/Cu MULTILAYERS

R.J. POLLARD*, M.J. WILSON and P.J. GRUNDY
Joule Laboratory, Department of Physics, University of Salford, Salford M5 4WT, UK.
*Now at Department of Pure and Applied Physics, the Queen's University of Belfast, Belfast BT7 1NN, UK.

ABSTRACT

This paper presents the results of giant magnetoresistance (GMR), magnetic and microstructural investigations of sputter-deposited Co/Cu multilayers. The multilayers were designed to be at the second maximum of the oscillatory exchange coupling and were deposited onto a range of ion beam irradiated (111) Si substrates giving a transition from a random orientation to a highly oriented (111) texture. The randomly oriented multilayers exhibited mixed coupling and ~20% GMR, which had components arising from irreversible and reversible magnetization changes, whereas the highly oriented (111) multilayers were almost completely ferromagnetically coupled and showed very little GMR.

INTRODUCTION

A topic of interest in the GMR literature has been the apparent difference in behaviour between MBE grown, epitaxial (111) Co/Cu multilayers and polycrystalline sputter-deposited systems. The latter [1,2,3] give large GMR values and oscillatory coupling with two or three GMR maxima at the spacer thickness, t_{Cu}, of ~0.9, 2.0 and 3nm. Early measurements [4] on MBE structures did not show oscillatory coupling, nor indeed much evidence of any antiferromagnetic coupling. Explanations for this difference have included the proposal [4] that a relatively small proportion of (100) oriented material was in fact responsible for the large effects in sputtered samples. The fact that multilayers sputter-deposited on iron buffer layers showed both large GMR and a lack of (111) texture gave some support to that reasoning [5].

The situation is somewhat clearer now, and (111) MBE multilayers grown with t_{Cu} at the first maximum have shown good GMR with 30-80% at 4.2K and >20% at room temperature [6,7,8,9]. The strength of antiferromagnetic coupling, however, appears to vary in the first maximum between investigations [8,10] and, surprisingly, the second and subsequent maxima are sometimes not observed [6,8] but sometimes are [11]. A recent [12] explanation for these different findings can be linked to the common observation [e.g. 2,7,8] that multilayers structured at the first antiferromagnetic maximum in the oscillation, in both MBE and sputtered samples, show large GMR but also a relatively large fraction of ferromagnetic coupling. It is suggested in [12] that the areal ferromagnetic fraction, linked to coupling across spacer "islands" increases with the order of the antiferromagnetic coupling maximum and can completely mask the higher order maxima. For spacer "islands" below a critical size ferromagnetic order persists despite antiferromagnetic coupling. In this paper we present some results on the magnetoresistive, magnetic and structural properties of sputter-deposited Co/Cu that link GMR with texture and orientation.

EXPERIMENT

The Co/Cu multilayers were fabricated with the structure 16x(1nm Co/2nm Cu) by sputter-

365

deposition on to nominally room temperature, i.e. 30∓10ºC, (111) Si substrates.   The samples
were prepared in a UHV-compatible system at a pressure of 3×10^{-1} Pa of Ar and at a deposition
rate of 0.1nm s^{-1}. The base pressure of the system was below 5×10^{-6} Pa.   No buffer layer was
used, but prior to deposition of the samples, some of the Si substrates were irradiated by a
Kaufman-type ion source at different beam energies from 50eV to 1keV. Using $t_{Cu} = 2$nm at the
second maximum [2] in the oscillatory coupling allowed us to fabricate reproducible structures
with potentially relatively large GMR and fairly low saturation fields.
    The multilayers were characterised by in-plane measurements at room temperature of their
initial magnetisation curves and magnetic hysteresis loops using an AGFM and their GMRs by the
conventional four point method. Two remanence curves for the samples were also measured both
magnetically and magnetoresistively, namely the d.c. demagnetising (DCD) and the isothermal
(IRM) remanence curves. For the DCD curve a saturating field is applied to the sample and the
remanent value of resistance or magnetization is measured after increasing values of a reverse field
are applied and then removed. In the IRM curve the sample is a.c. demagnetized and then the
remanent value measured after increasing values of a magnetizing field are applied and then
removed.   The microstructure of the multilayers was investigated by X-ray diffraction and by
high resolution electron microscopy.

RESULTS AND DISCUSSION

    Ion beam irradiation or "etching" of the Si substrates prior to deposition of the multilayers
resulted in remarkable effects on the GMR, magnetization behaviour and microstructure of the
samples. The GMR of multilayers deposited on the native oxide of the Si substrate approached
20%; this is a respectable value for multilayers containing 16 bilayers structured at the second
maximum in the oscillation with $t_{Cu} \simeq 2$ nm [1,2]. Increasing the energy of the ion beam at a flux
of $\sim 10^{16}$ ions $cm^{-2} s^{-1}$ caused a decrease in the GMR, to almost zero GMR above 300eV. This
effect was reproducible, as seen in results for several series of films. We surmise that the effect of
the substrate etching was to cause the nucleation and growth of samples with less and less apparent
antiferromagnetic coupling
    This effect could arise from accidental differences in thickness of the spacer layer in the
various samples which would take the multilayer structure away from the second maximum in the
coupling oscillation. However, measurements of satellite spacing in high angle x-ray diffraction
(HXRD) patterns confirmed that the bilayer periodicity was fairly constant. Alternatively, the
period of the oscillation may change with crystallographic texture of the multilayer. In sputtered
Co/Cu systems the period is generally found to be ~1nm [1,2,13]. In (100) MBE Co/Cu systems a
first maximum is found at about 1nm [10]. Recent work [14] on epitaxial sputter-deposited (100)
films shows the GMR to be higher than in (111) superlattices. A third possibility is that (111)
superlattices do not show oscillatory exchange coupling perhaps because of a discontinuous growth
process. In most cases the second and subsequent peaks are not observed [8,9] in (111) MBE
superlattices.
    Figure 1a shows hysteretic GMR loops for multilayers deposited on unetched and etched
substrates. They have a typical form, and the 17.5% GMR loop for the unetched sample shows
maxima at a finite field of about 40 Oe, a field well above the coercivity in the magnetization loop
(figure 1b).   For the 500eV etch the the $\Delta R/R$ value is about 1%. The loss of GMR is
accompanied by a change in form of the initial magnetization curve and hysteresis loop of the
multilayers, figure 1b.   The canted, wasp-wasted loop for the samples showing large GMR
suggests mixed ferromagnetic and antiferromagnetic coupled regions in the multilayers.   The
samples were isotropic in the plane of the sample.   The upright loop is representative of
multilayers deposited on substrates etched at energies greater than 300eV. It clearly represents
almost complete ferromagnetic coupling and switching.
    The GMR curve for the "unetched" multilayer in figure 1a contains only the hysteretic loop.
GMR values measured from an initially demagnetised or a.c. erased state can be larger than that
obtained on field cycling [15]. Figure 2a shows the results of such a measurement on one of our
multilayers. Figure 2b shows the d.c. demagnetising (DCD) and isothermal remanent (IRM)

magnetoresistance curves superimposed on the initial and demagnetizing parts of the loop of figure 2a. The significance of the remanence plots is that they reflect irreversible changes in magnetization and also magnetoresistance.

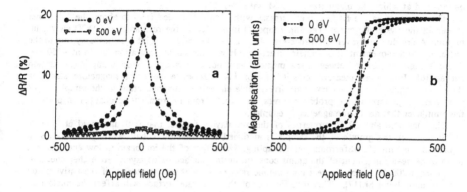

Figure 1.   (a) The dependence of giant magnetoresisitance in 16x(1nm Co/2nm Cu) multilayers on the Ar$^+$ ion beam energy used to "etch" the Si substrate (hysteretic GMR loops). (b) Magnetic hysteresis loops (in-plane) for the samples of figure 1b.

In the IRM curve we see a finite change in magnetoresistance, not what would be expected if the change in resistance was due only to antiferromagnetically coupled regions in the multilayer. The starting values for the IRM and the initial magetoresistance at zero field are equal as in both cases the sample has been a.c. demagnetised. The resistance of the IRM curve decreases on magnetization to that observed at zero field in the hysteretic magnetoresistance curve which is also the starting resistance of the DCD magnetoresistance curve. The "additional" component of magnetoresistance observed in the initial curve can be correlated to the resistance drop of the IRM curve. The IRM curve measures the resistance drop due to irreversible changes within the

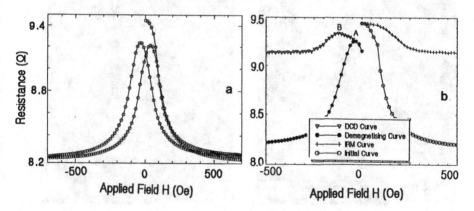

Figure 2.   (a) Initial magnetoresistance curve and hysteretic curve for a 16x(1nm Co/2nm Cu) multilayer deposited on untreated (111) Si. (b) Magnetoresistive DCD, IRM, initial and demagnetising curve for the sample.

material whereas the initial curve will have contributions from both reversible and irreversible changes. It is the reversible mutual rotation between cobalt layers antiferromagnetically coupled through the copper spacer layers that is responsible for the majority of the resistance change in the magnetoresistance cycle. The IRM magnetoresistance saturates at ~250 Oe and this is also found to be the field at which the magnetization IRM saturates [16].

The resistance in the DCD curve increases with negative applied field to above the value that is seen in the hysteretic curve. The superimposed initial rise in the resistance and DCD curves up to point A are due to irreversible processes. The further increase to point B is probably due to the nucleation and movement of a domain structure. It is followed by saturation, again at ~250 Oe. There is close correlation between these magnetoresistance curves and corresponding magnetization curves [16]. From these measurements it is reasonable to associate the "extra" magnetoresistance of the initial magnetoresistance curve with irreversible changes of magnetization in the sample. This additional magnetoresistance probably comes, therefore, from micromagnetic changes in parts of the multilayer that are ferromagnetically coupled.

The magnetic and GMR properties of the multilayers suggest that with increased etching of the Si substrate the growth of the sample is modified such that the cobalt layers gradually lose a significant amount of antiferromagnetic coupling. The effect of the ion beam at low energies will be to cause re-arrangement of the atoms constituting the surface oxide layer. At higher energies, above about 200eV, we can expect both atomic re-arrangement and sputtering of the native oxide and the underlying Si [17]. This modification of the substrate surface will affect the nucleation and subsequent growth of the multilayer.

Figures 3a shows a high resolution TEM micrograph of the cross-section of a multilayer deposited on a 500eV etched substrate imaged looking down the Si (110). In-set is an electron

 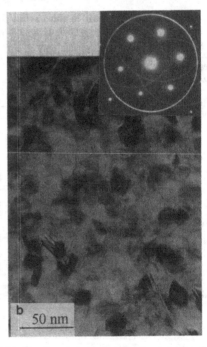

Figure 3. (a) HRTEM image and diffraction pattern from a cross-sectional specimen of a sample deposited on a 500 eV etched (111) Si substrate. (b) Plan view TEM image and diffraction pattern of the specimen in (a)

diffraction pattern from a region containing both the multilayer film and the substrate. The multilayer and the Si substrate are clear in the micrograph, as is the non-crystalline oxide at the interface between the two. The individual Co and Cu layers are not clear in these interference images because of the similar electron scattering factors of the two elements, but they are obvious in defocused Fresnel images and in diffraction contrast images. The figure shows a well defined lattice image showing almost continuous (111) lattice fringes in the superlattice that are oriented parallel to the (111) Si lattice fringes. These fringes pass through or end at the boundaries of columnar crystals which run through the layered structure from the substrate to the top of the structure.

The inset diffraction pattern to figure 3a shows the Si spot pattern, with short arcs of the multilayer (111) ring above and below the Si (111) spots. Although there does appear to be disordered material at the interface between the Si and the layers in figure 3a, close inspection shows that in some places continuity between the two lattices is obtained. The identity of the disordered material is unclear. In view of the reported [17] growth process of (111) Co films on (111) Si, it may very well be a disordered silicide which is a precursor to the growth of the oriented (111) multilayer after several monolayers. We intend to extend these observations by examination of the initial nucleation and growth of the layered structures using atomic force microscopy. Micrographs and electron diffraction patterns from cross-sectional samples of multilayers deposited on the untreated substrate show and suggest randomly oriented grains.

The information in figure 3a suggests a (111)-oriented Co/Cu superlattice with a high degree of texture. This is confirmed by HXRD patterns and plan-view HRTEM. As an illustration, figure 3b is a micrograph and in-set electron diffraction pattern from plan view specimen taken from the multilayer illustrated in figure 3a. It shows the microstructure of the multilayer as viewed in a direction normal to the multilayer. The multilayer is composed of small grains, ~50 nm across, which, as shown by dark field cross-sectional images, extend down through the superlattice in the form of a columnar microstructures. A similar microstructure is observed in the untextured multilayers grown on untreated substrates. The diffraction patterns of the plan view specimen in figure 3b show an extremely low intensity or absent (111) and (200) maxima and an extremely strong (220) ring. This result indicates a very strong (111) texture in the multilayer and confirms the existence of a highly oriented, columnar superlattice in multilayers deposited on ion etched (111) Si. Similar patterns from multilayers deposited on the untreated substrates show the ring pattern typical of a randomly oriented f.c.c. thin film.

CONCLUSIONS AND SUMMARY

From these results we may infer that GMR in our sputter-deposited (1 nm Co/2 nm Cu) multilayer samples is reduced as the microstructure of the samples is refined from a randomly oriented, small grained polycrystalline arrangement to a highly oriented (111) columnar superlattice. The GMR values, the magnetisation curves and the microstructural investigations all correlate to show that sputter-deposited (111) oriented Co/Cu multilayers fabricated with t_{Cu} = ~2nm do not show a GMR maximum and are apparently mainly ferromagnetically coupled.

This finding agrees with the often observed lack of a second, antiferromagnetic maximum in both the exchange coupling and GMR curves for MBE-deposited (111) superlattices. This lack of oscillatory coupling may arise in a lack of uniformity of the layered structure, in particular in discontinuities in the spacer layer. It could be argued that this is unlikely in sputtered samples, but the conditions leading to a highly oriented superlattice may cause island growth, as is often observed in f.c.c. metal films [18] and by us in Cu and Co deposited on high temperature Si substrates. Calculated critical spacer "island" dimensions are $\simeq 10$nm [12]. We intend to examine these possibilities by further TEM and also by AFM (atomic force microscopy) observations. Parallel experiments on multilayers structured at the first coupling maximum at $t_{Cu} \simeq 1$nm are also called for to extend these comparative studies with MBE-deposited superlattices.

ACKNOWLEDGEMENTS

The authors acknowledge support from the UK Engineering and Physical Sciences Research Council (EPSRC, previously SERC) through the Physics Committee (grant GR/F40870), the Materials Commission (grant GR/H65320) and the Magnetism and Magnetic Materials Initiative. The Royal Society is thanked for an equipment grant.

REFERENCES

1. S. S. P. Parkin, R. Bhadra and K. P. Roche, Phys. Rev. Lett. 66, 2152 (1991).
2. M. E. Tomlinson, R. J. Pollard, D. G. Lord and P. J. Grundy, J. Magn. Magn. Mater. 111, 79 (1992).
3. R. Mattheis, W. Andra, L. Fritzsch, A. Hubert, M. Ruhrig and F. Thrum, J. Magn. Magn. Mater. 121, 424 (1993).
4. W. F. Egelhoff and M. T. Kief, IEEE Trans. Magn. 28 (1992) 2742 and Phys. Rev. B 45, 7795 (1992).
5. P. J. Grundy, R. J. Pollard and M. E. Tomlinson, J. Magn. Magn. Mater. 126, 516 (1993).
6. M. T. Johnson, R. Coehoorn, J. J. de Vries, N. W. E. McGee, J. ann de Stegge and P. J. Bloemen, Phys. Rev. Lett. 68, 2688 (1992).
7. D. Greig, M. J. Hall, C. Hammond, B. J. Hickey, H. P. Ho, M. A. Howson, M. J. Walker, N. Weiser and D. G. Wright, J. Magn. Magn. Mater. 110, L239 (1992).
8. G. R. Harp, S. S. P. Parkin, R. F. C. Farrow, R. F. Marks, M. F. Toney, Q. H. Lam, T. A. Rabedeau, A. Cebollada and R. J. Savoy, MRS Symp. Proc. vol 313, p 41, 1993 and G. R. Harp, S. S. P. Parkin, R. F. C. Farrow, M. F. Toney, Q. H. Lam, T. A. Rabedeau and R. J. Savoy, Phys. Rev. B 47, 8721 (1993).
9. Y. Kobayashi, Y. Aoki, H. Sato, A. Kamijo and M. Abe, J. Magn. Magn. Mater. 126, 501 (1993).
10. J. J. de Miguel, A. Cebollada, J. M. Gallego, R. Miranda, C. M. Schneider, P. Schuster and J. Kirschner, J. Magn. Magn. Mater., 93, 1 (1991).
11. A. Schreyer, K. Brohl, J. F. Ankner, Th. Zeidler, P. Bodeker, N. Metoki, C. F. Majkrzak and H. Zabel, Phys. Rev. B 47, 15334 (1993).
12. U. Gradmann and H. J. Elmers, J. Magn. Magn. Mater. 137, 44 (1994).
13. S. J. Lenczowski, M. A. M. Gijs, J. B. Giesbers, R. J. M. van der Veerdonk and W. J. M. deJonge, Phys. Rev. B 50, 9982 (1994).
14. T. Kingetsu and F. Yoshizaki, Jpn. J. Appl. Phys. 33, 2041 and 6168 (1994).
15. P.A. Schroeder, S.F. Lee, P. Holody, R. Loloee, Q. Yang, W.P. Pratt jnr and J. Bass, J. Appl. Phys. 76, 6610 (1994).
16. R.J. Pollard, M.J. Wilson and P.J. Grundy, in press, J. Magn. Magn. Mater.
17. M.R. Cohen, R.J. Simonson, M.M. Altamirano, K.L. Critchfield, W.T. Kemp and J.A. Meinhardt, J. Vac. Sci. Tech. A11(4), 971 (1993).
18. C. Pirra, J.C. Peruchetti, G.Gewinner and J. Derrier, Phys. Rev. B 29, 3391 (1984).
19. A. Brodde and H. Neddermeyer, Ultramicroscopy 42-44, 556 (1992).

Part VII

Giant Magnetoresistance II
and Colossal Magnetoresistance

STM Studies of GMR Spin Valves

R. D. K. Misra,* T. Ha**, Y. Kadmon,*** C. J. Powell, M. D. Stiles, R. D. McMichael,
and W. F. Egelhoff, Jr.,
National Institute of Standards and Technology
Gaithersburg, MD 20899

Abstract

We have investigated the surface roughness and the grain size in giant magnetoresistance (GMR) spin valve multilayers of the general type: $FeMn/Ni_{80}Fe_{20}/Co/Cu/Co/Ni_{80}Fe_{20}$ on glass and aluminum oxide substrates by scanning tunneling microscopy (STM). The two substrates give very similar results. These polycrystalline GMR multilayers have a tendency to exhibit larger grain size and increased roughness with increasing thickness of the metal layers. Samples deposited at a low substrate temperature (150 K) exhibit smaller grains and less roughness. Valleys between the dome-shaped individual grainsare the dominant form of roughness. This roughness contributes to the ferromagnetic, magnetostatic coupling in these films, an effect termed "orange peel" coupling by Néel. We have calculated the strength of this coupling, based on our STM images, and obtain values generally within about 20% of the experimental values. It appears likely that the ferromagnetic coupling generally attributed to so-called "pinholes" in the Cu when the Cu film thickness is too small is actually "orange peel" coupling caused by these valleys.

Introduction

In the few years since the giant magnetoresistance (GMR) effect was discovered[1-3], cross-section transmission electron microscopy (TEM) has been used extensively to characterize of GMR structures,[4] but surprisingly little use has been made of scanning tunneling microscopy (STM). Cross-section TEM views a thin-film sample side-on, and presents an image that is in a sense an average over the width of the sample, which is thinned-down to a few tens of nanometers (nm) for TEM. Such images provide useful information about the layering structure in the interior of GMR multilayers. However, additional detail about the film morphology is available from STM, which views a sample surface from above and is not subject to the averaging effect just described. Our purpose in this work is to examine the interior of GMR spin valves by terminating the deposition at various stages of growth and examining the resulting surfaces by STM.

There has been a growing interest[5] in the "orange peel" coupling idea of Néel[6] largely as a result of its apparent manifestation in GMR spin valves[7]. This idea, which is illustrated in Fig. 1, is that if two magnetic films are separated by a nonmagnetic film then any bumps or protrusions in the magnetic films will have magnetic poles on them, and a dipole field will be set up (this model assumes the magnetization is in the plane of the film). If this roughness is conformal (i.e., if the same bumps occur in all three films one above another), then the dipole fields will interact in a manner that tends to produce parallel (or ferromagnetic) alignment in the magnetic films. This means, for example, that if one magnetic film is pinned by

373

FeMn, and one is unpinned, the hysteresis loop of the unpinned film will be shifted off center from zero field by an amount corresponding to the strength of the interaction.

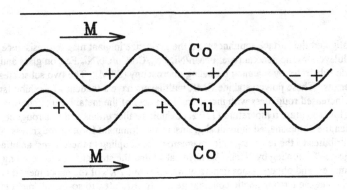

Figure 1 An illustration to the "orange peel" coupling idea of Néel[6] in which magnetostatic coupling occurs due to the interaction of magnetic poles in a magnetic/nonmagnetic/magnetic structure with conformal roughness.

Experimental

The two types of substrates used in this work were cover-glass slides and aluminum oxide films on Si wafers. The cover-glass slides and the Si wafers were cleaned ultrasonically, rinsed in distilled water, dried, and installed in the deposition chamber.

It is very important to remove the hydrocarbon contamination on the cover-glass slides (several tenths of a nm of which is accumulated on the glass substrate from exposure to the laboratory air) prior to the deposition of each spin valve in order to achieve the highest GMR values. Samples were Ar^+ sputtered with a neutralized-beam ion gun at a beam voltage of 500 eV until the carbon was removed, as judged by x-ray photoelectron spectroscopy (XPS) in a connected chamber.

A 50 nm film of aluminum oxide was deposited *in situ* on the Si wafers by reactive dc-magnetron sputtering of an metallic Al target in a 2 mTorr 85/15 mixture of Ar/O_2 at a rate of ~0.1 nm/s. The GMR films were deposited by sputtering without further treatment of the aluminum oxide.

The base pressure before depositing a spin valve was typically 2×10^{-8} Torr (~10^{-6} Pa) of which ~95% was H_2 and the remainder largely H_2O. The presence of H_2 during deposition has no apparent effect on spin valve properties unless the partial pressure exceeds ~10^{-6} Torr. The low base pressure is achieved, in part, by depositing a ~1.5 nm Ti film on the inside of the deposition chamber from a centrally mounted Ti filament just prior to deposition of each spin valve.

The metal films were deposited by dc-magnetron sputtering in 2 mTorr Ar at a rate of ~0.1 nm/s. During deposition the samples are subject to an in-plane field of ~20 mT (200 Oe) provided by permanent magnets mounted on either side of the sample on two quartz-crystal-oscillator holders. The magnetoresistance measurements were made *in situ* in the DC mode using a 4-point probe with a 5 1/2 digit ohm meter under computer control.

The scanning tunneling microscope (STM) is connected to the deposition chamber through a vacuum interlock so that samples could be transferred and investigated in a vacuum of better than 10^{-7} Torr. This vacuum appears to be adequate since we found little change in the roughness and no change at all in the grain size upon brief exposures (e.g., 1 min.) of these samples to air (by opening the STM to air during a scan). All images were recorded with a tunneling current of 0.2 nA with the tip biased at -50 mV with respect to the sample. The tips were prepared from 0.25 mm $Pt_{90}Ir_{10}$ wire clipped it under tension with a wire cutter. For the STM data discussed here, multiple images were taken at a variety of locations on each sample to ensure that the results were indeed typical of that sample. Care was taken to ensure that the data were not influenced by the use of different tunneling tips. The majority of data was recorded with a single tip, and great effort was devoted to repeated intercomparisons among the samples to ensure that changing tip conditions did not change the main features of the data. This approach is important because the STM image can be a convolution of the sample and tip morphologies. If the features on the sample are sharper than the tip, as could occur in pathological cases, the sample may even image the tip.

Results and Discussion

The present work was based on a rather common type of spin valve structure $FeMn/Ni_{80}Fe_{20}/Co/Cu/Co/,Ni_{80}Fe_{20}$, which often achieves a moderate GMR at a rather low coercivity.[6] The top two magnetic films ($Ni_{80}Fe_{20}$ and Co) are pinned by exchange bias from the FeMn, and the bottom two magnetic films (Co and $Ni_{80}Fe_{20}$) are free to switch at low applied fields (unpinned). The standard sample of this type used as a reference point in the present work is illustrated in Fig. 2. In our work such samples typically give a GMR of 8%, a coercivity of 0.5 mT (5 Oe), and a coupling field of 0.8 mT (8 Oe).[8] These studies will be published separately.[8]

The present studies concentrated on the STM images at the early and middle stages of deposition of films such as those illustrated in Fig. 2. The issues of principle interest in this work lay in two areas. One was that of nucleation and growth in the early stages of deposition, and the other was that of grain size and roughness at the Cu layer, which plays a key role in the GMR effect.

We found it impossible to obtain STM images on our bare substrates, which are highly insulating. After deposition of ~1 nm of $Ni_{80}Fe_{20}$, images of low quality could be obtained and it seems likely that a partially continuous metal film was present; nevertheless bare regions of the insulating substrate impaired the quality of the images. After deposition of 2.5 nm of $Ni_{80}Fe_{20}$, good images could be obtained, but this thickness was found to be too thin for optimum spin valve performance. If the bottom $Ni_{80}Fe_{20}$ film is 2.5 nm thick instead of the optimum value of 5 nm in the type of structure illustrated in Fig. 2, the coupling rises from 0.8 mT to 2 mT (8 Oe to 20 Oe) and the GMR drops from 8% to 6%.[8]

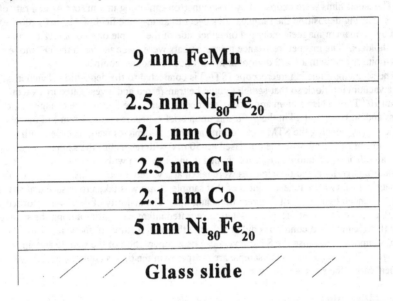

| 9 nm FeMn |
| 2.5 nm $Ni_{80}Fe_{20}$ |
| 2.1 nm Co |
| 2.5 nm Cu |
| 2.1 nm Co |
| 5 nm $Ni_{80}Fe_{20}$ |
| Glass slide |

Figure 2 An illustration of a standard type of spin valve structure on which the present investigations are based.

Figure 3 presents two typical STM images of samples after deposition of the optimum 5 nm of $Ni_{80}Fe_{20}$. Figure 3a presents the result for deposition at room temperature, and Fig. 3b presents the result for deposition at 150 K followed by warming to room temperature for the STM studies. Deposition at 150 K is of interest because in recent work we have found that the GMR can be increased to 9% or even 10%. [8]

The first thing to note about the images in Fig. 3 is that the vertical scale is 15 times smaller than the horizontal scales. The resulting vertical exaggeration is an important aid in visualizing the surface roughness since these films are actually quite smooth.

The most noticeable difference between these two images is that the $Ni_{80}Fe_{20}$ film deposited at room temperature (RT) may be seen to have somewhat sharper or more pronounced peaks associated with each grain than the $Ni_{80}Fe_{20}$ film deposited at 150 K. However, other differences between these images are not readily apparent by mere visual inspection, and it is very helpful to quantify the images.

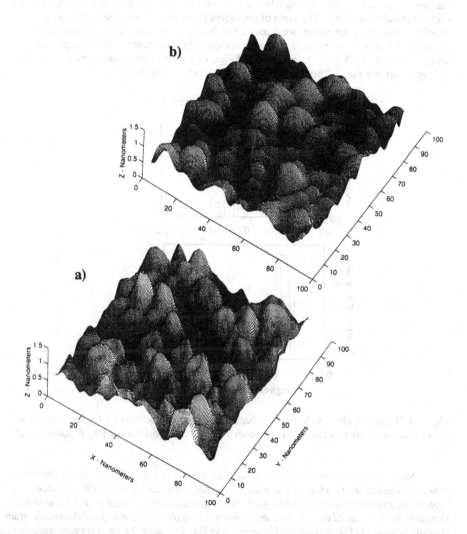

Figure 3 The STM images of an early stage of deposition of a typical spin valve structure. In a) 5 nm of $Ni_{80}Fe_{20}$ was deposited at room temperature, and in b) 5 nm of $Ni_{80}Fe_{20}$ was deposited at 150 K and the sample warmed to room temperature for the STM studies. The STM is *in situ*, and the samples were in high vacuum continuously.

Figure 4 presents the results of quantifying the surface roughness in images of the type illustrated in Fig. 3 (several locations were imaged on each sample). In this case, the roughness was defined as the difference in height between the maximum at the center of each grain and the depth of an adjacent valley. The depth of the valleys in the positive and negative x and y directions (from the grain center) were counted as four separate data points. This particular definition of roughness seems appropriate for assessing the importance of the "orange peel" coupling illustrated in Fig. 1. A comparison of Figs. 4a and 4b suggests that there is not a great difference between them. The film deposited at RT is only slightly rougher.

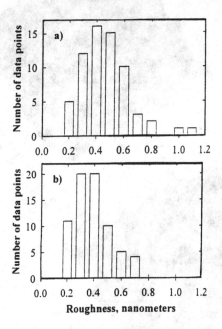

Figure 4 Histogram plots of the surface roughness (defined in the text) for the samples of Fig. 3, where a) corresponds to 5 nm $Ni_{80}Fe_{20}$ deposited at RT and b) to 5 nm $Ni_{80}Fe_{20}$ deposited at 150 K.

Measurements of the average grain diameter, presented in Fig. 5, show major differences between images of the two types of samples. We define the grain diameter as the distance between minima on opposite sides of a grain. Most grains are nearly circular. For those that are elongated, the average of the long and short distances was plotted. The major difference in grain diameter apparent in Fig. 5 is not readily apparent in Fig. 3 because the vertical exaggeration tends to hide the depths of the minima. This shortcoming illustrates the value of quantifying the images.

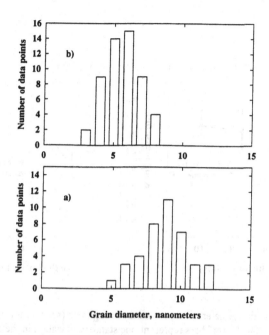

Figure 5 Histogram plots of the grain diameter (defined in the text) for the samples of Fig. 3, where a) corresponds to 5 nm $Ni_{80}Fe_{20}$ deposited at RT and b) to 5 nm $Ni_{80}Fe_{20}$ deposited at 150 K.

The smaller average grain size in Fig. 5b compared to Fig. 5a is attributable to less diffusion of deposited atoms at 150 K than at RT. At lower temperatures grains nucleate more closely together on the substrate and thus grain diameters are smaller when a film becomes continuous. Since grain growth in these systems tends to be columnar,[4] the grain size tends to propagate through the layers. Some increase in grain size with increasing film thickness is apparent, particularly in the early stages. Figure 6a illustrates this increase for $Ni_{80}Fe_{20}$ films of different thickness deposited at RT. The two points plotted in Fig. 6a for the 5 nm $Ni_{80}Fe_{20}$ films correspond to the data of Figs 5a and 5b. Each point in Fig. 6 represents the average of the corresponding histogram and the "error" bars represent one standard deviation in the distribution of values in the histogram.

As mentioned above, the coupling strength increases sharply if the bottom $Ni_{80}Fe_{20}$ layer in a spin valve of the type illustrated in Fig. 2 is 2.5 nm thick rather than the optimum value of 5 nm. From the data of Fig. 6, one can infer that this increase may be attributed to the smaller grain size. The surface roughness is rather similar for the 2.5 and 5 nm thicknesses of $Ni_{80}Fe_{20}$ for RT deposition, as seen in Fig. 6b. However, as seen in Fig. 6a, the grain size is significantly smaller at 2.5 nm $Ni_{80}Fe_{20}$, and thus there is a higher density of valleys. According to the Néel model, the greater the density of the waviness (illustrated in Fig. 1), the greater will be the

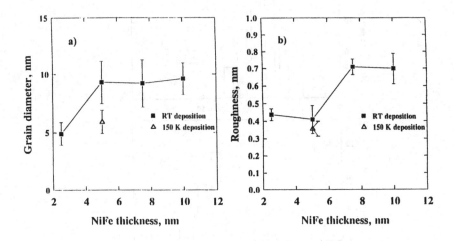

Figure 6 Plots of a) the grain diameter and b) the surface roughness (see text for definitions) versus $Ni_{80}Fe_{20}$ thickness. The "error" bars represent one standard deviation in the distribution of observed values. The data for the 5 nm $Ni_{80}Fe_{20}$ film correspond to the samples of Fig. 3.

coupling strength, all other things being equal. In the present structures, the valleys between grains are the only prominent form of waviness, and it seems reasonable to attribute the observed magnetostatic coupling to the effects of roughness.

In fact, we have calculated the magnetostatic coupling strength using our STM images and the values obtained are generally within about 20% of the observed values. Several assumptions are made in these calculations. It is assumed that the roughness is conformal, as in Fig. 1, so that an STM image taken after deposition of Cu is representative of both the Co/Cu and the Cu/Co interfaces. It is also assumed that the magnetization is rigidly in-plane right up to the interfaces. In reality, the demagnetizing field will tend to twist the magnetization slightly into parallel alignment with the interface. This twisting is probably responsible for the calculated values tending to be higher that the experimental ones. The calculation is performed by using the slope of the interface to determine the density of magnetic poles, and then by making a simple double sum (of the Coulomb interaction) over the top and bottom poles to get the coupling energy.[9]

Another example of the effect of the grain diameter on the coupling strength may be found in the 150 K deposition data of Fig. 6. The roughness is not very different from the value for RT deposition, but the grain size is significantly smaller for 150 K deposition. The observed coupling strength when the entire spin valve was deposited at 150 K was 1.2 - 1.3 mT (12-13

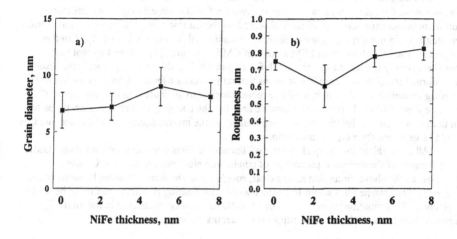

Figure 7 Plots of a) the grain diameter and b) the surface roughness (see text for definitions) versus $Ni_{80}Fe_{20}$ thickness for spin valves in which deposition was terminated after Cu deposition (see Fig. 2). The "error" bars represent one standard deviation in the distribution observed values. All depositions were at RT. The vertical scales are the same as in Fig. 6 to facilitate comparison.

Oe), or 50% larger than for RT deposition.[8] This increase may be attributed to the increased density of valleys.

Figure 7 presents STM data similar to that of Fig. 6, except that, after deposition of the $Ni_{80}Fe_{20}$, 2.1 nm Co and 2.5 nm Cu (illustrated in Fig. 2) were also deposited. By comparing Figs. 6 and 7, some insight can be obtained into the evolution of grain size and roughness during film growth. The most noticeable effect is that the plots are somewhat flatter in Fig. 7 than in Fig. 6. For example, the increase in grain diameter in Fig. 7a in going from 2.5 nm to 5 nm of $Ni_{80}Fe_{20}$ is smaller than that in Fig. 6a. The reason for this observation is that the most rapid change in grain size generally occurs in the early stages of film deposition. In Fig. 6, the images are recorded after $Ni_{80}Fe_{20}$ deposition, but in Fig. 7a the images are recorded after $Ni_{80}Fe_{20}$, Co , and Cu deposition so the film is thicker. Although the increase in grain diameter in Fig. 7a in going from $Ni_{80}Fe_{20}$ thicknesses of 0 to 2.5 to 5 nm is rather modes, there is a more pronounced decrease in coupling strength from >5 mT (>50 Oe) to 2 mT (20 Oe) to <1 mT (<10 Oe), respectively.[8] This nonlinear dependence of coupling strength on the density of valleys is expected on the basis of Fig. 1 since the interaction is dipolar.

No evidence was found in the present studies for the much discussed "pinholes" that are often invoked to explain the strong ferromagnetic coupling generally observed in GMR structures when the Cu layer is too thin. The roughness in Figs. 6b and 7b for a 5 nm $Ni_{80}Fe_{20}$

film is about 0.7 nm whereas the Cu film must be thinner than about 1.9 nm in our spin valves for the coupling to be so strong that it eliminates any GMR. This observation makes true pinholes in the Cu (direct contact between the two Co films) seem unlikely in our structures. Furthermore, no depressions at all were observed on the Cu that were as deep as 1.9 nm. Finally, the roughness in our structures seems likely to be conformal (as sketched in Fig.1). Conformal roughness is widely observed in TEM studies of GMR structures,[4] and the fact that the roughness is very similar in Figs. 6b and 7b for a 5 nm $Ni_{80}Fe_{20}$ thickness is probably the result of conformal roughness. Therefore, it seems that rather than attributing strong ferromagnetic coupling to pinholes in the Cu, it would be more plausible to attribute it to the "orange peel" coupling idea of Néel. In our calculations we find that the coupling strength rises steeply as the Cu thickness decreases below 2 nm, varying roughly as the inverse square of the Cu thickness, as would be expected for a dipolar interaction.

All other things being equal, the use of thinner Cu films tends to produce a larger GMR effect because of the increased proportion of conduction electrons crossing the Cu layer. However, as the above discussion makes clear, roughness in the form of valleys between grains in these polycrystalline films seems to be the major factor limiting how thin the Cu can be made in practice. This situation suggests that larger GMR values may be attained in the future if deposition methods can be found to suppress the depths of valleys.

Conclusions

The major conclusions of this work may be summarized as follows.
1) STM observations of GMR spin valves show that valleys between grains are the dominant form of roughness.
2) These valleys have about the right depth and occur in about the right concentration to explain the observed magnetostatic coupling using the "orange peel" coupling idea of Néel.
3) The deposition of spin valve structures at 150 K produces smaller grain size and similar values of the roughness. The resulting greater density of valleys produces increased coupling strength.
4) Additional increases in GMR may be possible if deposition methods can be found to reduce the depth of these valleys further so that thinner Cu films may be used.

*Present address: Defense Metallurgical Institute, Hyderabad, India
**Present address: 5248 Signal Hill Rd., Orlando, FL 32808
***Present address: NRCN, Beer-Sheva, Israel

Acknowledgements

The authors would like to acknowledge useful conversations with Drs. J. M. Daughton, S. S. P. Parkin, and V. S. Speriosu. This work has been supported in part (W.F.E. and R.D.McM.) by the NIST Advanced Technology Program.

References

[1]   E. Velu, C. Dupas, D. Renard, J. P. Renard, and J. Seiden, Phys. Rev. B37, 668 (1988).
[2]   G. Binasch, P. Grunberg, F. S. Sauerenbach, and W. Zinn, Phys. Rev. B 39, 4828 (1989).
[3]   M. N. Baibich, J. M. Broto, A. Fert, F. Nguyen van Dau, F. Petroff, P. Etienne, G. Creuzet, A. Friederich, and J. Chazelas, Phys. Rev. Lett. 61, 2472 (1988).
[4]   S. S. P. Parkin, Z. G. Li, and D. J. Smith, Appl. Phys. Lett. 58, 2710 (1991); R. J. Highmore, W. C. Shih, R. E. Somekh, J. E. Evetts, J. Mag. Mag. Mat. 116, 249 (1992); A. R. Modak, D. J. Smith, and S. S. P. Parkin, Phys. Rev. B 50 ,4232 (1994); T. Shinjo, Surface Sci. Rep. 12, 49 (1991); R. F. C. Farrow, R. F. Marks, G R. Harp, D. Weller, T. A. Rabedeau, M. F. Toney, and S.S.P. Parkin, Mat. Sci. Eng. R11, 155 (1993).
[5]   G. S. Almasi and K. Y. Ahn, J. Appl. Phys. 41, 1258 (1970); A. Layadi and J. O. Artman, J. Mag. Mag. Mat. 92, 143 (1990); A. Layadi, J. O. Artman, R. A. Hoffman, C. L. Jensen, D. A. Saunders, and B. O. Hall, J. Appl. Phys. 67, 4451 (1990); H. O. Grupta, H. Niedoba, L. J. Heyderman, I. Tomas, I. B. Puchalska, and C. Sella, J. Appl. Phys. 69, 4529 (1991); E. W. Hill, J. P. Li, and J. K. Birtwistle, J. Appl. Phys. 69, 4526 (1991); R. P. Erickson and J. R. Cullen, J. Appl. Phys. 73, 5981 (1993); M. R. Parker, S. Hossain, D. Seale, J. A. Barnard, M. Tan, and H.Fujiwara, IEEE Trans. Mag. 30, 358 (1994); D. Altbir, M. Kiwi, R. Ramirez, and I. K. Schuller, to be published.
[6]   L. Néel, Comp. Rend. Acad. Sci. (France) 255, 1545 (1962) and 255, 1676 (1962); J.-C. Bruyère, O. Massenet, R. Montmory, and L. Néel, Comp. Rend. Acad. Sci. (France) 258, 841(1964) and 258, 1423 (1964).
[7]   V. S. Speriosu, B. Dieny, P. Humbert, B. A. Gurney, and H. Lefakis, Phys. Rev. B 44 5358 (1991); B. Dieny, V. S. Speriosu, S. S. P. Parkin, B. A. Gurney, D. R. Wilhoit, and D. Mauri, Phys. Rev. B43, 1297 (1991); C. Meny, J. P. Jay, P. Panissod, P. Humbert, V.S. Speriosu, H. Lefakis, J. P. Nozieres, and B. A. Gurney, Mat. Res. Soc. Symp. Proc. 313, 289 (1993); and B. Dieny, J. Mag. Mag. Mat.136, 335 (1994).
[8]   W. F. Egelhoff, Jr., R. D. K. Misra, Ha, Y. Kadmon, C. J. Powell, M. D. Stiles, R. D. McMichael, L. H. Bennett, C.-L. Lin, J. M. Sivertsen, and J. H. Judy, to be published.
[9]   W. F. Egelhoff, Jr. and Y. Kadmon, to be published.

TIME DEPENDENT MAGNETIC SWITCHING IN SPIN VALVE STRUCTURES

J. B. RESTORFF, M. WUN-FOGLE, S. F. CHENG, AND K. B. HATHAWAY
Naval Surface Warfare Center, 10901 New Hampshire Ave., Silver Spring, MD 20903

ABSTRACT

We have observed time dependent magnetic switching in spin-valve sandwich structures of Cu/Co/Cu/Fe films grown on silicon and Kapton substrates and Permalloy/Co/Cu/Co films grown on NiO or NiO/CoO coated Si substrates. The giant magnetoresistance (MR) values ranged from 1 to 3 percent at room temperature. The films were grown by DC magnetron sputter deposition. Measurements were made on the time required for the MR to stabilize to about 1 part in 10^4 after the applied field was incremented. This time depends almost linearly on the amplitude of the time-dependent MR change with a slope (time / ΔMR) of 20 000 to 30 000 s. Some samples took as long as 70 s to stabilize. The time dependent effects may be important for devices operating in these regions of the magnetoresistance curve. In addition, measurements were made on the time history of the MR value for a period of 75 s following a step change in the field from saturation. We observed that the time dependent behavior of the MR values of both experiments produced an excellent fit to a function of the form $\Delta MR(t) = \alpha + \beta \ln(t)$ where α and β are constants. This time dependence was consistent with the behavior of the magnetic aftereffect.

INTRODUCTION

Layered structures consisting of alternating ferromagnetic and nonmagnetic materials are capable of exhibiting giant magnetoresistance[1]. A brief review in given in Ref. 2. Because the resistance of these devices depends on the relative orientation between the spins in the ferromagnetic layers, they are referred to as spin-valves. Consider a simple example. Fig. 1 shows a layered device (Sample C of this paper). The copper serves as a "magnetic insulator" to reduce the coupling between the ferromagnetic layers. When a large magnetic field is applied, the spins in both the Co and Fe layers line up in the same direction. When the magnetic field is reversed, the Fe spins will change direction first since Fe has a lower coercivity, ~ 4.5 kA/m (56 Oe). As the field intensity is increased further, the spins in the Co layers [coercivity ~ 15.9 kA/m (200 Oe)] will eventually change direction. This can be seen in Fig. 2(a), which shows the magnetization vs field loop of sample C. The resistance of the spin valve is given by[3]

$$R = R_p + \frac{(R_{AP} - R_P)(1 - \cos\theta)}{2} \qquad (1)$$

where R is the resistance, θ is the angle between the spins, R_{AP} is the resistance when the spins are antiparallel ($\theta = \pi$), and R_P is the resistance when the spins are parallel ($\theta = 0$). We measure the magnetoresistance (MR), which is defined as

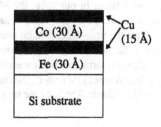

Fig. 1. Example of a spin-valve structure.

$$MR = \Delta R / R = (R - R_P) / R_P. \tag{2}$$

With equation (1), this gives

$$MR = \frac{R_{AP} - R_P}{2 R_P} (1 - \cos \theta) \tag{3}$$

The measured magnetoresistance of Sample C is shown in Fig. 2(b).

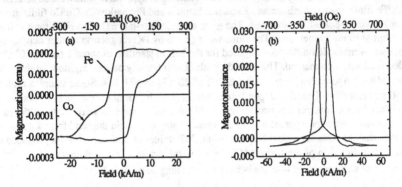

Fig. 2. (a) Magnetization and (b) magnetoresistance vs field of a
Cu (15 Å)/Co (30 Å)/Cu (15 Å)/Fe (30 Å)/Si sandwich structure (Sample C).

The angle θ depends not only on the applied field, but also on time. This is due to the magnetic aftereffect[4] which occurs when magnetization changes by domain wall motion. When a magnetic field is abruptly applied to a material, the magnetization M changes rapidly to a new value and then continues to change as a function of time until it eventually stabilizes. If the material can be described by a single time constant, M behaves as

$$M = M_s < \cos \theta > = I (1 - e^{-t/\tau}) \tag{4}$$

where M_s is the saturation magnetization, I is the amplitude of the time dependent part, τ is the time constant and $<\cos \theta>$ is the spin angle averaged over domains. This means that $<\cos \theta>$ changes with time. In our case, the MR is caused by M in *one* of the ferromagnetic layers changing while M in the other remains saturated, so the $<\cos \theta>$ in Eq. (3) is the same as $\cos \theta$ in Eq. (4). Thus the time-dependent magnetization leads to a time dependent MR. Usually the time dependence displays a range of time constants, from τ_1 to τ_2. If it is assumed that the distribution of the ln τ's is uniform, the time dependence breaks into three regions[4]: 1) if $t << \tau_1$, the change in M is linear in time; 2) if $t >> \tau_2$, the change is a decaying exponential, and 3) if $\tau_1 < t < \tau_2$, the change is logarithmic. The last case is of interest to us, and the detailed time dependence is given by[4]

$$\Delta M \cong \frac{I}{\ln(\tau_2 / \tau_1)} [\ln \tau_2 - 0.577 - \ln(t)]. \tag{5}$$

Using Eq. 5, we can write the time-dependent change in MR as

$$\Delta MR = \alpha + \beta \ln(t) \qquad (6)$$

where α and β are constants[5].

SAMPLE PREPARATION

Table I shows the details of the spin-valve sandwich structures used in this study. Cu/Co/Cu/Fe and Permalloy/Co/Cu/Co films were grown on various substrates by DC magnetron sputtering with a base pressure of 3×10^{-7} Torr and a working argon pressure of 2 or 3 mTorr. The oxide layers were prepared by oxidation (500°C, 0.5 hr) in air of a sputter deposited 500 Å Ni film or a multilayer $(10 \text{ Å Ni}/10 \text{ Å Co})_{25}$ film. They were to provide a bias field for the bottom Co layer. The films with Permalloy were grown in a magnetic field of 1.2 kA/m (150 Oe).

Table I. Details of the spin-valve sandwich structures. Samples containing Permalloy were grown in a magnetic field. The "orientation" is the relationship between the growth field direction and the measurement field direction.

Sample	A	B	C	D	E	F
Substrate	Si	Kapton	Si	Si	Si	Si
Layer 1	30 Å Fe	40 Å Fe	30 Å Fe	(Ni/Co)O	NiO	NiO
Layer 2	50 Å Cu	50 Å Cu	15 Å Cu	50 Å Co	50 Å Co	50 Å Co
Layer 3	30 Å Co	30 Å Co	30 Å Co	50 Å Cu	50 Å Cu	50 Å Cu
Layer 4	50 Å Cu	50 Å Cu	15 Å Cu	4 Å Co	4 Å Co	4 Å Co
Layer 5				50 Å Permalloy	50 Å Permalloy	50 Å Permalloy
Orientation				parallel	transverse	parallel
MR (%)	1.6	1.0	2.9	3.0	2.1	2.6

MEASUREMENTS AND ANALYSIS

After sputtering, the samples were cut into 1.2 cm x 5.2 cm rectangles and four copper wires were attached with silver paint. Each sample was held in place on an aluminum bar inside the field coil by a layer of silicon vacuum grease. Samples were oriented with the long axis parallel to the coil. Resistance measurements were made at room temperature by using an HP 3457A digital voltmeter in the four terminal measurement mode. A magnetic field H of up to 55.7 kA/m (700 Oe) was applied by a coil. The sample and coil were air cooled sufficiently to eliminate effects due to heating from the coil. To insure that the time dependence of the resistivity was not due to the measurement apparatus, the sample was replaced with a resistor. A second, larger resistor was added in parallel during the series of measurements to slightly change R. The resulting resistivity measurements showed a sharp step, without any indication of a time dependence. Two types of time-dependent measurements were made; details of which are given in the next two sections.

Stepped Field Measurement

This type of measurement more closely resembles the expected uses for these devices. In these measurements, the magnetic field was moved to 55.7 kA/m and the saturation resistance R_p was measured. After this, H was stepped to -55.7 kA/m and stepped back to 55.7 kA/m in varying increments, depending on the sample. After each step, measurements were taken until the change

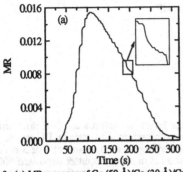 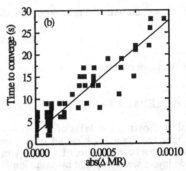

Fig. 3. (a) MR response of Cu (50 Å)/Co (30 Å)/Cu (50 Å)/Fe (30 Å)/Si sandwich structure (Sample A) vs time. The field was incremented [400 A/m (5 Oe)] after the resistance stabilized at each field. The inset shows the aftereffect. (b) Stabilization time vs absolute value of the time-dependent part of the change in MR for each step, also for Sample A. The straight line represents a linear fit to the data.

in R was 0.002 Ω or less. Since the resistance of the samples was about 25 Ω, this represents a precision of about one part in 1.2 x 10^4 in the MR. This is shown in Fig. 3(a) for Sample A; the steps were 400 A/m (5 Oe). The inset shows an expanded view of the time-dependent portion of the MR (curved region) between two field steps. The time-dependent portion of the MR is visible as the curved portion between the steps. This part can be fit rather well to Eq. 6, showing that the behavior is as expected. Unfortunately, the values of α and β are not particularly illuminating. Of more practical interest is the time the MR takes to settle to its equilibrium value. This time depends upon the change in amplitude of the time-dependent part of the MR for each field step in a more or less linear fashion. Fig. 3(b) shows settling time vs the absolute value of the change, along with a linear fit to the data. Note that when the change in MR is largest, this sample can take up 30 s to stabilize. The curves in Fig. 3 (a) and (b) have the same shape for the other samples. Sample D, which has an oxide bias layer, required as much as 70 s for the longest step and 1000 s for the same measurement. Table II shows the values of the slope of the linear fit, uncertainties in the slope (σ) and the correlation coefficient (r) for all of the samples.

Table II Summary of the experimental results for the stepped field measurements.

Sample	A	B	C	D	E	F
Slope (sec)	25 300	26 500	26 000	31 800	34 300	25 300
σ (sec)	670	1 000	1 400	440	1 000	800
r	0.94	0.92	0.81	0.98	0.92	0.91

Time Dependent Measurements

In the second type of measurement, the field was first moved to 55.7 kA/m and R_p was determined. Then the field was moved to the desired value and R was measured at 0.5 s intervals for 75 s, giving a time history of the MR. This is shown in Fig. 4(a), along with the fit to $\alpha + \beta \ln(t)$ for Sample A. In theory, the situation during these measurements could be complicated. If the measurement field was not large enough to move the Co spins, then only the Fe spins would be involved and the MR would increase as the angle between the Fe and Co spins increased. If the field was larger, *both* the Co and Fe spins would move and there could be two sets of time constants involved. In practice, the time dependence of all the measurements fit Eq. 6 very well, so we cannot determine if more than one set of time constants is involved. The jagged line in Fig. 4(a) shows the time dependence of the MR for Sample A where the measurement field is -10.3 kA/m (-130 Oe), near the coercivity of Co. The smooth line is the fit to Eq. 6. Fig. 4(b) shows the values of α and β for a series of measurements taken at different fields, along with the MR curve for comparison. The values of β for all the samples show the same shape and almost the same amplitude. The values of α show similar shapes but vary in amplitude by a factor of about 3.

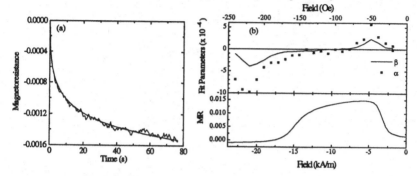

Fig. 4. (a) Time dependent response of the magnetoresistance after moving from 55.7 kA/m to -10.3 kA/m (jagged line) for Cu (50 Å)/Co (30 Å)/Cu (50 Å)/Fe (30 Å)/Si sandwich structure (Sample A). Also shown is the fit to $\alpha + \beta \ln(t)$ (smooth line). (b) Top, values of fit parameters α and β; bottom, magnetoresistance as a function of field, also for Sample A.

SUMMARY

Measurements of the magnetoresistance of spin-valve structures show time-dependent effects consistent with the behavior of the magnetic aftereffect. Up to 70 s were needed to reach equilibrium for some of the spin-valve structures studied. This could have consequences for use of spin-valve structures for some types of devices, while other devices should not be affected. The effect seen here is probably due to domain wall motion. Thin film spin-valve read heads for disk drives are single domain devices with the magnetization changing by rotation. This type of device should not be affected. Devices which apply only large fields, which are able to overcome the energy barriers impeding domain wall motion, would also remain unaffected. Devices that require switching from one domain state to another at small to moderate fields would be susceptible to time dependent effects.

We would like to thank J. R. Cullen for helpful discussions. We would also like to acknowledge the support of the Independent Research Program at Naval Surface Warfare Center, Carderock Division and the ARPA ULTRA Program.

REFERENCES

1. S. S. P. Parkin, N. More, and K. P. Roche, Phys. Rev. Lett. **64**, 2304 (1990).
2. Gary A. Prinz, Physics Today **48**, 58 (April 1995).
3. B. Dieny, J. Magn. Magn. Mat. **136**, 335 (1994).

4. Soshin Chickazumi, Physics of Magnetism (John Wiley & Sons, New York, 1964), p. 303.
5. It is tempting to try to use something like $\Delta MR = \alpha + \beta \ln(t/\tau)$ to find a characteristic time; however, due to the properties of the log function, this results in $\Delta MR = [\alpha - \ln(\tau)] + \beta \ln(t)$ where the quantity in square brackets is just a constant. Thus the "characteristic time" is folded into the constant a.

EFFECTS OF INTERFACE INTERMIXING ON THE MAGNETORESISTANCE OF SPIN VALVES WITH UNCOUPLED Co-LAYERS

M.M.H. WILLEKENS[1], TH.G.S.M. RIJKS[1,2], H.J.M. SWAGTEN[1] AND W.J.M. DE JONGE[1]
[1]Eindhoven University of Technology, P.O.Box 513, 5600 MB, The Netherlands.
[2]Philips Research Laboratories, Prof. Holstlaan 4, 5656 AA Eindhoven, The Netherlands.

ABSTRACT

We have studied the effect of an artificially intermixed region grown at the interfaces of Co/Cu spin valves with uncoupled layers. Two different structures are used: exchange-biased spin valves and engineered spin valves in which two layers are antiferromagnetically coupled and a third layer, on top of this system, is not coupled to the other two. It is shown that structural effects, induced by variation of the deposition parameters and by the intermixing can play an important role. Since the present study uses sputtered layers an intrinsic initial intermixing of 4-5Å is already present. For both types of spin valves G_p, ΔG and $\Delta R/R$ all show a gradual decrease when the nominal thickness of the total intermixed region is enlarged from 0 to 36Å. Also when the initial degree of intermixing is decreased by sputtering at higher Ar-pressure, G_p, ΔG and $\Delta R/R$ still show a gradual decrease as a function of intermixed layer thickness. Combined with the fact that there is no difference between an intermixed region of thickness t at one Co/Cu interface or intermixed regions of thickness $t/2$ at two interfaces, this indicates that the electron scattering in the intermixed region is predominantly spin *in*dependent, although this region preserves a magnetic moment.

INTRODUCTION

Apart from layer thicknesses and intrinsic material quantities, such as (spin dependent) conductivities of the layers, interface roughness is one of the parameters that determines the magnitude of the giant magnetoresistance (GMR). On the one hand it is known that dilute impurities in ferromagnetic materials such as Fe, Co and Ni can lead to spin dependent electron scattering and thus some intermixing at the interfaces can enhance the GMR. On the other hand, when the interfaces are too rough, the large amount of interface scattering will reduce the probability of electrons to cross the non-magnetic interlayer and therefore reduce the GMR. One could imagine that some optimum in the interface roughness exists as is reported for the Fe/Cr-system [1].

In the study of the effect of interface roughness on the GMR, antiferromagnetic (AF) coupling between the magnetic layers can be a complicating factor. When, due to roughness, the strength of the AF coupling changes, this can influence the degree of antiparallel alignment, which will in turn also affect the GMR-effect. Therefore this kind of experiment should be performed preferably on decoupled magnetic layers.

In this article we report on the effects of interface intermixing on the GMR in Co/Cu spin valves with uncoupled Co-layers. Two different structures are used: exchange-biased spin valves and spin valves in which one of the Co-layers of the Co/Cu spin valve is antiferromagnetically coupled to a third magnetic layer. Although the GMR can be influenced by roughness on a very small scale [2,3] and one would therefore like to start with perfectly flat interfaces, the present study uses sputtered samples in which an intrinsic intermixing of ca. 4-5Å is artificially increased to 36Å during the growth of the sample. Therefore the results of these samples will provide mainly information on the "bulk properties" of the intermixed regions. Currently measurements on MBE-grown samples with sharper interfaces are in progress.

SAMPLE PREPARATION

Two types of samples were used which we will refer to as type I and type II. The type I samples consist of a spin engineered structure of three magnetic layers: 100ÅCo(M_1)/6ÅRu/25ÅCo(M_2)/(40-t/2)ÅCu/tÅ CoCu/(100-t/2)ÅCo(M_3), as is shown in Fig.1A for t=0. The 6Å Ru-layer provides a strong AF-coupling

391

Mat. Res. Soc. Symp. Proc. Vol. 384 ° 1995 Materials Research Society

between M_1 and M_2 which causes an antiparallel alignment between M_2 and the other Co-layers at small applied fields. The thickness of the Cu-layer is chosen such that this layer decouples M_2 and M_3. The samples are HV-magnetron sputter deposited on SiO_2 substrates at room temperature and an Ar-pressure of 7 mTorr. The intermixed region of CoCu at the Cu/M_3 interface is introduced by alternately depositing 1Å Co and 1Å Cu. The nominal thickness, t, of this intermixed region varies from 0 to 36Å. In some cases this intermixed region was divided into two regions of thickness $t/2$, one at the M_2/Cu interface and one at the Cu/M_3 interface.

The type II samples are exchange biased spin valves which are DC magnetron sputtered on Si(100) substrates at room temperature. They consist of 30ÅTa/(100-t/4)ÅCo/(t/2)ÅCoCu/(40-t/2)ÅCu/(t/2)Å CoCu/(50-t/4)ÅCo/100ÅFeMn/30ÅTa. Again the thickness of the Cu spacer is such that the 100Å Co-layer is not coupled to the 50Å Co-layer. The magnetization direction of the 50Å Co-layer is pinned in a certain direction through direct exchange coupling to the antiferromagnetic FeMn-layer. In these samples the nominal thickness of the total intermixed CoCu-region, t, varies from 0 to 30Å. The CoCu-regions in these samples were deposited by co-sputtering of Co and Cu. To vary the intrinsic initial intermixing two series were grown in which the Co/CoCu/Cu/CoCu/Co-stack was deposited at Ar-pressures of 5 and 10 mTorr. All other layers are grown at an Ar-pressure of 5 mTorr.

SAMPLE CHARACTERIZATION

To characterize the samples magnetization measurements were performed. In Fig. 1A a typical magnetization loop is shown for the type I samples. As layer M_3 is not coupled to the other magnetic layers, the magnetization of this layer will always point in the direction of the field. Layers M_1 and M_2 are coupled strongly antiferromagnetically and therefore at small fields ($< H_1$) the magnetization directions of these layers will align antiparallel with the magnetization direction of the thinner layer, M_2, pointing opposite to the field direction. Note that now also the magnetization directions of layers M_2 and M_3 are aligned antiparallel although they are not coupled to each other. When the field is increased, the magnetiza-

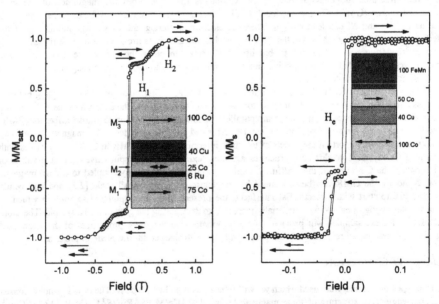

Fig. 1: *Typical M(H)-loops for A) type I samples and B) type II samples. The arrows indicate the magnetization directions of the Co-layers.*

tion direction of layer M_2 will gradually reverse to the field direction until saturation is reached at $H=H_2$.

In Fig. 1B a typical magnetization loop of a type II sample is shown. In fact the measurement shows two hysteresis loops. One around H=0 T for the "free" Co-layer and one around the exchange biasing field H_e for the exchange-biased Co-layer. Between H=0 T and $H=H_e$ a region exists of complete antiparallel alignment.

The most essential feature of both systems is the possibility to create a transition between parallel and antiparallel alignment of the magnetization directions of two uncoupled Co-layers through the application of a magnetic field. Both systems show a clear plateau in the magnetization, the moment of which corresponds to the moment one would expect from antiparallel alignment of the magnetization directions. Also when the intermixed regions are present we observe full antiparallel alignment of the constituents of the spin valve.

Now we would like to know the structure and magnetic behavior of the regions inserted at the interfaces. According to the bulk phase diagram of Co_xCu_{1-x} [4], no thermodynamically stable solid solutions exist at any temperature in the composition range $0.05 \leq x \leq 0.88$ due to immiscibility of the two components. However, it has been established that it is possible to produce a metastable Co_xCu_{1-x}-alloy over the whole concentration range by coevaporation [5] and cosputtering [6]. Both Childress [6] and Kneller [5] have reported a reduction of the magnetic moment of Co-atoms when intermixed with Cu. For a metastable alloy $Co_{0.5}Cu_{0.5}$ cosputtered at 77K on glass or mica, Childress reports a saturation magnetization of ≈ 125 emu/g_{Co} compared with 175 emu/g_{Co} for bulk fcc Co. In our samples however we did not measure any reduction of magnetic moment as a function of intermixed region thickness. Also in a separate series of samples 200ÅRu/6*[(40-x)ÅCu/x*(1ÅCo+1ÅCu)/(25-x)ÅCo]/10ÅCu/30ÅRu with x up to 23 such that almost all the Co is intermixed with Cu, we did not observe any loss of magnetic moment. This could indicate the existence of Co-clusters in the intermixed regions and might be due to the fact that our samples were not sputtered at 77K but rather at room temperature, which causes a higher mobility of the atoms reaching the substrate.

The initial, intrinsic intermixing was for some samples determined by quantitative fitting of low-angle X-ray spectra combined with X-ray fluorescence [7]. An initial intermixing of ca. 4 and 5Å was found for the samples of type I and type II (5 mTorr) respectively. For the samples of type II sputtered at 10 mTorr we expect the initial intermixing to be less, in the order of one monolayer. To investigate the thickness of the artificially intermixed regions we used the same techniques. Here we described the intentionally introduced intermixing as an extra layer between Co and Cu that consists of a CoCu-alloy with a density averaged from that of Co and Cu. The measured results are in reasonable agreement with the nominal thicknesses of the intermixed regions, although the conservation of magnetic moment seems to indicate that the intermixing of Co and Cu is not extending downto an atomic scale.

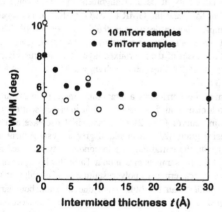

According to the X-ray measurements all samples are grown in the (111)-direction. However, rocking curves indicate that especially the samples of type I have a rather poor texture. For all samples of type I the full width at half maximum (FWHM) of the rocking curves amounts ≈ 13. The FWHM of the rocking curves of the type II samples is shown in Fig. 2 as a function of the total intermixed thickness t. We will return to this point in the next section.

Fig. 2: *FWHM of the XRD-rocking curves versus total intermixed thickness t, for the type II spin valves.*

RESULTS AND DISCUSSION

In Fig. 3 a typical resistance versus field curve is shown for the type I and type II samples. The magnetoresistance was measured at room temperature in a standard four-probe set-up with the current and the field in the plane of the sample (H⊥I). In both measurements the resistance is low at saturation and there is a clear plateau where the magnetization directions are aligned antiparallel corresponding to a high resistance. As noted before, the samples of type I actually consist of two spin valves. However, the contribution to the GMR from the Co/Ru/Co-system, needed to obtain the antiparallel alignment, is negligible compared to the contribution of the Co/Cu/Co-part, which is shown in Fig. 3A where the GMR of the system 200ÅRu/75ÅCo/6ÅRu/25ÅCo/150ÅCu/30ÅRu is reproduced. The peaks around H=0 T in the GMR of the type I samples are due to the magnetization reversal of all magnetic layers (see Fig. 1), which causes deviations of the perfect antiparallel alignment, and to the changing angles between the magnetization directions and the current (anisotropic magnetoresistance effect).

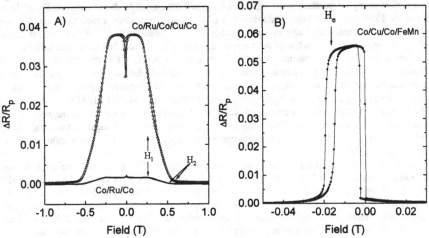

Fig. 3: *Typical current in plane resistance versus field measurements measured at room temperature for A) type I spin valves and B) type II spin valves.*

In Fig. 4 we show the results for the sheet conductivity at parallel alignment G_p, the change of conductivity between parallel and antiparallel alignment, ΔG, and the GMR (=$\Delta G/G_{ap}$=$\Delta R/R_p$). For large thicknesses of the intermixed regions all series of samples show a decrease in the G, ΔG and GMR. The decrease in conductivity is due to the fact that pure Co and pure Cu are substituted by a mixed region of Co and Cu with a higher resistivity. For these large thicknesses of the intermixed layer, the high resistivity of these layers reduces the flow of polarized electrons across the Cu-layer, resulting in a smaller value for ΔG. The quantity GMR can be simply deduced from G and ΔG.

At small thicknesses the samples of type II, grown at 5 mTorr, show a remarkable increase in ΔG and GMR. It would be tempting to relate this behavior to a change in the spin dependent interface scattering. However, for these samples the FWHM of the rocking curves (Fig. 2) show an initial decrease and a saturation at about 10Å. This could point at an increasing grain size or a smaller degree of misalignment at the grain boundaries which seems to be in agreement with the initial increase in conductivity observed in these samples. It is known from other experiments [8] that, for samples grown on a Ta buffer layer, a non-negligible part of the resistance is due to grain boundary scattering. In polycrystalline $Ni_{80}Fe_{20}$-films it is found [8] that the effective value of $\alpha = \lambda^\uparrow/\lambda^\downarrow$ is smaller than the bulk value due to grain boundary scattering of mainly the spin-up electrons. Because of their larger (bulk) mean free path spin-up electrons are more sensitive to grain boundary scattering than spin-down electrons. Therefore, an increase in the grain size or a reduction of the degree of misalignment near the grain boundaries can 1) decrease the scattering probability, leading to an increase in conductivity and 2) increase the spin dependence of the scattering, leading to a larger value of ΔG. This is just what is shown in Fig. 4 for the samples grown at 5mTorr for small intermixing. Possible effects of the artificial intermixing on the (spin dependent) interface scattering, if present with the present degree of initial intermixing, are not discernible from these structural effects.

The 10 mTorr samples of type II, that are expected to have a smaller degree of initial intermixing, and the samples of type I do not show large changes in structural quality when the artificial intermixed thickness is increased. In these samples (for type I also measured at T=10K) no dramatic changes are observed in G, ΔG and GMR when the thickness of the intermixed layers increases. They rather show a gradual decrease even when the thickness of the intermixed regions is comparable to the thickness of the other layers. Of course in this case it is difficult to think of this region as interface intermixing since the "interface" between Co and Cu has now become a layer of its own.

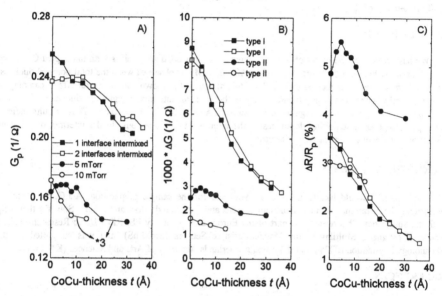

Fig. 4: G_p (a), ΔG (b) and $ΔR/R_p$ (c) as a function of the total intermixed region thickness t. The conductivity G_p of the type II samples is multiplied by a factor 3.

The present results are in marked contrast with Suzuki [2] who has reported a sharp decrease in GMR from 27 to 4% when only cosputtering 1.5Å Co and Cu at the interfaces of AF-coupled Co/Cu-multilayers. This difference could be explained either by a coupling effect in the samples of Suzuki, or to a different state of initial intermixing.

A decrease of GMR in spin valves with uncoupled layers is also reported by Parkin [9]. In NiFe/Cu spin valves with thin Co-layers at the interfaces and Co/Cu spin valves with thin NiFe-layers inserted at the interfaces, the GMR decreases upon annealing. These results are different from results obtained in the Fe/Cr-system. The GMR-effect in Fe/Cr is often, by analogy with bulk alloys [10], ascribed to spin dependent scattering from Cr-atoms dissolved in the Fe-layers. As a consequence one could expect that the GMR increases with increasing intermixing. This seems not to be the case in Co/Cu spin valves (Fig. 4).

For the Fe/Cr system it has been reported [11] that when ultrathin layers [0..4Å] of V, Mn, Al, Ir and Ge are inserted at the Fe/Cr interfaces, it makes no difference whether a thickness t of these layers is inserted at alternate interfaces or a thickness t/2 at each Fe/Cr-interface. What seems to be more important than the number of interfaces, is the overall number of additional (spin dependent) scatterers per multilayer period. Also for the type I samples two different series samples were grown. In the first series there was only an intermixed region between the Cu-layer and the free Co-layer (M_3). In the second series both interfaces of the Cu-layer were intermixed with Co. From Fig. 4 we can conclude that also in our Co/Cu spin valves there is no significant difference between an intermixed region of thickness t at one interface or intermixed regions of thickness $t/2$ at both interfaces. Both cases result in the same slope of G_P, ΔG and $\Delta R/R$ as a function of total intermixed thickness. The small difference in magnitude is already present in samples of the same composition (samples where $t_{CoCu} = 0$) and is therefore ascribed to non-perfect reproducibility.

This result, combined with the fact that the GMR decreases with increasing intermixing indicates that in the intermixed CoCu regions, at least for the thicknesses we have investigated here, the spin independent scattering is dominant over the spin dependent scattering. This conclusion seems to be in qualitative agreement with model calculations we performed recently with the well known Camley-Barnas model on a Co/Cu/Co-trilayer [7].

CONCLUSIONS

We have measured the effect of interface intermixing in Co/Cu spin valves with uncoupled Co-layers. In general a gradual, monotone decrease of G, ΔG and GMR is observed when the total nominal thickness of the intermixed region is increased from 0 to 36Å, although it is shown that structural effects can play an important role. There is no significant difference between an intermixed region of thickness tÅ at one Co/Cu-interface or intermixed regions of thickness $t/2$Å at both Co/Cu-interfaces. These results indicate that spin independent scattering dominates in the intermixed CoCu-regions. In the intermixed region no reduction of magnetic moment was observed.

ACKNOWLEDGEMENTS

The authors would like to thank H.T. Munsters for the sample preparation, A.J.G. Leenaers for performing the X-ray measurements and calculations and R.A. van de Roer and R.L.H. Sour for their help with measurements. This research is part of the European Community ESPRIT3 Basic Research Project, Study of Magnetic Multilayers for Magnetoresistive Sensors (SmMmS), and was supported by the Technology Foundation (STW) and the Royal Netherlands Academy of Arts and Sciences (KNAW).

REFERENCES

[1] F. Petroff, A. Barthélémy, A. Hamzic, A. Fert, P. Etienne, S. Lequien and G. Creuzet, Journ. o Magn. and Mag. Mat. **93**, 95-100 (1991).

[2] Motofumi Suzuki and Yasunori Taga, J. Appl. Phys., **74** (7), 4660 (1993).

[3] Randolph Q. Hood, L.M. Falicov and D.R. Penn, Phys. Rev. **B49**, 368 (1994).

[4] M. Hansen and K. Anderko, Constitution of Binary Alloys, McGraw-Hill Book Company, Inc., Ne York, (1958).

[5] E. Kneller, J. Appl. Phys., suppl. to vol **3**, 1355 (1962).

[6] J.R. Childress and C.L. Chien, Phys. Rev. **B43**, 8089 (1991).

[7] Willekens et al., submitted to J. Appl. Phys.

[8] Rijks et al., Paper submitted to the INTERMAG '95 San Antonio, Texas.

[9] S.S.P. Parkin, Phys. Rev. Lett. **71**, 1641 (1993).

[10] I.A. Campbell and A. Fert in Ferromagnetic Materials, Vol. **3**, p.747, edited by E.P. Wohlfart (North Holland, Amsterdam, 1982).

[11] Peter Baumgart, Bruce A. Gurney, Dennis R. Wilhoit, Thao Nguyen, Bernard Dieny and Virgil S Speriosu, J. Appl. Phys. **69** (8), 4792 (1991).

IMPROVED THERMAL STABILITY OF GMR SPIN VALVE FILMS

R. D. McMICHAEL, W. F. EGELHOFF, Jr. and MINH HA
National Institute of Standards and Technology, Gaithersburg, MD 20899

ABSTRACT

In order to improve the thermal stability of magnetic multilayer "spin valve" structures, we have measured the magnetic and magnetoresistive properties of a number of samples with the general structure of NiO/Co/Cu/Co/Cu/Co/NiO as a function of annealing time at 250 °C. The magnetoresistance (MR) of the samples annealed in air decreases proportionally to the square root of the annealing time. For samples annealed in a vacuum, the decrease in magnetoresistance is reduced, but not eliminated. Magnetometry of a vacuum annealed NiO/Co/NiO sample shows a magnetization reduction and a coercivity increase which suggest oxidation of the NiO-biased "outer" Co layers of the spin valve structure. For increasing NiO-biased Co layer thickness, we show enhanced thermal stability and even increasing MR with annealing time for samples with the thickest outer Co layers.

INTRODUCTION

Spin valves[1-3], which are based on the giant magnetoresistance effect, are rapidly finding applications in devices such as magnetic field sensors and read heads for high density magnetic recording. In recent work[4], symmetric spin valves[5-6] of the general structure NiO/Co/Cu/Co/Cu/Co/NiO were optimized for maximum MR as a function of the thicknesses of the Co and Cu layers. As-deposited MR values exceeding 21% in fields less than 10 mT (100 G) were obtained. In processing these films into devices, there is often a baking step where the temperature is raised to 240 °C for periods up to 24 hours. Therefore, it is important not only that the spin valve films should be able to endure the processing, but also that the films should be optimized to give the best performance after the anneal.

The basic symmetric spin valve structure is shown in fig. 1. The structure consists of three Co layers separated by two Cu layers, with an underlayer and an overlayer of antiferromagnetic NiO. Through an interfacial interaction between NiO and Co, the outer Co layers have a large coercivity, H_c^{outer}, typically on the order of 20-50 mT (200-500 G) and a smaller exchange bias field, which shifts the hysteresis loop of the outer films by 6-8 mT. The center Co layer has a much lower coercivity, H_c^{center}. The sign and magnitude of the coupling

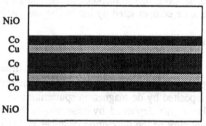

Fig. 1. Schematic structure of NiO-biased symmetric spin valve.

between the Co layers through the Cu layers is characterized by a coupling field, H_{ex}^{center}, and depends sensitively on the Cu layer thickness[5], which is chosen to produce the minimum coupling between the Co layers.

The MR response of a spin valve with the field and the current parallel and in the plane of the film is shown in fig. 2. The resistance of the film depends on the relative orientations of the magnetizations in the Co films, with the lowest resistance obtained when the magnetizations are parallel. In a negative saturating field, the magnetizations of the layers are

Mat. Res. Soc. Symp. Proc. Vol. 384 ©1995 Materials Research Society

aligned in the field direction and the electrical resistance is low. As the field is increased through zero to H_1, a high resistance state is obtained as the center film remagnetizes to the positive direction, while the outer films remain essentially unchanged. As the field is further increased through H_2, the outer Co films are remagnetized to the positive direction, and a low resistance state is obtained. The asymmetry in the MR response is due to the exchange biasing of the NiO/Co interfaces.

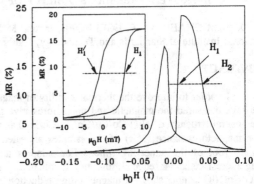

Fig. 2 Magnetoresistance loop for an as-deposited symmetric spin valve (sample #2). The minor loop, where only the central film switches, is shown in the inset.

The coercivity of the outer Co layers is determined primarily by coupling to the antiferromagnetic NiO. An energy balance argument shows that since the Co film's interaction with the NiO is a surface phenomenon while its interaction with the applied field is a volume interaction (proportional to the thickness of the film), the coercivity of the Co film should vary inversely with the film thickness.

$$H_c^{outer} \approx \frac{K_{NiO}}{\mu_0 M_s t} \tag{1}$$

where K_{NiO} is an effective surface anisotropy due to the NiO layers, M_s is the saturation magnetization of Co and t is the thickness of the Co layer.

In this paper we present results of a study to determine the mechanisms of thermal degradation in NiO biased symmetric spin valves. The results indicate that degradation of the outer Co layers is responsible for a large portion of the decrease of MR upon annealing which can be compensated by increasing the thickness of the outer Co layers.

EXPERIMENTAL

The substrates for these experiments were 3" Si wafers coated with ~50 nm of polycrystalline NiO by reactive magnetron sputtering. The metallic Co and Cu layers were deposited by dc-magnetron sputtering in 2 mTorr Ar at a rate of ~0.1 nm/s, and the top NiO layer was deposited by sputtering a Ni target with an 85/15 mixture of Ar/O_2. The base pressure of the deposition chamber is 1×10^{-8} Torr (~ 10^{-6} Pa).

A number of samples of the general structure given in fig. 1 were deposited with varying layer thicknesses which are presented in Table I. With the exception of sample #3 and sample #8, the primary difference between the samples is the thickness of the outer Co layers.

Annealing of the samples was carried out either in air or under a vacuum in the 10^{-6} Torr (10^{-4} Pa) range. All annealing was performed at 250 °C. To shorten the time required to cool the vacuum annealed samples, purified Ar was admitted to the furnace as a heat transfer agent.

Magnetoresistance measurements were made using an AC four-point probe technique using spring-loaded pin contacts. Occasionally, when difficulty was encountered making contact through the top NiO layer, an AC current of 100 mA peak-to-peak was passed through each of the pins. This procedure dramatically reduced the contact resistance from ~ $10^6 \Omega$ to ~ $10^1 \Omega$,

Table I.
Layer thicknesses of the samples.

sample	NiO (nm)	Co (nm)	Cu (nm)	Co (nm)	Cu (nm)	Co (nm)	NiO (nm)
#1	50	2.5	1.8	5.1	1.8	2.5	50
#2	50	2.8	1.9	4.0	1.9	2.8	50
#3	50	2.8	1.9	*	1.9	2.8	50
#4	50	3.2	2.0	4.0	2.0	3.2	50
#5	50	4.2	2.1	4.0	2.1	4.2	50
#6	50	5.0	1.8	4.0	1.8	5.0	50
#7	50	7.5	2.0	4.1	2.0	7.5	50
#8	50	-	-	6.0	-	-	50

* 2.0 nm Co/ 2.5 nm $Ni_{80}Fe_{20}$/ 2.0 nm Co

presumably by causing a dielectric breakdown of the NiO in a small area under the pin contact. Ferromagnetic resonance measurements were made in a 9.87 GHz (X-band) spectrometer with the field aligned in the plane of the sample. All measurements (MR, FMR, SQUID) were made at room temperature. All listed annealing times are cumulative.

RESULTS AND DISCUSSION

Center Co Film

Minor loops of the MR response were used to estimate the coercivity, H_c^{center} of the center film and the effective field of its interactions with the outer Co layers H_{ex}^{center}. Using the fields where the MR reaches one half of its maximum value as estimates of the fields where the magnetization of the center film passes through zero, (the fields H_1 and H_1' in fig. 2)

$$H_c^{center} = (H_1 - H_1')/2 \qquad (2)$$

$$H_{ex}^{center} = (H_1 + H_1')/2 \qquad (3)$$

H_c^{center} and H_{ex}^{center} are plotted in fig. 3 as a function of annealing time at 250 °C in air for sample #2. The interaction field, H_{ex}^{center} was ferromagnetic in all samples in this study.

Ferromagnetic resonance (FMR) spectra of spin valves were taken with the field aligned in the plane of the sample. In these spectra, only one resonance is observed, which is associated with the center Co film. The absence of a second (and third) resonance is attributed to a large inhomogeneous broadening and weakening of the resonances of the outer Co films due to their interaction with the NiO. FMR spectra were taken as a function of the orientation of the applied field in the plane of the films, and in-plane anisotropy fields, H_K, were calculated from the maximum and minimum resonance in-plane resonance fields,

$$H_K = (H_{res}^{max} - H_{res}^{min})/2. \tag{4}$$

The FMR linewidth and the in-plane anisotropy field, H_K are plotted in fig. 4.

Figs. 3 and 5 show that while coercivity and FMR linewidth increase rapidly during the initial 15 minutes of annealing, the coupling field and the in-plane anisotropy field remain nearly constant. None of the measured properties of the center film vary dramatically after the first 30 minutes of annealing.

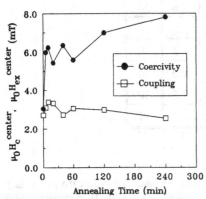

Fig. 3 Coercivity and coupling field of the center Co layer for sample #2 as a function of annealing time at 250 °C in air.

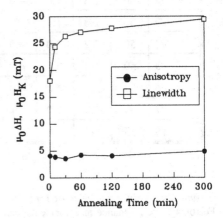

Fig. 4. FMR linewidth, and anisotropy field of the center Co layer for sample #2 as a function of annealing time at 250 °C in air.

Outer Co Film

In order to isolate the effect of annealing on the outer Co films and their interaction with the NiO, we deposited a sample with the central Co film and Cu spacer films eliminated. The remaining structure is a Co film sandwiched between thick layers of NiO, which we annealed at 250 °C in vacuum. SQUID magnetometry measurements shown in fig. 5 show that the moment per unit area of the as-deposited film is somewhat reduced from the nominal value for a 6.0 nm film with the magnetization of bulk Co. We expect that this is due, at least

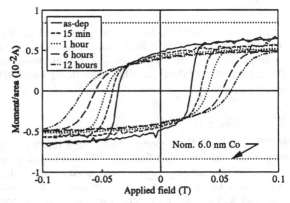

Fig. 5 Room temperature hysteresis loops of a NiO/Co/NiO sandwich sample for several annealing times. Annealing was done under vacuum at 250 °C .

in part, by oxidation of up to 1.5 nm of the Co when O_2 is admitted to the deposition chamber in preparation for reactive sputtering of the NiO. The SQUID measurements show that annealing in vacuum produces a further reduction in the remanence and saturation moment of the film as well as an increase in its coercivity. The observed increase in coercivity of the sample is greater than what would be expected from eq. 1 to arise via a reduction in moment per unit area of the sample, $M_s t$, alone. The origin of the increased coercivity may lie in the formation of a thin layer of CoO or mixed $Co_xNi_{1-x}O$ oxide at the Co/NiO interfaces. It has been shown that existence of a thin (≤ 2.0 nm) layer of CoO ($T_n = 293K$) between permalloy and NiO does not reduce the exchange bias dramatically[7], and studies of exchange biasing of permalloy films on $Co_xNi_{1-x}O$ have shown that the mixed oxide produces a greater effect than pure NiO[8].

Magnetoresistance

Magnetoresistance values plotted as a function of the square root of annealing time for anneals in air and in vacuum are shown in fig. 6. Comparing samples #1 and #2, annealed in air, it is clear that sample #2, which has slightly thicker outer Co layers, is more thermally stable. The fact that the MR values fall on a straight line when plotted vs. the square root of the annealing time was suggestive that diffusion of oxygen through the NiO is the mechanism by which the spin valve degrades. However, while switching to a vacuum anneal reduced the amount of MR degradation, the degradation was not eliminated.

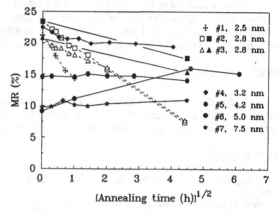

Fig. 6 Magnetoresistance of samples #1-#7 with the listed outer Co layer thicknesses as a function of annealing time at 250 °C. Dashed lines indicate annealing in air and solid lines indicate annealing in vacuum.

Increasing the outer Co layer thicknesses to 3.2 nm appeared to stabilize the MR value at about 20% after an initial decrease.

For the samples with thicker outer Co layers, the MR loops are more symmetric than the loop shown in fig. 2. In the as deposited state, samples #5 and #7 have symmetric MR loops with "shoulders" similar to the right side of fig. 2., but with reduced values of H_2. These samples change very little during annealing, sample #5 showing a stable MR value and sample #7 showing a slightly increasing MR value. Sample #6 has an MR loop with cusps similar to the left side of fig. 2, and is sensitive to annealing. The cusps are seen when the outer layers begin to reverse before the center layer has completed its reversal. In this case, the high resistance state with complete antiparallel alignment of adjacent layers is not obtained, and the maximum observable MR value is sensitive to changes in coercivity of the outer films during annealing. Sample #6 shows a dramatically increasing MR with annealing time, reaching a maximum MR near 20 hours.

CONCLUSIONS

The experimental results indicate that the thermal stability of the center Co layer differs strongly form that of the outer Co layers in a NiO-biased symmetric spin valve structure. The center layer experiences a rapid increase in coercivity and FMR linewidth in the first 15 minutes of annealing at 250 °C, but it is not strongly affected by further annealing. The in-plane anisotropy field and the coupling to the outer Co layers are only weakly affected by annealing.

It seems clear from the difference between air annealed and vacuum annealed samples that oxidation of the top Co layer by oxygen which diffuses through the NiO is a major contributor to the degradation of spin valves annealed in air. Annealing in vacuum significantly reduces the degradation, but the reduced moment and increased coercivity of the NiO/Co/NiO sandwich sample suggest oxidation of outer Co layers even under vacuum. We speculate that oxidation of vacuum annealed samples might occur through diffusion of excess nonstoichiometric oxygen deposited in the NiO layers or by reduction of NiO into Ni.

Using a spin-valve structure optimized for maximum MR in the as-deposited state as a baseline, increasing the thickness of the outer Co layers compensates for the oxidation that occurs during annealing. For maximum MR response after annealing at 250 °C for 20 hours, the optimum thickness for the outer Co layers was found to be about 3.2 nm.

ACKNOWLEDGEMENT

The authors thank K. Takano, A. E. Berkowitz, C.-L. Lin, J. M. Sivertsen, and J. H. Judy for providing substrate material, H. J. Brown for assistance with SQUID magnetometry, and P. R. Menon for assistance with MR measurements.

REFERENCES

[1] B. Dieny, V. S. Speriosu, S. S. P. Parkin, B. A. Gurney, D. R. Wilhoilt, and D. Mauri, Phys. Rev. B, 43, 1297 (1991).

[2] E. Velu, C. Dupas, D. Renard, J. P. Renard, and J. Seiden, Phys. Rev. B 37, 668 (1988).

[3] G. Binasch, P. Grunberg, F. S. Sauerenbach, and W. Zinn, Phys. Rev. B 39, 4828 (1994).

[4] W. F. Egelhoff, Jr., T. Ha, R. D. K. Misra, Y. Kadmon, J. Nir, C. J. Powell, M. D. Stiles, R. D. McMichael, C.-L. Lin, J. M. Sivertsen, J. H. Judy, K. Takano, A. E. Berkowitz, T. C. Anthony and J. A. Brug, to be published, J. Appl. Phys.

[5] T. C. Anthony, J. A. Brug, and S. Zhang, IEEE Trans. Mag. 30, 3819 (1994).

[6] P. M. Baumgart, B. Dieny, B. A. Gurney, J.-P. Nozieres, V. S. Speriosu, and D. R. Wilhoit, U. S. Patent 5 287 238 (Feb. 15, 1994).

[7] M. J. Carey and A. E. Berkowitz, J. Appl. Phys. 73, 6892 (1993).

[8] M. J. Carey and A. E. Berkowitz, Appl. Phys. Lett., 60, 3060 (1992).

IMPROVEMENT OF GMR IN NiFeCo/Cu MULTILAYERS
BY A LAYER - BY - LAYER MAGNETIC FIELD SPUTTERING

K.SAITO, Y. YANAGIDA*, Y. OBI, H. ITOHO* AND H. FUJIMORI
IMR, Tohoku University, Sendai 980, Japan
*Teikoku Tsushin Kogyo Co., Ltd., Kawasaki 211, Japan

ABSTRACT

NiFeCo/Cu multilayers fabricated by an improved magnetic field sputtering were investigated in order to achieve the soft GMR (giant magnetoresistance) with a high sensitivity at low magnetic fields. A magnetic field was applied to the film during sputter-deposition, and its field direction was changed alternately from layer to layer. Such an alternate field sputtering is called hereafter layer-by-layer magnetic field sputtering. The best GMR characteristics (large MR at low magnetic fields) were achieved when the angle between the directions of magnetic field applied to neighboring two magnetic layers was 90°. As one of the speculation, it has been considered that the result is attributed to the induced composite magnetic anisotropy which causes the magnetization to occur more dominantly by spin rotation than by domain wall movement.

INTRODUCTION

Giant magnetoresistance (GMR) multilayered thin films are now considered to be useful for various kinds of electric, magnetic and mechanical sensors, if one can improve the GMR so as to occur easily at low magnetic fields. The NiFeCo/Cu system has been anticipated to be the most suitable GMR multilayers [1,2]. Kanda et al. [3] and Jimbo et al. [4] have reported that the combination of elements of Ni, Fe and Co has opportunities to make both magnetocrystalline anisotropy and magnetostriction negligibly small and to induce a fairly large uniaxial magnetic anisotropy by magnetic field sputter-deposition, leading to a magnetically soft GMR in NiFeCo/Cu multilayers. However, the magnetic fields needed for exciting GMR are still large and a large hysteresis is observed in the MR vs. field curves. The bar-like 180° magnetic domains arising from the induced uniaxial magnetic anisotropy might be responsible for these disadvantages, particularly for the large hysteresis.

In order to solve this problem, we have attempted to control the magnetic domain structure by using an improved sputtering technique. That is, during sputter-deposition for fabricating GMR multilayers (NiFeCo/Cu, for example), a magnetic field has been applied to magnetic layers by changing the field direction alternately from layer to layer. The angle between the field directions changed from a given layer to the nearest neighbor layer has been selected to be 0°, 45° and 90°. In addition, we have also investigated the case that field is not applied. We have found that the MR vs. magnetic field curves are different in these four cases, and that the most prominent MR response with a considerably reduced hysteresis can be obtained in the case of the 90° alternate field sputtering. As one of the speculation, it may be considered that such an alternate magnetic field sputtering (layer-by-layer magnetic field sputtering, in another word) causes the multilayer to have a composite magnetic anisotropy consisting of the uniaxial magnetic anisotropies piled up spirally from layer to layer. This induced composite magnetic anisotropy may compete to the exchange coupling between magnetic layers and then affects to the domain structure, resulting in rotation of magnetization rather than domain wall movement. In the followings, the obtained MR behavior will be described in details.

Mat. Res. Soc. Symp. Proc. Vol. 384 © 1995 Materials Research Society

EXPERIMENTAL

An ion-beam sputtering apparatus with three gun-target systems was used for fabricating multilayer samples. The target materials used were $Ni_{63}Fe_{12}Co_{25}$ alloy disk and Cu metal disk, both with 10 cm diameter. Glass thin plates were used for the substrates. The sputter-deposition was done at room temperature with Ar pressure of 3 - 8 mTorr, ion-beam acceleration voltage of about 1 kV and ion-beam current of about 60 mA. The deposition rates were about 1.0 Å/sec for NiFeCo and about 1.7 Å/sec for Cu, respectively.

The compositions of the multilayers fabricated were $[Ni_{63}Fe_{12}Co_{25}(16 Å)/Cu(21 Å)]_{20}$ and $[Ni_{63}Fe_{12}Co_{25}(21 Å)/Cu(21 Å)]_{20}$. The Cu layer thickness was 21 Å. It corresponds to the second peak region in the oscillation curve of MR vs. Cu thickness, where considerably soft GMRs were confirmed to be obtained [2,3,5].

The multilayering of NiFeCo/Cu was made by applying a magnetic field (H') to NiFeCo magnetic layers during sputter-deposition. The direction of the field was changed for each magnetic layer as shown in Fig.1. The first magnetic layer was deposited under the magnetic field with a given direction and then the Cu layer was deposited. Next, the substrate was rotated by a certain angle (θ) with respect to the first field direction and then the deposition was done for the second magnetic layer. Such depositions were continuously repeated for all the layers. We call this method "layer-by-layer magnetic field sputtering". In the present study, we examined four cases, i.e., the zero field sputtering and the layer-by-layer field sputterings with H'=200 Oe of $\theta=0°$, $45°$ and $90°$. These are denoted as H'=0, H'(0 °), H'(45°) and H'(90°), respectively, as shown in Fig.1. For generating H', a couple of small SmCo permanent magnets was set in the rotatable substrate holder. The direction of the coupled magnets with respect to the multilayer sample was automatically changed by using a stepping motor.

(a)　　　　(b)　　　　(c)　　　　(d)

H'=0　　　H'(0°)　　H'(45°)　　H'(90°)

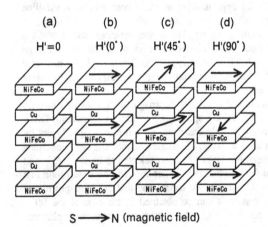

S ——→ N (magnetic field)

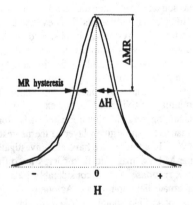

Fig.1. Schematic picture of the magnetic field sputter-deposition employed in the present study. a is the zero field sputtering, and b, c and d are the 0°, 45° and 90° layer-by-layer field sputtering, repectively.

Fig.2. Definition of the MR properties; hysteresis, field width ΔH and sensitivity ΔMR/ΔH.

The structure of the fabricated multilayers was analyzed by an X-ray diffraction and a cross section TEM-image. The magnetization (M) vs. magnetic field (H) was measured by a SQUID magnetometer. The electrical resistivity ρ vs. H was measured by a four-terminal electric resistive method at room temperature, and then the magnetoresistance MR $(=\Delta\rho/\rho)$ vs. H was obtained. For characterizing the properties of MR vs.H, we define 1) hysteresis for MR, 2) field width for MR (ΔH), 3) sensitivity for MR $(\Delta MR/\Delta H)$, as shown in Fig.2.

RESULTS AND DISCUSSION
1) Structure of multilayer

Figure 3 shows the examples of the X-ray profiles of [Ni63Fe12Co25(16 Å)/Cu(21 Å)]20 multilayer samples. As clearly seen, the diffraction peak around $2\theta = 44$ degree for fcc (111) is stronger than another peak around 50 degree for fcc (200), which indicates that the obtained NiFeCo/Cu multilayers have a <111> textile structure oriented normal to film plane. Figure 4 shows a cross-section TEM profile for the sample of H'(90°), which exhibits a clear layered structure with fairly flat layer boundaries.

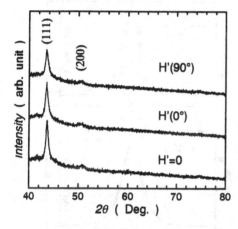

Fig.3. X-ray profiles for the [Ni63Fe12Co25 (16 Å)/Cu(21 Å)]20 multilayers fabricated by three different field sputterings. (See text for the notation of the samples).

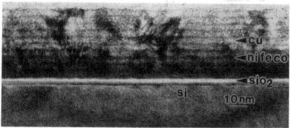

Fig.4. Cross section TEM profile for a [Ni63Fe12Co25(16 Å)/Cu(21 Å)]20 multilayer fabricated by the 90° layer-by-layer field sputtering.

2) Magnetoresistance and Magnetization

In Fig.5, ρ and MR for the H'=0, H'(0°), H'(45°) and H'(90°) multilayers with the same composition in Fig.3 are shown as a function of magnetic field, H. For the H'(0°) multilayer, two curves measured along the two principal axes of easy magnetization and hard magnetization are shown. The other cases are represented with one curve, because they are nearly isotropic. As seen in Fig.5, appreciable hysteresis is observed for all the MR vs. H curves, excepting for the H'(90°) multilayer. It should also be noted that the field width for MR, ΔH, is smallest and thus ΔMR/ΔH is largest for H'(90°); ΔMR/ΔH at around H=20 Oe is 0.12%, 0.21%, 0.12% and 0.23% for H'=0, H'(0°), H'(45°) and H'(90°), respectively. In the case of the H'(90°) [Ni63Fe12Co25(21 Å)/Cu(21 Å)]20 multilayer, much enhanced ΔMR/ΔH of about 0.35% at 20 Oe was obtained. The absolute values of ΔMR are 6% at 20 Oe and 10% at 100 Oe, as shown in Fig.6 . These magnitudes are similar to those of the soft GMR multilayers reported previously [3, 4]. But, the MR hysteresis is much reduced in the present H'(90°) multilayer, thus, the layer-by-layer field sputtering with θ=90° permits us to obtain soft GMR multilayers.

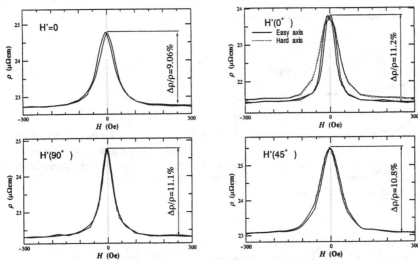

Fig.5. MR vs. H curves for [Ni63Fe12Co25(16 Å)/Cu(21 Å)]20 multilayers fabricated by four different field sputterings. (See text for the notation of the samples).

In order to understand the reason of this effect, we have investigated the magnetization behavior for the present multilayer samples. The magnetization curves obtained are shown in Fig.7, where two curves corresponding to easy magnetization and hard magnetization are shown for all samples of H'=0, H'(0°), H'(45°) and H'(90°). The magnetization for H'(0°) is strongly anisotropic, while the magnetizations for H'=0, H'(45°) and H'(90°) are nearly isotropic. However, the appreciable hysteresis can be seen in all cases, although it is significantly reduced in the case of H'(90). The slight difference between two magnetization curves, which can be seen even in the case of H'=0, seems to be due to somewhat geometrical asymmetry of the sputtering apparatus. On the contrary, the anisotropic magnetization for H'(0°)

is obviously due to an induced composite magnetic anisotropy with a uniaxial symmetry in the whole multilayer sample, which has been caused by the parallel layer-by-layer field sputtering. And hence, the large hysteresis for H'(0°) may be attributed to bar - like 180° magnetic domains associated with the uniaxial anisotropy. The value of the anisotropy energy (Ku) estimated from the two magnetization curves for H'(0°) is about 3.5 $\times 10^3$ erg/cc. On the other hand, the isotropic magnetization for H' (90°) is thought to be attributed to a cross-induced composite magnetic anisotropy caused by the 90° layer-by-layer field sputtering. Therefore, the magnetization by spin rotation becomes dominant and hence the hysteresis can be reduced. However, the directions of the uniaxial anisotropies induced in magnetic layers seem to be not completely crossed each other as shown in Fig.1 for H'(90), because the antiferromagnetic exchange coupling energy (Ex) estimated from

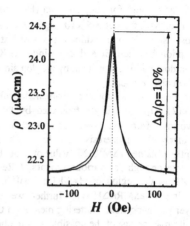

Fig.6. MR vs. H curve of a [Ni63Fe12Co25(21 Å)/Cu(21 Å)]20 multilayer fabricated by the 90° layer-by-layer field sputtering.

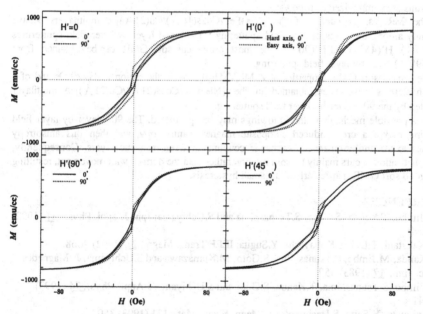

Fig.7. M vs. H curves for [Ni63Fe12Co25(16 Å)/Cu(21 Å)]20 multilayers fabricated by four different field sputterings. (See text for the notation of the samples).

the saturation field for magnetization (Hs) shown in Fig.7 is about -4 ×10^4 erg/cc being much larger than Ku of about 3.5 ×10^3 erg/cc estimated above. Nevertheless, the H'(90°) multilayer is most isotropic and exhibits the smallest hysteresis. This fact suggests that somewhat competition between Ex and cross-induced Ku can cause the magnetization to occur by spin rotation dominantly rather than by the domain wall movement. This speculation, however, is not based on clear evidence.

From the above mentioned experimental results, it may be concluded that the [Ni$_{63}$Fe$_{12}$Co$_{25}$(21 Å)/Cu(21 Å)]$_{20}$ multilayer fabricated by the 90° layer-by-layer magnetic field sputtering is noteworthy from the point of view of the magnetically soft GMR characteristics. Particularly, the hysteresis can be significantly reduced by this method. However, the presently obtained △MR/△H value of about 0.35% in the region of H=20 Oe (see Fig.6) is still not large enough for various sensors. (Recently, Kondo et al. [6] have reported △MR/△ H=0.5% in the region of H=30 Oe in [Ni$_{24}$Fe$_{31}$Co$_{45}$(15 Å)/Cu(23 Å)]$_{30}$ multilayer). For improving the △MR/△H value further, we need to reduce △H value. Application of the layer-by-layer magnetic field sputtering method to GMR multilayers with a weak interlayer exchange coupling may be one of the possible way in obtaining further advanced soft GMR multilayers.

SUMMARY

We have investigated on the improvement of soft GMR in NiFeCo/Cu multilayers by using a new magnetic field sputtering method, namely, layer-by-layer magnetic field sputter-deposition. The results are summarized as follows:

1) The MR vs. H curves of the [Ni$_{63}$Fe$_{12}$Co$_{25}$(16 Å)/Cu(21 Å)]$_{20}$ multilayers behave differently among the cases of zero field sputtering (H'=0) and layer-by-layer field sputterings of H'(0°), H'(45°) and H'(90°). The most prominent soft GMR can be obtained from the H'(90°) layer-by-layer field sputtering.

2) The typical soft GMR properties of △MR/△H=0.35% in the region of H=20 Oe and of a small hysteresis have been obtained in the [Ni$_{63}$Fe$_{12}$Co$_{25}$(21 Å)/Cu(21 Å)]$_{20}$ multilayer fabricated by the 90° layer-by-layer field sputtering.

3) As a possible mechanism, the followings may be speculated. The 90° layer-by-layer field sputtering makes a cross-induced composite magnetic anisotropy, and then this anisotropy competes to the antiferromagnetic exchange coupling between magnetic layers. Consequently, the magnetization occurs mainly by spin rotation rather than the domain wall movement, resulting in the good soft GMR, particularly in the small hysteresis.

REFERFENCES

[1] M.Jimbo, T.Kanda, S.Goto, S.Tsunashima and S.Uchiyama: Jpn. J. appl. Phys., 31 (1992) L1348.

[2] R.Nakatani, T.Dei, T.Kobayashi, Y.Sugita: IEEE Trans. Magn., 28 (1992) 2668.

[3] T.Kanda, M.Jimbo, S.Tsunashima, S.Goto, M.Kumazawa and S.Uchiyama; J. Magnetics Soc. Jpn., 17 (1983) 359.

[4] M.Jimbo, S.Tsunashima, T.Kanda, S.Goto and S.Uchiyama: J.Appl. Phys., 74 (1993) 3341.

[5] K.Inomata, Y.Saito, S.Hashimoto: J. Magn. Magn. Mat., 121 (1993) 350.

[6] J.Kondo, S.Kumagai, H.Kubota and T.Miyazaki: J. Magn. Soc. Jpn., 19 (1995) 385.

(100) EPITAXIAL AND (111) POLYCRYSTALLINE SPIN VALVE HETEROSTRUCTURES ON SI (100) : MAGNETOTRANSPORT AND THE IMPORTANCE OF INTERFACE MIXING IN ION BEAM SPUTTERING

Hyun S. Joo, Imran Hashim,[†] and Harry A. Atwater
Thomas J. Watson Laboratory of Applied Physics
California Institute of Technology, Pasadena, CA 91125

ABSTRACT

We have investigated magnetoresistance properties of (100) epitaxial, (111) textured and polycrystalline spin valve heterostructures of the form $Ni_{80}Fe_{20}/Cu/Ni_{80}Fe_{20}/Fe_{50}Mn_{50}$ on (100) Si substrates by ultra high vacuum (UHV) ion beam sputtering at room temperature. Magnetoresistance was measured as a function of Cu interlayer thickness (t_i) with 10 Å $\leq t_i \leq$ 100 Å and the maximum was found at 20 Å in the case of (100) epitaxial spin valves. Highly (111) textured spin valves with heterostructure configurations similar to the (100) spin valves were found to have a slightly lower magnetoresistance than the (100) heterostructures, but the functional dependence of the magnetoresistance on t_i was very similar.

Interface mixing during the sputtering process by energetic neutral bombardment was found to significantly affect the magnetoresistance. Samples were made under various sputtering conditions (gas pressure, ion beam energy, target and substrate configuration) that could enhance or suppress high energy neutral bombardment of the growing film surface. Samples made under the conditions that suppressed neutral bombardment showed higher magnetoresistance and more abrupt interfaces as confirmed by small angle X-ray diffraction analysis of interface mixing by energetic neutral bombardment during sputter deposition.

INTRODUCTION

Many investigators have reported magnetotransport properties related to crystalline texture and interface roughness in sputtered or molecular beam epitaxy (MBE) grown magnetic multilayers such as Fe/Cr, Co/Cr, Co/Cu and Fe/Cu.[1-3] In investigations of spin valve heterostructures, the effects of (111) and random polycrystalline film texture[4] and interdiffusion induced by post-deposition annealing have been reported.[5] The dependence of magnetoresistance on interface mixing and orientation of multilayers is not clearly understood, but previous work indicated that magnetoresistance is sensitive to the method of film deposition. For example, MBE-grown Co/Cu epitaxial films showed lower magnetoresistance than sputtered polycrystalline samples, even though they had been expected to have unmixed sharp interfaces and well controlled crystalline orientations.[1]

In order to understand the effects of interface mixing and interface orientation in NiFe/Cu/NiFe/FeMn spin valves, we have grown (100) epitaxial, (111) highly textured and polycrystalline spin valve structures by UHV ion beam sputtering with different seed layers, and measured their magnetotransport properties as a function of Cu interlayer thickness. In

---

[†] present address : Applied Materials, Santa Clara, CA 95054

addition, energetic neutral bombardment was found to significantly affect the extent of interface mixing during sputtering.

EXPERIMENTAL DETAILS

Samples were prepared at room temperature by UHV ion beam sputtering at a base pressure lower than 2×10^{-9} Torr. Samples discussed here have the structures of 150 Å $Fe_{50}Mn_{50}/50$ Å $Ni_{80}Fe_{20}/ t_i$ Å Cu/50 Å $Ni_{80}Fe_{20}$/seed layer|Si(100). 50 Å (100) epitaxial Cu layers and 1000 Å SiO_2/50 Å Cu, 1000 Å SiO_2 layers were used for (100) epitaxial and polycrystalline samples respectively. For (111) textured samples, seed layers were not used and multilayers were grown directly on (100) Si wafers. An in-plane 450 Oe magnetic field was applied during growth by a permanent magnet. The deposition rate was 0.3-0.4 Å/sec, as calibrated by Rutherford backscattering (RBS). Epitaxial (100) samples have been successfully grown on 50 Å (100) epitaxial Cu seed layers.[6] Epitaxial (100) Cu seed layers were grown on hydrogen terminated Si (100) wafers.[7] Si wafers were cleaned with a solution of $NH_4OH:H_2O_2:H_2O$ = 1:1:3 and dipped into 50:1 HF solution for 1 minute for hydrogen termination. (111) highly textured samples were be grown directly on hydrogen terminated (100) Si wafers. Polycrystalline samples were grown on 1000 Å SiO_2 and 50 Å Cu/1000 Å SiO_2.

The crystalline orientation of each layer was characterized by *in situ* Reflection High Energy Electron Diffraction (RHEED). X-ray diffraction was also used to confirm the textures of samples in large sample areas.

During sputtering, the Ar gas pressure, the ion beam voltage and the ion source-target-substrate angle were varied to enhance or to suppress interface mixing by energetic neutral bombardment. Multilayers of [20 Å Cu/20 Å Pd]$_{15}$ deposited under identical sputtering conditions and seed layers to those of spin valve heterostructures were used for small angle X-ray diffraction to survey the relative roughness of spin valve heterostructures. Cu/Pd multilayers were chosen because Pd has a better X-ray contrast relative to Cu than does NiFe and Pd (100) can be grown epitaxially on Cu (100).[8]

Resistance measurements were performed using a four point probe DC method. All resistance data in this article was measured at room temperature. A DC magnetic field was applied horizontally to the samples and parallel to the current, and varied from 100 to -100 Oe by means of a Helmholtz coil.

RESULT AND DISCUSSION

Magnetoresistance (MR) was measured as a function of Cu interlayer thickness, t_i, with 10 Å $\leq t_i \leq$ 100 Å and the maximum was found at t_i = 20 Å for both (100) epitaxial and (111) textured films. For t_i < 20 Å, the MR decreased very quickly with thickness (Fig. 1). At t_i = 10 Å, the spin valve effect vanished, possibly because of the ferromagnetic coupling of the free and pinned NiFe layers. Figure 2 shows the interlayer interaction field H_i (defined as H at $\rho_i(H) = 0.5*(\rho_{max}-\rho_{min})$ as a function of the Cu interlayer thickness t_i. H_i drops very quickly below t_i = 20 Å similar to MR. Above t_i = 50 Å, H_i approches 0 Oe and does not exceed 0 Oe up to t_i = 100 Å.

The MR of highly (111) textured films with layer structures identical to (100) films was found to be slightly lower than the (100) films, but exhibited a very similar functional dependence on t_i. The sample grown directly on SiO_2 showed very small MR ($\approx 0.1\%$) and spin valves grown on randomly oriented polycrystalline Cu seed layer structures identical to (100) epitaxial heterostructures did not show spin valve effects.

These results suggest that the crystallographic texture itself does not dominate spin valve effects in the NiFe/Cu/NiFe/FeMn heterostructures, although it can affect exchange bias between the FeMn and the top NiFe film.

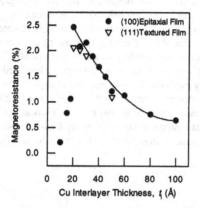

Fig. 1. Magnetoresistance as a function of Cu interlayer thickness, deposited at an ion beam voltage of 500V and 8×10^{-4} Torr Ar pressure for the ion source-target-substrate angle that inhibits substrate irradiation by high energy neutral species.

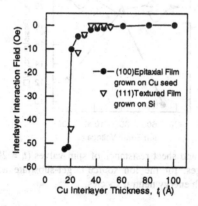

Fig. 2. Interlayer interaction fields as a function of Cu interlayer thickness, deposited at an ion beam voltage of 500V, 8×10^{-4} Torr Ar pressure for the ion source-target-substrate angle that inhibits substrate irradiation by high energy neutral species.

In ion beam sputtering, energetic neutral bombardment of the growing film surface can be controlled by varying the ion beam voltage, Ar gas pressure and the ion source-target-sample angle, while this cannot be done easily with DC or RF sputtering ,or MBE. In spin valve magnetic multilayers, we found that interface mixing by energetic neutral bombardment of growing heterostructures during ion beam sputtering significantly affected the MR. Samples deposited with higher ion energies (higher ion beam voltage and lower argon gas pressure) had high resistance and low MR. Figure 3 illustrates the variation of the MR and the resistance of samples deposited at various ion beam voltages. High energy neutral irradiation of growing film surfaces enhanced interface atomic mixing. The films grown with high energy neutral irradiation had high resistance and low MR due to interface mixing.

To minimize interface mixing by high energy bombardment, we located substrates off-axis with respect to the high energy neutral flux from the sputtering target. With this configuration, the MR was dramatically increased. A doubling of the maximum MR from 1.2 % to 2.45 % was observed when the target was oriented so as to suppress bombardment of the growing heterostructure by high energy neutrals. To measure the change in interface mixing, small angle X-ray diffraction experiments were carried out. We used [20 Å Cu/20 Å Pd]$_{15}$ multilayer samples deposited under identical sputtering conditions as the NiFe/Cu/NiFe/FeMn heterostructures. Diffraction data indicated higher intensity first-order satellite peaks from the samples grown at the off-axis position than for those grown under conditions which enhanced irradiation (Fig. 4).

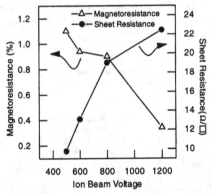

Fig. 3. Magnetoresistance and sheet resistance of spin valves (t_i = 20 Å) deposited at different ion beam voltages, for the ion source-target-substrate angle that permits substrate irradiation by high energy neutral species.

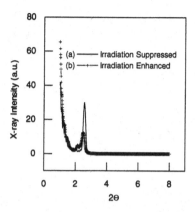

Fig. 4. Small angle X-ray diffraction data of [20 Å Cu/20 Å Pd]₁₅ multilayers deposited at an ion beam voltage of 500V, 8×10⁻⁴ Torr Ar gas pressure for the substrate position that (a) suppressed irradiation and (b) enhanced irradiation.

CONCLUSION

We found that interface mixing by energetic neutral bombardment during deposition significantly affected magnetotransport properties of NiFe/Cu/NiFe/FeMn heterostructures whereas the interface crystallographic orientation alone did not significantly affect spin valve heterostructures. Spin valves with different texture showed similar functional trends with Cu interlayer thickness. Polycrystalline heterostructures without strong preferred orientations did not show spin valve behavior, possibly due to rough interfaces or small exchange bias fields.

REFERENCES

1. W. F. Egelhoff and M. T. Kief, IEEE Trans. Magn., **28**, 2742 (1992).
2. R. Q. Hood, L. M. Falicov, and D. R. Penn, Phys. Rev. B, **49**, 368 (1994).
3. K. I. Min. and S. K. Joo, J. Appl. Phys., **75**, 4632 (1994).
4. R. Nakatani, K. Hoshino, and Y. Sugita, Jpn. J. Appl. Phys., **33**, 133 (1994).
5. V. S. Speriosu, J. P. Nozieres, B. A. Gurney, B. Dieny, T. C. Huang, and H. Lefakis, Phys. Rev. B, **47**, 11579 (1993).
6. I. Hashim, B. Park, and H. A. Atwater, Appl. Phys. Lett., **63**, 2833 (1993).
7. I. Hashim and H. A. Atwater, J. Appl. Phys., **75**, 6516 (1994).
8. C. A. Chang, J. Vac. Sci. Tech., **9**, 98 (1991).

GIANT MAGNETORESISTANCE IN HYBRID MAGNETIC NANOSTRUCTURES INCLUDING BOTH LAYERS AND CLUSTERS.

P.A.SCHROEDER, P. HOLODY, and R. LOLOEE.
Department of Physics and Astronomy, and Center for Fundamental Materials Research.
Michigan State University, East Lansing, MI 48824-1116.
J. L. DUVAIL, A. BARTHÉLEMY, L.B.STEREN, R.MOREL AND A.FERT.
Laboratoire de Physique des Solides, Université Paris Sud, 91405 Orsay, France.

ABSTRACT

Early experiments to define oscillations in the CIP magnetoresistance (CIP-MR) of Ag/Co analogous to those for $t_{Cu} < 5nm$ in Cu/Co were unsuccessful. The MR in this region was very small. Later experiments by Araki using thin (0.6nm) Co layers produced much larger MRs and lead us to look at the MRs of similar samples more closely. We conclude that the large MR of such samples is associated with the discontinuous nature of the Co layers. The object of the present paper is to combine the high MR associated with the thin Co layers with the field dependence governed by the magnetization reversal in thick, and magnetically soft permalloy (Py) layers. We have measured the CIP-MR of sputtered samples of the $[Co(0.4nm)/Ag(t_{Ag})/Py(t_{Py})/Ag(t_{Ag})]x15$ system with t_{Ag} ranging from 1.05 to 4nm and with $t_{Py} = 2$ or 4nm. We obtain MRs at 4.2K as large as 35% in less than 10Oe with slopes as high as 5%/Oe. With CPP measurements slopes as high as 10%/Oe have been obtained. Squid magnetometer measurements indicate that, as the temperature increases, there is a crossover to superparamagnetic behaviour and a resulting gross deterioration of the MR slopes at room temperature. Efforts to increase the room temperature sensitivity are described. Detailed measurements of the CPP-MR of the $[Co(0.4nm)/Ag(4nm)/Py(t_{Py})/Ag(4nm)]x20$ series of multilayers are consistent with a two spin band model modified to take account of the granular nature of the Co.

INTRODUCTION

To define the terms H_o, H_p and H_{sat} we show in Fig. 1 a typical plot of magnetoresistance (MR) versus magnetic field for a Ag(1.8nm)/Co(0.4nm) sample. We define MR(H) as

$$MR(H) = \{R(H) - R(H_{sat})\}/R(H_{sat}) \qquad (1)$$

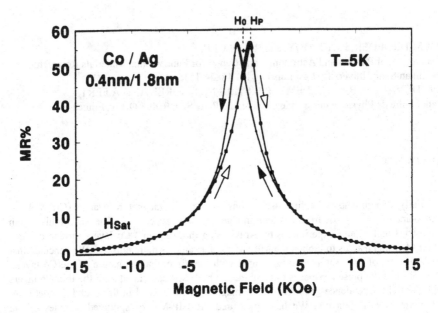

Figure 1. Magnetoresistance plotted against magnetic field for a [Ag(1.8nm)/Co(0.4nm)]x60 sample. The + represents the MR(H_0) point. H_p and H_{sat} are defined by this figure.

MR(H_0) and MR(H_p) will be the MRs for H = H_0 and H = H_p. In this paper we are primarily concerned with standard CIP-MR measurements in which the measuring current is in the plane of the layers. However we will look at some CPP-MR measurements with the measuring current perpendicular to the planes.

Early attempts to see the oscillations of MR as a function of t_{Ag}, the thickness of the Ag layers, analogous to those seen in the Cu/Co system[1,2] failed. Most of our work on the Ag/Co[3] system was then pursued on samples with t_{Ag} > 5nm which showed a substantial MR. More recently, Araki et al.[4] reported a large and oscillating MR in *evaporated* Ag/Co multilayers with very thin Co layers, t_{Co} = 0.6nm. This lead us to look more closely at the Ag/Co system in the region of very thin Co thicknesses for *sputtered* systems[5]. The properties of this system lead us to conjecture whether a hybrid (taken to mean a multilayer with more than one magnetic component) system making use of the high MR and coercive field of the Ag/thin Co system and the low coercive field of the permalloy might lead to a very sensitive sensor for magnetic fields. In this paper we give the results of our research on both the Ag/thin Co system, the hybrid Ag/thin Co/Ag/Py system[6], (Py ~ permalloy, in our case $Ni_{84}Fe_{16}$), and some similar multilayers with Cu substituting for Ag.

In section 2 we give some of the experimental details followed by our results for the Ag/thin Co system, and the hybrid system in sections 3 and 4 respectively. In section 5 we briefly mention some of our work on CPP-MR.

2. EXPERIMENTAL

Sample Preparation and Measurement

Our samples were sputtered onto Si(100) substrates at ~0°C, using argon sputtering pressure of 2.5mTorr and deposition rates ~ 0.8nm/s. Further details are given elsewhere[7]. CIP-MR measurements on the Ag/Co samples were made at MSU, at a variety of temperatures, using conventional 4-terminal methods, on samples mounted on a probe that was inserted into the magnet of a Quantum Designs MPMS magnetometer. The direction of the field was always parallel to the current. The M-H measurements were made on the same instrument. The CIP-MRs of the hybrids were mostly measured at Orsay, but all the CPP-MR and some of the CIP-MR measurements were made at MSU. The design of the CPP-MR samples and a description of CPP-MR is given elsewhere[7].

Nuclear magnetic resonance studies on similar samples made at MSU have been performed by van Alphen et al.[8] in Eindhoven.

3. RESULTS. Ag/THIN Co

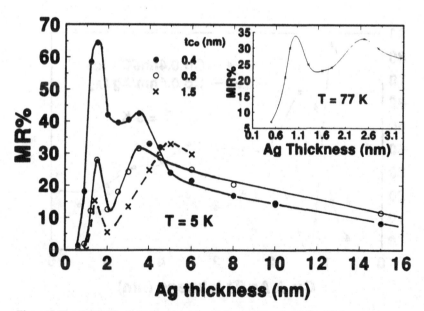

Figure 2. The CIP-MR plotted against Ag thickness for several Co thicknesses.
Inset: Results of Araki et al[4].

In Fig. 2, we show the MR% plotted as a function of t_{Ag} for t_{Co} = 0.4, 0.6, and 1.5nm. Araki's results for evaporated samples at 77K are shown in the inset. In both cases two MR peaks are observed, but the disagreement in position indicates that the results are strongly dependent on the mode of preparation. The salient points are a strong peak at t_{Ag}~ 1.4nm, followed by a minimum and a slow rise to a maximum between t_{Ag} =3.5 and 4.5 nm depending on the Co thickness. Only $MR(H_p)$ values are plotted in Fig. 2. For all our measurements with t_{Ag} > 6nm, $MR(H_0) > MR(H_p)$. The reverse is true for t_{Ag} < 5nm.

Between 1 and 5nm the MR for $Ag(t_{Ag})/Co(0.4nm)$ is substantially larger than that for $Cu(t_{Cu})/Co(0.4nm)$ (Fig.3). For these two systems, the graph indicates that there is a large difference between the periods of oscillation with non-magnetic (N) metal thickness, which would not be expected if the oscillations rose entirely from the oscillatory exchange interaction between the Co layers[9]. In a previous publication[5] we conclude that in the Co/Ag system there are two possible processes at work. First, since discontinuities occur in the Co, it is possible that the system takes on certain aspects of a granular system. Second, in samples with thicker Co layers, substantial ferromagnetic bridging occurs in the Ag/Co system for t_{Ag} < 5nm. Discontinuities in the Co layers localize the lateral spread of ferromagnetic bridging, so that the contribution from layering increases in the thin Co/Ag system. It is difficult to rigorously separate these two mechanisms. To support the granular picture we have recently measured both the CIP and CPP-MR of the Co(0.4nm)/Ag(1.5nm) sample, corresponding to the first peak, and find these to be approximately the same. For a granular effect we expect the MR to be isotropic, i.e. the CIP-MR and CPP-MR should be the same[10,11], whereas they should be significantly different if they are produced by the layer mechanism.

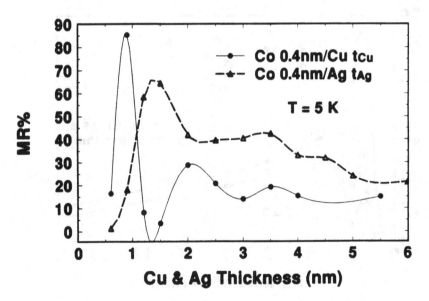

Fig. 3. Comparison of the CIP-MR for Cu/Co and Ag/Co samples with the same t_{Co}.

On the basis of the discontinuous layer picture, one can argue that the first peak arises from the exchange interaction associated with layering. It occurs at a larger nominal t_{Ag} than would be anticipated, because, in a discontinuous layer, t_{Co} may locally be appreciably greater than the nominal value, which implies that the local distance between layers may be considerably smaller than the nominal value.

The system we are studying is thus very similar to the Py/Ag system (Py=permalloy) investigated by Hylton et al.[12], but whereas they obtained discontinuous layers by annealing we produce them by using thin layers of Co.

A concommitant property is that the coercive force becomes very large as indicated in Fig4(a). The high field (~2500 Oe) needed to saturate the magnetization of the Co clusters reflects the slow alignment associated with these discontinuous layers. This is one of the essential properties we use when we come to consider the hybrid structures. Fig. 4(b) indicates this coercive force has disappeared at room temperature.

Plots of magnetization versus temperature, like those shown in Fig.5 for a Cu/Co sample, indicate the onset of superparamagnetism at blocking temperatures (T_b)~100K or less for the Ag/Co system, and these temperatures correspond with the disappearance of the high coercive fields. Such behaviour is typical of materials with magnetic granules. From the frequency dependence of the blocking temperature, an estimate of the size of the granules can be made[5]. For a Ag(3.5nm)/Co(0.2nm) sample we estimate that the radius of a spherical granule is ~ 2nm. Van Alphen et al.[8] have performed NMR measurements on samples prepared in our laboratory. They found a volume contribution to the NMR in samples with t_{Co} <1nm for which no volume contribution would be expected. Their conclusion was that the growth of

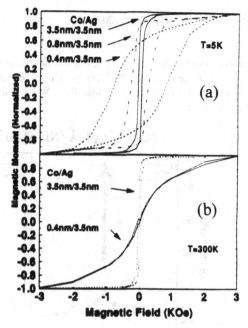

Figure 4. Hysteresis curves for several Ag/Co samples

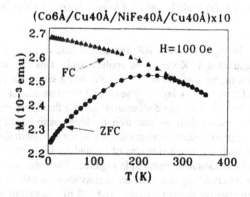

Figure 5. Blocking temperature curve for a hybrid multilayer

sputtered Ag/Co multilayers starts with a three dimensional island growth of Co. For $t_{Co} >$ ~1nm continuous Co layers are formed. For $t_{Co} <$ 1nm the layers are discontinuous and the large MR is associated with this discontinuous nature.

4. RESULTS. HYBRID SYSTEMS.

In these systems we seek to combine the advantages of discontinuous Co layers with a large coercive force with thick Py layers with high permeability and low coercivity. Data for the CIP-MR have already been published[6,13]. We will therefore concentrate on more recent results on CPP-MR.

(Co4Å/Ag40Å/NiFe100Å/Ag40Å)x20

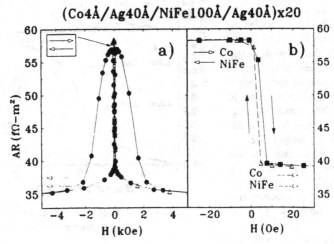

Figure 6. (a) Variation of AR (A=sample area) with H. (b) is the central part of (a) magnified.

In Fig.6 we show AR, the cross-sectional area of the sample times its CPP resistance measured at 4.2K versus the applied field. This can be understood in terms of Fig. 7 which gives the magnetization of a sample, and a schematic representation of the magnetization of the Co and the Py layers. The Co and Py plots indicate the large difference in coercive fields for these layers. The dark arrows indicate the directions of the magnetizations in the Co and Py layers. Initially we start from the high positive field situation. Decreasing the field results in a sharp decrease in magnetization of the sample in the vicinity of the Py coercive field. At that stage the magnetizations of Co and Py layers are antiparallel which is the condition for the high resistance observed in Fig.6(a). Thereafter the magnetization of the Co layers slowly reverses resulting in a decrease in resistance until the magnetizations become parallel again at high negative magnetic fields. The MR of this sample is ~62%, but more importantly almost all the change in resistance occurs in a field of less than 10Oe, as shown in Fig. 6(b). This

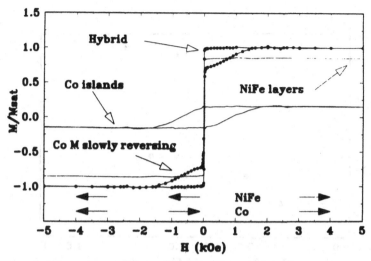

Figure 7. Schematic representation of the magnetizations associated with a hybrid sample. Darkened arrows represent the directions of the Py (NiFe) and Co layers.

corresponds to a sensitivity of ~10MR%/Oe. One reason for the large effects at low fields is that the volume of Co is so small that the dipolar field, which it produces, has very little effect on the coercive properties of the Py. Measurements of the coupling field have been made through the study of minor hysteresis loops[13]. Coupling fields as low as 4 Oe have been measured for a sample with $t_{Ag} = t_{Py} = 4nm$, and $t_{Co} = 0.4nm$. The Py is very effectively decoupled from the hard Co layers, and consequently retains its low coercive force.

The properties of these hybrid systems are well suited for application as a read head for extracting information from magnetic media. Unfortunately, the sensitivity decreases dramatically in going to room temperature. The reason is linked to the reduction of the Co coercive field as the temperature increases from 4.2K (Fig.4(a)) to 300K (Fig. 4(b)).

The critical quantity as far as the size of the blocking temperature is concerned is the product kV where k is the anisotropy energy per unit volume and V is the volume of the granule. To obtain higher blocking temperatures and hence extend the temperature range of high sensitivity, one might try to increase the volume of the granules or increase the anisotropy. We have performed preliminary experiments in both these directions.

The blocking temperature (T_b) graph in Fig.5 is for a sample with Cu rather than Ag as the N metal. This graph shows that granules also exist in the Cu/Co system. Generally speaking higher T_b's are observed for Cu/Co, which implies that larger magnetic granules exist in this system. In Fig.8 we show the variation of T_b as a function of t_{Co} for the Ag/Co and Cu/Co systems. Whereas T_b saturates at ~100K for the Ag based hybrids, for the Cu based hybrids, it continues to rise as t_{Co} increases until T_b becomes close to 300K for t_{Co} ~ 1.6nm. In Fig. 9 we show the sensitivity S, defined with units MR%/Oersted, and MR% as a function of t_{Co} for both hybrids of the form $[Co(t_{Co})/N(4nm)/Py(4nm)/N(4nm)]x10$ at 4.2 and 300K respectively. At both temperatures, S for the Ag hybrid peaks at t_{Co}~0.8nm, but it is an order of magnitude smaller at 300K. At 300K the MR for the Cu based hybrid is

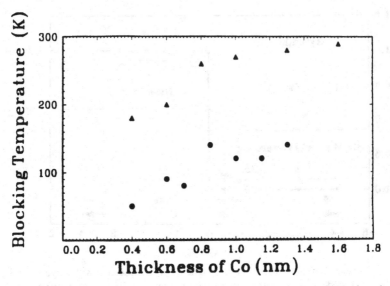

Figure 8. The blocking temperature of [Co(t_{Co})N(4nm)/Py(4nm)/N(4nm)]x10 plotted against t_{Co}. triangles N = Cu; circles N = Ag

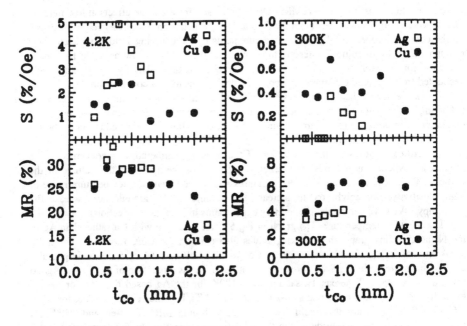

Figure 9. The sensitivity S measured in MR%/Oersted and the MR for the series [Co(t_{Co})/N(4nm)/Py(4nm)/N(4nm)]x10. N=Ag or Cu.

considerably larger than that for the Ag hybrid, but S is only nominally larger, despite the larger value of T_b.

We remark here that the large difference in coercive fields of the thin Co and the thicker Py layers makes this system an excellent one for studying the dependence of MR on the angle between magnetizations of the two magnetic layers. This has been treated elsewhere[13].

5.TWO SPIN CHANNEL MODEL

The theory of the CPP magnetoresistance has been well established[14] for multilayers in which there is only one magnetic metal, and for which the spin diffusion length is much larger than the bilayer thickness. It has been used in the analysis of CPP measurements on several

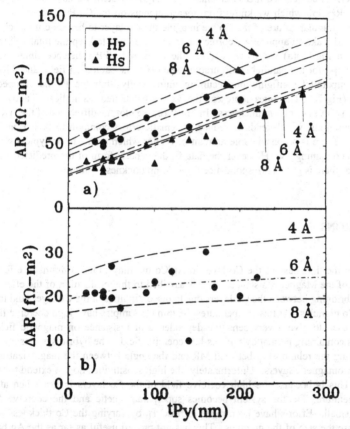

Figure 10. (a) AR=sample area x resistance plotted against t_{Co} for the series $[(Co(0.4nm)/Ag(4nm)/ Py(t_{Py})/Ag(4nm)]x20$. The lines are the calculated values assuming granule thicknesses indicated. (b) A similar plot of $A\Delta R = AR(H_p)-AR(H_s)$ vs t_{Co}.

systems[15] to determine, the important parameters $\alpha_F = \rho_F\downarrow/\rho_F\uparrow$ and the analogous $\alpha_{F/N} = R_{F/N}\downarrow/R_{F/N}\uparrow$ which describe the spin dependent scattering in the bulk F-resistivity and in the F/N interface resistance respectively.

For a hybrid system it is much more difficult to extract these parameters, and it is more satisfactory to calculate the resistance for a model system and compare it with the experimental results. Besides earlier CPP-MR measurements we have performed on the [Co/Cu/Py/Cu] system with thicker Co layers[16] , we have also studied the [Co(0.4nm)/Ag(4nm)/Py(t_{Py})/Ag(4nm)]x20 series (Fig.10).

In the model we adopt, we assume that the long spin diffusion length limit is appropriate as has been demonstrated for the Ag/Co system[17]. We consider two channels of spin up and spin down electrons. We add the resistances of N and F metals, and of the N/F interfaces in series for each channel. We then add the resistances of the two channels in parallel. We have assumed the α_{Co} value derived from measurements on the Ag/Co and Cu/Co systems. Since we have not performed detailed measurements on the Ag/Py system we have had to assume that $R_{Ag/Py} \approx R_{Cu/Py}$ which we know from measurements on the Cu/Py system. To take into account the granular nature of the Co we imagine the Co layer to be made up of clumps of thickness f and make an appropriate correction to the layer area to keep the total volume of Co constant. In Fig. 10(a) we show $AR(H_s)$, the measured CPP resistance times A, the sample area, vs t_{Py} for the parallel configuration when H > the saturation field, and $AR(H_p)$ for the true antiparallel configuration occurring immediately after the Py has flipped its magnetization (Fig.7). The lines represent the values calculated, as indicated above, for several assumed values of clump thickness. In Fig.10(b) we plot the difference of the ARs for the two configurations, which tends to reduce systematic errors common to both F and AF configurations. There is a considerable amount of scatter which seems to be typical of data for multilayers containing Py. However the data fit the general form of the predicted values and in particular there is good correspondence for a clump thickness ~ 6nm.

6. CONCLUSIONS

Decreasing the thickness of the Co layer in Ag/Co multilayers has a dramatic effect on the magnitude of the Magnetoresistance. This is ascribed to the localization of the effects of ferromagnetic bridging between the Co layers due to discontinuities in the Co layer, and to the formation of Co granules. At low temperatures the thin Co samples have high coercive fields which can be used to give a very sensitive dependence of resistance on magnetic field in hybrid systems containing permalloy with its low coercive field. The hybrid systems are also ideal for studying the relationship between MR and the angle between the magnetization of consecutive ferromagnetic layers. Unfortunately, the high sensitivities do not extend to room temperature. This is because the high coercive field of the Co layers is lost when at the blocking temperature, T_b, the system becomes superparamagnetic and the coercive field becomes very small. Efforts have been made to raise T_b by varying the Co thickness in an effort to increase the size of the granules. This has not proved useful as far as the Ag based hybrids are concerned, but T_b for the Cu based hybrid has been increased to ~ 300K. However the best sensitivity at 300K is still limited to ~ 0.5MR%/Oersted.

A series of CPP measurements is consistent with a model in which the Co layers nominally 0.4nm thick, consist of Co clumps ~0.6nm thick.

ACKNOWLEDGEMENTS

This research was supported in part by the US-NSF under Grant No. DMR-22614 and by the US-France cooperation program under grant No. INT-92-1609 for the US-NSF and Companion Grant No. AI 0693 from the French CNRS. It was also supported by the Esprit Basic Research Project No.6146 SMMS, and by the MSU Center for Fundamental Material Research and by the Ford Scientific Laboratory. The authors express their thanks to Professors J. Bass and W. P. Pratt Jr. for useful conversations and to Professor I. Schuller for the original suggestion to study the Ag/Co system.

REFERENCES

1. S. S. Parkin, R. Bhadra, and K. P. Roche. Phys. Rev. Lett. **66**, 2152 (1991).
2. D. H. Mosca, F. Petroff, A. Fert, P. A. Schroeder, W. P. Pratt Jr. and R. Loloee. J. Magn. Magn. Mat. **94**, 1 (1991).
3. S. F. Lee, W. P. Pratt Jr., R. Loloee, P. A. Schroeder, and J. Bass. Phys. Rev. **B46**, 548 (1992): W. P. Pratt Jr., S. F. Lee, Q. Yang, P. Holody, R. Loloee, P. A. Schroeder, and J. Bass. J. App. Phys.**73**, 5326 (1993).
4. S. Araki, K. Yasui and Y.Narumiya. J. Phys. Soc. Japan.**60**, 2827 (1991); S. Araki, J. Appl. Phys. **73**, 3910 (1993).
5. R. Loloee, P. A. Schroeder, W. P. Pratt Jr., J. Bass and A. Fert. Physica **B204**, 274 (1995).
6. P. Holody, L. B. Steren, R. Morel, A. Fert, R. Loloee and P. A. Schroeder. Phys. Rev. **B50**, 12999 (1994).
7. J. M. Slaughter, W. P. Pratt Jr. and P. A. Schroeder. Rev. Sci. Instr. **69**, 127 (1989).
8. E.A.M.van Alphen, P.A. A van Heijden and W. J. M. de Jonge. J. Appl. Phys. **76**, 6607 (1994).
9. R. Coehoorn, Phys. Rev. **B44**, 9331 (1993).
10. C. L. Chien, J. Q. Xiao, and J. S. Jiang. J. App. Phys. **73**,5309 (1993).
11. S. Zhang and P. M. Levy. J. App. Phys.**73**, 5315 (1993).
12. T. L. Hylton, K. R. Coffey, M. A. Parker, and J. K. Howard. Science **261**, 1021 (1993).
13. L. B. Steren, J. L. Duvail A. Fert R. Morel, F. Petroff,, P. Holody,R. Loloee and P. A. Schroeder. Phys. Rev. **B51**, 292 (1995).
14. S. Zhang and P. M. Levy, J. App. Phys.**69**, 4786 (1991): T. Valet and A. Fert. Phys. Rev **B48**, 7099 (1993).
15. P. A. Schroeder, J. Bass,.P. Holody, S. F. Lee,R. Loloee, W. P. Pratt Jr., and Q. Yang. NATO ASI series **B309**, 129 (1993).
16. Q. Yang, P. Holody, R. Loloee, L. L. Henry, W. P. Pratt Jr., P. A. Schroeder and J. Bass. Phys. Rev. **51**,3226 (1995).
17. C. Yang, P. Holody, S.-F. Lee, L. L. Henry, R. Loloee, P. A. Schroeder, W. P. Pratt Jr., and J. Bass. Phys. Rev. Lett. **72**, 3274 (1994).

COLOSSAL MAGNETORESISTANCE IN THICK La$_{0.7}$Ca$_{0.3}$MnO$_3$ FILMS

RANDOLPH E. TREECE,[†] P. DORSEY, M. RUBINSTEIN, J. M. BYERS, J. S. HORWITZ, E. DONOVAN, AND D. B. CHRISEY
Naval Research Laboratory, Washington, DC 20375.

ABSTRACT

Thick films (0.6 and 2.0 μm) of the colossal magnetoresistance (CMR) material, La$_{0.7}$Ca$_{0.3}$MnO$_3$ (LCMO), have been grown by pulsed laser deposition (PLD). The films were grown from single-phase LCMO targets in 100 mTorr O$_2$ pressures and the material deposited on (100) LaAlO$_3$ substrates at deposition temperatures of 800°C. The deposited films were characterized by X-ray diffraction (XRD), magnetic field-dependent resistivity, and Rutherford backscattering spectroscopy (RBS). The LCMO films were shown by XRD to adopt an orthorhombic structure. Brief post-deposition annealing led to ~50,000% and ~12,000% MR effect in the 0.6 μm and 2.0 μm films, respectively.

INTRODUCTION

The magnetoresistive (MR) properties of the substituted lanthanum manganates (La^{3+})$_x$(M^{2+})$_{1-x}$MnO$_3$, where M^{2+}=alkaline earth metal, recently have been rediscovered and greatly magnified by the deposition of these compounds in thin film form. The magnetoresistance ratio is defined as $\Delta R/R_H = (R_H - R_0)/R_H$, where R_H is the resistance in an applied magnetic field and R_0 is the resistance in zero field. Thin films (1000 Å<thickness<5000 Å) of (La,M)MnO$_3$ were coined as giantmagnetoresistive (GMR) when MR values as high as 150 % were observed in as-deposited materials.[1,2] When epitaxial thin films (thickness≈1000 Å) of these materials were briefly annealed at high temperatures, the MR values exceeded 100,000 % and led to the term colossal magnetoresistance (CMR).[3,4]

The observation of an increase in magnetoresistance ratios by several orders of magnitude in the epitaxial thin films led researchers to consider what differences in the films could have led to such large increases in MR. The ~1000 Å films that displayed ~10^2 MR were deposited as polycrystalline with a non-cubic structure type and were not annealed after deposition[1,2] while the films displaying >10^5 MR were grown as epitaxial, cubic materials and were annealed after deposition.[3,4,5] The fact that the CMR films grew as cubic, epitaxial films led investigators to propose that strain in the films could be key to observations of large magnetoresistance. In order to determine if epitaxial growth was required for >10^2 MR effect, we sought to investigate oriented, thick films. In short, we describe here that we have found >10^4 MR effect in thick, non-cubic, oriented thin films of La$_{0.7}$Ca$_{0.3}$MnO$_3$ (LCMO).

EXPERIMENTAL PROCEDURE

The LCMO films were grown by pulsed laser deposition (PLD) in a high vacuum chamber equipped with a turbomolecular and a cryopump. A pulsed KrF excimer laser beam (248 nm, 30 ns FWHM) operating at 10 Hz was focused with a 50 cm focal length lens at 45° onto a dithering and rotating (30 RPM) LCMO target prepared from the metal oxides using conventional ceramic techniques. The ambient O$_2$ input pressure was regulated to a dynamic equilibrium (~10 sccm) by a solenoid-activated leak valve controlled by a capacitance manometer with the chamber under gated or throttled pumping. The substrates were washed with ethanol, attached to the substrate

427

heater with silver paste and maintained ~6 cm from the target. The laser fluence was 1-2 J/cm^2. The film compositions and thicknesses were determined by Rutherford Backscattering Spectroscopy (RBS) with 6 MeV He^{2+}. X-Ray diffraction patterns were collected using a rotating anode source and a conventional θ–2θ geometry and indexed using a least squares fit of the data. The rocking curve widths, Γ (FWHM), for a particular reflection was measured by fixing 2θ at the peak maximum and scanning through θ. The resistivities were measured by four-point method in a superconducting magnet where the temperature (T) and applied field (H) could be controlled.

RESULTS AND DISCUSSION

Structural characterization

LCMO films were grown by PLD on (100) LaAlO$_3$ at substrate deposition temperatures of 800°C and O$_2$ pressures of 100 mTorr. The composition of the deposited films was found to be La$_{0.7}$Ca$_{0.3}$MnO$_3$ and thicknesses were determined to be 0.6 μm and 2.0 μm. The structure and magnetoresistive properties of the LCMO films were characterized before and after annealing in air. The structures of the LCMO films were determined using XRD. Representative XRD patterns of the 2.0 μm films before and after annealing are shown in Figure 1. The as-deposited films and the annealed films, with the annealing times andtemperatures, are indicated in the figure along with the Miller indices.

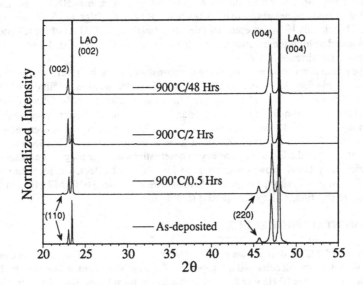

Figure 1. Representative XRD patterns of the 2.0 μm as-deposited and annealed films, with the annealing times and temperatures indicated along with the Miller indices.

The as-deposited and 0.5 hr annealed films were indexed as orthorhombic, isostructural with LaMnO$_3$.[6] The films annealed for 2 and 48 hrs were indexed as perovskite (cubic). The as-deposited films were highly oriented with crystallites growing in one of two orientations. The (220) and (004) peaks had rocking curve widths of Γ=0.15° (FWHM) and Γ=0.42° (FWHM), respectively. This is consistent with crystallites growing with either c-axis perpendicular or c-axis parallel to the substrate surface.

Annealing had two primary effects on the structures of the deposited LCMO films. The first effect is that the LCMO material underwent a structural transition from orthorhombic to cubic symmetry. This transition is evident from the disappearance of the (110) and (220) reflections. The annealed films retained a strong (00l) orientation. The rocking curve width of the (004) peak for the film annealed to 900°C for 48 hrs was Γ=0.56° (FWHM). The second effect is that the c-axis expanded from 7.784 Å in the as-deposited film to 8.113 Å in the 48 hr annealed film.

Structural phase transitions in ceramic samples of (La^{3+},M^{2+})MnO$_3$ materials have been documented previously.[7] Since LaMnO$_3$ is orthorhombic and CaMnO$_3$ is cubic, the solid solution undergoes a orthorhombic-to-cubic phase transition as Ca^{2+} is substituted for La^{3+} in La$_x$Ca$_{1-x}$MnO$_3$. The primary effect of substituting the Ca^{2+} for the La^{3+} is to oxidize Mn^{3+} to Mn^{4+} resulting in a mixed valence (Mn^{3+}/Mn^{4+}) material. However, the calcium content is not the only factor determining the oxidation state of the Mn ions. It has been shown that based on the annealing conditions, the Mn^{3+}/Mn^{4+} ratio can be manipulated tremendously. For instance, in LaMnO$_3$ all of the manganese should be in the Mn^{3+} oxidation state, but when annealed in pure O$_2$ for short times the Mn^{3+} can be oxidized to up to 35% Mn^{4+}.[7] The structural phase, orthorhombic or cubic, has been shown to depend soley on Mn^{3+}/Mn^{4+} ratio, as represented in Figure 2.[7] This figure, compiled from data tabulated in Ref. 7, plots lattice parameters versus %Mn^{4+}. The orthorhombic-to-cubic phase transition in La$_x$Ca$_{1-x}$MnO$_3$ occurs at a Mn^{4+} content ≈27%. Since the metal ratios in our samples do not change from La$_{0.7}$Ca$_{0.3}$MnO$_3$ on annealing, the likely reason our films go from orthorhombic to cubic is due to oxidation of some of the Mn^{3+} to Mn^{4+} so that the total Mn^{4+} content of the films exceeds 27%.

<u>Magnetoresistive characterization</u>

The MR properties of the LCMO films were characterized before and after annealing. A data set for a 2.0 μm film annealed at 900°C for 0.5 hr is presented in Figure 3. The MR ratio ($\Delta R/R_H$ %) is plotted against applied field (Oe) for runs at different sample temperatures. The peak MR for this 2.0 μm film was ~12,000 % at a temperature of 100 K. The sharpness of the peak in $\Delta R/R_H$ % is shown by the drop in MR by an order of magnitude with the application of just 10,000 Oe (1 Tesla). Assuming that the shape of the %MR versus field peak is symmetric about the $\Delta R/R$ axis, the full width at half maximum (FWHM) of the peak measured at 100 K would be 4,000 Oe. The figure of merit for magnetoresistive materials is %MR/Oe (FWHM). For the 2.0 μm material, this value is 3.0 % MR/Oe (=11,200 %/4,000 Oe) The maximum MR value observed for the 0.6 μm film was 50,000 % measured at 77 K. While the maximum MR value was ~4 times larger for the thinner film, the width of the MR peak also was larger. The %MR/Oe value for the 0.6 μm film was found to be 1.7.

Plots of the maximum MR ratios versus measurement temperature for LCMO films annealed under different conditions are presented in Figure 4, where the inset shows the full peak of the CMR material. The peak in the MR ratio versus temperature data for each film is indicated with an arrow. The MR ratio maxima were at 30 K, 100 K, 150 K and 200 K for the films as-deposited and annealed at 900°C for 0.5 hr, 2 hrs and 48 hrs, respectively. The movement of the peak in MR ratio maxima follows the movement of the maximum in resistance versus temperature

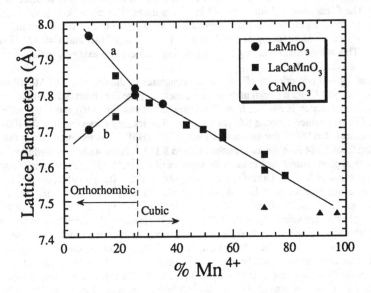

Figure 2. Plot of lattice parameters versus %Mn^{4+} in $La_xCa_{1-x}MnO_3$. The orthorhombic-to-cubic phase transition occurs at a Mn^{4+} content ≈27%.

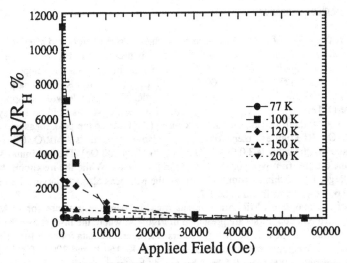

Figure 3. The MR ratio ($\Delta R/R_H$ %) is plotted against applied field (Oe) for runs at different measurement temperatures.

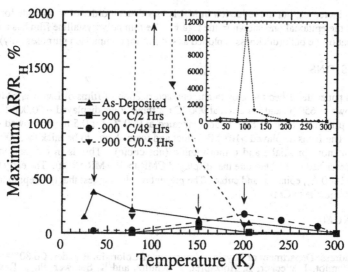

Figure 4. The maximum MR ratios versus measurement temperature for LCMO films annealed in air at 900 °C for different duration's.

(R(T)) and has been observed in other CMR systems. When MR ratios $>10^4$ % are observed, the peak in the R(T) curve occurs at temperatures well below the Curie temperature of the bulk ceramic. Likewise, when the peak in the R(T) curve is close to the Curie temperature, MR ratios are $< 10^3$ %.

The effect of thickness on MR properties is not clearly understood, but the films described here add more information. Figure 5 plots the log MR % versus sample thickness for the present materials, as well as, some films and bulk samples described in the literature. The MR decreases

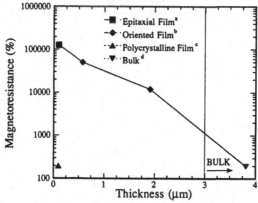

Figure 5. Plot of the log MR % versus sample thickness for the present materials (b), as well as, some films (a [Ref. 3], c [Ref. 1]) and bulk samples (d [Ref. 4]) described in the literature.

as thickness increases, but the reduction in MR is less than an order of magnitude for a film 20x thicker than the epitaxial one shown in the figure. The thin polycrystalline film has an MR value similar to that of the bulk (thickness is plotted as < 4 μm for comparison purposes only).

CONCLUSIONS

This paper described the growth and characterization of films of $La_{0.7}Ca_{0.3}MnO_3$. A 0.6 μm film grown at 650 °C and annealed at 900 °C for three hours displayed ~ 50,000% MR effect, while a 2.0 μm film grown at 800 °C and annealed at 900 °C for 0.5 hours showed ~12,000 % MR effect. The films displayed >10^4 MR effect and were shown to be thick (> 5000 Å), highly oriented (but not epitaxial), and orthorhombic (not cubic). This is in contrast to the films previously described in the literature that displayed CMR (> 10^2 MR effect). The early CMR films were thin (~1000 Å), epitaxial and cubic. The present results suggest that thin, epitaxial materials are not a prerequisite to CMR.

REFERENCES

† Present address: Department of Physics, University of Colorado, Boulder, Co 80309-0390.
1. R. von Hemlot, J. Wecker, B. Holzapfel, L. schultz, and K. Samwer, Phys. Rev. Lett. **71**, 2331 (1993).
2. K. Chahara, T. Ohno, M. Kasai, and Y. Kozono, Appl. Phys. Lett. **63**, 1990 (1993).
3. S. Jin, T. H. Tiefel, M. McCormack, R. A. Fastnacht, R. Ramesh, and L. H. Cohen, Science **264**, 413, (1994); M. McCormack, S. Jin, T. H. Tiefel, R. M. Fleming, Julia M. Phillips, and R. Ramesh, Appl. Phys. Lett. **64**, 3015 (1994).
4. S. Jin, M. McCormack, T. H. Tiefel, R. Ramesh, J. Appl. Phys. **76**, 6929 (1994).
5. G. C. Xiong, Q. Li, H. L. Ju, S. N. Mao, L. Senapati, X. X. Xi, R. L. Greene, and T. Venkatesan, Appl. Phys. Lett. **66**, 1427 (1995).
6. Powder Diffraction File (Joint Committee on Powder Diffraction Standards-International Center for Diffraction Data, Swarthmore, PA, 1986) file no. 35-1353 (LaMnO3).
7. H. L. Yakel, Acta. Cryst. **8**, 394 (1955).

GIANT MAGNETORESISTANCE BEHAVIOR IN EPITAXIAL Nd$_{0.7}$Sr$_{0.3}$MnO$_{3-\delta}$ AND La$_{0.67}$Ba$_{0.33}$MnO$_{3-\delta}$ THIN FILMS

XIONG G.C.*, LI Q., JU H.L., WU J., SENAPATI L., GREENE R.L., VENKATESAN T.
Center for Superconductivity Research, University of Maryland,
College Park, MD 20742-4111
* also Department of Physics and National Mesoscopic Physics Laboratory,
Peking University, 100871 Beijing, P.R. China.

ABSTRACT

Epitaxial Nd$_{0.7}$Sr$_{0.3}$MnO$_{3-\delta}$ and La$_{0.67}$Ba$_{0.33}$MnO$_{3-\delta}$ thin films with large magnetoresistance ratios have been prepared by pulsed laser deposition. Huge negative magnetoresistance ratios of $-\Delta R/R_H$ $> 1 \times 10^6$ % were obtained at 60 K and a magnetic field of 8 T in a Nd$_{0.7}$Sr$_{0.3}$MnO$_{3-\delta}$ film. The influence of sample preparation conditions on the resistivity behavior of these films has been studied. Results suggest that oxygen stoichiometry and diffusion are important factors in causing the behavior observed in doped manganese oxide films.

INTRODUCTION

Recently, a large negative magnetoresistance (MR) effect was reported in doped manganese oxide, (Re,B)MnO$_3$ thin films, where Re is a trivalent rare earth element such as La, Nd and Pr and B represents a divalent element such as Sr, Ba, Ca and Pb.[1-6] The largest reported values of magnetoresistance ratio, $-\Delta R/R_H$ in the doped manganese oxide films are obtained in Nd-Sr-Mn-O film ($>10^6$ % at 60 K and 8 T)[6] and La-Ca-Mn-O film ($>10^5$ % at 70 K and 6 T)[3,4]. The La-Ba-Mn-O thin films gave a $-\Delta R/R_H$ value of ~150 % at room temperature.[1]

From technological point of view, it is desired to prepare films with large GMR. It is also necessary to know the effects of heat-treatments in obtaining large GMR. In this paper, we report the influence of film preparation on the properties of epitaxial Nd$_{0.7}$Sr$_{0.3}$MnO$_z$ (NSMO) and La$_{0.67}$Ba$_{0.33}$MnO$_{3-\delta}$ (LBMO) thin films. Huge negative magnetoresistance changes of $-\Delta R/R_H >$ 1×10^6 % were obtained at 60 K and a magnetic field of 8 T in a Nd$_{0.7}$Sr$_{0.3}$MnO$_{3-\delta}$ film.

EXPERIMENT

Doped manganese oxide films with thickness of 200-300 nm were grown on (100) LaAlO$_3$ single crystal substrates by pulsed laser deposition. Ceramic targets with nominal composition of Nd$_{0.7}$Sr$_{0.3}$MnO$_3$ and La$_{0.67}$Ba$_{0.33}$MnO$_3$ was prepared by standard solid state reaction from metallic oxides and carbonates at 1300°C to 1450°C. During the deposiion a 200-300 mTorr N$_2$O atmosphere was used. The as-grown films were directly cooled down to room temperature in 400 Torr oxygen after deposition. Fig. 1 shows (a) X-ray diffraction patterns of ϕ scan for the (112) peak of a NSMO film and (b) Rutherford random backscattering and ion channeling spectra (3

Mat. Res. Soc. Symp. Proc. Vol. 384 ° 1995 Materials Research Society

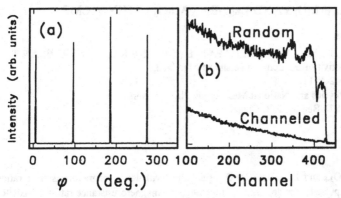

Fig. 1. (a) X-ray diffraction patterns of ϕ scan for (114) peak of a NSMO film and (b) Ion beam channeling and random Rutherford backscattering patterns (3 MeV He^+ ion) for an as-grown NSMO film.

MeV He^+) for an as-grown NSMO film. The results of X-ray diffraction and ion channeling indicated a perfect epitaxial crystalline structure.

A standard four-probe method was used to measure the dc resistivity of the samples. The magnetoresistance measurements were performed in a superconducting magnet with the applied magnetic field parallel to the film surface and the current direction.

Magnetoresistivty Behavior of NSMO and LBMO Films

The magnetoresistance ratio is defined here as $-\Delta R/R_H = -(R_H-R_0)/R_H$, where R_H is the resistance in an applied magnetic field of H and R_0 is the resistance at zero field. Fig. 2 shows temperature dependencies of resistivity with and without magnetic field for (a) an as-grown NSMO film and (b) a post-annealed LBMO film. Under our deposition conditions the films show resistivity maxima, ρ_{MAX}, at a peak temperature, T_P. Below the peak temperatures, the resistivity of the as-grown films exhibits dR/dT > 0, a metallic temperature dependence behavior.

As can be seen in Fig. 2, the resistivity maximum of the LBMO film (ρ_{MAX} = 1.46 Ω·cm at 100 K) is much lower than that of the NSMO film (ρ_{MAX} = 16.6 Ω·cm at 110 K). In an applied magnetic field the resistivity is repressed. Below the peak temperature large negative MR effect

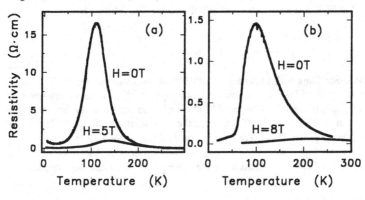

Fig. 2 Resistivity versus temperature for (a) an as-grown NSMO film in zero field and a magnetic field of 5 T and (b) a post-anealed LBMO film in zero field and magnetic field of 8 T.

was observed in these doped manganese oxide thin films. In a magnetic field of 5 T a maximum MR ratio of 3230 % was observed at 95 K for this as-grown NSMO film. A maximum MR ratio of 9100 % was observed at 85 K for this LBMO film in a magnetic field of 8 T.

Influence of Substrate Temperature

The temperature dependence of the resistivity for four as-grown NSMO thin films is shown in Fig. 3 (a). These films were deposited at 615, 710, 730 and 815 °C respectively. It is clear that the film deposited at 615 °C had the lowest resistivity maximum of ρ_{MAX} = 3.5 $\Omega\cdot$cm at a highest peak temperature of T_P ~ 175 K. By increasing the deposition temperature, the peak temperature shifts to a lower temperature and ρ_{MAX} increases. The film deposited at 815 °C had a highest ρ_{MAX} = 61.2 $\Omega\cdot$cm at T_P = 95 K. In an applied magnetic field, large MR ratios were observed in the as-grown films. In a magnetic field of 8 T for the film deposited at 815 K the giant magnetoresistance ratio of $-\Delta R/R_{H=8\,T}$ is over 7000 % at 60 K and for the film deposited at 615 K the $-\Delta R/R_{H=8\,T}$ is about 3500 % at 165 K.

The influence of increasing the deposition temperature on the resistivity behavior of epitaxial LBMO films is similar to that observed in NSMO films except that the resistivity of epitaxial LBMO films is lower. Fig. 3 (b) shows the resistivity versus temperature of three as-grown LBMO thin films, which were deposited at 640 °C, 700°C and 760 °C. The ρ_{MAX} and the T_P were 0.013 $\Omega\cdot$cm at 320 K, 0.22 $\Omega\cdot$cm at 170 K and 1.97 $\Omega\cdot$cm at 70 K, respectively. A large MR ratio of 6500 % is observed at ~50 K for the film deposited at 760 °C in a magnetic field of 8 T.

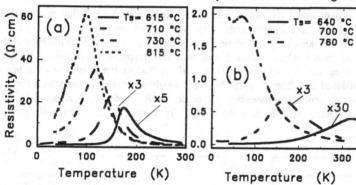

Fig. 3 R-T for (a) as-grown NSMO films deposited at 615, 710, 730 and 815 °C; and (b) for *in-situ* LBMO films deposited at 640, 700 and 760 °C.

Annealing Effect

The temperature dependence of the resistivity for the NSMO film deposited at 815 °C before and after several annealing steps are shown in Fig. 4. The heat-treatment process is annealing in O_2 atmosphere at 900 °C for 1 hr and subsequent annealing for 2 hr and 3 hr. The actual annealing times were 1, 3 and 6 hours respectively. The annealing treatment lowers the resistivity maximum from ρ_{MAX} = 61.2 $\Omega\cdot$cm (before annealing) to ρ_{MAX} =2.42 $\Omega\cdot$cm (after an one hour annealing) and then to ρ_{MAX} =0.526 $\Omega\cdot$cm (after a three hour annealing) and finally to ρ_{MAX} =0.386 $\Omega\cdot$cm (after a six hour annealing). After heat treatments the T_P was raised from about 95 K (before annealing) to about 195 K (after 3 and 6 hr annealing). It can be seen that with increasing annealing time the increase of T_P seems to saturate to about 195 K.

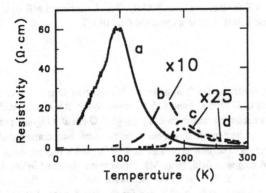

Fig. 4. R-T for the NSMO film deposited at 815 °C (a) before and after annealing at 900 °C in O_2 for (b) 1 hr, (c) 3 hr and (d) 6 hr.

A significant change of the resistivity below the T_P can also be observed, for example from 13.0 Ω·cm at 35 K (before annealing) to about 0.004 Ω·cm at 130 K (after a three hour annealing). After annealing the resistivity in low temperature region dropped more quickly than before.

Large Giant Magnetoresistance Ratio

Since the resistivity properties of the doped manganese oxide films were influenced by the substrate temperature and the annealing procedure, the MR ratios were affected by the sample preparation also. If a suitable annealing procedure can only reduce the R_H and keep the R_0 still at a high value, a large MR ratio will be expected. Normally, a heat treatment can effectively reduce the resistivity in the temperature range below T_P as well as R_H. At the beginning of a heat treatment, the R_H is reduced first. Therefore, the MR ratio would probably increase after a short-time annealing. However, prolonged annealing will cause the R_0 to decrease substantially as shown in Fig. 4 and hence the MR ratio decreases. From our results, a film with good

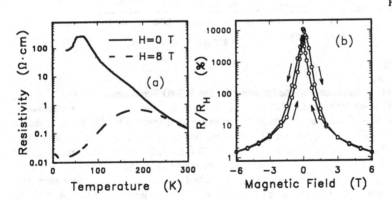

Fig. 5 (a) R-T in zero field and a magnetic field of 8 T and (b) R/R_H ver. magnetic field at 60 K for a post-anealed NSMO film.

crystalline structure and high resistivity maximum is needed to have a large MR ratio. It appears that a way to get large magnetoresistance is to prepare films with perfect structure and proper composition by adjusting the deposition temperature with or without a heat-treatment.

The largest MR ratio we obtained was in a NSMO film after annealing at 900 °C for 30 minutes. Fig. 5 (a) shows the resistivity versus temperature for this film, in which the $-\Delta R/R_H$ excesses 10^6 % at 60 K and in a magnetic field of 8 T. The $R/R_{H=8\,T}$ versus temperature at 60 K is shown in Fig. 5 (b) for this film.

DISCUSSION

Two processes, deposition at a high temperature and annealing at a high temperature have opposite influences on the resistivity behavior of NSMO and LBMO films. A higher deposition temperature under a fixed gas pressure results in a larger resistivity maximum at a lower T_P. Annealing at 900 °C in oxygen atmosphere reduces the resistivity maximum and moves the T_P to a higher temperature. The opposite effects can hardly be explained by film-substrate chemical reactions, lose of cation elements or structure variation, as in such cases a similar effect was expected for the two processes.

The opposite effects of sample preparation on the resistivity behavior of the NSMO and LBMO thin films can possibly be understood by considering the variation of the oxygen content during deposition and subsequent heat-treatment. It was proposed that in doped (La,A)-Mn-O (A=Ba, Ca, Sr) compounds the mixed Mn^{3+}/Mn^{4+} valence gives rise to both, ferromagnetism and metallic behavior.[7,8] The mixed Mn^{3+}/Mn^{4+} ratio can be changed by changing the doping level and by varying the oxygen content. Due to the thermodynamics, during the deposition a high deposition temperature under our deposition pressure may result in more oxygen deficiency. An annealing process can refill oxygen into oxygen-deficient films. The effect of oxygen deficiency on MR behavior has also been observed in La-Ba-Mn-O bulk samples,[9] which is similar to that of increasing the deposition temperature in the film preparation.

In addition to the oxygen content, the heat treatment can affect the defect and vacancy motion and recombination which can also have an effect on the resistivity and the GMR ratio. A small composition change during the depositon also can not be totally ruled out, especially at high deposition temperature. In our experiments, after a long-time annealing the T_P of the films saturated to slightly different values, which implies that the microstructure, defect density or composition of the films may be slightly different in different films.

CONCLUSION

Epitaxial $Nd_{0.7}Sr_{0.3}MnO_{3-\delta}$ and $La_{0.67}Ba_{0.33}MnO_{3-\delta}$ thin films with large magnetoresistance ratios have been prepared by pulsed laser deposition. Huge negative magnetoresistance changes of $-\Delta R/R_H > 1 \times 10^6$ % were obtained at 60 K and a magnetic field of 8 T in a $Nd_{0.7}Sr_{0.3}MnO_{3-\delta}$ film. $Nd_{0.7}Sr_{0.3}MnO_{3-\delta}$ and $La_{.67}Ba_{.33}MnO_{3-\delta}$ thin films show a maximum in resistivity as a function of temperature. Two processes, deposition at a high temperature and annealing at a high temperature have opposite effects on the resistivity behavior. Higher substrate temperature results in a larger resistivity maximum at a lower temperature. However, annealing in oxygen atmosphere reduces the magnitude of the resistivity maximum and moves it to a higher temperature. Our results suggest that oxygen stoichiometry and diffusion are important factors in causing the behavior observed in doped manganese oxide films.

ACKNOWLEDGMENTS

We appreciate the discussions with Prof. S.M. Bhagat, Prof. D. Drew and helps of X.G. Jiang, X.X. Xi, C. Kwon and S.N. Mao. G.C.X. would like to thank the international university exchange program committee of University of Maryland.

REFFERENCES

1. R. von Helmolt, J. Wecker, B. Holzapfel, L. Schultz and K. Samwer, Phys. Rev. Lett., **71**, 2331 (1993).
2. Ken-ichi Chahara, T. Ohno, M. Kasai and Y. Kozono, Appl. Phys. Lett., **63**, 1990 (1993).
3. S. Jin, T.H. Tiefel, M. McCormack, R.A. Fastnacht, R. Ramesh and L.H. Chen, Science **264**, 413 (1994).
4. M. McCormack, S. Jin, T.H. Tiefel, R.M. Fleming, J.M. Phillips and R. Ramesh, Appl. Phys. Lett., **64**, 3047 (1994).
5. H.L. Ju, C. Kwon, Qi Li, R.L. Greene and T. Venkatesan, Appl. Phys. Lett., **65**, 2109 (1994).
6. G.C. Xiong, Q. Li, H.L. Ju, S.N. Mao, L. Senapati, X.X. Xi, R.L. Greene and T. Venkatesan, Appl. Phys. Lett. **66**, 1427 (1995).
7. G.H. Jonker and J.H. van Santen, Physica, **16**, 337 (1950).
8. G.H. Jonker, Physica, **22**, 707 (1956).
9. H.L. Ju, J. Gopalakrishnan, J.L. Peng, Qi Li, G.C. Xiong, T. Venkatesan and R.L. Greene, Phys. Rev. **B51**, 6143 (1995).

Calculated Electronic Structure and Transport Properties of La$_{.67}$Ca$_{.33}$MnO$_3$

W. H. Butler * , X.-G. Zhang * and J. M. MacLaren **
*Metals and Ceramics Division, Oak Ridge National Laboratory, Oak Ridge, Tennessee
37831-6114
**Department of Physics Tulane University New Orleans, LA 70118

ABSTRACT

We have calculated the electronic structure, total energy, magnetic moments and electrical resistivities of LaMnO$_3$ and La$_{.67}$Ca$_{.33}$MnO$_3$ using mean field band theory. The magnetic and structural properties seem to be in good agreement with experiment. The calculations predict that La$_{.67}$Ca$_{.33}$MnO$_3$ is metallic for the majority spins and semiconducting for the minority spins.

Introduction

The recent observations[1, 2, 3] of a very large negative magnetoresistance effect in perovskite compounds of composition La$_{1-x}$A$_x$MnO$_3$ where A is an alkaline earth element have elicited considerable interest because of possible magnetic sensor applications and because of their possible similarity to materials which show giant magnetoresistance (GMR). In contrast to GMR materials which have a large magnetoresistance because of their inhomogeneous structure, the magnetoresistance of these manganite materials seems to be a property of the homogeneous, bulk material and to be associated with a ferromagnetic phase transition. The details of the transport and magnetic properties, however, vary dramatically from sample to sample and depend on the oxygen content and annealing history of the sample. Recently Hundley et al[3] have observed that the resistivities of their samples are a simple exponential function of the magnetization.

In this paper we present calculations of the mean field electronic structure of La$_{.67}$Ca$_{.33}$MnO$_3$ and LaMnO$_3$. The calculations are performed within the local spin density approximation to density functional theory implemented using the Layer Korringa Kohn Rostoker technique[4]. The atomic potentials were assumed to be spherically symmetric about each nucleus. Space filling spheres were used with the radii chosen proportional to the ionic radii. The coherent potential approximation[5] was used to treat the disorder associated with the calcium substitution for the lanthanum. We also report on some preliminary calculations of the electrical resistivity arising from the substitutional disorder.

Electronic Structure of LaMnO$_3$

The calculated density of states of nonmagnetic LaMnO$_3$ in the cubic perovskite structure is shown in Figure 1. The primary features in the density of states are a large complex of oxygen-p states centered near 0.2 Hartree, and a large peak consisting of manganese-d states with a cusp that lies exactly at the Fermi energy. These manganese-d states have an admixture of oxygen-p. The lanthanum-f states lie about 2 eV above E_F.

Because of the large number of manganese-d states at the Fermi energy, it is not surprising that this electronic structure is unstable against spin polarization. We find that the

439

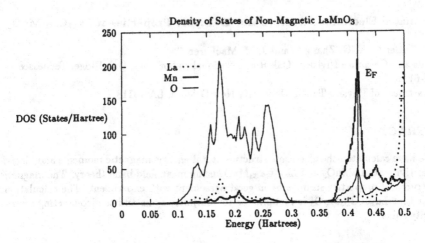

Figure 1: Calculated density of states for non-magnetic LaMnO₃.

ferromagnetic phase and the anti-ferromagnetic (layered A-type using the nomenclature of [6]) phases are lower in energy than the non-magnetic phase by 50 and 43 mHartrees respectively. Figure 2 shows the calculated density of states for the ferromagnetic phase of the cubic perovskite structure. For clarity we have omitted the lanthanum f states which still lie 2 eV above the Fermi Energy. The interesting feature of this density of states is that the states at the Fermi energy consist entirely of hybridized manganese-d and oxygen-p *of majority spin only*. Thus ferromagnetic LaMnO₃ in the cubic perovskite phase is predicted to be a ferromagnetic "half-metal". It is metallic for the majority spins and semi-conducting for the minority spins. The spin polarization of the hybridized manganese-d oxygen-p complex is sufficient to move the minority bands above the Fermi energy. There is a much smaller but still perceptible spin polarization of the oxygen-p complex centered near 0.2 Ha. The calculated magnetic moment is $3.9\mu_B$ per manganese atom.

It is known that the stable phase of LaMnO₃ is not the cubic one, but the Pnma structure[7] obtained by rotating and distorting the oxygen octahedra. Pickett and Singh[8] have shown that the local spin density approximation correctly yields an anti-ferromagnetic insulator as the ground state when this distortion of the structure is included.

Electronic Structure of La.₆₇Ca.₃₃MnO₃

Most of the recent interest in these materials has centered on the the compounds which are alloyed with alkaline earth elements. It is these that show the extraordinary magnetoresistance. In order to investigate these materials we used the coherent potential approximation to treat the disorder associated with calcium substituting for lanthanum. Figure 3 shows the calculated electronic structure of ferromagnetic La.₆₇Ca.₃₃MnO₃. Our calculations correctly predicted that the ferromagnetic phase is stable with respect to the anti-ferromagnetic phase. The calculated energy difference was 6 mHa per formula unit. Our calculated moment per manganese atom was $3.31\mu_B$, in good agreement with the experimental value of 3.4[6].

The major changes in the density of states from Figure 2 are that the Fermi energy has

440

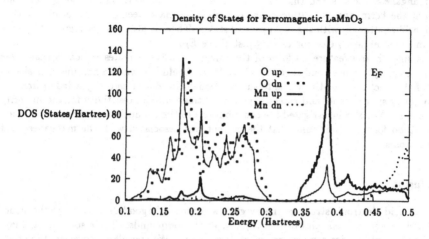

Figure 2: Calculated density of states for ferromagnetic LaMnO$_3$. The lanthanum states have been omitted for clarity.

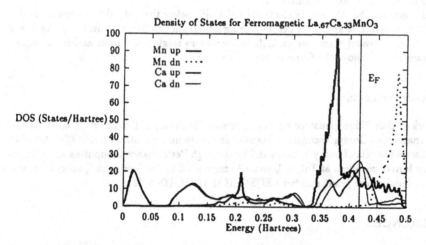

Figure 3: Calculated Density of States for La$_{.67}$Ca$_{.33}$MnO$_3$. Oxygen and lanthanum states have been omitted for clarity.

moved lower relative to the sharp peak in the density of states associated with the majority spin manganese-d states and that calcium states of both minority and majority spins are found at the Fermi energy. Although the oxygen states have been omitted for clarity they still hybridize strongly with the manganese-d states. There is also a larger region of little or no manganese majority electron density just above E_F.

The smooth and featureless form of the calcium density of states indicates that these states are highly localized and that $La_{.67}Ca_{.33}MnO_3$ should behave as a half metal. Calculations of the electrical resistivity due to substitutional disorder at zero K yielded $6\mu\Omega$cm for the majority spin channel and essentially an essentially infinite resistivity for the minority spin channel. We also investigated the contributions to the conductivity through evaluation of the Kubo formula and found that the current was associated with the manganese and oxygen atoms.

Conclusions

Mean field electronic structure theory gives a surprisingly good description of the structural and magnetic properties of the $La_{1-x}A_xMnO_3$ compounds. Their spectacular magnetotransport properties remain somewhat mysterious. We speculate, however, that they will eventually be understood in terms of the half-metallic band structure predicted by electronic structure calculations. The resistivity calculations described here only included substitutional disorder. It is likely, if indeed these systems are half-metals, that spin-disorder will give a much larger contribution to the electrical resistivity. The large variation in the transport and magnetotransport properties from sample to sample leads one to believe that these materials may not be completely homogeneous magnetically and that they may be similar to the GMR materials after all.

We speculate that the rise in electrical resistivity below the critical temperature may be explained in terms of spin disorder scattering which should be extremely large in a half-metal. The activated transport at high temperatures might arise from anti-ferromagnetic short-range order above the Curie temperature.

Acknowledgements

Work at Oak Ridge sponsored by DOE Assistant Secretary of Defense Programs, Technology Management Group, Technology Transfer Initiative under contract DEAC05-84OR21400 with Martin Marietta Energy Systems and by the High Performance Computing and Communication Initiative. Work at Tulane University supported by the Louisiana Quality Education Support Fund under grant number LEQSF (1991-1994)-RD-A-30.

REFERENCES

1. R. von Helmolt, J. Wecker, B. Holzapfel, L. Schultz and K. Samwer Phys. Rev. Lett. **71** 2331 (1993).

2. S. Jin, M. McCormack, R. A. Fastnacht, R. Ramesh, and L. H. Chen, Science **264** 413 (1994).

3. M. F. Hundley, M. Hawley, R. H. Heffner, Q. X. Jia, J. J. Neumeier, J. Tesmer, J. D. Thompson, and X. D. Wu *preprint*

4. J. M. MacLaren, S. Crampin, D. D. Vvednsky, R. C. Albers, and J. B. Pendry, Computer Physics Communications, **60** 365 (1990).

5. G. M. Stocks, W. M. Temmerman and B. L. Gyorffy, Phys. Rev. Lett. **41**, 339 (1978).

6. E. O. Wolland and W. C. Koehler, Phys. Rev. **100**, 545 (1955).

7. J. B. A. A. Elemans, B. van Laar, K. R. van der Veen, and B. O. Loopstra, J. Solid State Chem **3**, 238 (1971).

8. W. E. Pickett and D. J. Singh, Magnetoelectronic and Magnetostructural Coupling in the $La_{1-x}Ca_xMnO_3$ System (*preprint*).

Part VIII

Spectroscopies, Magneto-Optical Properties

ATOM SPECIFIC SURFACE MAGNETOMETRY WITH LINEAR MAGNETIC DICHROISM IN DIRECTIONAL PHOTOEMISSION

GIORGIO ROSSI,[1,2] FAUSTO SIROTTI,[2] and GIANCARLO PANACCIONE[2]

[1] *Laboratorium für Festkörperphysik, ETH-Zürich, CH-8093*
[2] *Laboratoire pour l'Utilisation du Rayonnement Electromagnetique, CNRS, CEA, MESR F-94305 Orsay*

ABSTRACT

A practical atom specific surface magnetometry can be based on the measure of magnetic dichroism in the angular distribution of core photoelectrons using linearly polarized synchrotron radiation. The magnetic dichroism effect on the photoemission intensity of 3p core levels of the ferromagnetic transition elements is as large as 46% in the case of Fe(100). The most efficient scheme for measuring the magnetic dichroism in photoemission requires two mirror experiments in chiral geometry, i.e. only two times more experiments than standard core level photoemission for surface chemical analysis. We describe the dichroism magnetometry and show examples for Fe, Co, Ni and Cr surfaces and interfaces, including the measurement of the temperature dependence of the Fe(100) surface magnetization and of the effect of S-segregation on the surface magnetic moment of iron.

INTRODUCTION

The understanding of the magnetic properties of surfaces and low dimensional solids requires accurate measurements of all the structural, chemical, and magnetic parameters of the surface atoms.

The field of surface magnetism has been opened when the measurement of the magnetization of surface and near-surface atoms became practical by the application of spin-polarimetry to the photoelectrons and secondary electrons ejected from surfaces.[1] The measure of spin polarization (SP) of secondaries is intrinsically surface sensitive due to the short escape depth for low energy photoelectrons in ferromagnets, and can be understood semi-quantitatively, but gives an average magnetization values, i.e. does not resolve contributions from individual atomic species constituting the magnetic surface. On the other hand the basic surface science spectroscopies, Auger electron spectroscopy and photoelectron spectroscopy, which allow a full chemical characterization and sensitivity to local order in the diffraction of the ejected electrons, are not directly sensitive to magnetism. To gain magnetic sensitivity AES and PES have been measured in the spin-resolved mode, i.e. by measuring the number of Auger electrons or core photoelectrons with their spins aligned or counteraligned to a macroscopic magnetization direction.[2-4] These are close to ideal tools for surface magnetism, but suffer for a great technical handicap: the low efficiency of spin-detection which is only some 10^{-3} and severely reduces the applications of these techniques. An alternative is to exploit the fact that polarized atoms, like the atoms in a magnetically ordered ferromagnetic material, can be recognized by their directional anisotropies in the photoionization matrix elements.

Angular photoemission experiments on core levels about the magnetization direction (vector) show dichroism and therefore allow to probe magnetism in a highly efficient way. A large magnetic dichroism effect is shown in figure 1 for the Fe 3p core levels from Fe(100). The

447

dichroism is obtained as the differences of two photoelectron spectra of the exchange split-ted Fe 3p magnetic sublevels, measured in the mirror chiral geometries which are shown in the inset of the figure 1, using linearly polarized synchrotron radiation at soft X-ray ener-gies. This effect was observed first by Roth and coworkers.[5] The special case of magnetic dichroism in photoemission which is measured as a difference in the angular distribution of photoelectrons between two mirror experiments, is called linear magnetic dichroism in the angular distribution of photoelectrons (LMDAD).[7-11] We show below that a practical surface magnetometry can be based on the LMDAD effect, combining all the power of photoelectron spectroscopy with magnetometric information at a total cost of doubling the experimental effort of a standard core level photoemission surface analysis.

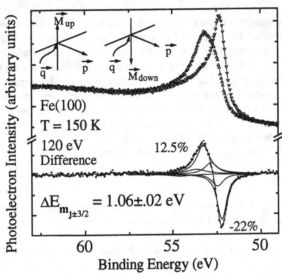

Fig.1: Fe 3p photoemission mirror experiments done in the chiral geometries shown at the top, and the LMDAD dichroism curve for Fe(100). The dichroism is fitted by a sextuplet of magnetic sublevels which are weighted by the angular matrix elements.[ref. 8] The extrema of the LMDAD curvve coincide with the mj=+-3/2 sublevels.

PRINCIPLES

In a core level photoemission experiment of a ferromagnetic material the photoexcited core hole state of total angular momentum J is split by the spin orbit and exchange interac-tions with the spin polarized valence electrons into sublevels with a given projection m_j on the magnetic quantization axis. Exchange dominates over orbital interaction and $J = L + S$ is not a good quantum number so that the ordering of the m_j sublevels is different with respect to the Zeeman effect. By disregarding altogether the manybody processes involved in the photoemission from atoms and from solids, as well as the possible splitting of the high energy final states, one expects each m_j sublevel to contribute intensity with a character-istic angular distribution.[12] Within this one-electron atomic picture the intensity and the angular distribution of the photoemission spectrum is completely determined by the core hole multiplet. The angular dependence of photoionization of a particular m_j core sublevel can be measured if the energy splitting of the core hole is at least of comparable size with respect to the intrinsic width of the core hole. The relative variations of lineshape measured in the mirror experiments of figure 1 represent the relative variations of intensity from a grid of effective m_j sublevels for the Fe 3p core hole.[8,11] The experimental variations can be compared with calculated photoionization cross sections.[12,13,10]

LMDAD experiments on Fe, Co, Ni and Cr 3p photoemission.

The experimental LMDAD spectra for the ferromagnetic transition metals and for chromium (layered antiferromagnet) measured at 150 K are collected in figure 2. The top curves in each panel represent the as measured spectra at the indicated photon energies. Up full triangles indicate the spectra measured with the external magnetic field direction in the up-vertical direction, and open down-oriented triangle indicate the experimental points measured after reversal of the surface magnetization, by the external field directed in the down-vertical direction. The chiral geometry is the same for all the experiments but the angular acceptance of the analyzer is different in the case of Fe at 170 eV. The second panel represents the spectra after background substraction (integral background plus exponential decay of secondaries). This procedure allows to estimate the amount of photoemission intensity at final state energies lower than the main 3p peak, i.e. the deviation from the simple one-electron reference model.[11] Finally the bottom panels show the experimental difference curves i.e. the LMDAD spectra. The photoemission magnetic asymmetry is defined as $A = \frac{I_{up} - I_{down}}{I_{up} + I_{down}}$, where $I_{up \, (down)}$ are the photoelectron spectral intensities obtained with the magnetization in the upward (up) or downward ($down$) directions.

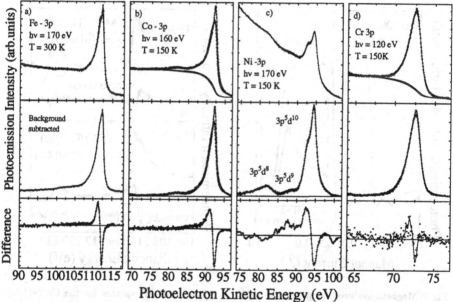

Fig.2: 3p LMDAD spectra of Fe, Co, and N polycrystalline layers, and of the Cr(100) surface.

The maximum LMDAD asymmetry was 46% of the Fe 3p intensity (corresponding to a 22% variation of the total photoemission including background) measured at $140K$ and for $h\nu = 120eV$ for a clean Fe(100) surface.

The three ferromagnetic 3d metals all show similar dichroism curves under the main peak, with the magnitude of the asymmetry scaling with the element and with the photon energy. The LMDAD asymmetry result are related to the magnetic moment of the metals, but in a complex way. By inspecting the spectra one finds that the photoemission intensity is basically all concentrated under the main peak in the case of Fe 3p, but it shows tails and well resolved satellites in the cases of Co and Ni. This is a consequence of the increasing correlation of the d-bands with d-filling, which implies a higher probability of multiplet final state configurations. The Ni 3p peak and satellites correspond to $3p^5d^{10}$, $3p^5d^9$ and $3p^5d^8$ final state configurations respectively. The photoemission intensity of the satellites amounts to 20% of the total intensity in nickel. Since every configuration is different, it will display in principle a different magnetic dichroism. Such result was previously suggested by a circular dichroism experiment on Ni.[14] In the 4th panel we see the Cr 3p photoemission spectra from an epitaxial overlayer on Fe(100). The ferromagnetic order of the Cr surface can be seen by the presence of a LMDAD curve. Its orientation with respect to the substrate Fe can be directly observed from the sign of the LMDAD curve (reversed in this case indicating antiferromagnetic coupling of the Cr surface to the substrate).

LMDAD as a Kerr-like diagnostic tool

A direct application of LMDAD is to probe the magnetic order at surfaces in a similar qualitative way as the magneto-optic Kerr effect probes the bulk magnetic order, but with the important advantage of chemical sensitivity.

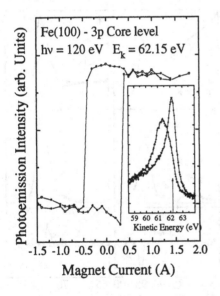

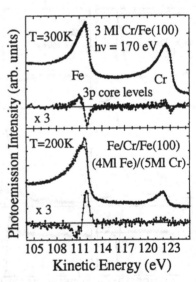

Fig.3: Magnetic hysteresis loop measured with LMDAD on Fe(100).

Fig.4: LMDAD spectra for the Cr/Fe(100) interface and Fe/Cr/Fe(100) trilayer.

Figure 3 shows the LMDAD spectra at two stages during the growth of a Fe/Cr/Fe trilayer structure. The top panel shows that the ferromagnetic order of the Fe(100) substrate is not disturbed by the epitaxial growth of 3 ML of Cr. The Cr 3p also shows LMDAD indicating ferromagnetic order in plane. Cr is known to grow along the [100] direction as a layered antiferromagnet,[15] the measured Cr 3p LMDAD is due to the uncompensated signal from the surface layer. The lower panel shows the completion of the trilayer structure by epitaxial overgrowth of Fe on a 5 ML thick Cr interlayer. The LMDAD signal shows that the top Fe film is antiferromagnetically coupled to the substrate iron across the 5 monolayers of chromium.

The LMDAD asymmetry can be measured at the optimum final state energy (i.e. where the asymmetry is maximum) while varying the external magnetic field to obtain magnetic hysteresis curves for the selected atoms. A LMDAD magnetic hysteresis curve for Fe(100) is shown in figure 3. There is a practical limitation of photoemission hysteresis curves which is due to the quite small value of external field that can be present *during* the photoemission measurement.

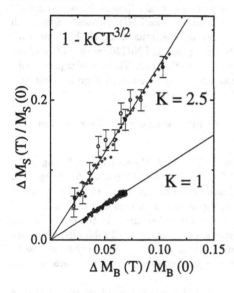

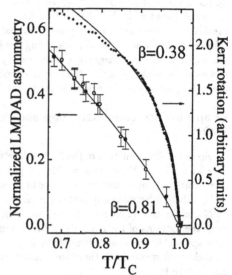

Fig.5: Thermal decrease of the relative surface magnetization as measured by Fe 3p LMDAD asymmetry.

Fig.6: LMDAD (circles) and Kerr rotation (diamonds) as afunction of reduced temperaturein the near Tc region..

Testing the Temperature Dependence of LMDAD in The Spin Wave Regime and Near T$_{Curie}$.

The temperature dependence of the surface magnetization in the spin wave regime has been measured for Fe-Ni-B and for Fe layers and interfaces in SP experiments. The results indicate that the exchange interaction within the surface and along a path perpendicular to the surface are different from the bulk exchange, and can be obtained by fitting model calculations to the slope of the $M_S(T)$ curve.

A crucial test for the LMDAD photoemission magnetometry is the dependence of the LMDAD asymmetry on temperature for a clean Fe(100) surface, and to compare the results with spin polarization (SP) measurements. The temperature dependent results of the Fe 3p LMDAD asymmetry, SP of the secondary electrons as measured with a 100KV Mott-scattering experiment, and of bulk magnetization sensitive Kerr rotation data are shown in figures 4 and 5.[16]

For $T \leq .4T_C$, the thermal decrease of the relative surface magnetization $\Delta M_S(T)/M_S(0)$ of atomically clean Fe(100) is consistently measured by LMDAD and by SP. The results are described by the law $M(T)/M(0) = 1 - kCT^{3/2}$ of spin-wave theory.[17] The Kerr data measure the bulk thermal decrease of relative magnetization with $k = 1$ which determines the bulk constant C. A surface enhancement factor $k \simeq 2.5$ is obtained by fitting the LMDAD and SP data to the $T^{3/2}$ law: it represents the reduction of the exchange interaction of the Fe(100) surface atoms along a path perpendicular to the surface.[18] In the critical region (figure 3) the Kerr-rotation and LMDAD signals vanish with, respectively, bulk and surface critical exponents according to $M_B \propto (1 - T/T_{CB})^{\beta_B}$ and $M_S \propto (1 - T/T_{CB})^{\beta_S}$.[19] The Fe 3p LMDAD results for Fe(100) are described by a surface critical exponent $\beta_S = 0.81 \pm 0.01$, to be compared with $\beta_B = 0.38 \pm 0.01$ from Kerr rotation. The LMDAD results agree with previous SP measurements of energy selected secondary electrons on Fe (110),[20] and with spin-dependent elastic electron scattering results for Ni(100) and Ni(110) surfaces,[21] and are consistent with a ferromagnetic surface weakly coupled to the bulk. These results provide direct experimental proofs that the measure of photoemission magnetic dichrism gives directly the order parameter of magnetization for the Fe surface atoms $< M_S >$.

Fe 3p Fine Structure: Exchange and Spin-Orbit Parameters, Magnetochemical Shifts

From the fine structure of the 3p core levels one can evaluate the exchange splitting of the core hole. The width of the $J = 3/2$ multiplet is found here to be $1.06 \pm .02eV$ for Fe 3p. The spin orbit splitting of the 3p core holes in Fe is known from silicide data[22] to be $1.05 \pm .05eV$. The splitting between adjacent m_j sublevels corresponds to a value of exchange field of 3.5 10^7Gauss for the 3p core hole in bcc-Fe. This analysis is rather direct in the case of Fe 3p since the extrema of the asymmetry curve coincide with the $m_j = \pm 3/2$ sublevels, which are pure spin-orbit states and display the largest magnetic asymmetry. The "width" of the Fe 3p dichrism spectrum is therefore a direct measure of the atomic exchange interaction for the 3p core hole.

The atomic exchange depends upon the details of the spin-polarized electron states of the valence band which is determined by crystal structure and chemistry at the surface. We compare in figure 6 the 3p dichrism spectra of clean Fe(100) and S segregated Fe(100). Sulfur segregation occurs at temperatures higher then 500 C and saturates when an ordered c(2x2) superstructure is completed (clearly observed in LEED) for annealing above 600 C.[23] The 3p LMDAD dichrism spectra of Fe(100) and c(2x2)S/Fe(100) at 300K are shown in figure 4. The narrowing of the Fe 3p dichrism spectrum for the sulfurated surface shows that the $m_j = \pm 3/2$ splitting is reduced to $.99 \pm .01eV$. This indicates the at the magnetic moment of the Fe atoms bonded to the S is reduced by a large amount. The photoemission measurement averages the information of the top three layers of material , so if the reduction of magnetic moment is only occurring at Fe surface sites, this reduction is of the order of 20%. From the inspection of the magnetization dependent spectra one observes a clear "magnetochemical shift" of the peak dominated by the (minority spin) $m_j = +3/2$ sublevel when sulfur is present at the surface. The reduced magnetic moment of the Fe(100) surface and subsurface atoms when S atoms occupy the fourfold surface sites corresponding to the c(2x2) superstructure can be attributed to hybridization of the Fe 3d bands with the S sp valence electrons.[24] This fact has implications on the valence configuration of the c(2x2) S atoms. In figure 7 we show the LMDAD spectra for the S 2p core level for the segregated c(2x2)S surface of Fe(100): the presence of dichroism shows that bonding with Fe implies a transfer of magnetic moment on the sulfur. The $m_j = \pm 3/2$ splitting of the S 2p 3/2 peak is $.45 \pm .03 eV$.

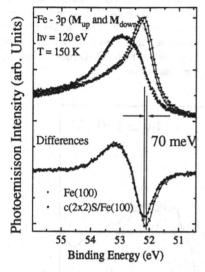

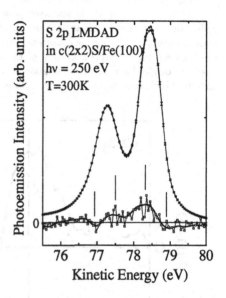

Fig.7: Fe 3p LMDAD spectra for Fe(100) and c(2x2)S/Fe(100) surface. The reduce width of the dichroism spectrum reflects the reduced magnetic splitting of the core hole.

Fig.8: S 2p LMDAD spectra measured. The dichroism indicated that a magnetic moment is present on the S atoms.

A large magnetochemical shift can be measured in chromium at the interface with iron. The first epitaxial layer of Cr on Fe(100) is ferromagnetically ordered and antiferromagnetically coupled to the iron substrate. The LMDAD spectrum measured at 150 K for one monolayer (identical spectra were measured for half monolayer) of Cr on Fe(100) is shown in figure 8 and compared with the LMDAD spectrum measured from the surface of a thicker epitaxial layer. The width of the interface LMDAD spectrum is 30% larger then the width of the Cr(100) surface LMDAD for the epitaxial (100) film. The relationship between the magnetic splitting of the Cr 3p core level in the two cases, shows the relative values of the magnetic moment of chromium atoms at the interface with iron, or at the Cr(100) surface.

CONCLUSIONS AND OUTLOOK

LMDAD in chiral photoemission experiments on $l > 0$ initial states is a large effect. The measure of LMDAD on core levels is a diagnostic of ferromagnetic order which cost only a double effort with respect to the standard photoemission lineshape inspection. It is clear that photoemission magnetic dichroism provides a powerful magnetometer. The measure of the order parameter $< M >$ is directly obtained from the asymmetry. Relative changes in the magnetic moment as a function of the atomic environment can be obtained from the LMDAD spectral widths. All the advantages of surface sensitivity and chemical sensitivity of the photoemission technique are transferred to the dichroism magnetometry. The surface analysis of ferromagnetic alloys, heterostructures, impurity terminated surfaces can be done with the full power of core level spectroscopy and magnetic order sensitivity. The fact that only two times more measurements are needed in photoemission LMDAD experiments with respect to standard photoemission, implies that the technique does not suffer for the intensity limitations which severely affects spin-resolved photoemission and even circular polarization dichroism photoemission measurements.[25,26] This implies that all the high resolution implementation of the photoemission technique made possible by new linearly polarized synchrotron radiation sources can be extended to LMDAD-magnetometry, including high energy resolution, lateral resolution, time resolution.

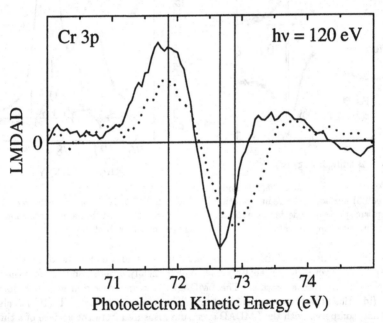

Fig.9: Cr 3p LMDAD for interface Cr atoms on Fe(100), and for the surface of an epitaxial Cr(100) film.

This work was partially supported by the Swiss National Fund for Research, under program 24. We are in debt with N.A. Cherepkov and F. Combet Farnoux for their collaboration, and to H.C. Siegmann for stimulating comments and continuous support. One of us, G.P. acknowledges the EC for a grant under the Human Capital and Mobility Program.

References

[1] H.C. Siegmann; J. Phys.: Conden. Matter **4**, 8395 (1992) and references quoted therein

[2] M. Landolt, in *Polarized Electrons in Surface Physics*, ed. by R. Feder, World Scientific, Singapore, 1985; R. Allenspach, D. Mauri, M. Taborelli, and M. Landolt, Phys. Rev. **B35**, 4801 (1987)

[3] C. Carbone, and E. Kisker; Solid State Commun. **65**, 1107 (1988)

[4] E. Kisker, in *Polarized Electrons in Surface Physics*, ed. by R. Feder, World Scientific, Singapore, 1985

[5] Ch. Roth, F.U. Hillebrecht, H. Rose, and E. Kisker, Phys. Rev. Lett. **70**, 3479 (1993)

[6] Ch. Roth, H. Rose, F.U. Hillebrecht, and E. Kisker, Solid State Commun. **86**, 647 (1993)

[7] F. Sirotti and G. Rossi, Phys. Rev. **B49**, 15682 (1994)

[8] G. Rossi, F. Sirotti, N.A. Cherepkov, F. Combet Farnoux, and G. Panaccione, Solid State Commun. **90**, 557 (1994)

[9] D. Venus; Phys. Rev. **B48**, 6144 (1993), ibid. **49**, 8821 (1994))

[10] G. van der Laan; Phys. Rev. **B51**, 240 (1995)

[11] G. Rossi, F. Sirotti and G. Panaccione; in *Core Level Spectroscopies For Magnetic Phenomena: Theory and Experiment* ed. by P.S. Bagus, G. Pacchioni, and F. Parmigiuani, Plenum ASI-NATO series 1995.

[12] N.A. Cherepkov, Phys. Rev. **B50**, 13813 (1994)

[13] E. Tamura, G.D. Waddill, J.G. Tobin, and P.A. Sterne; Phys. Rev. Lett. **73**, 1533 (1994)

[14] G. van der Laan, M.A. Hoyland, M. Surman, C.F. J. Flipse, and B.T. Thole, Phys. Rev. Lett **69**, 3827 (1993)

[15] J. Unguris, R.J. Celotta, and D. Pierce, Phys. Rev. Lett. **69**, 1125 (1992)

[16] F. Sirotti, G. Panaccione, and G. Rossi, to be published.

[17] J. Mathon and S.B. Ahmad, Phys. Rev. **B37**, 660 (1988).

[18] D. Mauri, D. Scholl, H.C. Siegmann, and E. Kay; Phys. Rev. Lett. **61**, 758 (1988); D. Scholl. M.Donath, D. Mauri, E. Kay, J. Mathon, R.B. Muniz, H.C. Siegmann; Phys. Rev.**B43**, 13309 (1991)

[19] K. Binder, in *Phase Transitions and Critical Phenomena* vol 8, ed C. Domb and J. Lebowitz (New York, Academic)

[20] M. Taborelli, O. Paul, O. Zuger, and M. Landolt; Journal de Physique **49**, C8-1659 (1988).

[21] S.F. Alvarado, M. Campagna, and H. Hopster, Phys. Rev. Lett. **48**, 51 (1982)

[22]F. Sirotti, M. De Santis, and G. Rossi, Phys. Rev. **B48**, 8299 (1993)

[23]K.O. Legg, F. Jona, D.W. Jepsen, and P.M. Marcus; Surf. Sci. **66**, 25 (1977).

[24]S.R. Chubb, and W.E. Pickett; Phys. Rev. **B38**, 10227 (1988).

[25]L. Baumgarten, C.M. Schneider, H. Petersen, F. Schafers, and J. Kirschner; Phys. Rev. Letters **65**, 492 (1990)

[26]H. Ebert, L. Baumgarten, C.M. Schneider, and J. Kirschner, Phys. Rev.B **44**, 4406 (1991)

THE CONNECTION OF SUM RULE AND BRANCHING RATIO ANALYSES OF MAGNETIC X-RAY CIRCULAR DICHROISM IN 3d SYSTEMS

J.G. Tobin[A], G.D. Waddill[B], A.F. Jankowski[A], P.A. Sterne[C], and D.P. Pappas[D]

[A]Lawrence Livermore National Laboratory, Chemistry and Materials Science Department, Livermore, CA 94550 • USA
[B]University of Missouri-Rolla, Department of Physics, Rolla, MO 65401-0249 • USA
[C]University of California-Davis, Department of Physics, Davis, CA 95616 • USA
[D]Virginia Commonwealth University, Department of Physics, Richmond, VA 23284-2000 • USA

In the recent past, Carra, et al.[1-3] derived a sum rule for electric dipole transitions in a single ion model that could be used to extract an elementally-specific spin-magnetic-moment (μ_{SPIN}) from magnetic x-ray circular dichroism (MXCD) spectra. Earlier, we proposed the utilization of a branching ratio analysis[4] for the determination of μ_{SPIN}, based upon a simplified one-electron, atomic picture which assumed complete orbital quenching. Here, it will be shown that these two approaches are essentially related in the case of 3d ferromagnetic materials. Both methods are based upon a comparison of the integrated intensity in the $L_3(J=3/2)$ white line peak versus the sum of the intensities in the $L_3(J=3/2)$ and $L_2(J=1/2)$ peaks, after background removal. An error estimate will also be presented. A more complete description of our work is under preparation[5].

Consider Equations 1 and 2, taken from references 1, 2, and 3.

$$\rho = \frac{\int_{j^+ + j^-} d\omega\left(\mu^+ - \mu^-\right)}{\int_{j^+ + j^-} d\omega\left(\mu^+ + \mu^- + \mu^\circ\right)} = \frac{1}{2}\frac{\ell(\ell+1)+2-c(c+1)}{\ell(\ell+1)(4\ell+2-n)}\langle L_Z\rangle \tag{1}$$

$$\delta = \frac{\ell(\ell+1)-2-c(c+1)}{3c(4\ell+2-n)}\langle S_Z\rangle + A(c,\ell,m)\langle T_Z\rangle = \frac{\int_{j^+} d\omega\left(\mu^+ - \mu^-\right) - \frac{c+1}{c}\int_{j^-} d\omega\left(\mu^+ - \mu^-\right)}{\int_{j^+ + j^-} d\omega\left(\mu^+ + \mu^- + \mu^\circ\right)} \tag{2}$$

For the case of 3d magnetic materials and using the 2p → 3d transition, $\ell = 2$ and c = 1. The number of 3d electrons (holes) is n (10–n). In our notation:

$$I_{3/2}^+ = \int_{j^+} d\omega\, \mu^+,\ I_{3/2}^- = \int_{j^+} d\omega\, \mu^-,\ I_{1/2}^+ = \int_{j^-} d\omega\, \mu^+ \text{ and } I_{1/2}^- = \int_{j^-} d\omega\, \mu^-.$$

Switching to our notation and using $\mu^\circ = 1/2(\mu^+ + \mu^-)$ as in Reference 2, a combination and rearrangement of Equations 1 and 2 gives us Equation 3, shown below.

$$2\langle S_Z \rangle + 3\langle L_Z \rangle = 6(10-n)\frac{\left[I_{3/2}^+ - I_{3/2}^-\right]}{\left[I_{3/2}^+ + I_{1/2}^+ + I_{3/2}^- + I_{1/2}^-\right]} \tag{3}$$

Here, we have also taken the liberty of dropping $\langle T_Z \rangle$ as done previously in References 1, 2, and 3. Again, Equation 3 is merely a restatement of the sum rules of Carra, et al., using our notation and explicitly showing the spin moment $\left(\mu_{SPIN}^{SR} = 2\langle S_Z \rangle\right)$ and orbital moment $\left(\mu_{ORB}^{SR} = \langle L_Z \rangle\right)$. The super script SR stands for sum rule.

Now consider the branching ratio (BR) analysis previously proposed in Reference 3.

$$\mu_{SPIN}^{BR} = \frac{4(10-n)}{P_{h\nu}}\left[\frac{BR^{+,-} - BR_{LIN}}{BR_{LIN}}\right] = \frac{4(10-n)}{P_{h\nu}}\left[\frac{BR^+ - BR^-}{BR^+ + BR^-}\right] \tag{4}$$

Here, we use $BR^+ + BR^- = 2\,BR_{LIN}$. Again LIN, +, and − denote polarization: linear, left circular, and right circular. ($BR^{+,-} = BR^+$ or BR^-, with a sign change in $P_{h\nu}$.) $P_{h\nu}$ is the circular polarization (+1 for left, 0 for linear, and −1 for right circular). (Note: If $P_{h\nu} = 0$, then $BR^+ = BR^- = BR_{LIN}$ and this equation is meaningless.) For the remainder of this work, $|P_{h\nu}| = 1$, to be consistent with Carra, et al. As noted earlier[4]:

$$BR = I_{3/2}/(I_{3/2} + I_{1/2}) \tag{5}$$

The BR has the advantage of being internally normalized for each spectrum: Cross normalization between spectra is not required. By doing a series expansion of Equation 5 and rearranging, we can get Equation 6. (This requires dropping terms of 1% magnitude or smaller.)

$$\mu_{SPIN}^{BR} = \frac{\dfrac{4(10-n)}{a}\left[I_{3/2}^+ - I_{3/2}^-\right]}{\left[I_{3/2}^+ + I_{1/2}^+ + I_{3/2}^- + I_{1/2}^-\right]} - 3\langle L_Z \rangle \tag{6}$$

here,

$$a = \frac{I_{3/2}^+ + I_{3/2}^-}{I_{3/2}^+ + I_{1/2}^+ + I_{3/2}^- + I_{1/2}^-} \cong \frac{1}{2}\left(BR^+ + BR^-\right) = BR_{LIN} \tag{7}$$

thus,

$$\mu_{SPIN}^{SR} - \mu_{SPIN}^{BR} = \left(6 - \frac{4}{a}\right)(10-n)\frac{\left[I_{3/2}^+ - I_{3/2}^-\right]}{\left[I_{3/2}^+ + I_{1/2}^+ + I_{3/2}^- + I_{1/2}^-\right]} \qquad [8]$$

In a single electron picture with complete orbital quenching and a statistical branching ratio for linear polarization ($BR_{LIN} = 2/3$), a = 2/3 and $\mu_{SPIN}^{SR} = \mu_{SPIN}^{BR}$. Now, it is well known[5-7] that $BR_{LIN} \neq 2/3$, with the causes including $\langle L_Z \rangle \neq 0$, band dispersion and multi-electronic effects[7].

However, for the 3d magnetic materials, $\langle L_Z \rangle / \langle S_Z \rangle$ is often quite small (References 1–3) and the BR_{LIN} value and a are close to 2/3. The deviation of a from 2/3 is a measure of the error. To be more specific, we can apply Equation 9.

$$\% \text{ error } = \left|\frac{\mu_{SPIN}^{SR} - \mu_{SPIN}^{BR}}{\mu_{SPIN}^{SR}}\right| \cong \left|\left(\frac{a - 2/3}{a}\right)\left(1 + \frac{3\langle L_Z \rangle}{2\langle S_Z \rangle}\right)\right| \qquad [9]$$

Using the values from Reference 1, $1+3\langle L_Z \rangle / 2\langle S_Z \rangle \leq 1.25$. As an example, consider Fe/Cu(001), where a $\cong 0.73$ and $\langle L_Z \rangle / \langle S_Z \rangle = 0.1$: Here the % error would be 10%. For Ni and Co the error would probably be higher: To be conservative, a value 20% can be used as a rule of thumb.

Thus the approximate analysis procedure using branching ratios, based upon an atomic, single-electron picture and assuming strong orbital quenching, can give the same value of μ_{SPIN} as the sum rule approach, to within 20% or better. This can be done without explicit knowledge of $\langle L_Z \rangle$ and without cross-normalization between spectra. Furthermore, the BR approach permits a proper correction for non-ideal ($|P_{hv}|<100\%$) polarization. Finally, it is necessary to note that both of these models are grounded in localized atomic or ionic pictures. Band dispersion and multi-electronic effects ultimately must be included for a full analysis[7].

Acknowledgments

Work performed under the auspices of the U.S. Department of Energy by Lawrence Livermore National Laboratory under contract number W-7405-ENG-48. This work is based upon work supported by the National Science Foundation under grant number DMR 94-58004. The measurements were performed at the Stanford Synchrotron Radiation Laboratory (SSRL), which is supported by the Chemical Sciences Division of the Office of Basic Energy Sciences in the U.S. Department of Energy. Special thanks to Michael Rowen of SSRL for his work on

Beamline 8–2. We also wish to thank Ms. Karen Clark for her clerical support of this work. Conversations with P. Carra and T. Thole were enlightening and enjoyable.

References
1. P. Carra, B.T. Thole, M. Altarelli, and X. Wang, Phys. Rev. Lett. 70, 694 (1993).
2. P. Carra, Syn. Rad. News. 5, 21 (1992).
3. P. Carra, Jpn. J. Appl. Phys. Preprint.
4. J.G. Tobin, G.D. Waddill, and D.P. Pappas, Phys. Rev. Lett. 68, 3642 (1992).
5. J.G. Tobin, G.D. Waddill, A.F. Jankowski, D.P. Pappas, and P.A. Sterne submitted to Phys. Rev. B, 1995.
6. G. Vanderlaan and B.T. Thole, Phys. Rev. B 43, 13401 (1991); 42, 6670 (1990); 38, 3158 (1988).
7. N.V. Smith, C.T. Chen, F. Sette, and L.F. Mattheiss, Phys. Rev. B 46, 1023 (1992).

SOFT X-RAY OPTICAL ROTATION
AS ELEMENT-SPECIFIC MAGNETO-OPTICAL PROBE

J.B. KORTRIGHT AND M. RICE
Center for X-ray Optics, Lawrence Berkeley Laboratory, Berkeley, California 94720 USA

ABSTRACT

Continuously tunable multilayer linear polarizers extend magneto-optical rotation techniques into the extreme ultraviolet and soft x-ray regions. Resonant magneto-optical rotation is large at the Fe $L_{2,3}$ edges. Magneto-optical Kerr rotation can be either bulk or surface sensitive depending on incidence angle. A SMOKE hysteresis loop measured from the Fe in an Fe/Cr multilayer reveals high sensitivity and a complex magnetic response of Fe.

INTRODUCTION

Several different element specific extreme ultraviolet and soft x-ray spectroscopic techniques have been developed recently as probes of magnetic materials. Large resonant magneto-optical effects are associated with the 3p and 2p levels of, e.g., Fe, Co and Ni (ranging in photon energy from 50 to 900 eV) because of the strong dipole transition matrix elements coupling these states directly and exclusively to the 3d states responsible for magnetism. Magnetic circular dichroism (MCD) using total electron yield and requiring a large degree of circular polarization has revealed large signals at the $L_{2,3}$ edges of these elements[1-3] and has provided a means of distinguishing orbital and spin contributions to magnetism.[4,5] MCD with fluorescence detection has been used to measure hysteresis loops of different elements in multicomponent samples.[6] Intensity measurements of Kerr effects (no polarization analysis) at the Co $L_{2,3}$ edges have shown large signals in transverse and longitudinal geometries using linear and circular polarization, respectively.[7,8] Photoemission spectroscopy techniques are sensitive to magnetic properties, both through spin-resolved measurements of the emitted electrons[9] and through dichroism in the angular distribution of photo-electrons emitted from a remnantly magnetized sample in certain geometries.[10]

We report on the development and use of linear polarizers to extend into the soft x-ray magneto-optical rotation techniques common in the IR, visible and UV ranges. A previous study demonstrates that soft x-ray Faraday rotation can be larger in the soft x-ray than at these longer wavelengths.[11] Here we present the first measurement of the magneto-optical Kerr effect (MOKE) in the soft x-ray and demonstrate its application in measuring hysteresis loops. Like the above techniques, x-ray MOKE is element specific and offers the ability to study element-resolved contributions to the net magnetic properties of homogeneous or heterogeneous multicomponent materials. In the x-ray range, MOKE can be either bulk or surface sensitive depending on the incidence angle.

Mat. Res. Soc. Symp. Proc. Vol. 384 ©1995 Materials Research Society

BACKGROUND

Magneto-optical effects result from differential interaction of the net electron angular momentum with circular components of light and manifest themselves through various measurement channels. Since light is equivalently decomposed into orthogonal circular or linear components, linearly polarized light is equivalent to the coherent superposition of equal amounts of left (l) and right (r) circularly polarized light. The propagation of these different components is described by a refractive index which in the x-ray range is expressed as $n(\lambda) = 1 - \delta(\lambda) - i\beta(\lambda)$ were $\delta(\lambda)$ and $\beta(\lambda)$ are related through a Kramers-Kronig dispersion relation. Magnetic circular dichroism, $\beta_l - \beta_r$, can be measured directly if circular polarization is available, or indirectly through the ellipticity induced in an initially linearly polarized beam, if linearly polarizing optical elements are available. Magnetic circular birefringence, $\delta_l - \delta_r$, does not produce directly observable effects using circularly polarized light, but rotates the plane of polarization of initially linearly polarized light. Linear polarizers can easily measure this rotation, yielding commonly used magneto-optical probes in the IR, visible and UV.

The extension of magneto-optical rotation techniques into the soft x-ray thus requires high quality linear polarizers, which are obtained in the form of multilayer interference structures designed to position the first order Bragg interference peak at 45° (the Brewster angle in the x-ray region). Figure 1 shows the calculated ratio of the s to p component reflectances at this condition for multilayers of different constitution. Generally the extinction ratio is quite high, and increases with hv. The extinction ratio is relatively insensitive to multilayer imperfections, and can be increased beyond those values in the figure by multiple reflections or by working closer to the precise Brewster angle, which deviates significantly from 45° at lower energies.[12] Peak reflectance for the s-component is expected to range from roughly 0.5 at 50 eV to 0.01 at 700 eV for the materials designated in the figure. Continuously tunable linear polarizers are obtained by

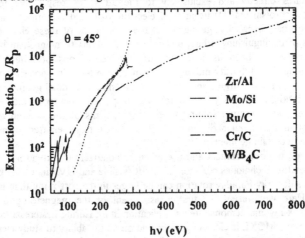

Figure 1. Calculated extinction ratios of s to p component peak reflectances for a series of multilayers designed to position the first order Bragg peak at 45°. These different materials combinations as noted are typical of those used in practice at these energies. Calculations assume ideal structures.

laterally grading the multilayer period and translating the optic along the gradient to shift the 45° Bragg peak spectral position according to $\lambda = 2d\sin(45°)$. Such continuously tunable polarizers have been incorporated into a spectro-polarimeter[13] which was used in the measurements reported here.

Faraday rotation on transmission through an Fe/Cr multilayer across the Fe $L_{2,3}$ edges was measured using this spectro-polarimeter as reported in ref. 11. Here we note that the specific rotation values observed at the L_3 resonance are larger than for Fe in the infrared, visible, ultraviolet, and hard x-ray regions. Such large rotation is consistent with the large MCD measured at the 2p edges of Fe, Co and Ni,[1-3] and implies that soft x-ray magneto-optical rotation studies can generally be expected to yield useful element-resolved effects.

SOFT X-RAY MAGNETO-OPTICAL KERR ROTATION

Kerr magneto-optical rotation (on reflection) is perhaps of greater value than Faraday rotation for materials studies in the soft x-ray, since ultrathin transmission samples are generally less available than reflection samples. In the soft x-ray range the magneto-optical Kerr effect can be either bulk sensitive or surface sensitive depending on the incidence angle and photon energy as indicated in Figure 2. Figure 2 (a) and (b) show calculated reflectance and penetration depth (s component) for Fe at two energies just below and above the L_3 ($2p_{3/2}$) and L_2 ($2p_{1/2}$) edges, respectively. Calculations use optical constants of ref. 14. For incidence angles below the critical angle for total external reflection the penetration depth is limited to several nm. At greater angles the penetration depth is proportional to the angle, with proportionality constant dependent on β. Figure 2 (c) and (d) show similar calculations with hv below and above the Fe $M_{2,3}$ (3p) edge at 52.7 eV, revealing similar trends in depth sensitivities. At these lower photon energies absorption is a more significant limit to penetration depth above the critical angle, which occurs at significantly higher angles. These calculations demonstrate that it is generally possible to obtain surface or bulk sensitivity in MOKE measurements across a broad energy range by tuning the incidence angle with respect to the critical angle for total external reflection. Thus magnetic properties of buried interfaces or layers can be studied.

Our first measurement of Kerr rotation used a surface sensitive longitudinal geometry to measure hysteresis loops from Fe in an Fe/Cr multilayer. The multilayer consisted of 40 periods of Fe (20 Å) and Cr (19 Å) magnetron sputter deposited onto a Si wafer. Cr was the top layer. This structure is consistent with ferromagnetically coupled Fe layers.[15] With a 1° incidence angle only the top one or possibly two Fe layers contribute to the measured signal (see Fig. 2(b)). Measurements were made using bending magnet beamline 6.3.2 at the Advanced Light Source at LBL. A vertical aperture selected an incident beam having 0.99 degree of linear polarization as determined from measurements with the spectro-polarimeter. The sample was mounted with its surface on the axis of a solenoidal electromagnet, both of which were positioned to reflect the beam with a 1° incidence angle into the polarimeter, which was aligned to accept the reflected beam. Based on prior Faraday rotation measurements we tuned the incident energy and the polarizer to the low energy inflection point of the Fe L_3 white line (706.5 eV) where magneto-optical rotation associated with Fe is maximum.[11] Measurements of SMOKE (surface MOKE) hysteresis loops were made with the polarizer at several different azimuthal angles with respect to the crossed position. Figure 3 shows the hysteresis loop measured at an azimuth 30° from the crossed position and averaged over roughly 1.5 seconds of measurement per point.

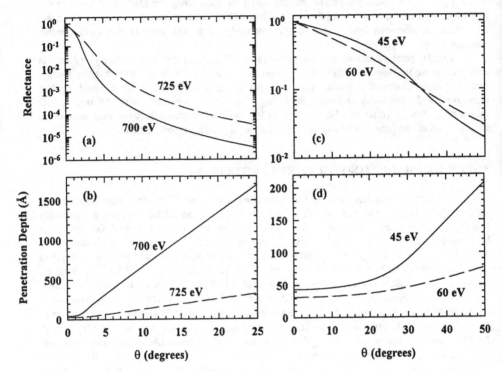

Figure 2. (a) and (b) show, respectively, reflectivity and penetration depth (s component) calculated for an ideal Fe surface at two energies just above and below the Fe $L_{2,3}$ edges as noted. (c) and (d) show similar calculated data with photon energies on either side of the Fe $M_{2,3}$ edge. Surface or bulk sensitivity can be obtained with the incidence angle below or above the critical angle for total external reflection, respectively.

Several features of this soft x-ray SMOKE hysteresis loop are noteworthy. First, it represents the magnetic response of only the Fe atoms in the Fe/Cr multilayer sample, because of the resonant nature of soft x-ray magneto-optical effects. Second, its shape contains a ferromagnetic component and an apparent paramagnetic or antiferromagnetic contribution represented by the sloping data beyond saturation of the ferromagnetic loop. Similarly shaped bulk hysteresis loops for Fe/Cr multilayers coupled antiferromagnetically have been observed,[16] and were interpreted to result from a residual ferromagnetic component in samples having an odd number of Fe layers. The sample we studied is thought to be ferromagnetically coupled. Since our measurement is primarily sensitive to just the top Fe layer, the origin of the shape of the the hysteresis loop in Fig. 3 may or may not be similar to that in this previous work, which is also sensitive to any magnetic response of the Cr in the samples. Further investigations will include measurements at both Fe and Cr L_3 edges of samples with differing layer thicknesses using different penetration depths and higher applied fields. Third, the resonant rotation is large. The

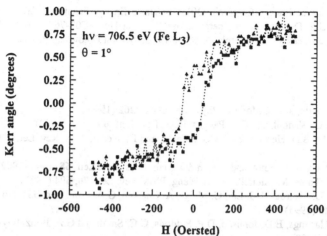

Figure 3. Longitudinal SMOKE hysteresis loop from an Fe/Cr multilayer reveals the magnetic response of just the Fe in the sample to include a ferromagnetic loop and an apparent paramagnetic component.

remnant magnetization corresponds to a Kerr rotation angle jump of 0.6°. If we assume that this ferromagnetic component results from the uppermost 20 Å thick Fe layer and that the Fe is pure ferromagnetic bcc, then this corresponds to roughly 0.06° or 1 mrad rotation per monolayer of Fe atoms, which is comparable to the noise level in the data in Figure 3. With improvements to increase the signal-to-noise ratio beyond that in this first measurement, this large soft x-ray resonant magneto-optical response will yield monolayer sensitivity.

SUMMARY

Multilayer optical elements extend magneto-optical rotation techniques into the extreme ultraviolet and soft x-ray range. These techniques will be useful over at least the 50 - 900 eV range including the 3p and 2p levels of most 3d transition metal as well as 4p levels of rare earth elements of interest in magnetism. Both Faraday and Kerr rotation have been observed at the Fe $L_{2,3}$ edges, where they yield large effects compared to other spectral regions. Soft x-ray MOKE can be either bulk or surface sensitive depending on incidence angle, and the measurement of hysteresis loops is easily accomplished. An initial measurement of the hysteresis loop from just the Fe in an Fe/Cr multilayer reveals a complex magnetic response of the Fe in the sample which is under further investigation. Thus soft x-ray optical rotation techniques using optics to measure rotation of linear polarization are clearly useful to obtain element-specific information from complex homogeneous or heterogeneous magnetic samples, and as such are applicable to a wide variety of materials of fundamental and practical interest.

ACKNOWLEDGEMENTS

We acknowledge assistance from J. Underwood, E. Gullikson, R. Steele, D. Kemp, S. Klingler and R. Delano. D. Humphries designed the electromagnet. This work was supported by the Director, Office of Energy Research, Office of Basic Energy Sciences, Materials Sciences Division, of the U.S. Department of Energy under Contract No. AC03-76SF00098. M.R. was supported by the University of California President's Postdoctoral Fellowship Program.

REFERENCES

1. C.T. Chen, F. Sette, and S. Modesti, Phys. Rev. B **42** 7262 (1990).
2. J.G. Tobin, G.C. Waddill, and D.P. Pappas, Phys. Rev. Lett. **68**, 3642 (1992).
3. Y. Wu, J. Stohr, B.D. Hermsmeier, M.G. Samant, and D. Weller, Phys. Rev. Lett. **69**, 2307 (1992).
4. B.T. Thole, P. Carra, F. Sette, and G. van der Laan, Phys. Rev. Lett. **68**, 1943 (1992).
5. P. Carra, B.T. Thole, M. Altarelli, and X.Wang, Phys. Rev. Lett. **70**, 694 (1993).
6. C.T. Chen, Y.U. Idzerda, H.-J. Lin, G. Meigs, A. Chaiken, G.A. Prinz, and G.H. Ho, Phys. Rev. B **48**, 642 (1993).
7. C. Kao, J.B. Hastings, E.D. Johnson, D.P. Siddons, G.C. Smith and G.A. Prinz, Phys. Rev. Lett. **65**, 373 (1990).
8. C.-C. Kao, C.T. Chen, E.D. Johnson, J.B. Hastings, H.J. Lin, G.H. Ho, G. Meigs, J.-M. Brot, S.L. Hulbert, Y.U. Idzerda, and C. Vettier, Phys. Rev. B **50**, 9599 (1994).
9. F.U. Hillebrecht, R. Jungblut, and E. Kisker, Phys. Rev. Lett. **65**, 2450 (1990).
10. L. Baumgarten, C.M. Schneider, H. Petersen, F. Schafers, and J. Kirschner, Phys. Rev. Lett. **65**, 492 (1990). See also G. Rossi, these proceedings, and references therin.
11. J.B. Kortright, M. Rice, and R. Carr, Phys. Rev. B **51**, 10240 (1995).
12. J.B. Kortright, Proc. SPIE **2010**, 160 (1993).
13. J.B. Kortright, M. Rice, and K. Franck, Rev. Sci. Instrum. **66**, 1567 (1995).
14. B.L. Henke, E.M. Gullikson, and J.C. Davis, At. Data Nucl. Data Tables **54**, 181 (1993).
15. S.S.P. Parkin, N. More, and K.P. Roche, Phys. Rev. Lett. **64**, 2304 (1990).
16. S.S.P. Parkin, A. Mansour, and G.P. Felcher, Appl. Phys. Lett. **58**, 1473 (1991).

[197]Au MÖSSBAUER STUDY OF
Au/Fe, Au/Co AND Au/Ni MAGNETIC MULTILAYERS

SABURO NASU AND YASUHIRO KOBAYASHI
Department of Material Physics, Faculty of Engineering Science, Osaka University, Toyonaka, Osaka 560, Japan

ABSTRACT

[197]Au Mössbauer measurements have been performed for Au/Fe, Au/Co and Au/Ni magnetic multilayers. [197]Au Mössbauer spectra observed from multilayers consist of at least 4 components having different magnitude of hyperfine fields and isomer shift values those depend on the local environments of the Au probe-atoms in multilayers. Rather large electron spin polarization at Au atoms has been observed in the interface with adjacent ferromagnetic layers, but nearly no magnetic hyperfine field has been observed to the interior of Au layer. It implies that the largest hyperfine field observed is not due to the conduction-electron polarization but induced by the hybridization in the interface with ferromagnetic layer.

INTRODUCTION

Artificial multilayered films are prepared by alternately depositing two elements using the vacuum deposition or sputtering techniques. A modulation in composition is constructed along the direction normal to the film plane. The structure and physical properties of such films have been expected unique because these films do not exist in nature. Many researchers have been devoted to investigate the physical property of artificial multilayered films [1]. To understand the microscopic properties, especially magnetic properties of the multilayers, various investigations using nuclear methods like Mössbauer spectroscopy and nuclear magnetic resonance technique have been reported [1, 2]. In applications of Mössbauer spectroscopy, the most common isotope [57]Fe was usually used as a probe atom for multilayers. Other isotopes like [119]Sn and [197]Au are also quite useful as probe atoms for multilayers constructed by two alternated layers of non-magnetic and magnetic one, because these non-magnetic elements might make it possible to search the magnetic properties of conduction electrons in the multilayers. Anyhow the results from the microscopic methods like Mössbauer spectroscopy may offer a new sight to elucidate the physical properties of multilayers.

For Au/Ni multilayer, the enhancement of the elastic modulus so called "supermodulus" effect has been reported and a recent X-ray diffraction study using an external pressure has supported the existence of supermodulus effect in Au(1 nm)/Ni(1 nm) multilayer [3]. In previous investigation, we performed [197]Au Mössbauer measurements for Au(1 nm)/Ni(1 nm) multilayer and determined the recoilless-fraction of [197]Au Mössbauer resonance in order to clarify its supermodulus effect [2]. Results obtained from [197]Au Mössbauer measurements show the larger recoilless-fraction and the higher Debye temperature of Au comparing to the bulk Au metal supporting the existence of the supermodulus effect in this multilayer.

467

Mat. Res. Soc. Symp. Proc. Vol. 384 ° 1995 Materials Research Society

In the present investigation, at first the [197]Au Mössbauer measurements have been performed for the Au-Ni, Au-Co and Au-Fe alloy specimens those were prepared by melt. Subsequently we performed the measurements for Au/Ni, Au/Co and Au/Fe multilayers in order to clarify the magnetic properties of Au spacer layers sandwiched by ferromagnetic Ni, Co and Fe layers. Ferromagnetic Fe, Co and Ni layers are 0.8 nm, 2 nm and 2 nm in thickness for each multilayer and thickness of the Au layers varies from 0.5 to 3 nm. [197]Au Mössbauer spectra observed from multilayers consist of at least 4 components having different magnitude of hyperfine fields and isomer shift values that depend on the local environments of the Au probe-atoms in multilayers. Rather large electron spin polarization at Au atoms has been observed in the interface with adjacent ferromagnetic layers, but nearly no magnetic hyperfine field has been observed to the interior of Au layer. Magnitude of the largest magnetic hyperfine field has been determined to be 115 T, 69 T and 23 T in the interfaces of Au(1 nm)/Fe(0.8 nm), Au(1 nm)/Co(2 nm) and Au(1 nm)/Ni(2 nm) multilayers, respectively. It implies that the largest hyperfine field observed is not due to the conduction-electron polarization but induced by the hybridization in the interface with ferromagnetic layer.

EXPERIMENTAL

The multilayers used for the present investigation are prepared by the electron-beam evaporation technique in an ultrahigh vacuum chamber. Thicknesses of the Au/Ni, Au/Co and Au/Fe multilayers, number of stacked layers and the Ag buffer and overlayer are as follows;

Au(0.5 nm)[Ni(2 nm)/Au(0.5 nm)]$_{59}$,
Au(1 nm)[Ni(2 nm)/Au(1 nm)]$_{29}$,
Au(2 nm)[Ni(2 nm)/Au(2 nm)]$_{14}$,
Ag(100 nm)[Au(0.5 nm)/Co(2 nm)]$_{100}$Ag(5 nm),
Ag(100 nm)[Au(1 nm)/Co(2 nm)]$_{70}$Ag(5 nm),
Ag(100 nm)[Au(2 nm)/Co(2 nm)]$_{40}$Ag(5 nm),
Ag(100 nm)[Au(0.5 nm)/Fe(0.8 nm)]$_{120}$Ag(5 nm),
Ag(100 nm)[Au(1 nm)/Fe(0.8 nm)]$_{60}$Ag(5 nm),
Ag(100 nm)[Au(2 nm)/Fe(0.8 nm)]$_{50}$Ag(5 nm) and
Ag(100 nm)[Au(3 nm)/Fe(0.8 nm)]$_{30}$Ag(5 nm).

Schematic drawing of the Au/Co multilayers is shown in Fig. 1. These multilayer specimens have been used for the [197]Au Mössbauer measurements. [57]Fe Mössbauer measurements have been also performed for Au/Fe multilayers using same specimens as [197]Au Mössbauer measurements. Structural analysis of the Au/Ni superlattice has been performed using X-ray diffraction technique and the direct observation of cross sectional view by a high resolution transmission electron microscope [4]. Lattice relationships have been determined as $(111)_{Ni}//(111)_{Au}$ and $[2\bar{2}0]_{Ni}//[2\bar{2}0]_{Au}$ between Au and Ni layers and $(110)_{Fe}//(111)_{Ni}$.between Au and body-centered cubic Fe layers. Structural analysis of Au/Co multilayers have not yet been completed.

For [197]Au Mössbauer measurements, [197]Pt gamma-ray source was prepared by the reaction of [196]Pt(n,γ)[197]Pt in Pt metal foil using Kyoto University Reactor (KUR). Superlattice specimens were used as gamma-ray absorbers in a transmission geometry. Source and absorber were cooled by a closed cycle refrigerator and usual measurements were performed at 16 K and 11 K.

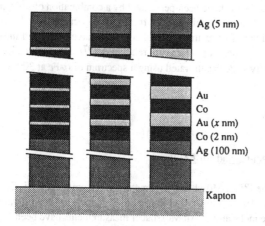

Fig. 1 Schematic Drawing of Au/Co multilayers stacked on Kapton film with Ag buffer and cover layers.

Ag (5 nm)

Au
Co
Au (*x* nm)
Co (2 nm)
Ag (100 nm)

Kapton

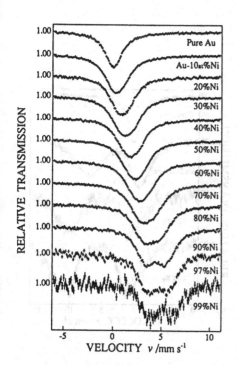

Fig. 2. Typical ^{197}Au Mössbauer spectra obtained from Au-Ni alloys at 16 K. Specimens were annealed and quenched from 1173 K.

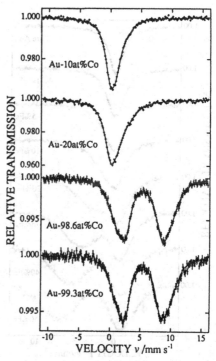

Fig. 3. Typical ^{197}Au Mössbauer spectra obtained from Au-Co alloys at 11 K. Specimens were annealed and quenched from 1253 K.

469

Transmission ^{57}Fe Mössbauer measurements have been performed by a combination of ^{57}Co in Rh source and absorber specimen. Conventional velocity transducer operated by a constant acceleration mode and data acquisition system using personal computer have been used in time mode for Mössbauer measurements. Isomer shift values are relative to ^{197}Au in Au metal at 16 K and ^{57}Fe in α-Fe at 300 K. Velocity scale is calibrated using a spectrum of α-Fe at 300 K.

RESULTS AND DISCUSSION

^{197}Au Mössbauer effects of Au-Ni, Au-Co and Au-Fe alloys

In order to determine precisely the ^{197}Au Mössbauer parameters in ferromagnetic Ni, Co and Fe and also in ferromagnetic alloys, Au-Ni, Au-Co and Au-Fe alloy specimens having various compositions were prepared from the melts and ^{197}Au Mössbauer measurements have been

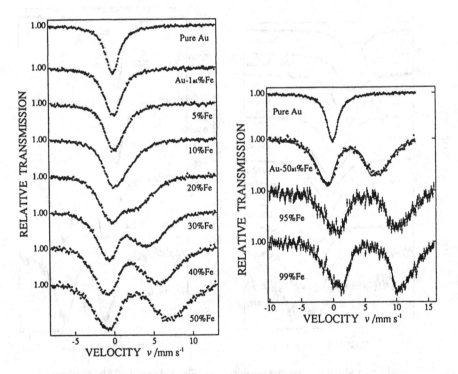

Fig. 4. Typical ^{197}Au Mössbauer spectra obtained from Au-Fe alloys at 11 K. Solid lines in the spectra shown in right are the results from least-square fits using a distribution of the Hi and IS for Au-50 at.%Fe and single sets of the hyperfine interaction parameters for Au-95%Fe and Au-99%Fe.

performed for these alloy specimens. For Au-Ni alloys whose Ni concentrations are from 10 % to 99 %, the specimens were annealed at 1173 K for 24 hours in vacuum and then quenched into water in order to get random solid solutions, because the Au-Ni alloy system decomposed into 2 phases of Au-rich and Ni-rich phases at low temperature. For Au-Co alloys, the specimen compositions are 10 %, 20 %, 98.6 % and 99.3 % Co in Au because of the limited solid solubility of each elements. Au-Co alloy specimens were annealed at 1253 K and then quenched into water. For Au-Fe alloys, specimen composition was ranged from 1 % to 50 % Fe and two Fe-rich specimens of Au-95 % and 99 % Fe were prepared from the melts. These specimens were annealed at 1173 K and then quenched into water to get random solid solutions. ^{197}Au Mössbauer measurements have been performed by a combination of absorber specimen and a gamma-ray source of ^{197}Pt in Pt metal.

Figure 2 shows typical ^{197}Au Mössbauer spectra obtained from Au-Ni alloys as a function of Ni concentration. At the top of the figure, spectrum obtained from pure Au metal foil is shown and used as a reference spectrum. As an increase of the Ni concentration, the center of the gravity of the spectrum shifts toward the positive velocity side indicating an increase of the isomer shift (IS) values. The spectrum obtained from the specimen whose Ni concentration exceeds 40 % became to be broad and magnetic. The spectra from the specimens whose Ni concentrations are above 90 % are magnetically split doublets. Hyperfine magnetic field (Hi) and isomer shift (IS) value of Au in Ni-1 % Au have been determined to be 29 T and 5.0 mm/s relative to Au at 16 K. IS value is relative to pure Au metal at 11 K. Relationships between IS value and Ni concentration (Ni%) have been determined as follows;

$$IS = (0.046 \pm 0.001) \cdot Ni\% - (0.02 \pm 0.02) \quad \text{for } 0 \leq Ni\% \leq 40 ,$$
$$IS = (0.052 \pm 0.001) \cdot Ni\% - (0.24 \pm 0.09) \quad \text{for } 40 \leq Ni\% \leq 100.$$

Above relation shows clearly the linear dependence of the charge density at 197Au nucleus on the concentration of the alloy. Linear relationship between Hi and Ni% also determined from the spectra obtained for the specimen whose Ni concentration is above 60 %.

Figure 3 shows typical ^{197}Au Mössbauer spectra obtained from various Au-Co alloys. Hi and IS values of Au in Co-0.7 % Au have been determined to be 86 T and 5.5 mm/s at 11 K. Figure 4 shows typical ^{197}Au Mössbauer spectra obtained from Au-Fe alloys. In Au-Fe alloy system, the Fe atom has local magnetic moment even in dilute alloy concentration range and for Au-20 % Fe alloy specimen the ^{197}Au Mössbauer spectrum shows the existence of rather large hyperfine magnetic fields. Hi and IS values at ^{197}Au nucleus in Fe-1 % Au are 126 T and 5.8 mm/s at 11 K. IS value is relative to pure Au metal at 11 K. From the above measurements for the alloy systems, we can determined the relation between IS, Hi and alloy concentration for each alloy

Table I. Hyperfine magnetic fields (Hi) and Isomer shifts (IS) at ^{197}Au nucleus in ferromagnetic Ni, Co and Fe matrixes. IS is relative to pure Au at each temperature.

	IS, mm/s	Hi, T	Temperature, K
Ni-1 at.%Au	5.0	29	16
Co-0.7 at.%Au	5.5	86	11
Fe-1 at.%Au	5.8	126	11

systems. Table I shows the hyperfine magnetic fields Hi and isomer shift IS values for the isolated Au impurities in ferromagnetic Ni, Co and Fe.

^{197}Au Mössbauer effects of Au/Ni, Au/Co and Au/Fe multilayers

Figure 5 shows typical ^{197}Au Mössbauer spectra obtained from Au/Ni multilayers with a spectrum of pure Au metal. The ^{197}Au Mössbauer spectra in Fig. 5 have been obtained from the specimens that were stacks of about 150 layers of the [Au(2 nm)/Ni(2 nm)]$_{14}$/Au(2 nm), [Au(1 nm)/Ni(2 nm)]$_{29}$/Au(1 nm) and [Au(0.5 nm)/Ni(2 nm)]$_{59}$/Au(0.5 nm) multilayers corresponding to about 8.7 mgAu/cm^2 (4.5 mm in thickness). In Fig. 5 the dots show the experimental results at 16 K and solid lines are results from the least-square fit using several Lorentz functions. Dotted lines in the spectra are the partial components of the spectra. Three spectra for multilayers show rather broad absorption lines indicating non-symmetrical lineshape that suggests the

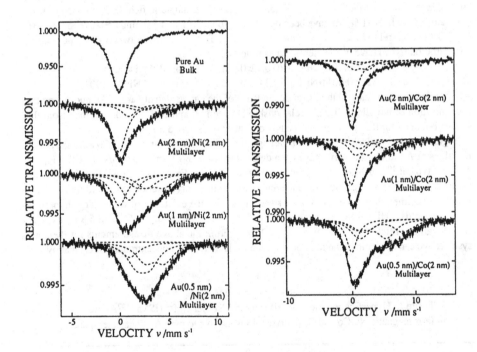

Fig. 5 Typical ^{197}Au Mössbauer spectra at 16 K obtained from Au/Ni multilayers with a spectrum of pure Au

Fig. 6 Typical ^{197}Au Mössbauer spectra at 11 K obtained from Au/Co multilayers.

472

existence of the components at positive velocity side, that is, positive isomer shift values. These spectra can be decomposed mainly into two components. One component is a rather sharp single line having a zero isomer shift. This component is due to the Au atoms having an identical electronic and magnetic environment to pure Au bulk metal. The other components show rather broad and complicated components having positive isomer shift (IS) values. Since their broadness and intensity depend on the thickness of the Au layers, the components observed at positive velocity side are due to the Au atoms whose electronic states are perturbed by the existence of the ferromagnetic Ni layers. Thickness of the Au layer in Au(1 nm)/Ni(2 nm) is correspond to 4 layers of (111) Au plane and all of 12 nearest neighbor sites of Au atoms at interior two layer could be occupied by Au atoms when the interface between Au and Ni layer is perfectly flat. From the area ratio between perturbed and non-perturbed components in spectra the region perturbed by the Ni layer is determined to be about one Au layer from interface assuming all of the Au atoms have an identical Debye-Waller factor in multilayer, which means two Au layers are affected at both sides by Ni layers. As shown in the lowest part of Fig. 5, the spectrum for Au(0.5 nm)/Ni(2 nm) multilayer that corresponds to only two Au layers is entirely perturbed one and shows the correctness of the above interpretation. To analyze the spectra for Au/Ni multilayers, four different models are adopted. First one is two-components model mentioned above. The second and third models are to analyze a spectrum using discrete components as 4 and 8 subspectra having different isomer shifts and magnetic hyperfine interactions. Fourth model is to analyze the spectrum using a continuous distribution of isomer shift (IS) and magnetic hyperfine field (Hi) values. Dotted lines in Fig. 5 are the partial components obtained by the analysis using a model of 4 discrete components. The component having the largest positive isomer shift has the isomer shift of 3.61 mm/s and the hyperfine magnetic field of 23 T at 16 K. This component might correspond to the Au atoms that occupy the interface sites and associated with Ni atoms being perturbed most strongly. Existence of the magnetic components in ^{197}Au Mössbauer spectrum of multilayers directly indicates the existence of the spin-polarized electrons at Au atoms associated with the ferromagnetic Ni layer.

Figure 6 shows typical ^{197}Au Mössbauer spectra obtained from Au(2 nm)/Co(2 nm), Au(1 nm)/Co(2 nm) and Au(0.5 nm)/Co(2 nm) multilayers at 11 K. Dotted lines are the partial components obtained by the analysis using a model of 4 discrete components. As shown in Fig. 5, the intensities of the components having larger IS values increase as an decrease of the thickness of the Au layer. Similarly to the case of Au/Ni multilayers, the components having positive IS values correspond to the Au atoms which locate at or near interface and perturbed strongly with the ferromagnetic Co layer. The magnitude of the hyperfine magnetic fields Hi of these components for Au/Co multilayers are larger than the case of Au/Ni.

Figure 7 shows typical ^{197}Au Mössbauer spectra obtained from Au/Fe multilayers. As shown in Fig. 7, components having rather large hyperfine magnetic fields have been observed superimposed to the single non-magnetic component having a zero IS value. Magnitude of the largest hyperfine magnetic field was determined to be 115 T from the spectrum of Au(1 nm)/Fe(0.8 nm) and the component having the largest Hi corresponds to the Au at or near interface with ferromagnetic Fe layer. Single non-magnetic component having a zero IS value corresponds to the Au whose electronic state is identical to the pure Au metal and associates to the interior atoms in the Au layer which may have no-influence from the ferromagnetic Fe layers. Distribution of the hyperfine magnetic fields at ^{197}Au nucleus has been observed for Au(0.5 nm)/Fe(0.8 nm) multilayer, which might be interpreted by the atomic mixing between Au and Fe

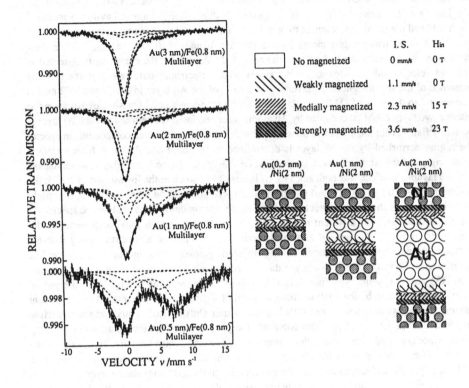

Fig. 7 Typical [197]Au Mössbauer spectra at 11 K obtained from Au/Fe multilayers.

Fig. 8 Schematic representation of the magnetic perturbations of the Au spacer layer sandwiched by ferromagnetic layers for the case of Au/Ni multilayers.

at the interface.   Using same Au/Fe multilayer specimens, we measured [57]Fe Mössbauer spectrum at 300 K and observed well-defined magnetically split sextet suggesting no-appreciable atom-mixing at the interface.   The above results suggest that the largest hyperfine magnetic field observed is not due to the conduction-electron polarization but induced by the hybridization at the interface with ferromagnetic Fe layer.   Magnitude of the largest hyperfine magnetic fields has been determined 69 T and 23 T in the interfaces of Au(1 nm)/Co(2 nm) and Au(1 nm)/Ni(2 nm) multilayers, respectively. Schematic representation of the magnetic perturbations for the Au spacer layer sandwiched by ferromagnetic layers is shown in Fig. 8 for the case of Au/Ni multilayers.

SUMMARY

Using [197]Au Mössbauer spectroscopy, magnetic properties of Au spacer layers sandwiched by the ferromagnetic Ni, Co and Fe layers have been investigated. Mössbauer spectra obtained from magnetic multilayer specimens consist of at least two components. One is a non-magnetic component having same isomer shift value as bulk Au. The other components are magnetically split components having the positive isomer shift values. From the above results, we interpreted that the magnetically split components are due to the Au atoms those are strongly perturbed at/near interface by the ferromagnetic layer. Origin of the hyperfine magnetic field is due to the direct hybridization with ferromagnetic 3d electrons. Contribution of the conduction electron polarization to the hyperfine magnetic field at [197]Au nucleus is small.

ACKNOWLEDGMENTS

The authors thank to Professor T. Shinjo of Kyoto University for his collaboration and Prof. Y. Nakayama and Messrs. T. Shibatani, H. Dohnomae and N. Mima for their help in experiments and to Profs. Y. Maeda and H. Sakai, Dr. M. Seto and Mr. S. Kitao at Research Reactor Institute of Kyoto University (KUR) for their kind cooperation in [197]Au Mossbauer spectroscopy.

REFERENCES

1. T, Shinjo, in Metallic superlattices, edited by T. Shinjo and T. Takada, (Elsevier, Amsterdam, 1987).pp. 1
2. Y. Kobayashi and S. Nasu, J. Phys. Soc. Jpn, (1995) in press.
3. H. Konishi, Y. Fujii, N. Hamaya, H. Kawada, Y. Ohishi, N. Nakayama, L. Wu, H. Dohnomae, T. Shinjo and T. Matsushita, Rev. Sci. Instrum., 63 1035 (1992).
4. H. Dohnomae, N. Nakayama and T. Shinjo, Materials Trans. JIM, 31 615 (1990).

A NEW MAGNETOOPTICAL EFFECT
DISCOVERED ON MAGNETIC MULTILAYERS :
THE MAGNETOREFRACTIVE EFFECT

J. C. JACQUET AND T. VALET
THOMSON CSF, Laboratoire Central de Recherches, Domaine de Corbeville, 91 404 Orsay
Cedex, France

ABSTRACT

We show theoretically that the change in the magnetization structure of magnetic metallic multilayers under the application of a magnetic field shall be generally associated with a significant change of the refractive index. This constitutes a new magnetooptical effect : the magnetorefractive effect.

Optical transmission measurements under an applied magnetic field through [Ni$_{80}$Fe$_{20}$/Cu/Co/Cu] multilayers, in the light wavelength region between 2 μm and 20 μm, clearly demonstrate the existence of the predicted effect and are found in reasonnable agreement with the theoretical calculations.

INTRODUCTION

Metallic magnetic multilayers have been a subject of great interest in the last few years, since they exhibit unique physical properties : giant magnetoresistance [1,2], oscillatory interlayer exchange coupling [3], giant magnetothermal conductivity and giant magnetothermopower [4]. Their optical and magnetooptical properties at visible wavelengths have been also investigated, mainly in connection with magnetic data-storage applications. At such wavelengths the interband contribution to the optical constants is very significant. Thus, considering interband optical properties as resulting from a weighted average over all the electronic transitions between initial and final states, one expects some influence of the artificial layering on the optical and magnetooptical constants through quantum size effects. Reports of oscillations of the saturation Kerr rotation with the interlayer thickness in Fe/Cu/Fe [5], Fe/Au/Fe and Fe/Ag/Fe [6] sandwiches, due to the appearance of quantum well states, are among the recent demonstrations of such quantum size effects.

Here we will be interested in a totally different kind of size effects appearing in the optical properties of metallic surfaces, thin films and multilayers, when either the skin depth (SD, the decay length of the optical electric field in the metal), or the layer thicknesses, become of the order of the conduction electron mean free path (MFP). These are quasiclassical effects, related to the non locality of the intraband optical conductivity on length scales shorter than the MFP. The anomalous skin effect, which was thoroughly analyzed by Reuter and Sondheimer [7], is a prototype of this second kind of size effects. When considering the specific case of an optical wave incident on a metallic multilayer, even in the normal skin effect regime MFP << SD, it is clear that the non locality cannot be neglected if the layer thicknesses are shorter than the MFP. This leads to effective optical constants for the multilayer which depend in an intricate way upon the thicknesses and material parameters characterizing the stacking [8]. It is clear also that such effects show up more clearly at infrared wavelengths, where the main contribution to the optical constants comes from the intraband conduction electron response (Drude).

In the present paper we report on the extension of these concepts to the case of magnetic metallic multilayers made of transition or noble metals, which means to take into account properly

Mat. Res. Soc. Symp. Proc. Vol. 384 ° 1995 Materials Research Society

spin dependent scatterings and two spin channel conduction [9] in the calculation of the optical constants. We show that the non locality of the electron response to an external electric field, which is responsible for the giant magnetoresistance (GMR) under static conditions [1,10-11], shall lead at optical frequencies to a dependence of the dielectric constant upon the relative orientation of the layer magnetizations. This constitutes a new magnetooptical effect, that we propose to name the magnetorefractive effect. Extensive experimental confirmations of this prediction are given, based on optical transmission measurements under an applied magnetic field through [$Ni_{80}Fe_{20}$/Cu/Co/Cu] multilayers. Finally these experimental results are confronted with a theoretical calculation of the optical properties, relying on numerical integration of the Boltzmann-Maxwell coupled system of equations, and we found a reasonable agreement between the calculated and measured quantities.

THE PHYSICS OF OPTICAL CONDUCTIVITY IN MAGNETIC MULTILAYERS - THE MAGNETOREFRACTIVE EFFECT

All along this paper we will describe the intraband dynamic of the conduction electrons of metallic multilayers submitted to an harmonic electromagnetic field according to the semi-classical linearized Boltzmann equation. This basically means that our description is valid if the involved length scales remain larger than $k_F^{-1} \approx 1$ Å, and the frequency smaller than $E_F/h \approx 10^{15}$ Hz, where k_F and E_F are typical values of respectively the Fermi wavevector and energy of the considered metals. The latter restriction is reinforced by the obvious neglect of the interband transitions in such semi-classical description, which further limit the spectral domain of validity of our approach. Owing to the known results on the onset wavelength of interband transitions in transition metals [12], it means that we are dealing with infrared light above 5 μm of wavelength. Furthermore, we will make the extra assumptions that the electronic structure of all the metal of interest is well approximated by the same spin degenerated parabolic conduction band (no potential step at the metal-metal interfaces), and that one can define isotropic relaxation times in the bulk of the layers and specular transmission coefficients at the interfaces to take into account the scattering processes.

In this section, we will first briefly review the problem of optical wave propagation in an homogeneous metal, putting the emphasis on the non locality of the optical conductivity, in order to clarify the role of the various length scales which will be involved in the more complex case of metallic multilayers. Secondly, to obtain a physical insight into the kind of original optical properties one can expect from magnetic metallic multilayers, we will derive explicit expressions of the "effective" optical constants for a magnetic bimetallic multilayer, in what was called in the context of GMR theory the "self-averaging" limit (SAL) [13]. This will allow us to show that the optical constants of magnetic multilayers shall depend upon their magnetization arrangement. This can be considered as a new magnetooptical effect. Finally, we will discuss the general features of the effect out of the SAL, and we will compare this new effect to the well known first order (Kerr-Faraday effects) and second order (Cotton-Mouton or Voigt effect) magnetooptical effects. We will also make some predictions on the kind of experiments suitable to observe the effect.

Non local optical conductivity and transverse wave propagation in an homogeneous metal

Here we consider an homogeneous infinite metal characterized by an effective mass m, a Fermi velocity v_F, an electron density n, and an isotropic relaxation time τ at the Fermi energy; and we are looking for monochromatic transverse plane wave solutions of the Boltzmann-Maxwell coupled system of equations, propagating in the z direction and polarized along the x direction. Assuming a common time dependence $e^{i\omega t}$ for all the fields of interest, the Maxwell equations reduce to the scalar wave equation :

$$\frac{d^2E_x}{dz^2} + \varepsilon_{st}\,\kappa_0^2\,E_x = i\,\omega\,\mu_0\,j_x(z) \tag{1}$$

where ε_{st} is the residual dielectric constant which takes care of the displacement current, atomic polarization, virtual interband transitions, etc, and will be assumed independent of the frequency in the spectral domain of interest from here on [13]; $\kappa_0 = (\omega/c)$ is the vacuum wavevector, μ_0 the vacuum magnetic permeability, and j_x the current density. At the same time, the linearized Boltzmann equation reduces to :

$$\left(\frac{1}{\tau} + i\,\omega\right)g + v_z\,\frac{\partial g}{\partial z} = q\,v_x\,\frac{\partial f^0}{\partial \varepsilon}\,E_x(z) \tag{2}$$

where we have obviously neglected the Laplace force due to the oscillating magnetic field of the wave; q, v and ε stand respectively for the charge, velocity and energy of the electrons. In Eq.(2), we have also implicitly split the conduction electron velocity distribution function $f(v,z)$ in the usual manner : $f(v,z) = f^0(\varepsilon) + g(v,z)$; with $g(v,z)$, the departure from the equilibrium Fermi-Dirac distribution $f^0(\varepsilon)$, assumed to be linear in E_x. The needed link between Eqs (1) and (2) is provided by the expression of the current density as a function of $g(v,z)$:

$$j_x(z) = -\,q\left(\frac{m}{h}\right)^3 \int d^3v\ v_x\,g(v,z) \tag{3}$$

By using standard mathematical techniques, one can cast the sytem of Eqs (1)-(3) into an integro-differential equation in the field E_x [7,13] :

$$\frac{d^2E_x}{dz^2} + \varepsilon_{st}\,\kappa_0^2\,E_x = i\,\omega\,\mu_0 \int dz'\ \sigma_{x,x}(\omega;z\text{-}z')\,E_x(z') \tag{4}$$

where $\sigma_{x,x}(\omega;z\text{-}z')$ is the two-point transverse conductivity at angular frequency ω, which relates the current at z to the electric field at z' [7,13] :

$$\sigma_{x,x}(\omega;z\text{-}z') = \frac{3}{4}\frac{\sigma_0}{\Lambda}\left\{Ei_1\left[(1+i\omega\tau)\frac{|z\text{-}z'|}{\Lambda}\right] - Ei_3\left[(1+i\omega\tau)\frac{|z\text{-}z'|}{\Lambda}\right]\right\}$$

$$\text{with } Ei_n(u) = \int_1^\infty ds\,\frac{e^{-u\,s}}{s^n} \tag{5}$$

where $\sigma_0 = (n\,q^2/m)\,\tau$ is the usual bulk static conductivity and $\Lambda = v_F\tau$ the bulk MFP.

The expression (5) of $\sigma_{x,x}(\omega;z\text{-}z')$ allows to identify the relevant length scales of the problem. When $\omega\tau \ll 1$, i.e. the quasi-static regime where the conductivity is limited by the collisions, the argument of the exponential integrals reduces to $|z\text{-}z'|/\lambda$ which is real. The two point conductivity is a positive real and decreases monotonously on the length scale of the MFP. When $\omega\tau \geq 1$, one enters gradually into the collisionless regime where the conductivity becomes independent of the relaxation time, the argument of the exponential integrals $[1/\lambda + i\,(\omega/v_F)]\,|z\text{-}z'|$ is complex. The two point conductivity is complex and its real and imaginary parts oscillate on the length scale of $\omega\,\tau/\lambda = (v_F/\omega)$ which is smaller than the MFP, with envelopes still decaying on the length scale of the MFP.

Let us assume for the moment that the electric field varies on a length scale much larger than the MFP. Following the above analysis of the range of $\sigma_{x,x}(\omega;z-z')$, it is clear that in this case we can sort $E_x(z')$ from the integral in the right hand side of Eq.(4), recovering the more classical wave equation for a conductive medium with a local conductivity :

$$\frac{d^2E_x}{dz^2} + \left(\varepsilon_{st} - i\,\frac{\sigma(\omega)}{\varepsilon_0\,\omega}\right)\kappa_0^2\,E_x = 0 \qquad (6)$$

as one can easily demonstrate by integration of $\sigma_{x,x}(\omega;z-z')$ over z'. In Eq.(6), $\sigma(\omega) = \sigma_0/(1 + i\,\omega\,\tau)$ is the local conductivity at angular frequency ω, and ε_0 the vacuum dielectric permitivity. The general solution of Eq.(6) is then obtained by elementary calculus as a linear combination of upward $A_{(-)}\,e^{-(1/\delta\,+\,2\,i\,\pi/\lambda)z}$ and downward $A_{(+)}\,e^{+(1/\delta\,+\,2\,i\,\pi/\lambda)z}$ travelling waves, with :

$$\frac{2\,\pi}{\lambda} - \frac{i}{\delta} = \tilde{n}\,\kappa_0 = \left(\varepsilon_{st} - i\,\frac{\sigma(\omega)}{\varepsilon_0\,\omega}\right)^{\frac{1}{2}}\kappa_0 \qquad (7)$$

where δ is the SD, λ the wavelength (WL) in the metal, and $\tilde{n} = n - i\,k$ the complex optical index.

We can now specify the domain of validity of Eq.(6) as the frequency domain for which both the SD and the WL in the metal are much larger than the MFP. If one of these conditions fails, one enters the so-called anomalous skin effect regime, and we shall go back to Eq.(4) to describe the propagation of transverse electromagnetic waves in the metal of interest. As one can see on Fig. 1, these conditions are well satisfied in the near to mid-infrared regions at room temperature for transition metals like Ni with quite short mean free path. However, for noble metals of reasonnable purity, like Cu with $\rho \approx 3\,\mu\Omega.cm$, one has SD $\approx$ MFP in the whole spectral domain of interest.

Effective optical constants of a magnetic metallic multilayers in the "self-averaging" limit - The magnetorefractive effect

In the case of a periodic metallic multilayer, the actual expression of the two-point conductivity will be quite cumbersome and much more difficult to obtain than Eq.(5), because we loose the translationnal invariance in the direction of the lamination, and a new length scale enters the problem : L, the length scale of inhomogeneity, i.e. the period of the multilayer. However, we will still be able to derive simple and compact expressions in a certain limit, the SAL, which is reached when the range of $\sigma(\omega;z,z')$ as a function of $|z-z'|$ is much larger than L. In this limit one can define an average scattering rate $1/\tau_{SAL}$ according to the Matthiessen's rule [16] which is simply the spatial average of the scattering rates over a period, as it has been demonstrated at zero frequency in Ref. 11 :

$$1/\tau_{SAL} = \frac{1}{L}\int_{period}\frac{dz}{\tau(z)} \qquad (8)$$

Then, we can define an average two-point optical conductivity $\sigma SAL(\omega;z-z')$, which is a scalar independent of the direction of propagation and of the polarization because the translationnal invariance has been restored by averaging. It will be simply given by Eq.(5), but with τ replaced by τ_{SAL}, Λ by $\Lambda_{SAL} = v_F\,\tau_{SAL}$ and σ_0 by $\sigma^0_{SAL} = (n\,q^2/m^*)\,\tau_{SAL}$. The consistency of this approximation is insured if the actual length scales upon which $\sigma_{SAL}(\omega;z-z')$ varies, Λ_{SAL} and (v_F/ω) as it has been shown in the preceeding sub-section,

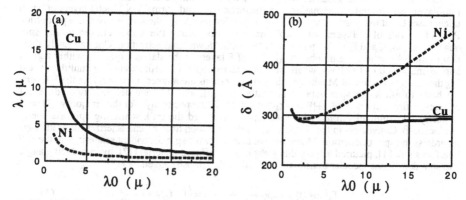

Fig.1 (a) Wavelength and (b) skin depth as a function of $\lambda 0$ the wavelength in vacuum, according to Eq.(7), for Ni and Cu at room temperature. We used $\varepsilon_{st} = 5.5$, $\Lambda_{Ni} = 60$ Å and $\Lambda_{Cu} = 300$ Å, assuming for both metals $\rho \Lambda = 1000$ $\mu\Omega$.cm.Å and $v_F = 10^6$ m.s^{-1} [15].

are effectively much larger than L. The appearance of the extra condition $L \ll (v_F/\omega)$, on the top of $L \ll \Lambda_{SAL}$ already present at zero frequency [11], marks the difference between the static and dynamic conductivity behaviours. Within the SAL, we can also define the wavelength and skin depth for light propagation, as well as an effective index of refraction, respectively λ_{SAL}, δ_{SAL}, and $\tilde{n}_{SAL}$. This is achieved by using Eq.(7), whith $\sigma(\omega)$ replaced by $\sigma_{SAL}(\omega)$:

$$\sigma_{SAL}(\omega) = \frac{\sigma_{SAL}^0}{1 + i\,\omega\,\tau_{SAL}} \tag{9}$$

In order to decide if one shall use Eq.(5) or Eq.(6) to describe light propagation in the multilayer, it is clear that one has to compare λ_{SAL}, δ_{SAL} and Λ_{SAL}. Due to the usual strong scattering at the interface in metallic multilayer, Λ_{SAL} is generally significantly shorter than the MFP of the constituting metals. For this reason, the anomalous skin effect regime $\Lambda_{SAL} \geq \delta_{SAL}$ will be hardly reached at room temperature in the spectral domain of interest. From here on, we will thus assume that it is not the case. Consequently, we will further discuss the optical properties of multilayers as those of effective homogeneous metals of refractive index $\tilde{n}_{SAL}$. At first sight this implies a local relationship between the electric field and the induced current, nevertheless the underlying non locality of the conductivity remains a deciding factor through the spatial average of Eq.(8).

For a magnetic metallic multilayer, all the above results are still valid providing that the spin is taken into account properly. For colinear magnetization arrangements, where one can define unambiguously a unique spin quantization axis along the common magnetization direction, this simply means to treat the electrons with (+) and (-) spin as two independent channels carrying the current in parallel (if one neglects the spin-mixing scatterrings) [9]. Thus, combining Eq.(7) and Eq.(9), we obtain for the effective refractive index $\tilde{n}_{SAL}$ and the corresponding dielectric constant ε_{SAL} :

$$\tilde{n}_{SAL} = \varepsilon_{SAI}^{\frac{1}{2}} = \left[\varepsilon_{st} - \frac{i}{\varepsilon_0\,\omega} \left(\frac{\sigma_{SAL,+}^0}{1 + i\,\omega\,\tau_{SAL,+}} + \frac{\sigma_{SAL,-}^0}{1 + i\,\omega\,\tau_{SAL,-}} \right) \right]^{\frac{1}{2}} \tag{10}$$

In order to be more specific, we consider now the simple case of a bimetallic magnetic multilayer, constituted of alternating ferromagnetic (F) and normal (N) metal layers of equal thicknesses t. Let us assume that within a slab of thickness Λ_{SAL}, there is on average a proportion $P_+ = (1 + p)/2$ of F layers with "up" magnetization along the z axis chosen as the spin quantization axis, and $P_- = (1 - p)/2$ of F layers with "down" magnetization. One can easily show that in the SAL where the number $\Lambda_{SAL}/(2\ t)$ of F layers in the slab is large, p is within a good approximation equal to M/M_S; where M is the macroscopic magnetization of the multilayer, and M_S the saturation value of M when all the magnetizations are aligned. Obviously, in the F layers, we have to allow for spin dependent scatterings [9], and thus we define two different relaxation times $\tau_\uparrow^{(F)} = \tau^{(F)}/(1+\beta)$ and $\tau_\downarrow^{(F)} = \tau^{(F)}/(1-\beta)$, respectively for the majority ($\uparrow$) and minority ($\downarrow$) spin electrons, where we have introduced the bulk scattering spin assymetry coefficient β. Conversely in the N layers, a single relaxation time $\tau^{(N)}$ characterizes the scatterings regardless the spin orientation. Moreover, we also allow for spin dependent scatterings at the F/N interfaces [10-11], pictured as very thin slabs of interdiffused material [17], by introducing angle and spin dependent specular transmission coefficients $T_s(\cos\theta)$ given by [18]:

$$T_s(\cos\ \theta) = \exp(-\frac{t_i}{\cos\ \theta\ \Lambda_{i,s}}) = \left(1 - D_s^0\right)^{\frac{1}{\cos\theta}} \tag{11}$$

where s stands for ($\uparrow$) or ($\downarrow$) spin, t_i and $\Lambda_{i,s}$ are respectively the thickness and MFP for spin s of the interfacial layer, and θ is the angle between the electron velocity and the interface normal [19]. To be consistent with the SAL, we assume that the scattering probabilities when crossing the interfaces at normal incidence D^0_s are not too large, leading to :

$$D_{\uparrow(\downarrow)}^0 = D\ (\ 1 +(-)\ \gamma) \approx \frac{t_i}{\Lambda_{i,\uparrow(\downarrow)}} \tag{12}$$

Where we have introduced γ the spin assymetry coefficient for interfacial scatterings. Applying Eq. (8) independently for spin (+) and spin (-), we obtain for $\tau_{SAL,\pm}$ the very simple expression :

$$\frac{1}{\tau_{SAL,\pm}} = \frac{1 \pm \beta_{SAL}\ (M/M_S)}{\tau_{SAL}} \tag{13}$$

where the effective relaxation time and spin assymetry coefficient, respectively τ_{SAL} and β_{SAL}, are given by :

$$\frac{1}{\tau_{SAL}} = \frac{1}{2}\left(\frac{1}{\tau^{(F)}} + \frac{1}{\tau^{(N)}} + 2\frac{v_F}{t}D\right) \tag{14.a}$$

$$\beta_{SAL} = \frac{\dfrac{\beta}{\tau^{(F)}} + 2\gamma\dfrac{v_F}{t}D}{\dfrac{1}{\tau^{(F)}} + \dfrac{1}{\tau^{(N)}} + 2\dfrac{v_F}{t}D} \tag{14.b}$$

Finally, Eq.(13) together with Eq.(10) give for the effective dielectric constant $\varepsilon_{SAL}(\omega)$ of the considered magnetic multilayer :

$$\varepsilon_{SAL} = \varepsilon_{st} - \frac{\omega_p^2}{\omega^2}\frac{i\ \omega\ \tau_{SAL}}{1 + i\ \omega\ \tau_{SAL}}\left(1 + \frac{\beta_{SAL}^2\ (M/M_s)^2}{(1 + i\ \omega\ \tau_{SAL})^2 - \beta_{SAL}^2\ (M/M_s)^2}\right) \tag{15}$$

In Eq.(15) we introduce the plasmon frequency $\omega_p = [n q^2/(\varepsilon_0 m)]^{1/2}$, in order to show more clearly the similarity with the standard Drude formula of metal dielectric constant.

One can easily see that Eq.(15) predicts a dependence of the effective dielectric constant of magnetic multilayers upon their magnetic state through the factor $(M/M_S)^2$, which have to be considered as measuring the degree of alignement of neighbouring layer magnetizations. This dependence which can be considered as a magnetooptical effect, because of the possibility to influence the relative orientation of the magnetizations by applying an external magnetic field, is described here for the first time and we propose to name it the magnetorefractive effect. It appears because of the two spin channel conduction in ferromagnetic metals, and of the existence of spin dependent scatterings, combined with the possibility to change the magnetization structure on a length scale smaller than the range of the non local conductivity. As a matter of fact, if one puts $\beta_{SAL} = 0$ into Eq.(15) (no spin dependent scattering on average), the effect disapears and one simply recovers a Drude dielectric constant.

In order to obtain a better understanting of the practical consequences of Eq.(15), we will consider the case of a $[F(5 \text{ Å})/N(5 \text{ Å})]_{20}$ multilayer whith the following set of parameters : $\rho \Lambda = 1630\ \mu\Omega.\text{cm.Å}$, $v_F = 10^6$ m.s^{-1}, $\tau^{(F)} = \tau^{(N)} = 10^{-14}$ s, $\beta = 0.5$, $D = 0.05$, $\gamma = 0.5$; i.e. equal spin dependence of the scattering in the bulk and at the interfaces. And we will assume that we can put the multilayer into two extreme magnetic configurations : the parallel (P) one with all the magnetizations aligned ($M = M_S$), and the antiparallel one with the magnetizations alternatively up and down along a common direction ($M = 0$). Using Eqs (12)-(15), one can easily compute the SAL dielectric constant in the (P) and (AP) configurations. We can derive from this (using for instance a standard interference matrix method [19]) the intensity transmission and reflection coefficient at normal incidence as a function of the wavelength, $T^{(P,AP)}(\lambda)$ and $R^{(P,AP)}(\lambda)$, assuming a self supported film in air. One can see on Fig.2 the relative change in the transmission and reflection coefficients, respectively defined as $\Delta T/T = (T^{AP}-T^P)/T^P$ and $\Delta R/R = (R^{AP}-R^P)/R^P$, as functions of λ.

As one can see, the magnetorefractive effect shall manifest itself as a significant dependence at infrared wavelengths of the transmission and reflection coefficients of magnetic multilayers upon their magnetization arrangements, but which should be independent of the polarization of the incident light beam. Moreover, we predict a smooth and slowly oscillatory wavelength dependence characteristic of a relaxation intraband effect, in strong contrast with the usual resonant behaviour of magnetooptical effects with strong interband contributions.

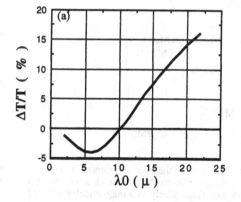
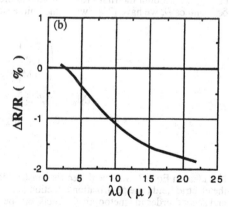

Fig.2 (a) $\Delta T/T$ and (b) $\Delta R/R$ at normal incidence in air, as functions of λ_0 the wavelength in vacuum, for a $[F(5 \text{ Å})/N(5 \text{ Å})]_{20}$ multilayer whith the following set of parameters : $\varepsilon_{st} = 1$, $\rho \Lambda = 1630\ \mu\Omega.\text{cm.Å}$, $v_F = 10^6$ m.s^{-1}, $\tau^{(F)} = \tau^{(N)} = 10^{-14}$ s, $\beta = 0.5$, $D = 0.05$, $\gamma = 0.5$.

The effect out of the SAL - Comparison between the magnetorefractive effect and the first and second order magnetooptical effects

Out of the SAL which will be hardly satisfied in practice (because one has certainly $\Lambda_{SAL} \leq 50$ Å in most of the cases and $v_F/\omega \approx 50$ Å at 10 μm of wavelength for $v_F = 10^6$ m.s^{-1}), it is clear that Eq.(15) is no longer valid. In fact, it will no longer be possible to derive an explicit expression of the effective dielectric constant, and one will loose the equivalence between the field in plane (the plane of lamination) and field out of plane. This latter point is well known for transport properties at zero frequency, where it was shown that the GMR in the so-called CPP (current perpendicular to the plane) [20] geometry scale with the spin diffusion length l_{sf} [21] instead of the MFP for the CIP (current in plane) geometry. However, as long as the scale of inhomogeneity will remain of the order of the range of the non local conductivity, the magnetorefractive effect will remain non zero. Consequently, we can guess a general phenomenological expression of the dielectric constant tensor for a magnetic multilayer at infrared wavelength even out of the SAL :

$$
\varepsilon = \begin{pmatrix} \varepsilon_{\parallel} + \delta\varepsilon_{\parallel}\left[\langle(M/M_S)^2\rangle_{\parallel}\right] & 0 & 0 \\ 0 & \varepsilon_{\parallel} + \delta\varepsilon_{\parallel}\left[\langle(M/M_S)^2\rangle_{\parallel}\right] & 0 \\ 0 & 0 & \varepsilon_{\perp} + \delta\varepsilon_{\perp}\left[\langle(M/M_S)^2\rangle_{\perp}\right] \end{pmatrix} \quad (16)
$$

where we put the z axis perpendicular to the plane of lamination. $\delta\varepsilon_{\parallel}$ and $\delta\varepsilon_{\perp}$ are complex numbers, which will decrease more or less exponentially respectively with t/Λ for the field in plane and t/l_{sf} for the field out of plane. Their functionnal dependence upon $\langle(M/M_S)^2\rangle_{\parallel,\perp}$, the spatial average of $(M/M_S)^2$ over the relevant length scales for either field in plane or out of plane, shall not be considered as rigourous. However it rightly indicates the fact that the dielectric constant tensor depends upon the relative orientation of neighbouring magnetizations, but does not depend on the relative orientation between the electric field (the polarization direction) and the magnetization.

If one considers now the dielectric tensor in case of a magnetized isotropic medium when taking into account the first order and second order magnetooptical contributions, respectively $\delta\varepsilon^{(1)}$ and $\delta\varepsilon^{(2)}$, we have [22] (with $\mathbb{I}$ the unit tensor) :

$$
\varepsilon = \varepsilon\,\mathbb{I} + \delta\varepsilon^{(1)}\begin{pmatrix} 0 & -i\,M_z/M_S & i\,M_y/M_S \\ i\,M_z/M_S & 0 & -i\,M_x/M_S \\ -i\,M_y/M_S & i\,M_x/M_S & 0 \end{pmatrix}
$$
$$
+ \delta\varepsilon^{(2)}\begin{pmatrix} (M_z^2 + M_y^2)/M_S^2 & 0 & 0 \\ 0 & (M_x^2 + M_z^2)/M_S^2 & 0 \\ 0 & 0 & (M_x^2 + M_y^2)/M_S^2 \end{pmatrix} \quad (17)
$$

It is clear in Eq.(17) that the dielectric constant depends now on the relative orientation between the electric field (the polarization direction) and the magnetization. As it is well known, the first and second order magnetooptical effects can be viewed respectively as a magnetically induced circular and linear birefringence (and dichroïsm) [22].

These profound differences in the symmetry of the effects, which shows that the magnetorefractive effect has no connection with the classical magnetooptical effects, is in fact a phenomenological counterpart of their totally different microscopic origins. The classical magnetooptical effects in magnetic materials are closely linked to the spin-orbit coupling [23], who couples the velocity (direction of the current, field) and spin (magnetization direction) degrees of freedom. On an other hand, the spin dependent scatterings and two spin channel conduction mechanisms which are at the origin of the magnetorefractive effect, does not induce any of such coupling on average. One recovers here at optical frequencies the differences existing between the classical magnetoresistive effects in magnetic metals (extraordinary Hall effect and anisotropic magnetoresistance) and the GMR. In some sense, one can say that the magnetorefractive effect is related to the GMR in the same way as the intraband part of the Kerr-Faraday effect is linked with the extraordinary Hall effect [23].

OPTICAL TRANSMISSION MEASUREMENTS UNDER AN APPLIED MAGNETIC FIELD THROUGH [$Ni_{80}Fe_{20}/Cu/Co/Cu$] MULTILAYERS

In order to test the existence of the predicted magnetorefractive effect, we performed optical transmission measurements at normal incidence under an applied magnetic field through a series of multilayers with the following structure : [50 Å $Ni_{80}Fe_{20}$/tCu Cu/20 Å Co/tCu Cu]$_{x3}$, with tCu ranging from 7 Å to 36 Å. The multilayers were deposited by RF diode sputtering on 200 μm thick high resistivity (300 Ω.cm) Si(100) substrates polished on the two sides, with a 50 Å Fe buffer layer. One can find elsewhere detailed results on the structural, magnetic and DC transport properties of these samples [24].

We performed two different kinds of measurements. First, we measured the transmission spectra at fixed magnetic field with a fourier transform infrared spectrometer Perkin Elmer 1760, used at 64 cm^{-1} of spectral resolution. We accumulated typically few thousand scans to obtain a good signal over noise ratio. The incorporated light source of the spectrometer is obviously not polarized, and the light was incident on the multilayer side of the samples. Furthermore we verified that the absoption of the Si substrates is very small between 2 μm to 20 μm, and that the interference fringes which could possibly appear on the spectra due to internal multiple reflections in the substrates were averaged out due to the moderate spectral resolution we used. Secondly, we measured the field dependence of the transmission at fixed wavelength, using a CO_2 laser operating at 10.6 μm of wavelength as light source incident on the multilayer side, and a liquid nitrogen cooled HgCdTe detector. The transmission hysteresis loops were recorded on a digital oscilloscope in XY mode, by averaging over 256 cycles of the magnetic field which was in this case modulated at few ten Hertz. We used in both cases approximatively 1/4 cm^2 pieces cut from the original 2" wafers, put into a special holder incorporating a pair of magnetic coils which allowed us to apply staticor low frequency modulated magnetic fields when measuring the spectra, in the plane of the films and ranging from 0 up to 1500 Oe.

We first focused our attention on the three samples with tCu = 9.5, 22 and 34.5 Å, corresponding to the three first maxima of antiferromagnetic interlayer exchange coupling of the $Ni_{80}Fe_{20}$/Cu/Co system [24]. This means that at some field H_{max} around zero field the magnetization of neighbouring $Ni_{80}Fe_{20}$ and Co layers are almost antiparallel in these samples, which is indicated by a maximum value of the DC resistivity; and above some saturation field $H_{sat} \leq 1500$ Oe sufficient to overcome the antiferromagnetic coupling, all the magnetizations are aligned [24]. One can see on Fig.3(a) the transmission spectra under saturating applied field for the three samples, and on Fig.3(b) the relative change in the transmission defined as $\Delta T/T_{sat} = [T(H_{max}) - T(H_{sat})]/T(H_{sat})$. The predicted variation of the transmission coefficient under the change of the magnetization structure, from the AP to the P state, is clearly observed. One has also to note that in the experimental geometry used (normal incidence, magnetization in the plane of the sample, unpolarized incident light, detector insensitive to the polarization) neither the Faraday nor the Cotton-Mouton effects could induce any change in the measured transmitted intensity. In order to further substantiate our point, we performed the same measurements on the whole series of samples. One can see part of these data on Fig.4(a) where we plot $\Delta T/T_{sat}$ as measured at two different wavelengths (9 μm and 20 μm) as a function of tCu, together with the plot of the relative change in the DC resistivity $\Delta \rho/\rho$ [24]. One clearly recovers in the optical measurements the oscillation in the magnitude of the effect observed in the DC transport

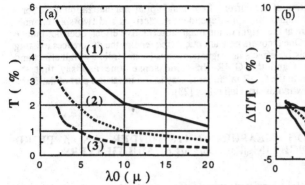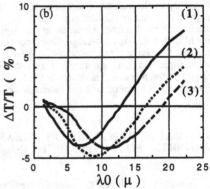

Fig.3 (a) The transmission T (b) and the relative change ΔT/T, measured as functions
of λ0 the wavelength in vacuum, for the three [50 Å Ni$_{80}$Fe$_{20}$/tCu Cu/20 Å Co/tCu Cu]$_{x3}$
multilayer samples with tCu = 9.5 Å (1), 22 Å (2) and 34.5 Å (3).

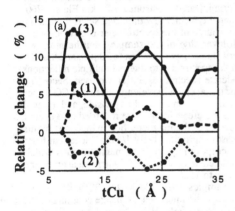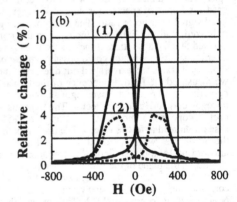

Fig.4 (a) ΔT/T$_{sat}$ at two different wavelengths 9 μm (1) and 20 μm (2), ploted with Δρ/ρ (3)
as a function of tCu, for the multilayers [50 Å Ni$_{80}$Fe$_{20}$/tCu Cu/20 Å Co/tCu Cu]$_{x3}$.
(b) - ΔT/T(H) = - [T(H)-T(H$_{sat}$)]/T(H$_{sat}$) (1) measured at 10.6 μm of wavelength, ploted with
Δρ/ρ(H) (2), for the sample [50 Å Ni$_{80}$Fe$_{20}$/22 Å Cu/20 Å Co/22 Å Cu]$_{x3}$.

measurements, which is due to the oscillatory interlayer exchange coupling [3,24]. This clearly
demonstrates that the observed effect requires a change in the relative orientation of the
magnetizations to be observed, because it almost disapears for tCu values (15 Å and 27 Å)
where the ferromagnetic coupling suppress the possibility of such a change. We also observed
for all the samples the change in sign of ΔT/T between the measurements made at 20 μm and
those made at 9 μm of wavelength, previously visible on the spectra given in Fig.3(b). Finally,

we also measured the field dependence of the effect at 10.6 μm of wavelength using the CO_2 laser on the sample with tCu = 22 Å, as one can see in Fig.4(b) where we also plot the field dependence of the change in the DC resistance. Despite a somehow larger hysteresis at low field, the two curves have almost the same shape. More strikingly, it is clear that once the sample is magnetically saturated there is no further change in the transmission, ruling out any possible spurious "paramagnetic" effect (direct influence of the magnetic field on any part of the experimental set up). Finally, when changing the polarization of the incident laser beam from parallel to perpendicular to the applied field we observe absolutly no change.

In our view, the presented experimental results constitute alltogether a strong demonstration of the existence of the predicted magnetorefractive effect, and confirm its essential predicted features : the variation of the index of refraction at infrared wavelength is polarization independent and is observed when a change in the relative orientation of adjacent layer magnetizations occurs. Furthermore, we clearly observed the predicted smooth and oscillating wavelength dependence with a disappearance of the effect at short wavelength when $\omega\tau \gg 1$. This is a strong argument in favor of the proposed microscopic mechanism based on spin dependent scatterings.

COMPARISON BETWEEN THE EXPERIMENTAL RESULTS AND NUMERICAL CALCULATIONS

Apart from the very simple model derived in the SAL, we also developed a Mathematica™ notebook in order to solve numerically the Maxwell-Boltzmann coupled set of equations, Eqs. (1)-(3), in the case of a monochromatic plane wave incident at normal incidence on an arbitrary magnetic multilayer with colinear magnetization structure deposited on a dielectric substrate of finite thickness. We solve self consistently for two velocity distribution functions g_s for the two spin directions s = ± and for the electric field, neglecting the spin mixing, and using angle dependent transmission probabilities for the electrons at the interfaces [18]. We used a standard interference matrix method to compute the transmission and reflection coefficients. The detail of the method and extensive numerical results will be given elsewhere [25]. Our aim here is just to give the calculated results correspondings to the experimental measurements shown on Fig.3, in order to demonstrate that one can reach a reasonnable agreement between the model and the experiments.

Thus, we considered the following stacking : 200 μm Si/50 Å Fe/[50 Å $Ni_{80}Fe_{20}$/tCu Cu/ 20 Å Co/tCu Cu]x3, in the three cases : tCu = 9.5 Å (1), 22 Å (2) and 34.5 Å (3). We take into account that approximatively the top 30 Å of the stacking is oxidized, as it was assessed by TEM [26], treating this part as a dielectric cap layer of dielectric constant ε_{ox} = 9. The set of parameters used for all the calculations is (following the previously defined notations) : ε_{Si} = 11.56 at all wavelengths [26]; ε_{st} = 1; $\rho \Lambda$ = 1000 μΩ.cm.Å, v_F = 0.75 10^6 m.s^{-1}, $\tau^{(Fe)}$ = 8 10^{-15} s, $\beta^{(Fe)}$ = 0, D(Fe/NiFe) = 0.06, γ(Fe/NiFe) = 0; τ(NiFe) = 2.3 10^{-15} s, $\beta^{(Fe)}$ = −0.73, D(NiFe/Cu) =0.06, γ(NiFe/Cu) = 0; τ(Co) = 2.4 10^{-15} s, β(Co) = −0.82, D(Co/Cu) =0.48, γ(Co/Cu) = − 0.88. We also assumed that the outer boundaries of the multilayer are perfectly diffusing for the electrons. Despite we did not perform a rigourous fit, this particular choice of parameters is found to reproduce quite well the experimental DC resistivities in the saturated state $\rho_{sat}^{(exp)}$ as well as the GMR ratio $\Delta\rho/\rho^{(exp)}$ for the three samples, as one can see in Tab.1. Using these parameters, we computed the transmission coefficients $T^{(P)}$ in the parallel state and the relative change $\Delta T/T = [T^{(AP)}-T^{(P)}]/T^{(P)}$. One can see these two quantities plotted as a function of the wavelength, respectively in Fig.5 (a) and (b). Thus, one can directly compare Fig. 3 and Fig. 5. It is clear that although not quantitative, the agreement is quite good, particularly when considering that we adjust the parameter to reproduce the DC measurements and then we compute with this same set the optical properties.

	Sample (1)	Sample (2)	Sample (3)
ρ_{sat}(exp) $\mu\Omega$.cm	29.	23.	17.
ρ_{sat}(calc) $\mu\Omega$.cm	28.1	20.6	16.3
$\Delta\rho/\rho$(exp) %	13.8	11.3	8.1
$\Delta\rho/\rho$(calc) %	13.9	10.4	8.1

Tab.1 . Experimental DC resistivities in the saturated state ρ_{sat}(exp) and GMR ratio $\Delta\rho/\rho$(exp), for the Si/50 Å Fe/[50 Å $Ni_{80}Fe_{20}$/tCu Cu/ 20 Å Co/tCu Cu]$_{x3}$ samples with tCu = 9.5 Å (1), 22 Å (2) and 34.5 Å (3), compared to the calculated values.

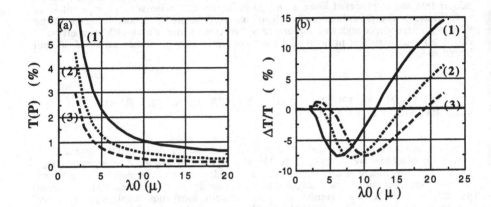

Fig.5 Calculated transmission $T^{(P)}$ (a), and relative change $\Delta T/T$ (b), as functions of λ0 the wavelength in vacuum, for the three [50 Å $Ni_{80}Fe_{20}$/tCu Cu/20 Å Co/tCu Cu]$_{x3}$ multilayer samples with tCu = 9.5 Å (1), 22 Å (2) and 34.5 Å (3).

In our view, this quite good agreement between a free electron (no potential steps at the interfaces) Boltzmann equation based model of the linear response functions and the exposed experimental results, from zero frequency up to 10^{15} Hz in [$Ni_{80}Fe_{20}$/Cu/Co/Cu] multilayers, is a strong and new support of the currently most popular interpretation of magnetotransport properties in multilayers based on spin dependent scattering mechanisms.

CONCLUSIONS

We have shown for the first time, both theoretically and experimentally, that a change in the relative orientation of adjacent F layer magnetizations in magnetic multilayers shall be generally associated with a significant change of their refractive index, which is mostly prominent at infrared wavelengths. This effect appears because of the non locality of the intraband optical conductivity which average the electron scattering rates over the shorter of the following length scale : Λ the electron mean free path, or v_F/ω the ratio of the Fermi velocity by the optical wave angular frequency.

This effect was shown to have completely different symmetry properties compared to the well known Kerr-Faraday or Cotton-Mouton magnetooptical effect : it does not depend on the relative orientation of the optical electric field (the polarization) and of the magnetizations. It is thus clearly a new magnetooptical effect that we propose to name the magnetorefractive effect.

This effect is closely linked to the GMR effect, and in some sense, one can say that the magnetorefractive effect is related to the GMR in the same way as the intraband part of the Kerr-Faraday effect is linked with the extraordinary Hall effect [23].

Quite good agreement is obtained between numerical calculations based on the Maxwell-Boltzmann coupled system of equations and optical transmission measurements between 2 μm and 20 μm of wavelength on [$Ni_{80}Fe_{20}$/Cu/Co/Cu] multilayers, using the same set of parameters which allows to reproduce the DC resistivities and GMR. In our view, this agreement from zero frequency up to 10^{15} Hz is a strong and new support of the interpretation of magnetotransport properties in multilayers based on spin dependent scattering mechanisms.

This newly observed effect seems to open many new experimental possibilities regarding to conventionnal DC transport measurements : easy in-situ characterizations, reflection measurements on multilayers deposited on metallic substrates ...

Finally, we stress that even though this paper is dealing with multilayers, one can certainly expect similar effects in non laminated structures like granular materials [28]. Quite independently of the spatial distribution and nature of the magnetized entities involved (granules, layers), the magnetorefractive effect as well as the GMR just require that one can induce some change in the magnetization arrangements on a length scale not too large compared to the range of the non local conductivity.

ACKNOWLEDGEMENTS

We would like to thank the technical assistance of D. Delacourt and D. Papillon, from the Groupe d'Optique of the LCR, for the optical transmission measurements. We also acknowledge the support from the European Union through the Esprit project BRA 6146.

REFERENCES

1. M.N. Baibich, J.M. Broto, A. Fert, F. Nguyen Van Dau, F. Petroff, P. Etienne, G. Creuzet, A. Friederich and J. Chazelas, Phys. Rev. Lett. **61**, 2472 (1988).

2. G. Binach, P. Grunberg, F. Saurenbach, and W. Zinn, Phys. Rev. B, **31**, 4828 (1989).

3. S.S.P. Parkin, N. More, and K.P. Roche, Phys. Rev. Lett. **64**, 2304 (1990).

4. J. Sakurai et al, J. Magn. Magn. Mat. **126**, 510 (1993).

5. W. R. Bennet, W. Schwarzacher, and W. F. Egelhoff, Phys. Rev. Lett. **65**, 3169 (1990).

6. Y. Suzuki and T. Katayama, Mater. Res. Soc. Symp. Proc. **313**, 153 (1993).

7. G. E. H. Reuter and E. H. Sondheimer, Proc. R. Soc. London, Ser. A, **195**, 336 (1948).

8. R. Dimmich, Phys. Rev. B **45**, 3784 (1992).

9. I. A. Campbell and A. Fert, in Ferromagnetic Materials, edited by E. P. Wohlfarth (North-Holland, Amsterdam, 1982), Vol. 3, p. 747.

10. R.E. Camley and J. Barnas, Phys. Rev. Lett. **63**, 664 (1989).

11. P. M. Levy, S. Zhang, and A. Fert, Phys. Rev. Lett. **65**, 1643 (1990).

12. M.M. Kirillova, Zh. Eksp. Teor. Fiz. 61, 336 (1971) [Sov. Phys. JETP, 34, 178 (1972)].

13. J. Szczyrbowski, K. Schmalzbauer and H. Hoffmann, Phys. Rev. B, 32, 763 (1985).

14. In the original work of Reuter and Sondheimer, Ref. 8, the notion of two-point conductivity is not explicitly stated. However, some relations equivalent to Eqs (4) and (5) of the text are explicitly derived.

15. H. Ehrenreich and H.R. Philipp, Phys Rev. 128, 1622 (1962); H. Ehrenreich, H.R. Philipp, and D.J. Olechna, ibid., 131, 2469 (1963).

16. J.M. Ziman, Electrons and Phonons, (Oxford University Press, London, 1962), pp. 285-287.

17. B.L. Johnson and R.E. Camley, Phys. Rev. B, 44, 9987 (1991).

18. H.E. Camblong and P.M. Levy, Phys. Rev. Lett. 69, 2835 (1992).

19. M. Born and E. Wolf, Principles of Optics, 5th ed. (Pergamon, Oxford, 1975), § 1.6.

20. W.P. Pratt Jr, S.F. Lee, J.M. Slaughter, R. Loloee, P.A. Schroeder and J. Bass, Phys. Rev. Lett. 66, 3060 (1991).

21. T. Valet and A. Fert, Phys. Rev. B, 48, 7099 (1993).

22. L.D. Landau and E.M. Lifshitz, Electrodynamics of Continuous Media, (Pergamon, New York, 1960), § 82.

23. S. Doniach, Optical Properties and Electronic Structure of Metals and Alloys, edited by F. Abeles (North-Holland, Amsterdam, 1966), pp. 471-483.

24. T. Valet et al, Appl. Phys. Lett. 61, 3187 (1992).

25. T. Valet and J. C. Jacquet, submitted to Phys. Rev. B.

26. P. Galtier, private communication.

27. C.D. Salzberg and J. Villa, J. Opt. Soc. Am. 47, 244 (1957).

28. A. E. Berkowitz et al, Phys. Rev. Lett. 68, 3749 (1992) ; .Q. Xia, J.S. Jiang and C.L. Chien, Phys. Rev. Lett. 68, 3749 (1992).

SECOND ORDER MAGNETO-OPTIC EFFECTS IN EPITAXIAL FE(110)/ MO(110) BILAYERS

R.M. OSGOOD III, R.L. WHITE and B.M. CLEMENS
Department of Materials Science and Engineering, Stanford University, Stanford, CA 94305

ABSTRACT

The signal measured during a Magneto-Optic Kerr Effect (MOKE) experiment is usually assumed to be linear in the magnetization (or the magnitude of the magneto-optic coupling vector Q that is proportional to the magnetization) so that a plot of the magnetization versus applied field can be obtained. We have observed an appreciable contribution from the Q^2 term in the magneto-optic response of epitaxial Fe(110)/Mo(110) bilayers. The Q^2 term in the magneto-optic response is much larger than that predicted by existing theory. We re-derive and modify the existing theory to fit the Q^2 term.

INTRODUCTION

MOKE has proven its usefulness in a number of magnetic thin film systems where it has been used to investigate the thickness dependence of the Curie temperature, search for magnetic ordering, and determine the anisotropy[1]. The latter measurement requires a magnetization curve in order to determine the anisotropy or saturation field. To obtain such a curve, the MOKE signal must be proportional to the magnetization $\vec{M}$ (the absolute value of which is proportional to Q, the absolute value of the *magneto-optic coupling vector*[2]). Most workers in magneto-optics have assumed that the higher order terms in the magneto-optic response are negligible due to the smallness of Q and because for purely in-plane magnetization the second order term is proportional to the product of the components of $\vec{M}$ lying parallel and perpendicular to the plane of incidence[1,3]. In an isotropic thin film, one of these two components is usually zero because the external field $\vec{H}$ is applied parallel or perpendicular to the plane of incidence so that the Q^2 terms vanish. In films with in-plane magnetic anisotropy, however, there can be a non-zero component of $\vec{M}$ perpendicular to $\vec{H}$; the anisotropy forces $\vec{M}$ to reverse by coherent rotation when $\vec{H}$ is close to parallel to the hard axis. We shall show that in a magnetically anisotropic Fe(110)/Mo(110) bilayer, the large size of the component of $\vec{M}$ perpendicular to $\vec{H}$ creates a measurable Q^2 term in the magneto-optic response.

EXPERIMENTAL

A (110)-oriented Fe/Mo bilayer was grown following a well-known procedure that guaranteed epitaxy[4]. A seed layer of Mo was deposited onto a Al_2O_3 substrate heated to above 650° C which was subsequently allowed to cool to below 100° C before deposition of the Fe layer. Torque magnetometry determined the uniaxial and biaxial anisotropy constants to be respectively 9.0×10^4 ergs/cm^3 and 1.6×10^4 ergs/cm^3, with an easy axis along the in-plane [001] direction. The magnetization curve (measured with vibrating sample magnetometry (VSM)) along the hard axis could be simulated assuming pure rotation (no domain wall motion). The demagnetizing field confined $\vec{M}$ to the plane of the sample.

A He-Ne laser beam (633 nm) polarized parallel to the plane of incidence ('p'-polarized) was incident on the sample at an angle from the film normal between 5° and 29°. Both rotation and ellipticity (θ_k and ϵ_k, respectively) were measured as functions of $\vec{H}$. The dimensionless longitudinal (m_l) and transverse (m_t) components of the unit vector in the direction of $\vec{M}$ were respectively parallel and perpendicular to $\vec{H}$. In this small angle limit, the magneto-optic response is given approximately by:

$$\theta_k + i\epsilon_k = \frac{-iQ\theta m_l}{(n_0^2 - 1)} - \frac{n_0}{(n_0^2 - 1)}\left(\frac{1 + 4\pi\alpha}{n_0^2 - 1 - 4\pi\alpha}\right)Q^2 m_l m_t, \tag{1}$$

where θ, n_0, and $4\pi\alpha$ are respectively the angle of incidence in radians, the index of refraction, and the contribution to the on-diagonal term of the dielectric tensor from electrons whose equations of motion are not affected by the force from the magetization (see below for an explanation of the $4\pi\alpha$ term). Time-reversal symmetry considerations dictate that the magnetization curve will go into itself if reflected through the origin[5]; therefore, the term proportional to $Q m_l$ is *symmetric* (i.e., can be reflected through the origin into itself), while the term proportional to $Q^2 m_l m_t$ is *asymmetric* (cannot be reflected through the origin into itself).

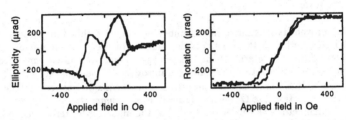

Figs. 1a and 1b. Observed ellipticity (left) and rotation (right) as a function of $\vec{H}$.

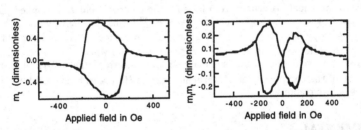

Figs. 2a. and 2b. m_t (left) and product of m_l and m_t (right) as a function of $\vec{H}$.

Displayed in Figs. 1a and 1b are the observed ellipticity and rotation for $\vec{H}$ close to parallel to the [1$\bar{1}$0] direction and an angle of incidence of 11°. The rotation and particularly the ellipticity are asymmetric and exhibit the presence of the $Q^2 m_l m_t$ term (from Eq. 1). Our magneto-optic measurement of m_t is shown in Fig. 2a, and the product of m_t and m_l (also determined magneto-optically) which determined the shape of the Q^2 term is shown in Fig. 2b (as in Figs. 1a - b, $\vec{H}$ is applied close to parallel to the [1$\bar{1}$0] direction and the laser's angle of incidence is 11°). An explanation of how m_t was measured magneto-optically is given in a related article[6]; m_l was obtained by using the same dependence on $\vec{H}$ as the rotation signal at an angle of incidence of 29° (this signal was almost perfectly symmetric and was very similar to the hysteresis loop measured by VSM) and assuming $m_l = \pm 1$ at saturation. As the angle of incidence is reduced, the coefficient of the term proportional to $Q\, m_l$ becomes smaller, while the coefficient of the Q^2 term remains constant: thus, the hysteresis loops become even more asymmetric at small angles of incidence, degenerating into the structure of Fig. 2b as the angle of incidence is reduced to zero. This has been confirmed by our experiments.

The existing theory of magneto-optics by Metzger et al.[3] assumed $\alpha = 0$. This gave a coefficient of the Q^2 term that was much too small to explain our observations. In the following section, we calculate explicitly the coefficient of the Q^2 term for the case of normal incidence (this is a fairly good approximation to our experiments because the coefficient of the Q^2 term is relatively constant with angle at small angles of incidence). We show how the existing theory predicts a very small effect while our modification to the theory is large enough to explain the observed $Q^2\, m_l m_t$ term.

THEORY

Let us consider the case of the laser beam at normal incidence to the sample surface (i.e. parallel to the z-axis so that $\theta = 0$), with the in-plane magnetization along the in-plane x-axis. We derive the size of the Q^2 term in the case of normal incidence because the physics is much more straightforward than in the case of an arbitrary angle of incidence and because exact calculations and our experiments show that the Q^2 term is approximately independent of angle of incidence for small angles of incidence[7]. Fitting α to the expression for the Q^2 term at normal incidence gives the correct value of α for small angles of incidence.

The incident electric field is $\vec{E}\exp(i\omega t - i\,|\,\vec{k}\,|\,z)$, where $\vec{k}, \vec{E}$ and ω are the wavevector, electric field vector, and frequency of the light wave, respectively. The absolute value of the wavevector $|\,\vec{k}\,|$ equals ω/c outside the sample and $n_0\omega/c$ inside the sample in the absence of magnetization (c is the speed of light and n_0 is the index of refraction in the absence of magnetization). Note that in the derivation below, the absolute value of the magnetization, $|\,\vec{M}\,|$, is constant.

The dielectric tensor enters the wave equation in the following manner:

$$\vec{k} \times \vec{k} \times \vec{E} + \frac{\omega^2}{c^2}\tilde{\epsilon}\vec{E} = 0. \tag{2}$$

where $\tilde{\epsilon}$ is the dielectric tensor of the sample. Following Yariv and Yeh[8] (although in CGS units), we write:

$$\begin{pmatrix} \epsilon_{xx} - n^2 & 0 & 0 \\ 0 & \epsilon_{zz} - n^2 & \epsilon_{yz} \\ 0 & \epsilon_{zy} & \epsilon_{zz} \end{pmatrix} \begin{pmatrix} E_x \\ E_y \\ E_z \end{pmatrix} = \begin{pmatrix} 0 \\ 0 \\ 0 \end{pmatrix}. \tag{3}$$

This allows us to find the eigenmodes of propagation with their respective eigenvalues or indices of refraction. There are two eigenmodes, polarized parallel and perpendicular to the magnetization, and therefore two indices of refraction: $n_{\parallel}$ and $n_{\perp}$ (given in terms of n_0, the index of refraction at zero magnetization):

Eigenmode # 1:

$$n_{\perp}^2 = \epsilon_{zz} + \frac{\epsilon_{yz}^2}{\epsilon_{zz}} \tag{4}$$

$$\begin{pmatrix} E_x \\ E_y \\ E_z \end{pmatrix} = \begin{pmatrix} 0 \\ \epsilon_{zz} \\ -\epsilon_{zy} \end{pmatrix}. \tag{5}$$

Eigenmode # 2:

$$n_\parallel^2 = \epsilon_{xx} = n_0^2 \tag{6}$$

$$\begin{pmatrix} E_x \\ E_y \\ E_z \end{pmatrix} = \begin{pmatrix} 1 \\ 0 \\ 0 \end{pmatrix}. \tag{7}$$

The sample reference frame has basis vectors parallel ($\hat{x}$) and perpendicular ($\hat{y}$) to the magnetization direction and perpendicular ($\hat{z}$) to the sample. In this reference frame, the reflectivity matrix is given by:

$$\begin{pmatrix} r_{yy} & 0 \\ 0 & r_{xx} \end{pmatrix} = \begin{pmatrix} \frac{1-n_\perp}{1+n_\perp} & 0 \\ 0 & \frac{1-n_\parallel}{1+n_\parallel} \end{pmatrix}. \tag{8}$$

where r_{xx} and r_{yy} are reflection coefficients of the sample that give the amplitudes of respectively reflecting light polarized parallel to the $\hat{x}$ and $\hat{y}$ directions into the same polarization state. Note that there is no coupling between different polarizations.

We can rotate the reflectivity matrix an angle θ about the $\hat{z}$ axis, where $\vec{M}$ is at an angle from the transverse direction. $\vec{M}$ (which lies along the x-axis in the film frame) can be written: $\vec{M} = |\vec{M}| \hat{x} = |\vec{M}| (\cos\theta \hat{s} + \sin\theta \hat{p}) = |\vec{M}| (m_t \hat{s} + m_l \hat{p})$, where $\hat{s}$ and $\hat{p}$ are unit vectors in the transverse and longitudinal directions, respectively. In the laboratory frame, the reflectivity matrix can be written:

$$\begin{pmatrix} r_{pp} & r_{ps} \\ r_{sp} & r_{ss} \end{pmatrix} = \begin{pmatrix} r_{yy}\cos^2\theta + r_{xx}\sin^2\theta & (r_{yy}-r_{xx})\sin\theta\cos\theta \\ (r_{yy}-r_{xx})\sin\theta\cos\theta & r_{xx}\cos^2\theta + r_{yy}\sin^2\theta \end{pmatrix}. \tag{9}$$

where r_{ps} and r_{sp} are reflection coefficients of the sample that give the amplitudes of respectively reflecting 's' polarized light (along the transverse direction) into 'p' polarized light (along the longitudinal direction) and vice versa.

The measured response is:

$$\theta_k + i\epsilon_k = r_{ps}/r_{ss} \cong \frac{(r_{yy}-r_{xx})\sin\theta\cos\theta}{r_{yy}} = \frac{(r_{yy}-r_{xx})m_l m_t}{r_{yy}}. \tag{10}$$

Substituting the reflection matrix from Eq. 8 gives (valid to third order in Q):

$$\theta_k + i\epsilon_k = \frac{(\frac{1-n_\perp}{1+n_\perp} - \frac{1-n_\parallel}{1+n_\parallel})m_l m_t}{\frac{1-n_0}{1+n_0}} = \frac{(n_\perp^2 - n_\parallel^2)m_l m_t}{n_0(n_0^2-1)} = \frac{(\epsilon_{xx}-\epsilon_{zz}-\frac{\epsilon_{yz}^2}{\epsilon_{zz}})m_l m_t}{n_0(n_0^2-1)} \tag{11}$$

Now let us consider the dielectric tensor used by Metzger et al.[3] (in the film frame, with $\vec{M} = |\vec{M}| \hat{x}$):

$$\begin{pmatrix} \epsilon_{xx} & 0 & 0 \\ 0 & \epsilon_{zz} & \epsilon_{yz} \\ 0 & \epsilon_{zy} & \epsilon_{zz} \end{pmatrix} = \begin{pmatrix} n_0^2 & 0 & 0 \\ 0 & n_0^2 + \frac{n_0^4 Q^2}{n_0^2-1} & in_0^2 Q m_x \\ 0 & -in_0^2 Q m_x & n_0^2 + \frac{n_0^4 Q^2}{n_0^2-1} \end{pmatrix}. \tag{12}$$

494

Note that the second order term in the dielectric tensor is equal to the first order term squared and divided by $(n_0^2 - 1)$; in a sense, these two terms are not independent of each other. Substituting the values for the dielectric components given in Eq. 12 into Eq. 11, we get:

$$\theta_k + i\epsilon_k = -\frac{Q^2 n_0 m_l m_t}{(n_0^2 - 1)^2},$$
(13)

which agrees with Metzger et al.[3]. Using values for Q and n_0 fitted to our experimental results ($Q = 0.018 + i\,0.004$, $n_0 = 2.5 - i\,3.5$)[5], we obtain:

$$\theta_k + i\epsilon_k \cong 3 \times 10^{-6} m_l m_t,$$
(14)

which is at the limits of detection of our system.

We can go further than Metzger et al.[3] and make the terms in the dielectric tensor first and second order in Q linearly independent. Metzger et al.[3] obtained the dielectric tensor using Bolotin and Sokolov's approach of calculating the conductivity from classical electronic equations of motion in the presence of an effective magnetic field equal to $\vec{M}$[9]. We follow this approach, except that we postulate the existence of two groups of electrons: one group (which we identify with bound electrons) whose equations of motion don't include a force from $\vec{M}$, and another (which we identify with the conduction electrons) whose equations of motion do include a force from $\vec{M}$. This division of electrons into two groups was inspired by Argyres[10], who in his quantum mechanical calculation of the dielectric tensor split integrals over states into two groups: integrals over states occupied by 'magnetic' and 'non-magnetic' electrons. In Argyres' calculation, both groups of electrons contributed to $4\pi\alpha$; for simplicity, we have assumed that only the 'non-magnetic' electrons contribute to $4\pi\alpha$[10]. Essentially, we are fitting the second order term independently of the first, not assuming the two to be related in the manner of Eq . 12. Note that our method gives the same first order term as Metzger et al.[3]. The force equations for the two groups of electrons are:

$$\text{Bound electrons}: \vec{F_1} = m\ddot{\vec{r_1}} = -e\vec{E} - m\omega_0^2 \vec{r_1} - m\dot{\vec{r_1}}\gamma$$
(15)

$$\text{Conduction electrons}: \vec{F_2} = m\ddot{\vec{r_2}} = -e\vec{E} - m\dot{\vec{r_2}}\gamma - \frac{e}{c}\dot{\vec{r_2}} \times 4\pi\vec{M},$$
(16)

where $\vec{r_{1,2}}$ is the displacement of the bound and conduction electron from the origin, ω_0 is the frequency of the bound electron, m is the electron effective mass, $(\vec{M} =| \vec{M} |)$ is the magnetization, e and c are the electronic charge and speed of light respectively, and γ is a damping factor (equal to $-i\times$ the scattering rate in the zero frequency limit).

Substituting $\vec{r_{1,2}}e^{i\omega t}$ for $\vec{r_{1,2}}$ and ω_c for the product $\frac{4\pi|\vec{M}||e|}{mc}$, we write down the following equation for the displacement of the conduction electrons:

$$-m\omega^2 \begin{pmatrix} x_2 \\ y_2 \\ z_2 \end{pmatrix} = \begin{pmatrix} -i\gamma m\omega & 0 & 0 \\ 0 & -i\gamma m\omega & -i\omega\omega_c m_x \\ 0 & i\omega\omega_c m_x & -i\gamma m\omega \end{pmatrix} \begin{pmatrix} x_2 \\ y_2 \\ z_2 \end{pmatrix} - e \begin{pmatrix} E_x \\ E_y \\ E_z \end{pmatrix}.$$
(17)

Recalling that $\tilde{\epsilon} = \tilde{1} + 4\pi\tilde{\chi}$ and the definition of the polarization, $\vec{P} = \tilde{\chi}\vec{E}$, we write $\vec{P} = -\beta n e \vec{r_1} - (1 - \beta)n e \vec{r_2}$, where β is the fraction of bound electrons and n is the total electron density. This gives for the total dielectric tensor $\tilde{\epsilon}$:

$$\tilde{\epsilon} = \tilde{1}(1 + 4\pi\alpha) + \frac{4\pi\tilde{\sigma}}{i\omega}$$
(18)

where $4\pi\alpha = \frac{-4\pi ne^2\beta}{m(\omega^2 - i\gamma\omega - \omega_0^2)}$ is the contribution to the dielectric tensor from the bound electrons and the conductivity tensor $(\tilde{\sigma})$ is given by:

$$\frac{4\pi\tilde{\sigma}}{i\omega} = \frac{4\pi\sigma_0}{i\omega}\begin{pmatrix} 1 - (\omega_c\tau)^2 & 0 & 0 \\ 0 & 1 & -i\omega_c\tau m_x \\ 0 & i\omega_c\tau m_x & 1 \end{pmatrix}\frac{1}{1 - (\omega_c\tau)^2}, \qquad (19)$$

where $\frac{\sigma_0}{i\omega} = \frac{-ne^2(1-\beta)}{m(\omega^2 - i\gamma\omega)}$ and τ is a frequency-dependent factor given by: $\tau = \frac{-\omega}{(\omega^2 - i\gamma\omega)}$. Note that $\frac{4\pi\sigma_0}{i\omega} = (n_0^2 - 1 - 4\pi\alpha)$. Writing the conductivity in this manner allows us to make the dielectric tensor a function of n_0, Q, and $4\pi\alpha$ only; the latter two variables can be fit to the magnitude of the terms in the magneto-optic response first and second order in Q, respectively (see Eq. 1 and Ref. 6).

Substituting Eq. 19 into Eq. 18 gives the complete expression for our dielectric tensor:

$$\begin{pmatrix} \epsilon_{xx} & 0 & 0 \\ 0 & \epsilon_{zz} & \epsilon_{yz} \\ 0 & \epsilon_{zy} & \epsilon_{zz} \end{pmatrix} = \begin{pmatrix} n_0^2 & 0 & 0 \\ 0 & n_0^2 + \frac{(n_0^2 - 1 - 4\pi\alpha)(\omega_c\tau)^2}{1 - (\omega_c\tau)^2} & \frac{4\pi\sigma_0}{i\omega}\frac{(i\omega_c\tau)m_x}{1-(\omega_c\tau)^2} \\ 0 & \frac{4\pi\sigma_0}{i\omega}\frac{(-i\omega_c\tau)m_x}{1-(\omega_c\tau)^2} & n_0^2 + \frac{(n_0^2 - 1 - 4\pi\alpha)(\omega_c\tau)^2}{1-(\omega_c\tau)^2} \end{pmatrix}. \qquad (20)$$

The first order term of the dielectric tensor is given by:

$$\epsilon_{yz} = in_0^2 Q m_x, \qquad (21)$$

where Q is of order 10^{-2}. Comparing our dielectric tensor (Eq. 20) to Eq. 21 and using the equality $\frac{4\pi\sigma_0}{i\omega} = (n_0^2 - 1 - 4\pi\alpha)$, we write down the following equality:

$$(n_0^2 - 1 - 4\pi\alpha)\frac{(i\omega_c\tau)m_x}{1 - (\omega_c\tau)^2} = in_0^2 Q m_x \qquad (22)$$

and solve for $\omega_c\tau$, which gives (valid to second order in the quantity $\frac{n_0^2 Q}{n_0^2 - 1 - 4\pi\alpha}$, which is of order 10^{-1} using the values of n_0 and Q given earlier and a fitted value of $4\pi\alpha = -3.6 - 14.6\ i^5$):

$$\omega_c\tau = \frac{n_0^2 Q}{n_0^2 - 1 - 4\pi\alpha}, \qquad (23)$$

which agrees with Metzger et al.[3] in the limit of $\alpha \to 0$. Metzger et al. use a different notation; they use the variables λ and A for the quantities $\omega_c\tau$ and $n_0^2 - 1$ (valid to first order in λ), respectively[3].

Having solved for $\omega_c\tau$, we can then write down an expression for ϵ_{zz} in terms of n_0, Q, and $4\pi\alpha$:

$$\epsilon_{zz} = n_0^2 + \frac{(n_0^2 - 1 - 4\pi\alpha)(\omega_c\tau)^2}{1 - (\omega_c\tau)^2} \cong n_0^2 + \frac{n_0^4 Q^2}{(n_0^2 - 1 - 4\pi\alpha)}, \qquad (24)$$

which is valid to third order in $\omega_c\tau$. This expression agrees with Metzger et al.[3] in the limit of $\alpha \to 0$. Our theory therefore has the following dielectric tensor:

$$\begin{pmatrix} \epsilon_{xx} & 0 & 0 \\ 0 & \epsilon_{zz} & \epsilon_{yz} \\ 0 & \epsilon_{zy} & \epsilon_{zz} \end{pmatrix} = \begin{pmatrix} n_0^2 & 0 & 0 \\ 0 & n_0^2 + \frac{n_0^4 Q^2}{(n_0^2 - 1 - 4\pi\alpha)} & in_0^2 Q m_x \\ 0 & -in_0^2 Q m_x & n_0^2 + \frac{n_0^4 Q^2}{(n_0^2 - 1 - 4\pi\alpha)} \end{pmatrix}. \qquad (25)$$

Plugging the tensor components $\epsilon_{xx}, \epsilon_{zz}$, and ϵ_{yz} into Eq. 11 and using the values of n_0 and Q given earlier, we obtain:

$$\theta_k + i\epsilon_k = -\frac{Q^2 m_l m_t (n_0^2 - \frac{n_0^4}{n_0^2 - 1 - 4\pi\alpha})}{n_0(n_0^2 - 1)} = -\frac{Q^2 m_l m_t n_0(\frac{1+4\pi\alpha}{n_0^2 - 1 - 4\pi\alpha})}{(n_0^2 - 1)} \cong 5 \times 10^{-4} m_l m_t, \quad (26)$$

which is easily observable by our system. We fit $4\pi\alpha$ to the magnitude of the Q^2 term. Note that our results coincide with those of Metzger et al.[3] in the limit of $\alpha \to 0$.

CONCLUSION

We have shown how the MOKE signal from an epitaxial Fe(110)/Mo(110) bilayer contains a significant contribution from a term proportional to Q^2. This term can appear in the magneto-optic response from any sample with a non-zero m_t (and therefore an in-plane anisotropy). This term is proportional to the product $m_l m_t$ and therefore does not go into itself if reflected through the origin. We have modified the existing theory of magneto-optics to explain the magnitude of the Q^2 term in the magneto-optic response.

REFERENCES

1. S. D. Bader, J. Mag. Magnetic Mats., **100**, 440 (1991).

2. W. Voigt. *Handbuch der Elektrizität und des Magnetismus.* (Barth **4:2**, Leipzig, 1915).

3. G. Metzger, P. Pluvinage and R. Torguet, Annales de Physique, **10**, 1965, p.5.

4. B. M. Clemens, R. M. Osgood, A. P. Payne, B. M. Lairson, S. Brennan, R. L. White, and W. D. Nix, J. Mag. Magnetic Mats., **121**, 37 (1993).

5. R.M. Osgood III, B.M. Clemens, and R.L. White. Asymmetric magneto-optic response in anisotropic thin films, 1995 (unpublished).

6. R. M. Osgood III, B. M. Clemens, and R. L. White. *This volume.* (Mats. Res. Soc. Proc., Pittsburgh, PA, 1995), edited by B. Heinrich.

7. R.M. Osgood III. PhD thesis, Stanford University, 1995.

8. A. Yariv and P. Yeh. *Optical Waves in Solids.* (John Wiley and Sons, 1984).

9. G. A. Bolotin and A. V. Sokolov, Fiz. Metall. i Metallov., **12**, 493,625,785 (1961).

10. P. N. Argyres, Phys. Rev., **97**, 334 (1955).

SURFACE ENHANCED MAGNETO-OPTICS
IN NOBLE METAL / FERROMAGNETIC METAL MULTILAYERS

V. I. SAFAROV[*], V. A. KOSOBUKIN[*], C. HERMANN[*], G. LAMPEL[*], J. PERETTI[*], AND
C. MARLIÈRE[**],

[*]Laboratoire de Physique de la Matière Condensée, Ecole Polytechnique,
91128 Palaiseau Cedex, France.
[**]Institut d'Optique Théorique et Appliquée, Centre Universitaire, Bât. 503,
91403 Orsay Cedex, France.

ABSTRACT

We present an electromagnetic enhancement mechanism for the magneto-optical response of noble metal / ferromagnetic metal multilayer thin films. When such a structure is illuminated in total reflection condition, the resonant coupling of light with the noble metal surface plasmons gives rise to an amplification of the magneto-optically induced component of the light electric field. The experimental results obtained on a 30nm-thick Au / Co / Au model system show that this resonant feature observed in the Kerr rotation and ellipticity corresponds to a strong enhancement of the magneto-optical figure of merit and signal-to-noise ratio.

INTRODUCTION

Enhancing magneto-optical (MO) effects stimulates a wide interest as it may promote new capabilities of optical techniques for studying magnetic material properties and because of the possible applications to information storage. Any approach to enhance the MO response (i. e. the MO figure of merit) of a structure involving a given magnetic material must account for two necessary conditions : the media must provide a high optical response (reflectivity) and a large (Kerr) rotation or ellipticity.[1,2]

Here we show that these two required conditions can be satisfied in simple systems such as noble metal / ferromagnetic metal multilayer thin films which are already known to exhibit large MO effects.[3] Our experimental and theoretical results obtained on a model Au / Co / Au structure[4] evidence that a resonance-like characteristic feature is observed in the MO effects at the angle of incidence beyond the total reflection limit where the noble metal (gold) surface plasmon modes[5] couple with the p-component of the evanescent light electric field. This feature results in a strong enhancement of the Kerr rotation and ellipticity and above all of the MO figure of merit and of the signal-to-noise ratio.

EXPERIMENT

The sample is a Au / Co / Au film of total thickness d = 30 nm, grown on a glass plate under UHV conditions.[6] The thicknesses of the substrate and cap gold layers are respectively d' = 25 nm and d" = 4 nm. The ℓ = 1 nm-thick cobalt layer has its easy-magnetization axis perpendicular to the film plane and exhibits a square hysteresis loop.

In our experiment,[4] the glass plate is optically coupled to a half-cylindrical glass prism with adapting refractive index liquid. Linearly polarized light of wavelength 647 nm (the red line of a krypton-ion laser) illuminates the sample through the prism at a variable incident angle θ. The intensity and polarization of the reflected beam are analyzed. The MO measurements are performed by specific modulation technics. The saturated magnetization of the Co-layer is periodically flipped by applying an alternative pulsed magnetic field perpendicular to the film plane and larger than the Co-layer coercive field (≈ 800 Oe). The resulting modulation of the reflected light polarization induced by the MO effect is detected as an intensity modulation through a polarization analyzer. The ac component of the detector output signal at the modulation frequency, measured with a lock-in amplifier, is the amplitude of the MO hysteresis loop at zero external magnetic field.

For a unit excitation intensity, the dc detector output signal is half of the reflectivity R and the ac signal is taken as the definition of the MO figure of merit.[1] The polarization analyzer is either a linear analyzer oriented at 45° to the incident polarization axis, or a circular analyzer, i. e. a quarter wave plate inserted in front of a linear analyzer oriented at 45° to the plate axis. For linear (resp. circular) detection, we note S^l (resp. S^c) the value of the MO figure of merit and K^l (resp. K^c) the Kerr rotation (resp. ellipticity) which is the ratio of the ac component to twice the dc component of the detector output signal.

When considering the MO effect as the appearance of a component perpendicular to the incident polarization direction, the complex amplitudes, E_R^s and E_R^p, of the s- and p-components of the reflected light electric field are related to the incident ones, E_0^s and E_0^p, by the following tensorial expression :[7]

$$\begin{pmatrix} E_R^s \\ E_R^p \end{pmatrix} = \begin{pmatrix} \rho^{ss} & \rho^{sp} \\ \rho^{ps} & \rho^{pp} \end{pmatrix} \begin{pmatrix} E_0^s \\ E_0^p \end{pmatrix} \tag{1}$$

This relation defines the complex reflection coefficients ρ of the whole structure for each component of the reflected electric field. Writing these coefficients in the general form $\rho = r\, e^{i\phi}$, the different quantities measured in the MO experiment, for instance in the case of a p-excitation, are given by :

$$R = \left(r^{pp}\right)^2 \tag{2}$$

$$K^l = \frac{r^{sp} \cos\left(\Delta\phi^{sp}\right)}{r^{pp}} \quad \text{and} \quad K^c = \frac{r^{sp} \sin\left(\Delta\phi^{sp}\right)}{r^{pp}} \tag{3}$$

$$S^l = 2\, r^{pp}\, r^{sp} \cos\left(\Delta\phi^{sp}\right) = 2\, R\, K^l \quad \text{and} \quad S^c = 2\, r^{pp}\, r^{sp} \sin\left(\Delta\phi^{sp}\right) = 2\, R\, K^c \tag{3}$$

where $\Delta\phi^{ps} = \phi^{ss} - \phi^{ps}$. Similar expressions hold for s-excitation with permutation of the superscripts s and p.

RESULTS

In Fig.1 are plotted the variations, with the incidence angle, of R (curves a and b), K^l (curves c and d) and K^c (curves e and f), that we measured for both incident polarizations. For p-excitation, the pronounced minimum of reflectivity (curve b) characteristic of the excitation of the gold surface plasmon[5] is observed beyond the total reflection limit (θ > 41.5°) at the

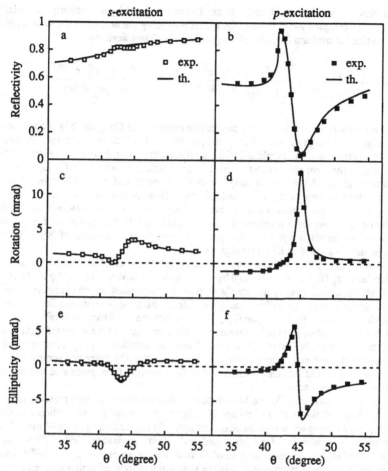

Figure 1 : Reflectivity (curves a and b), Kerr rotation (curves c and d) and ellipticity (curves e and f) as a function of the incidence angle θ, for s- (left column) and p- (right column) excitation. the Symboles represent the experimental points and the full lines are obtained from theory.

resonance angle $\theta_R \approx 44.5°$ (for the light energy that we uses). In the vicinity of θ_R, the MO quantities all exhibit a resonance behavior for both s- and p-excitations.

The precise theoretical description of the MO properties of magnetic metallic multilayer thin films, from which we obtained the theoretical curves plotted in Fig.1 (full lines), is not detailed here.[4,8] The main feature occuring at resonance may be understood when considering the analytical expressions, to first order in ζ of the MO components of the reflected light electric field, i. e. of ρ^{ps} and ρ^{sp} which are involved in all the measured MO quantities. In these expressions, we use the complex reflection coefficients ρ and transmission coefficients τ of the two light electric field components,[9] as indicated by the superscripts s or p, at the different interfaces between two adjacent media designated by double-number subscripts where 1 refers to glass, 2 to gold, and 3 to air. Denoting ε_2 and k_2 respectively the dielectric constant and the complex z-component (the z-direction being the normal to the surface) of the light wave vector in medium 2 (gold), ρ^{ps} is given, in this framework, by :

$$\rho^{ps} = \frac{k_2\,\varepsilon_b\,\ell}{2\varepsilon_2\,\cos(\theta)}\;\frac{e^{ik_2d'}\,\tau_{12}^s\,(\,1 + \rho_{23}^s\,e^{2ik_2d''}\,)}{1 + \rho_{12}^s\,\rho_{23}^s\,e^{2ik_2d}}\;\frac{e^{ik_2d'}\,\tau_{21}^p\,(\,1 + \rho_{23}^p\,e^{2ik_2d''}\,)}{1 + \rho_{12}^p\,\rho_{23}^p\,e^{2ik_2d}} \tag{5}$$

The first factor of Eq.(5) is the s→p conversion factor in the Co layer. It is proportional to the magnetization through ε_b, the off-diagonal part of the cobalt dielectric tensor. The second factor is the expression of the amplitude of the total s-excitation field in the Co layer. The third factor describes the propagation and interface transmission, between the Co layer and the observation in glass, of the p-wave magneto-optically generated in the Co layer. Both the second and third factors account for multiple reflections at the Au / air and Au / glass interfaces. The plasmon resonance arises here through $\rho^p{}_{23}$, in the numerator of the third factor (the denominator term involving $\rho^p{}_{23}$ is attenuated), which shows around θ_R a strong amplitude enhancement (responsible for the maximum in the variation of K^l) and phase variation (which explains that K^c go through zero).

For p-excitation, ρ^{sp} is obtained by replacing the first factor of Eq.(5) by the p→s conversion factor $\varepsilon_b\,\ell k_0{}^2\cos(\theta)/2k_2$ and by exchanging the superscripts s and p in the second and third factors. One can verify that for the considered magnetization orientation $\rho^{sp} = \rho^{ps}$. It is now the second factor, describing the amplitude of the total p-excitation field in the Co layer, which contains the resonant quantity $\rho^p{}_{23}$ in exactly the same form as in ρ^{ps}.

Thus, for any incident light polarization, the MO component of the reflected light electric field has the same complex amplitude $\rho^{sp} = \rho^{ps}$ and the coupling of the p-component of the light electric field, either magneto-optically induced in the Co layer or coming from the excitation, with the gold surface plasmon at the Au / air interface, gives rise to a resonant behavior of the MO component.

According to Eqs.(2) and (3), we have deduced from the measurements presented in Fig.1 the experimental variation with θ of r^{sp} and r^{ps} (Fig.2 curves a and b). As predicted by theory (full lines), these curves show that the magneto-optically induced component of the reflected light electric field is identical for both s- and p-incident polarizations [$r^{ps} = r^{sp}$ within the accuracy, around θ_R, on the determination of r^{pp} from the measurements of $R = (r^{pp})^2$] and is about three times larger at resonance than in the standard non-total reflection geometry.

Nevertheless, important differences appear around θ_R in the behavior of the measured MO quantities according as the incident light is s- or p-polarized. The Kerr rotation, for instance, is enhanced at θ_R by about a factor ten for p-excitation, and by a factor three for s-excitation. This larger enhancement of Kerr rotation for p-excitation is in fact "artificial". It is not due to the MO properties of the system but features the abrupt variation of the reflectivity. Indeed, as can be seen from curve b in Fig.1, r^{pp}, which enters as a denominator in the expression of K^l [see Eq.(4)], goes through a deep minimum (close to zero) at resonance. On the contrary, for s-excitation, the amplification (by a factor 3) of K^l at θ_R is directly related to the one of r^{ps} because r^{ss} (see curve a in Fig.1) does not show any particular feature and remains almost equal to 1 in the whole range of incidence angle. This emphasizes that when the reflectivity varies, the Kerr rotation and ellipticity are not significant enough to describe the MO properties of a system.

As an important consequence, although the MO component of the reflected light electric field are identical for both incident polarizations, the two conditions (large reflectivity and

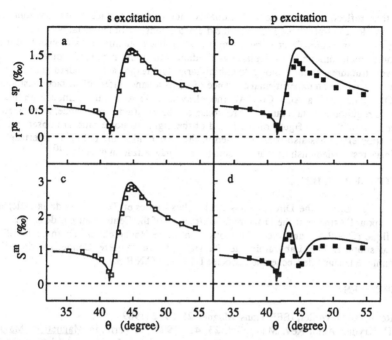

Figure 2 : Variation, as a function of the incidence angle θ, of r^{ps} and r^{sp} (curves a and b), and of the maximum figure of merit S^m for s- and p-excitation (curves c and d). the Symboles represent the experimental points and the full lines are obtained from theory. The variation of r^{ps} and r^{sp} also represents the variation of the signal to noise ratio (in arbitrary units) related to the measurement of S^m respectively for s- and p-excitation.

large rotation or ellipticity) required to obtained a large MO response are satisfied at resonance for s-excitation but not for p-excitation. This is evidenced when considering the variation with incidence angle of the maximum figure of merit S^m (= 2 r^{ss} r^{ps}, for s-excitation) that one can deduce from the measurements presented in Fig.1 or directly measure by compensating at detection the phase difference ($\Delta\phi^{ps}$ or $\Delta\phi^{sp}$) between the s- and p-components of the reflected light electric field. For s-excitation (Fig.2.c), S^m shows a strong enhancement directly related to the amplification of r^{ps} (Fig.2.a), while for p-excitation (Fig.2.b), the same resonance effect on r^{sp} (Fig.2.b) is strongly attenuated by the low value of the reflectivity, i. e. of r^{pp}.

Finally, one can roughly consider that the noise in the MO signal is proportional to the square root of the reflectivity, i. e. to r^{ss} or r^{pp}. Then, the signal-to-noise ratio related to the measurement of S^m is proportional to r^{ps} (Fig.2.a) or r^{sp} (Fig.2.b) and also shows a large enhancement at resonance. Therefore, for s-excitation, the situation is particularly favourable as both the true MO response, i. e. the figure of merit, and the signal-to-noise ratio are enhanced at resonance.

CONCLUSION

We have shown that in magnetic metallic systems the coupling of light with the surface plasmon modes may provide surface enhanced MO effects. This coupling occurs in a very simple total reflection arrangement. The quality factor of the surface plasmon resonance is high when $|Re(\varepsilon)/Im(\varepsilon)| \gg 1$, ε being the dielectric constant of the metal, which is the case for noble metals in a wide spectral range including the near infrared and visible domain. Therefore, this technique is particularly well adapted to noble metal / ferromagnetic metal multilayer structures, already known to exhibit large MO response in standard geometry. We have presented here an experimental evidence of the resonance effect of surface plasmons on the MO properties of a Au / Co / Au model system. These results demonstrate that the conditions required to obtain large MO response can be satisfied at resonance, where a strong enhancement of the MO figure of merit and of the signal to noise ratio are observed. This amplification effect has also been observed in the near-field at the gold / air interface[4] and preliminary results demonstrate that it occurs for any magnetization direction.[10]

ACKNOWLEDGEMENTS

We are grateful to the Direction des Recherches Etudes et Techniques de la Délégation Générale pour l'Armement, the Programme Ultimatech of the Centre National de la Recherche Scientifique (C.N.R.S.), and the Laboratoire Central de Recherche of Thomson C.S.F. for financial support. The Laboratoire de Physique de la Matière Condensée de l'Ecole Polytechnique is Unité de Recherche Associée D1254 au C.N.R.S..

REFERENCES

1. W. Reim and D. Weller, IEEE Trans. Magnet. **25**, 3752 (1989).
2. M. H. Kryder, Annu Rev. Mater. Sci. **23**, 411 1993 ; J. Ferré, in Magnetism, Magnetic Materials and their Applications, eds. F. Leccabue and J. L. Sanchez Llamazares (IOP Publishing, Bristol, 1992), p. 167.
3. See for instance, J. Ferré, G. Pénissard, C. Marlière, D. Renard, P. Beauvillain, and J. P. Renard, Appl. Phys. Lett. **56**, 1588 (1990).
4. V. I. Safarov, V. A. Kosoboukin, C. Hermann, G. Lampel, J. Peretti and C. Marlière, Phys. Rev. Lett. **73**, 3584 (1994).
5. See for instance, H. Raether, Surface Plasmons, in Springer Tracts in Modern Physics, vol. 11 (Springer Verlag, Berlin, 1988).
6. C. Marlière, D. Renard, and J. P. Chauvineau, Thin Solids Films **201**, 317 (1991) ; and references therein.
7. G. Metzger, P. Pluvinage, and R. Torguet, Ann. Phys. **10**, 5 (1965).
8. We have developped two theoretical approaches for describing the MO properties of metallic magnetic multilayer thin films. These treatments will be published elsewhere. Both show the same agreement with our experimental results. The first one is a MO generalization of the following work : V. A. Kosobukin, Zh. Tekh. Fiz. **56**, 1481 (1986) [Sov. Phys. Tech. Phys. **31**, 879 (1986)]. The second, after which are obtained the analytical expressions of the reflection coefficients ρ^{ps} and ρ^{sp} presented here, may be related to the following works : J. Zak, E. R. Moog, C. Liu, and S. D. Bader, J. Mag. Mag. Mat. **89**, 107 (1990) ; V. M. Agranovich, in Surface Polaritons, ed. by V. M. Agranovich and D. L. Mills (North Holland, Amsterdam, 1982), p. 187.
9. beyond the total reflection limit, the reflection and transmission coefficients are here formally defined by the same expressions than those generally used for propagating waves.
10. V. I. Safarov, V. A. Kosoboukin, C. Hermann, G. Lampel, J. Peretti, and C. Marlière (unpublished).

Part IX

Granular Nanostructures

EVOLUTION OF NANOSCALE FERROMAGNETIC PARTICLES IN Co-Cr AND Cr-Fe ALLOYS OBSERVED BY ATOM PROBE FIELD ION MICROSCOPY

K. HONO*, R. OKANO**, K. TAKANASHI**, H. FUJIMORI** Y. MAEDA*** and T. SAKURAI**
*National Research Institute for Metals, 1-2-1 Sengen, Tsukuba 305, Japan
**Institute for Materials Research, Tohoku University, Sendai 980-77, Japan
***NTT Basic Research Laboratories, Atsugi, Japan

ABSTRACT

With appropriate processing conditions, nanoscale ferromagnetic particles precipitate from nonmagnetic matrix phase in the Co-Cr and Cr-Fe systems. In these heterogeneous alloys, unique magnetic properties are observed. In order to correlate such magnetic properties with the microstructures, we have employed an atom probe field ion microscope (APFIM) and a three dimensional atom probe (3DAP). In the Co-22Cr thin film sputter-deposited at elevated temperatures (~500 K), both APFIM and 3DAP data convincingly showed that the film was composed of lamellae-like ferromagnetic and paramagnetic phases of approximately 8 nm in thickness. On the other hand, it was shown that the films sputter-deposited at ambient temperature was composed of ε-Co single phase without significant compositional heterogeneity. Based on these observations, we conclude that phase separation progresses during the growth of the film on a heated substrate. In the Cr-Fe alloy, large negative MR was observed in the as-quenched alloy at liquid helium temperature. However, the MR behavior changes as the phase decomposition progresses by annealing. The change in the MR behavior observed in this alloy with various heat treatment conditions will be discussed based on the microstructural characterization results by APFIM and 3DAP.

INTRODUCTION

Magnetic properties of materials are sensitive to the microstructure. Hence, in order to understand the mechanism of the emergence of magnetic properties, it is essential to characterize the microstructures in less than a nanometer scale. In many magnetic materials, properties arise when ferromagnetic particles evolve from a nonmagnetic matrix. Such examples can be found in some alloy permanent magnets such as Alnico, where nanoscale ferromagnetic particles precipitate from the matrix phase by phase decomposition. Similar phenomena were recently observed in some high density recording media and giant magnetoresistance (MR) materials. This paper present two such examples, i.e. decomposition in Co-Cr thin films and Cr-Fe alloys. The former is the fundamental system for high density recording media such as Co-Cr-Ta/Cr and Co-Cr-Pt/Cr, and the compositional heterogeneities in these films have been a subject of controversy. The latter alloy is interesting because it was recently found that large negative MR appeared in this system. In this study, we have analyzed local concentration changes in these materials in a subnanometer scale, and try to understand the correlation between the magnetic properties and the microstructure.

EXPERIMENTAL

For characterizing microstructures of these alloys, we have employed a time-of-flight atom probe field ion microscope (APFIM) and a three dimensional atom probe (3DAP). In an atom probe, atoms are field evaporated from the surface of a sharp needle shape specimen [1]. The mass-to-charge ratio is determined by measuring the time-of-flight of individual ions. By this way, atoms are sampled from a region which is covered by a small aperture (probe hole) of less than 2 nm. Since atoms are always ionized from the surface of the specimen, it is possible to obtain

507

concentration depth profiles by collecting an extended number of ions. The lateral spatial resolution of such analysis is determined by the probe hole diameter and a monoatomic layer resolution is obtained in the depth direction. In a three dimensional atom probe, the mass to charge ratios and the coordinates of individual ions are determined simultaneously [2]. By collecting such information from an extended number of atoms, two dimensional element mappings can be constructed with a subnanometer spatial resolution. As atoms always ionized from the specimen surface, atom number corresponds to the depth of the specimen. Scaling appropriate depth distance as a function of the number of detected ions, three dimensional element mappings can be constructed [3]. The details of atom probes used in the present study are described elsewhere [4,5].

The Co-Cr film was deposited by rf magnetron sputtering from a Co-22at.%Cr alloy target on a Cu coated Si wafer substrate at an Ar pressure of 1.1 Pa. The film thickness was estimated to be approximately 3 μm. During deposition, the substrate temperature was kept at approximately 500 K. The specimens for atom probe analyses were prepared by using a combination of photolithography and micro-electropolishing. A detailed description of this method is given elsewhere [4]. It should be noted that the FIM tips were prepared in a planar direction of the film, and hence atom probe analyses were conducted in this direction. The specimens were etched symmetrically during electropolishing so that the analysis region corresponded approximately to the middle of the film, i.e., about 1.5 μm from both substrate and the surface.

Cr-Fe alloy ingots with different compositions were prepared by argon arc melting. Small square rods of approximately 0.5 x 0.5 x 12 mm were cut out of the button ingot. These samples were solution heat treated in a vacuum sealed quartz tube at 1300 K for 1 h, then quenched into water (referred to as ASQ). Solution treated specimens were subsequently annealed in vacuum at 773 K for periods of 400 and 2600 h (hereafter, referred to as 400 h and 2600 h, respectively). Magnetoresistance (MR) was measured at 5.8 kOe at three different temperatures of 4.2, 77 and 300 K by the standard four-point probe method. Magnetic field is applied parallel to the electric current for all the MR measurements.

PHASE DECOMPOSITION IN Co-Cr SPUTTERED THIN FILMS

Co-Cr alloy is the fundamental system employed in current longitudinal high density recording media, such as Co-Cr-Ta and Co-Cr-Pt [7,8]. Co-Cr alloy thin film are also considered to be one of the most promising media for perpendicular recording [9]. It is well known that the magnetic properties of the thin film recording media depend on sputtering parameters, such as bias voltage, gas pressure and substrate temperature. One such example is the increase in coercivity and the decrease in media noise when Co-Cr based films are deposited on heated substrates [10,11]. The underlying mechanisms of such magnetic property changes, however, are not well understood.

Magnetic measurements indicate that there may be compositional heterogeneities in Co-Cr binary alloy thin films [12-16]. For example, Fisher et al. [13] found that the saturation magnetization and Curie temperature of sputtered thin films were higher than those of bulk materials with the same mean composition. Subsequently, many investigations have attempted to confirm the presence of compositional heterogeneities in these alloy thin films [14-16]. It has been suggested that such compositional heterogeneities may be explained by grain boundary segregation of Cr. In this model, each columnar grain is thought to be magnetically isolated from neighboring grains by a Cr enriched grain boundary phase and thus could function as the minimum unit of the recording bit [17]. In order to test this model, several analytical electron microscopy studies using x-ray energy dispersive spectroscopy (EDS) were performed [18-20]. However, limitations in the spatial resolution of these methods, which employ a focused electron beam, make it difficult to obtain quantitative results that provide evidence for the grain boundary segregation or chemical heterogeneities within the small grains in sputtered thin films.

It is impossible to observe compositional heterogeneities in these films by structure factor contrast using transmission electron microscopy (TEM), since the atomic scattering factor of Cr is very similar to that of Co. Maeda et al. [21-23], however, observed a chrysanthemum-like pattern (CP) structure within the grains of Co-Cr thin films deposited on heated substrates by chemically etching the Co in the Co-Cr films. They proposed that the bright contrast in this CP structure represented preferential dissolution of regions which were Cr depleted prior to etching. Based on this interpretation, they concluded that fine compositional fluctuations existed *within* each grain. Many subsequent results were published supporting the presence of compositional heterogeneities within the grains. These included analytical TEM studies [24], nuclear magnetic resonance (NMR) measurements [25,26], thermomagnetic analysis [27,28], small angle neutron scattering (SANS) studies [29] and atom probe investigations [30-33].

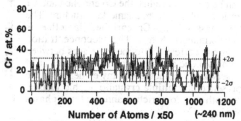

Figure 1 shows a one dimensional atom probe concentration depth profile of Co-22at.%Cr thin films sputter deposited at 473 K, which convincingly showed that the films prepared at elevated temperatures were composed of ferromagnetic and paramagnetic phases of comparable dimensions. The specimen was prepared along the direction of the film, so the concentration depth profile represents

Fig. 1 One dimensional atom probe concentration depth profile of Co-22at.%Cr alloy thin film sputter-deposited at 573 K.

fluctuations in the Cr concentration in the planar direction of the film. One phase is rich in Cr (~30 - 40 at.%Cr), and the other is depleted in Cr (~5 - 10 at.%Cr). The sizes of Cr and Co enriched regions are comparable (~8 nm). The Curie temperature becomes lower than room temperature when the Cr concentration exceeds 25 at.%, hence the Cr enriched phase is paramagnetic and the Co enriched region is ferromagnetic. As the size of the paramagnetic phase is several nanometers, each ferromagnetic phase is believed to be isolated magnetically. The scale of the compositional fluctuation is significantly smaller than that of the grains, hence it is concluded that compositional fluctuations are present within the grains. Such compositional inhomogeneity cannot be explained by equilibrium grain boundary segregation and it is suggested that this is caused by phase separation which occurred during the sputtering process. The thin films deposited at the ambient temperature did not show such long range fluctuation of Cr concentration [32]. This indicates that sputter deposition on a heated substrate enhances the kinetics of decomposition, because high energy ions can migrate on the growing surface by a surface diffusion mechanism. In this case, the two phase state, which will never be achieved in the bulk specimen, can be realized by the sputter deposition at elevated temperatures.

While the conventional atom probe data convincingly showed the presence of the compositional heterogeneity, it does not provide any information on the morphology of the phase. Hence, in order to visualize the morphology of the ferromagnetic particles, three dimensional atom probe analysis was performed. Figures 2 (a) and (b) show a three dimensional concentration mapping and isoconcentration

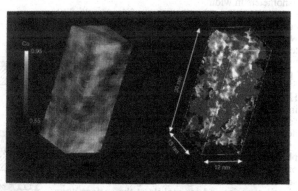

Fig. 2 (a) Three dimensional atom probe concentration mapping of Co-22at.%Cr alloy sputter-deposited at 573 K. The color scale indicates the local concentration of Co in the analyzed volume. By TEM observation of the specimen, it was confirmed that the data was obtained from an interior of a grain. (b) Isoconcentration contour of 75at.%Co. This surface corresponds to interfaces between the paramagnetic and ferromagnetic phases.

509

contours which were constructed by connecting concentration data points of 75at.%Co, respectively. The three dimensional concentration mapping (Fig. 2(a)) shows there is significant fluctuation in Co concentration. It can be clearly seen that the Co enriched and Cr enriched phases form a lamellar structure. The thickness of the Cr enriched lamellae is approximately 5 nm. The isoconcentration contours correspond to the interfaces between paramagnetic and ferromagnetic phases. TEM observations did not reveal any evidence for the presence of multiple crystallographic phases, so both the Co enriched and depleted phases are hcp ε-Co.

This work directly reveals compositional heterogeneity within grains of the Co-Cr thin films sputter-deposited at elevated temperatures. In addition, the three dimensional compositional mapping shows convincingly that the ferromagnetic and paramagnetic phases both exist in a lamellar structure. The lamellae of the Co enriched phase are also completely separated from each other by the lamellae of the Cr enriched paramagnetic phase, whose width is 5 - 10 nm. This thickness is believed to be sufficient to suppress exchange interactions between neighboring ferromagnetic regions. It also should be noted that if the ferromagnetic phase were composed of spherical particles less than ~10 nm in diameter, the film would exhibit superparamagnetic behavior. These phases, however, form a lamellar structure within the grain, as shown in Fig. 2, and hence the ferromagnetic phase will be more resistant to thermal fluctuations due to shape anisotropy even though they are only a few nanometers in width.

We believe that the origin of the lamella-like compositional heterogeneity is phase separation which progresses during film deposition at elevated temperatures. Based on the phase diagram calculated by Hasebe et al. [34], the isostructure two phase field appears at a temperature region below approximately 500 K for the Co-22at.%Cr alloy as shown in Fig. 3. Hence, phase decomposition is expected to take place in the low temperature region for the Co-22at.%Cr solid solution. In the bulk, however, such phase decomposition would never occur, due to sluggish volume diffusion in this temperature range. The phase decomposition would be kinetically possible only during the film deposition process at the film surface, because during sputtering atoms have high kinetic energies and are mobile at the surface upon deposition. By elevating the substrate temperature to around 600 K, phase separation would progress two dimensionally by surface diffusion as schematically shown in Fig. 4. When the second monolayer develops, Cr atoms would segregate above the Cr enriched underlayer and Co atoms would segregate above the Co enriched underlayer. Then a lamellar structure composed of the alternating ferromagnetic

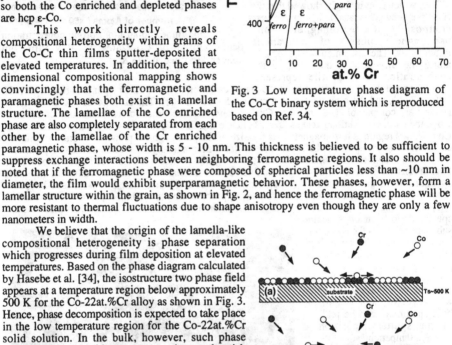

Fig. 3 Low temperature phase diagram of the Co-Cr binary system which is reproduced based on Ref. 34.

Fig. 4 Schematic drawing of the deposition process of Co-Cr thin film at an elevated temperature.

and paramagnetic phases would develop as the film grows (Fig. 4). When the substrate temperature is lower, the kinetics for surface diffusion are not fast enough and a compositionally uniform film should grow. For a better understanding of magnetic properties, it is also important to characterize the chemical inhomogeneities at grain boundaries.

MAGNETORESISTANCE AND PHASE DECOMPOSITION IN Cr-Fe ALLOY

Giant magnetoresistance (GMR) was reported for the first time in Cr/Fe thin films with the multilayer structure [35]. Since then, this phenomenon has been observed in many metallic multilayers which are composed of nonmagnetic and ferromagnetic thin layers of a few nanometers [36-39]. A recent discovery of GMR in alloy thin films with heterogeneous structure, in which nanoscale ferromagnetic particles are distributed in the nonmagnetic matrix phase, has stimulated research into non-multilayer type GMR [40,41]. This has been reported in the alloy systems composed of nonmagnetic and ferromagnetic elements with negligible solubilities like Cu-Co, Ag-Fe, Ag-Co etc. [42]. Moreover, it has been recently shown that a similar phenomenon is observed in melt spun ribbons of Cu-Co [43-45] as well as in bulk Cu-Ni-Fe alloy [46]. The origin of MR in these heterogeneous alloy films is believed to be similar to that of multilayer type GMR, i.e. the spin dependent electron scattering by the ferromagnetic particles. Although this type of GMR has been reported in many binary alloy systems, the Cr-Fe system has not been examined in the thin film or bulk alloy form with heterogeneous structure. The Cr-Fe system is quite different from the other systems where GMR has been reported in the heterogeneous structure in the following points: (1) the equilibrium solid solution exists above 1094 K in the entire concentration region, (2) it has a large miscibility gap below 1094 K, but the solubility limit of each side is as much as 10 at.% even at 700 K. Phase separation is expected to occur by aging the supersaturated solid solution at temperatures below

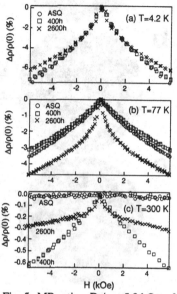

Fig. 5 MR ratios, Dr/r at 5.8 kOe, of as-quenched Cr-25at.%Fe (referred to as ASQ), the same alloy annealed at 773 K for 400 h. (referred to as 400 h) and for 2 600 h. (referred to as 2600 h). Measured at (a) 4.5 K, (b) 77 K and (c) 300 K.

the miscibility gap, although the kinetics of the decomposition is expected to be very slow below 800 K due to the low bulk diffusivity. In order to attain a heterogeneous structure in the Cr-Fe system, it is necessary to anneal the supersaturated solid solution for an extended period of time. We have recently observed large MR in sputtered Cr-Fe thin films as well as in the bulk Cr-Fe alloys [47,48].

Figure 5 (a-c) shows the variation of the MR ratios of Cr-25at.%Fe alloy with three different heat treatment conditions (as-quenched, aged for 400 h and 2600 h at 773 K). In this figure, the values of $\Delta\rho/\rho$ at 4.2, 77 and 290 K were plotted as a function of applied magnetic field. Large negative magnetoresistance is observed in all specimens at 4.5 K. MR decreases monotonously to 5.8 kOe, but the magnetoresistance change does not seem to saturate even at 5.8 kOe. The MR dependence on the heat treatment is not prominent at 4.5 K and all samples show similar behavior. However, it may be noted that the MR ratio at 5.8 kOe becomes smaller as the annealing time becomes longer. The tendency is entirely different at 77 K as shown in Fig.5 (b). The highest MR ratio was obtained in the 2600 h specimen and the as-quenched and 400 h

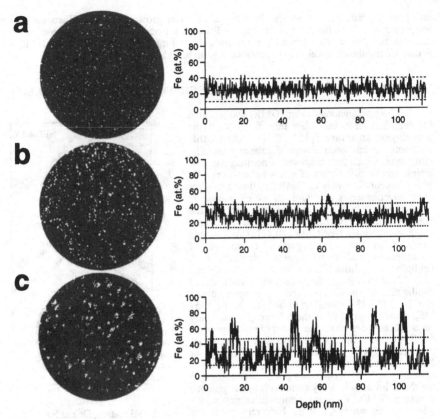

Fig. 6 Field ion micrographs and atom probe concentration depth profiles of the Cr-25at.%Fe alloy (a) as-quenched, (b) annealed at 773 K for 400 h and (c) annealed at 773 K for 2600 h. Images were observed with a mixture of Ne and He as an imaging gas. The probe aperture was located on the (011) pole. The dotted line shows the statistical error expected from 2σ, two times of the standard deviation.

specimens show similar MR behavior. The MR change at the low field region is more noticeable in the 2600 h specimen. At 300 K, the difference becomes more prominent. Although MR does not appear in the as-quenched specimen, the sample aged for 400 h. shows unsaturated monotonous MR changes. The specimen aged for 2600 h. shows a small but steep MR change at lower than 2 kOe, but it almost saturates at that field.

In order to understand these MR observations, it is essential to characterize the nature of Fe clusters in the alloy as a function of annealing time. Figures 6 (a-c) show field ion micrographs of Cr-25at.%Fe alloy in the (a) as-quenched, (b) annealed at 773 K for 400 h. and (c) annealed at 773 K for 2600 h. and their corresponding one dimensional atom probe concentration depth profiles. In the FIM image of the as-quenched alloy, bright spots are uniformly distributed, suggesting that this is a homogeneous solid solution. In the specimen aged for 400 h., regions of brighter contrast are observed. However, these brightly imaging regions are not well defined and appear somewhat interconnected. This suggests that a limited degree of concentration fluctuation is present. The specimen aged for 2600 h. clearly shows many isolated Fe particles. From the continuity of the concentric rings of the small particles to those of the matrix, it can be seen that these precipitates are coherent with the Cr matrix. The average size of Fe particles is estimated to be ~4 nm and the interparticle distance is ~ 6 nm. The volume fraction of the ferromagnetic particles which is estimated from the area fraction of the brightly imaging region is ~ 14 %. The atom probe concentration depth profiles demonstrate the changes in compositional fluctuations during the decomposition process quantitatively. In the as-quenched specimen, the concentration fluctuation is almost within the statistical errors, suggesting that the distribution of Fe atoms is homogeneous. No statistically significant long range concentration fluctuations are detected. On the other hand, small concentration fluctuations are recognized in the specimen aged for 400 h. There are Fe enriched regions with a diameter of ~2 nm and a concentration of 50 at%Fe. The interface between the Fe enriched particle and the matrix is not well defined, suggesting that the interface may be diffuse. In the specimen aged for 2600 h, the Fe concentration of the Fe-rich particles reaches approximately 80%, which is still slightly lower than the equilibrium concentration estimated from the binary phase diagram. Since the solubility limit of Fe to Cr is high, a large amount of Fe (~10at.%) still remains in the Cr matrix. The diameter of the particles is estimated to be ~ 4 nm and the interparticle distance is comparable or slightly larger than this.

Fe concentration   ~20 nm

0   20   40   60   80   100

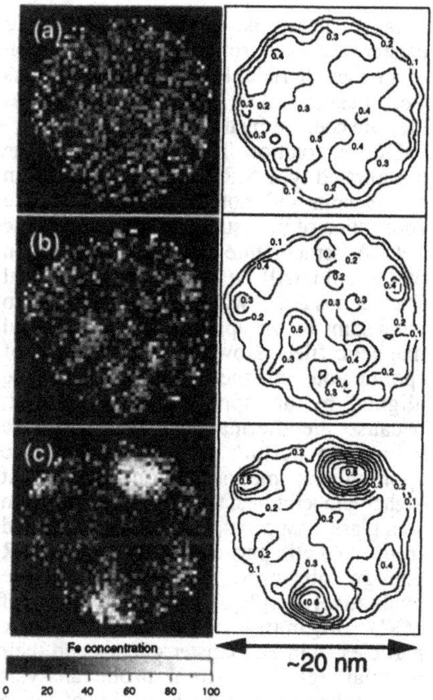

Fig. 7 Two dimensional concentration mapping fo Cr-25at.%Fe alloy (a) as-quenched, (b) annealed at 773 K for 400 h and (c) annealed at 773 K for 2600 h. The gray scale corresponds to th concentration of Fe.

The evolution of these ferromagnetic particles can be more clearly seen in the two dimensional concentration mapping obtained by the three dimensional atom probe. Fig. 7 shows two dimensional concentration mappings and their corresponding isoconcentration contours of (a) as-quenched specimen and of specimens (b) annealed at 773 K for 400 h and (c) annealed at 773 K for 2600 h. In this figure, local concentrations of Fe are displayed with a gray scale. The concentration mapping of the as-quenched sample show only statistical concentration fluctuations and no Fe rich particles are observed. In the specimen annealed at 773 K for 40 h., fluctuations in Fe concentration can be clearly recognized. However, the interface between the Fe-enriched region and the matrix is still not clear. The isoconcentration contours indicate that the maximum concentration of the Fe-enriched region is only 50at.%Fe, which is significantly lower than the equilibrium value, 80at.%Fe. The concentration mapping indicates that the Fe-enriched regions are interconnected. This type of concentration modulation is expected in the early stage of spinodal decomposition. In the specimen annealed for 2600 h., discrete Fe rich particles are recognized. The concentration of Fe in the particles are ~80at.%Fe, which is close to the thermal equilibrium value. Isoconcentration contours indicate that the concentration change at the particle/matrix interfaces are much more discrete that those in the previous two specimens. Note that ~20at.%Fe are still remain in the matrix.

Based on these atom probe results, the distribution of spins of Fe atoms as well as the spontaneous magnetization of ferromagnetic particles can be schematically represented as Fig. 8, where small arrows indicate the spins of independent Fe atoms and large arrows indicate the spontaneous magnetization of the single domain ferromagnetic particles. The aggregates of arrows represent ferromagnetically coupled spin clusters. The as-quenched alloy is spin cluster glass, in which a mixture of ferromagnetically and antiferromagnetically coupled spin clusters is present, while the chemical composition is homogeneous on a macroscopic scale but fluctuate in a statistical sense. The spins of the ferromagnetic clusters are easily rotated by the application of a magnetic field, but the individual spins frozen as spin glass do not rotate easily as they are coupled by the RKKY interaction. The latter results in the unsaturated type MR, because the direction of the spin will not be aligned with the application of the usual strength of magnetic field. As temperature increases, the direction of the spin will be disturbed by thermal fluctuations and it will become impossible to align spins by the application of an external magnetic filed. However, the direction of spontaneous magnetization will still be aligned by an applied magnetic field, because the thermal energy is not large enough to perturb the direction of the spontaneous magnetization. Hence, at higher temperatures, the MR originated from spin glass disappears, but the MR originated from magnetic particles remains. The MR change observed in Fig. 8 can be qualitatively explained by the interplay of MR's originated from ferromagnetic particles, spin and cluster glasses and their thermal agitation. Atom probe analysis clearly showed how much fraction of solute

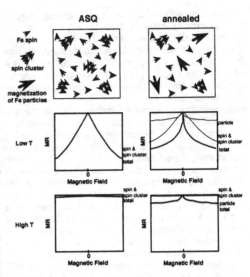

Fig. 8 Schematic drawings of MR curves expected from three types of scattering sources, i.e. individual spins of Fe atoms, spin clusters and ferromagnetic particles.

are dissolved in the matrix phase which cause spin glass effect at low temperatures and how much Fe are consumed for the formation of ferromagnetic particles. Also, it offered valuable information regarding the particle size, interparticle distance, density , and composition of ferromagnetic particles, all of which are key factors in controlling the MR in the heterogeneous alloy structure. Nanostructure characterization by atom probe analysis of these heterogeneous magnetic alloys will yield crucial information on the mechanisms of various magnetic properties.

REFERENCES

1. M. K. Miller and G. D. W. Smith, Atom-Probe Microanalysis: Principles and Applications to Materials Problems, MRS, Pittsburgh, 1989.
2. A. Cerezo, T. J. Godfrey and G. D. W. Smith, Rev. Sci. Instrum, **59**, 862 (1988).
3. A. Cerezo, M. G. Hetherington, J. M. Hyde, M. K. Miller, G. D. W. Smith and J. S. Underkoffler, Surf. Sci. **266**, 471 (1992).
4. N. Hasegawa, K. Hono, R. Okano, H. Fujimori and T. Sakurai, Appl. Surf. Sci. **67**, 407 (1993).
5. K. Hono, T. Hashizume and T. Sakurai, Surf. Sci. **266**, 506 (1992).
6. K. Hono, R. Okano, T. Saeda and T. Sakurai, Appl. Surf. Sci. (1995) in press.
7. T. Yogi and T. A. Nguyen, IEEE Trans. Mag. **MAG-29**, 307 (1993).
8. Y. Shiroishi, K. Yoshida, M. Futamoto and H. Aoi, J. Mag. Soc. Jpn. **17**, 784 (1993).
9. S. Iwasaki and K. Ouchi, IEEE Trans. Mag. **MAG-14**, 849 (1978) .
10. T. M. Coughlin, J. H. Judy and E. R. Wuori, IEEE Trans. Mag. **MAG-17**, 3169 (1981).
11. K. E. Johnson, J. B. Mahlke, K. J. Schulz and a. C. Wall, IEEE Trans. Mag. **MAG-29**, 215 (1993).
12. J. Zhu and H. N. Bertram, J. Appl. Phys. **63**, 3248 (1988).
13. R. D. Fisher, V. S. Au-Yeung and B. B. Sabo, IEEE Trans. Mag. **MAG-20**, 806 (1984).
14. J. W. Smits, S. B. Luitjens and F. J. A. den Broeder, J. Appl. Phys. **55**, 2260 (1984).

15. S. Iwasaki, K. Ouchi and T. Hizawa, J. Mag. Soc. Jpn. **9**, 57 (1985).
16. M. Sagoi, R. Nishikawa and T. Suzuki, IEEE Trans. Mag. **MAG-22**, 1335 (1986).
17. K. Ouchi and S. Iwasaki, IEEE Trans. Mag. **MAG-18**, 1110 (1982).
18. J. N. Chapman, I. R. McFadyen and J. P. C. Bernards, J. Mag. Mag. Mater. **62**, 359 (1986).
19. A. K. Jhingan, J. Mag. Mag. Mater. **54-57**, 1685 (1986).
20. D. J. Rogers, J. N. Chapman, J. P. C. Bernards and S. B. Luitjens, IEEE Trans. Mag, **MAG-25**, 4180 (1989).
21. Y. Maeda, S. Hirono and M. Asahi, Jpn. J. Appl. Phys. **24**, L951(1985).
22. Y. Maeda, M. Asahi and M. Seki, Jpn. J. Appl. Phys. **25**, L668 (1986).
23. Y. Maeda and M. Asahi, J. Appl. Phys. **61**, 1972 (1987).
24. Y. Maeda and M. Takahashi, Jpn. J. Appl. Phys. **28**, L248 (1989).
25. K. Yoshida, H. Kakibayashi and H. Yasuoka, J. Appl. Phys. **68**, 705 (1990).
26. K. Takei and Y. Maeda, Jpn. J. Appl. Phys. **30**, L1125 (1991).
27. Y. Maeda and M. Takahashi, J. Appl. Phys. **68**, 4751 (1990).
28. J. E. Snyder and M. H. Kryder, J. Appl. Phys. **73**, 5551 (1993).
29. J. Suzuki, Y. Morii, K. Takei and Y. Maeda, J. Mag. Mag. Mater. (1994).
30. K. Hono, S. S. Babu, Y. Maeda, N. Hasegawa and T. Sakurai, Appl. Phys. Lett. **62**, 2504 (1993).
31. K. Hono, Y. Maeda, J-L. Li and T. Sakurai, IEEE Trans. Mag. **29**, 3745 (1993).
32. K. Hono, Y. Maeda, S. S. Babu and T. Sakurai, J. Appl. Phys. **76**, 8025 (1995).
33. K. Hono, K. Yeh, Y. Maeda and T. Sakurai, Appl. Phys. Lett. **66**, 1686 (1995).
34. M. Hasebe, K. Oikawa and T. Nishizawa, J. Jpn. Inst. Metals, **46**, 577 (1982).
35. M. N. Baibich, J. M. Broto, A. Fert, F. Ngyen van Dau, F. Petroff, P. Eitenne, G. Creuzet, A. Friederich and J. Chazelas, Phys. Rev. Lett. **61**, 2473 (1988).
36. S. S. P. Parkin, R. Bhadra and K. P. Roche, Phys. Rev. Lett. **66**, 2152 (1991).
37. F. Petroff, A. Barthelemy, D. H. Mosca, D. K. Lottis, A. Fert, P. A. Schroeder, W. P. Pratt, Jr., R. Loloee and S. Lequien, Phys. Rev. **B44**, 5355 (1991).
38. S. Araki, K. Yasui and Y. Narumiya, J. Phys. Soc. Jpn. **60**, 2827 (1991).
39. C. S. Santos, B. Rodmacq, M. Vaezzadeh and B. George, Appl. Phys. Lett. **59**, 126 (1991).
40. A. E. Berkowitz, J. R. Mitchell, M. J. Carely, A. P. Young, S. Zhang, F. E. Spada, F. T. Parker, A. Hutten and G. Thomas, Phys. Rev. Lett. **68**, 3745 (1992).
41. J. Q. Xiao, J. S. Jian and C. L. Chien, Phys. Rev. Lett. **68**, 3749 (1992).
42. C. L. Chien, J. Q. Xiao and J. S. Jiang, J. Appl. Phys. **73**, 5309 (1993).
43. N. Kataoka, H. Endo, K. Fukamichi and Y. Shimada, Jpn. J. Appl. Phys. **32**, 1969 (1993).
44. J. Wecker, R. von Helmolt, L. Schulz and K. Sawer, Appl. Phys. Lett. **62**, 1985 (1993).
45. H. Takeda, N. Kataoka, K. Fukamichi and Y. Shimada, Jpn. J. Appl. Phys. **33**, 102 (1994).
46. L. H. Chen, S. Jin and T. H. Tiefel, J. Mater. Res. **9**, 1134 (1994).
47. K. Takanashi, T. Sugawara, K. Hono and H. Fujimori, J. Appl. Phys. **76**, 6790 (1994).
48. R. Okano, K. Hono, T. Takanashi, H. Fujimori and T. Sakurai, J. Appl. Phys. (1995) in press.

MICROSTRUCTURE AND MAGNETIC PROPERTIES OF NANOCRYSTALLINE
$Fe_{93-x-y}Zr_7B_xCu_y$ ALLOYS

M. KOPCEWICZ, A. GRABIAS, P. NOWICKI*
Institute of Electronic Materials Technology, Wólczyńska 133, 01-919 Warszawa, Poland
*Department of Materials Science and Engineering, Warsaw University of Technology,
Narbutta 85, 02-524 Warszawa, Poland

ABSTRACT

An unconventional technique combining Mössbauer spectroscopy with the effects induced by magnetic radio-frequency fields (rf collapse and rf sidebands) is employed to study the microstructure and magnetic properties of nanocrystalline clusters of bcc Fe formed by annealing amorphous $Fe_{93-x-y}Zr_7B_xCu_y$ (x=6, 8, 12, y=0, 2) alloys at 500-600°C. The rf-Mössbauer experiments allow us to distinguish magnetically soft nanoclusters from magnetically harder microcrystalline phases. The dependence of the bcc Fe phase formation on the alloy composition is discussed. The Mössbauer results are supplemented by DSC measurements.

INTRODUCTION

Recently, a new class of soft magnetic materials with high saturation magnetization and low magnetostriction has been developed by utilizing the first stage of crystallization of amorphous Fe-based alloys. It was found that by annealing the amorphous FeSiB-based alloys containing Cu and Nb the nanocrystalline bcc Fe(Si) phase is formed which exhibits excellent soft magnetic properties [1]. Addition of Cu decreases the crystallization temperature and increases the nucleation rate, and a small amount of Nb limits the grain growth. The good soft magnetic properties of nanocrystalline alloys are well explained by the reduction of the effective magnetic anisotropy due to refinement of the grain size [2].

A nanocrystalline bcc Fe phase was obtained in ternary FeZrB alloys which reveal superior magnetic properties (higher saturation magnetization and permeability) as compared with FeCuNbSiB alloys [3]. Annealing of the amorphous precursor causes formation of nanoscale grains of bcc Fe which exhibit high saturation magnetization combined with low anisotropy and coercive fields and vanishing magnetostriction. Addition of 1-2% Cu to FeZrB alloys decreases the crystallization temperature and increases the nucleation rate. Boron enhances thermal stability of the nanocrystalline bcc phase [4] and affects the homogeneity of the bcc precipitates [5].

The structure and magnetic properties of nanocrystalline alloys have been extensively investigated by various experimental techniques including the Mössbauer spectroscopy which was successfully used for phase identification. However, information regarding the grain size and magnetic anisotropy is not available by conventional Mössbauer measurements. Therefore in the study of the microstructure and magnetic properties of the FeZrBCu alloys we applied an unconventional technique which combines the Mössbauer effect with the phenomena induced by an external radio-frequency field (rf collapse and sideband effects) [6]. The collapse of the magnetic hyperfine splitting occurs due to fast magnetization reversal induced by an external radio-frequency (rf) magnetic field. If the frequency of the rf field is larger than the Larmor precession frequency and the rf field is strong enough to overcome local magnetic anisotropy, then the magnetic hyperfine field is averaged to zero at the Mössbauer nuclei. The rf-collapsed spectra consist of a single line or a quadrupole doublet in the place of the magnetically split

517

Mat. Res. Soc. Symp. Proc. Vol. 384 ©1995 Materials Research Society

six-line pattern, though the sample remains in the ferromagnetic state. The rf collapse is very sensitive to small changes of local magnetic anisotropy, what allows distinction of the soft nanocrystalline phase from the magnetically harder microcrystalline phases formed by annealing of the amorphous precursor. The rf sideband effect, directly related to magnetostriction, allows identification of the nanocrystalline phase thanks to vanishing magnetostriction.

The rf Mössbauer technique was recently successfully used in the study of nanocrystalline FeCuNbSiB alloys [7]. Here we present the results obtained for $Fe_{93-x-y}Zr_7B_xCu_y$ alloys in amorphous, nano- and microcrystalline states.

EXPERIMENTAL

Amorphous $Fe_{93-x-y}Zr_7B_xCu_y$ (x= 6, 8, 12; y= 0, 2) alloys were prepared by the melt-spinning technique. The ribbons were 4 mm wide and about 25 μm thick. In order to obtain the nanocrystalline phase the amorphous alloys were annealed in an Ar atmosphere for 1 hour at 430°C, 500°C, 550°C, 600°C and 780°C in a furnace heated with halogen lamps. The differential scanning calorimetry (DSC) measurements at the heating rate of 20 K/min were performed to determine the first and second crystallization peaks. Conventional Mössbauer measurements were performed for the as-quenched and annealed samples prior and after the rf exposure. The unconventional rf-Mössbauer measurements were performed during the exposure of samples to the rf field of 20 Oe at 60.8 MHz, frequency about three times higher than the Larmor one. The rf field was applied in the plane of the samples which played the role of an absorber. Adequate cooling of the samples preventing excessive heating was applied [6,7].

RESULTS AND DISCUSSION

DSC measurements

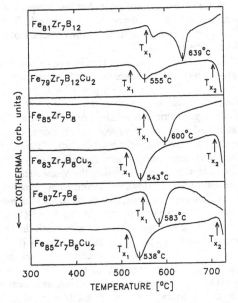

Fig. 1. The DSC curves for $Fe_{93-x-y}Zr_7B_xCu_y$ alloys.

The crystallization behaviour of $Fe_{93-x-y}Zr_7B_xCu_y$ alloys was studied by DSC. All alloys crystallized through two steps as revealed by two exothermic peaks in DSC curves (onset temperatures T_{x_1} and T_{x_2} are indicated in Fig. 1). The crystallization temperatures depend on the boron content and for a given boron level decrease markedly upon addition of 2% Cu. The first stage with the crystallization temperature (T_{x_1}) corresponds to the structural transformation from the amorphous to the nanoscale bcc phase. The second exothermic peak (T_{x_2}) reflects the transformation of the remaining amorphous phase to a mixture of FeZr compounds and microcrystalline α-Fe. The nanocrystalline bcc Fe phase can be formed in the temperature range between T_{x_1} and T_{x_2}. The 2% Cu content not only decreases the T_{x_1} temperature but also extends the temperature range between T_{x_1} and T_{x_2} (Fig. 1).

Conventional Mössbauer measurements

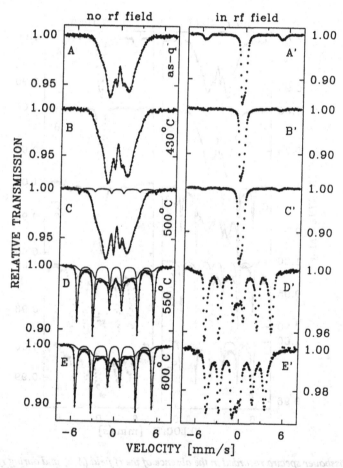

Fig. 2. Mössbauer spectra recorded in the absence of the rf field (A-E) and during rf exposure (A'-E') for $Fe_{85}Zr_7B_8$ alloy in the as-quenched state and after annealing at 430-600°C.

The Mössbauer measurements clearly revealed the changes in the microstructure of the amorphous $Fe_{93-x-y}Zr_7B_xCu_y$ alloys induced by annealing. Typical examples are shown in Figs. 2A-2E and Figs. 3A-3E for $Fe_{85}Zr_7B_8$ and $Fe_{83}Zr_7B_8Cu_2$, respectively. The spectra measured for the as-quenched alloys are typical for an amorphous structure (Figs. 2A and 3A). Annealing at T<500°C induces the structural relaxation in the amorphous phase, evidenced in the spectra by the increase of the average hyperfine field (Figs. 2B and 3B). At 500°C the bcc Fe phase appears; the spectral component with sharp lines and the characteristic hyperfine field H_{hf}=32.95 T and isomer shift δ=0.00 mm/s is observed in the spectra (Figs. 2C and 3C). Its spectral contribution increases with increasing annealing temperature (Figs. 2D-2E and 3D-3E). Beside the bcc Fe phase, the $Fe_3(Zr,B)$ and $Fe_2(Zr,B)$ phases can be detected. The increase of annealing temperature to 780°C causes complete crystallization of the amorphous alloys. The microcrystalline phase α-Fe dominates, however, another phase with a characteristic paramagnetic subspectrum (quadrupole splitting of 0.45 mm/s and δ=0.22 mm/s), corresponding

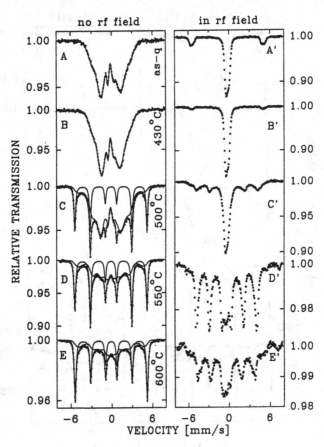

Fig. 3. Mössbauer spectra recorded in the absence of the rf field (A-E) and during rf exposure (A'-E') for $Fe_{83}Zr_7B_8Cu_2$ alloy in the as-quenched state and after annealing at 430-600°C.

probably to the $FeZr_2$-type phase, appears in the spectrum. Similar results were obtained for the remaining alloy compositions. The conventional Mössbauer measurements allow determination of the relative spectral contribution of the bcc Fe phase to the total spectral area. However, information regarding the grain size cannot be obtained.

The relative abundance of the bcc Fe phase for various alloy compositions is shown in Fig. 4. As can be seen, the increase of boron content in the FeZrBCu alloy increases the T_{x_1} temperature resulting in the enhanced thermal stability of the amorphous phase. The relative abundance of the bcc Fe phase increases with decreasing boron content (Figs. 4A-4C). The presence of 2% Cu decreases the T_{x_1} temperature and dramatically promotes the precipitation of the bcc Fe phase. This effect is clearly seen especially for temperatures close to T_{x_1} (e.g. 500°C) (Figs. 4B-4C). The highest relative amount of the bcc Fe phase formed at 600°C is observed for the lowest boron content in the alloy (Fig. 4C).

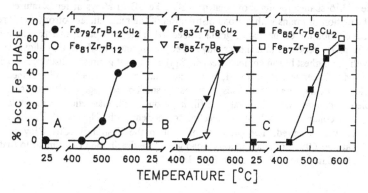

Fig. 4. The relative abundance of the bcc Fe phase determined from the conventional Mössbauer measurements.

The rf-Mössbauer results

The typical Mössbauer spectra measured during exposure to the rf field of 20 Oe at 60.8 MHz for $Fe_{85}Zr_7B_8$ and $Fe_{83}Zr_7B_8Cu_2$ alloys are shown in Figs. 2A'-2E' and Figs. 3A'-3E', respectively. A complete collapse of the magnetic hyperfine structure to a quadrupole doublet is observed in the amorphous state (Figs. 2A'-2B' and 3A'-3B'). The rf sidebands, clearly seen in the spectra of the as-quenched alloys (Figs. 2A' and 3A'), decrease markedly due to structural relaxation in the amorphous phase (Figs. 2B' and 3B') which causes a decrease of the magnetostriction constant [3]. The sidebands disappear almost completely for the samples annealed at 500°C (Figs. 2C' and 3C') in which the bcc Fe phase begins to form. The formation of the nanocrystalline bcc Fe phase dramatically affects the shape of the collapsed spectra [8]. The central collapsed part consists here of a superposition of a single line corresponding to the cubic nanocrystalline magnetically soft Fe phase and a quadrupole doublet corresponding to the remaining amorphous matrix (Figs. 2C' and 3C'). In addition to the completely collapsed central part a well resolved partially collapsed hyperfine structure appears (Figs. 2D'-2E' and 3C'-3E'). The partially collapsed six-line spectral component corresponds to the bcc Fe grains with magnetic anisotropy sufficiently large to limit the magnetization reversal. The hyperfine split pattern reveals an average hyperfine field markedly reduced as compared to α-Fe, suggesting that the magnetic anisotropy is much smaller than in the bulk α-Fe, for which no rf-induced narrowing is observed. Thus, the partially collapsed component corresponds to the nanoscale bcc

Fe grains having a magnetic anisotropy considerably larger than for the amorphous phase, for which a complete rf collapse was observed. The amount of such grains markedly increases with increasing annealing temperature, as seen by the dramatic increase of the magnetically split spectral component and the decrease of the rf collapsed component at 550°C (from 90% of the total spectral area, Fig. 2C', to 15% , Fig. 2D', and from 60%, Fig. 3C', to 20%, Fig. 3D'). However, annealing at 600°C causes partial restoration of the fully collapsed spectral component (26%, Fig. 2E', and 30%, Fig. 3E') thus revealing an increase of the relative abundance of the magnetically soft nanoscale bcc Fe grains (Figs. 2E', 3E'). It seems that in these alloys annealing at 600°C is most favourable for the formation of nanocrystalline Fe phase with low anisotropy. A further increase of annealing temperature (to 780°C) causes complete crystallization of the alloys and the formation of microcrystalline α-Fe with magnetic anisotropy large enough to prevent any rf-narrowing of the spectra.

The rf-Mössbauer results show that annealing FeZrBCu alloys at temperatures 550-600°C causes formation of grains with a broad distribution of the magnetic anisotropy fields related to a broad size distribution of the bcc Fe grains embedded in the amorphous matrix. The formation of the bcc Fe phase proceeds in a similar way in $Fe_{87}Zr_7B_6$ and $Fe_{85}Zr_7B_6Cu_2$ alloys. However, in alloys with the highest boron content ($Fe_{81}Zr_7B_{12}$) a different grain size distribution was detected by the rf-Mössbauer technique [8]. It was found that in a $Fe_{81}Zr_7B_{12}$ alloy annealed at 500-600°C the bcc Fe appears in two kinds of grains: (i) small ones, magnetically very soft, for which a complete rf collapse was observed, and (ii) large ones with magnetic anisotropy large enough to prevent rf collapse, for which the spectral component with $H_{hf}=32.95$ T, characteristic for bulk α-Fe, was observed. The presence of Cu in the $Fe_{79}Zr_7B_{12}Cu_2$ alloy results in the formation at 500-600°C of a continuous broad grain size distribution [9], similar to that proposed to explain the data in Figs. 2A'-2E' and Figs. 3A'-3E'.

CONCLUSIONS

Conventional Mössbauer measurements allow identification of phases formed by annealing of the starting amorphous alloys but do not provide information concerning the magnetic anisotropy of the grains. The rf-Mössbauer experiments, in which the rf collapse and sideband effects are observed, permit us to distinguish the soft nanocrystalline bcc phase from magnetically harder microcrystalline α-Fe. Some information concerning grain size distribution can be obtained. The nanocrystalline α-Fe phase is identified by the appearance of a single line component in the rf collapsed spectrum and by the disappearance of rf sidebands due to vanishing magnetostriction.

References

1. Y. Yoshizawa, S. Oguma, K. Yamauchi, J. Appl. Phys. **64**, 6044 (1988).
2. G. Herzer, Mat. Sci. Eng. **A133**, 1 (1991).
3. K.Suzuki, A.Makino, A.Inoue, T.Masumoto, J.Appl. Phys. **70**, 6232 (1991); **74**, 3316 (1993).
4. K.Suzuki, A.Makino, A.Tsai, A.Inoue, T.Masumoto, Mat. Sci. Eng. **A179/A180**, 501 (1994).
5. K. Y. Kim, T. H. Noh, I. K. Kang, Mat. Sci. Eng. **A179/A180**, 552 (1994).
6. M. Kopcewicz, Structural Chem. **2**, 313 (1991).
7. M. Kopcewicz, J. Jagielski, T. Graf, M. Fricke, J. Hesse, Hyperfine Inter. **94**, 2223 (1994).
8. M. Kopcewicz, A. Grabias, P. Nowicki, JMMM **140-144**, 461 (1995).
9. M. Kopcewicz, A. Grabias, P. Nowicki, Nanostructured Mater. (1995) in print.

MAGNETIC STUDIES OF PARAMAGNETIC CLUSTERS ENCAPSULATED WITHIN THE SODALITE CAGE

Lee J. Woodall*, P. A. Anderson*, A. R. Armstrong** and P. P. Edwards*
* The School of Chemistry, The University of Birmingham, Birmingham, B15 2TT, U.K.
** The School of Chemistry, The University of St. Andrews, Fife, Scotland, KY16 9ST, U.K.

ABSTRACT

The magnetic properties of alkali-doped zeolites have been investigated using magnetic resonance spectroscopy. Although paramagnetic centres have been detected, the predominant species, in many cases, is spin-paired.

INTRODUCTION

Zeolites are a well defined class of crystalline aluminosilicates, with structures based on tetrahedral networks which encompass channels and cavities of molecular dimensions. The general formula for the composition of a zeolite is:

$$M_{x/n}[(AlO_2)_x(SiO_2)_y] \cdot mH_2O$$

where cations M of valence n neutralise the negative charges on the aluminosilicate framework. The framework is made up from corner sharing SiO_4 and AlO_4 tetrahedra, with the cations found within the cavities and channels of the framework co-ordinated to the framework oxygens or water molecules. The framework structures of three zeolites, based on a common sodalite cage are shown in figure 1.

The idea of using the zeolite as a host in which to dissolve alkali metals is not a new one[1]. On contact with the dehydrated zeolite, the incoming alkali atoms are spontaneously ionised by the intense electric fields within the host matrix. This approach focuses on the deliberate modification of the host material to generate new solid-state compounds that have interesting electronic and magnetic properties[2]. We have studied a number of zeolite-based systems, containing the sodalite cage as a basic building block, in an attempt to understand and perhaps control the electronic properties of these types of materials.

EXPERIMENTAL

The syntheses of all of these powdered materials was carried out through the controlled reaction of the dehydrated host with alkali metal. The reactions were carried out between 100 and 250°C, in a sealed, evacuated quartz tube for between 12 to 48 hours, as described elsewhere[3].

The electronic properties of these materials were probed using ESR (Electron Spin Resonance) Spectroscopy. ESR studies were performed using a Bruker ESP300 Spectrometer, operating at X-Band frequencies (9.5GHz) between 4K and room temperature. To quantify the ESR measurements, the ESR signal from each sample was compared to that of a known $CuSO_4.5H_2O$ standard in a dual-mode cavity. Errors in this technique are estimated at ±20%.

Preliminary Solid-State NMR measurements have also been performed on some samples.

RESULTS

With regard to the composition of these materials, the following nomenclature will be adopted. The addition of x-mole equivalents of metal (M) to a host zeolite (Z) containing cations (N) results in the formation of a product labelled as M_x/N-Z.

On addition of sodium metal to sodalite, the sample undergoes a remarkable colour change. The white host first turns blue, then purple and finally black as the concentration of reacted metal increases[4]. The ESR spectra, recorded at room temperature, of samples of sodalite containing x mole-equivalents of sodium are shown in figure 2 (Table I). A similar set of spectra, recorded on samples of metal-loaded zeolite -Y are presented in figure 3 (Table II).

Potassium has been reacted with zeolite K_{12}-A. At the lowest loadings ($x\sim0.5$) the sample appears blue and at the highest (x=5) brown. The ESR spectra, recorded at room temperature, of samples of K_x/K_{12}-A containing x mole-equivalents of potassium are shown in figure 4 (Table III). The structures of a number of these samples have previously been determined[5].

For comparative purposes alone the ESR Intensity is expressed as:
i) number of electron spins per formula unit of sodalite cage

ii) % spins = $\dfrac{\text{number of visible spins (ESR Intensity)}}{\text{number of injected electron spins}}$ x 100%

The room temperature ^{23}Na NMR spectrum of K_5/Na_{12}-A is shown in figure 5.

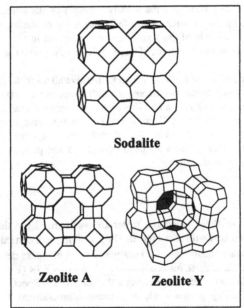

Sodalite

Zeolite A   **Zeolite Y**

**FIGURE 1. THE FRAMEWORK STRUCTURES OF
ZEOLITES -A, -Y AND SODALITE**

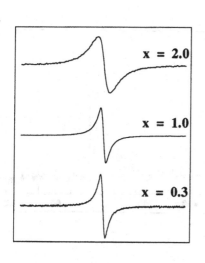

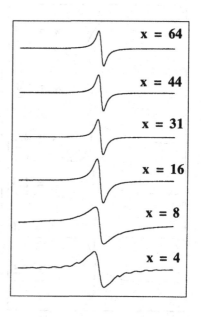

FIGURE 2. THE ROOM TEMPERATURE
ESR SPECTRA OF Na$_x$/Sodalite (x = 0.3-2)

FIGURE 3. THE ROOM TEMPERATURE
ESR SPECTRA OF Na$_x$/Na$_{56}$-Y (x = 4-64)

TABLE I

Sample	g	Linewidth (G)	Spins per Sodalite Cage	% Spins
Na$_{0.3}$/Sodalite	2.0000	2.1	0.012	4
Na$_1$/Sodalite	2.0000	2.2	0.15	15
Na$_2$/Sodalite	2.0000	5.3	0.28	14

TABLE II

Sample	g	Linewidth (G)	Spins per Sodalite Cage	% Spins
Na$_4$/Na$_{56}$-Y (291K)	2.0013 (singlet)	32.0 A=33 (13 lines)	0.6	120
Na$_8$/Na$_{56}$-Y (291K)	2.0012 (singlet)	27.5	0.79	79
Na$_{16}$/Na$_{56}$-Y (291K)	2.0011	20.6	1.00	50
Na$_{64}$/Na$_{56}$-Y (291K)	2.0011	16.1	0.6	7
Na$_8$/Na$_{87}$-X (291K)	2.0019	A = 26	0.16	16
Na$_{13}$/Na$_{87}$-X (291K)	2.0011	40.0	0.21	13
Na$_{31}$/Na$_{87}$-X (291K)	2.0011	4.4	0.27	7
Na$_{63}$/Na$_{87}$-X (291K)	2.0011	13.4	0.1	1.3

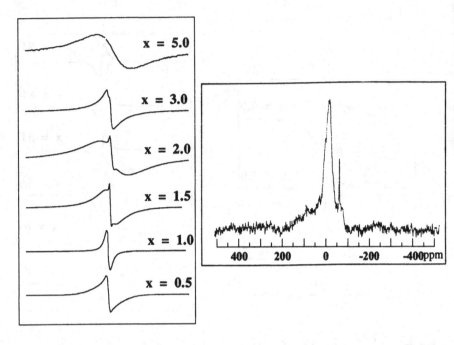

FIGURE 4. THE ROOM TEMPERATURE ESR SPECTRA OF K_x/K_{12}-A (x = 0.3-5)

FIGURE 5. THE ROOM TEMPERATURE ^{23}Na NMR SPECTRUM OF K_5/Na_{12}-A

TABLE III

Sample	g	Linewidth (G)	Spins per Sodalite Cage	% Spins
$K_{0.5}/K_{12}$-A (291K) (4K)	1.9996 (singlet)	2.6 (singlet) A=14 (10 lines)	0.05	9 (291K)
K_1/K_{12}-A (291K)	1.9997 (singlet)	complex	0.25	25
$K_{1.5}/K_{12}$-A (291K)	1.9987 (narrow) 1.9987 (broad)	1.2 (narrow) 6.4 (broad)	0.53	35
K_2/K_{12}-A (291K)	1.9987 (narrow) 1.9979 (broad)	1.4 (narrow) 17 (broad)	0.74	37
K_3/K_{12}-A (291K)	1.9978	4.2 (broad)	0.78	26
K_5/K_{12}-A (291K)	1.9974 (singlet)	22.0 (singlet)	0.95	19
Na_5/Na_{12}-A (291K)	1.9992 (singlet)	5.1 (singlet)	0.15	3.1
K_5/Na_{12}-A (291K)	2.0012 (singlet)	12.5 (singlet)	0.025	0.5
Rb_5/Na_{12}-A (291K)	1.9990 singlet)	4.5 (singlet)	0.035	0.7
Rb_5/Rb_{12}-A (291K)	1.9888	74 (narrow) 1000 (broad)	0.75	15
Cs_2/Cs_9Na_3-A (291K)	2.0003(narrow) 2.034 (broad)	4.0 (narrow) 250 (broad)	0.12 (10K)	6 (10K)

DISCUSSION

Low Density Regime

Two extreme may be noted in the ESR results for low metal loadings. The ESR spectrum of $K_{0.3}/K_{12}$-A (at low temperatures) and Na_x/Na_{56}-Y, for $x=4,8$ show hyperfine structure (as does Na_8/Na_{87}-X). These may be attributed to the K_3^{2+} and Na_4^{3+} paramagnetic centres respectively[6-8] whose formation may be represented as:

$$M^0 + nN^+_z \rightarrow M^+_z + N_n^{(n-1)+}_z$$

$$eg: Na^0 + 4Na^+_z \rightarrow Na^+_z + Na_4^{3+}_z$$

Although the %spins visible is high for the Na-Y system, it is relatively low for the Na-X, K-A and sodalite systems. The structure of Na_8/Na_{56}-Y has been determined[9]. The sodalite cages contain sodium cations (site I' occupancy 73%) that are separated by 4.4Å. These are thought to be bound together as local Na_4^{3+} states, in good agreement with the high %spin. This is in good agreement with the model of Ursenbach and Madden[10]. Interactions between clusters is believed to give rise to the central ESR singlets in these materials; an intercluster distance of 5.4Å in Na_8/Na_{56}-Y is consistent with this line of thought. However, for the other systems the %spins is considerably smaller than that expected for a complete set of independent localised moments. The Na_5^{4+} and Na_6^{5+} clusters have been observed in Na_3/Na_{87}-X[3]. However, in Na_8/Na_{87}-X only a small portion of the injected spins (16%) give rise to paramagnetic states. Therefore a correspondingly large portion of the spins (84%) must exist in spin-paired, diamagnetic states. It is conceivable that sodium cations located between sodalite cages could facilitate spin-pairing. Similarly, in $Na_{0.3}$/Sodalite only 4% of the injected spins appear visible. The distance between the centres of adjacent sodalite cages (6.5Å) (and also the average distance between sodiums in adjacent sodalite cages) in sodalite is smaller than the equivalent distance in zeolites -X,-Y or -A. Interactions between electrons in sodalite may be enhanced relative to those in zeolites -X,-Y or -A, due to a larger density of sodium cations.

Intermediate Density Regime

The composition dependence of the room temperature ESR spectra of K_x/K_{12}-A is shown in figure 4. The evolution of both a narrow singlet and broad singlet is to be noticed; each perhaps may be associated with a different spin environment. For x=1-5 the % spins visible is relatively large, suggesting a stable environment for localised spin states. In K_5/K_{12}-A each sodalite cage can formally be associated with a localised K_4^{3+} state. At least 80% of the spins exist in spin-paired states in this material at room temperature.

The ESR spectra of a number of similar systems have also been investigated. Within zeolite-A systems, the most intense ESR spectra appear to come from systems where the host zeolite contains the same cations as those injected. The ESR spectrum of K_5/Na_{12}-A can be associated with <1% of the injected electrons. Further evidence for diamagnetic states in this material has come from NMR studies[11] (figure 5). The narrow, diamagnetic shifted signal at 65ppm has been seen in metal-amines[12] and alkalides[13]. It has been assigned to the Na$^-$ ion.

High Density Regime

Beyond Na_1/sodalite, each sodalite cage can nominally be regarded as saturated with Na_4^{3+} centres[14]. However, our ESR data suggests that the excess electrons are predominantly spin-paired. For Na_2/sodalite, the %spins is lower than in the equivalent -Y and -A system. Again, a higher cation density may be responsible for the enhanced spin-pairing.

For zeolites -X and -Y with the largest concentrations of injected electrons, the %spins starts to fall as the systems move closer towards metallic behaviour.

REFERENCES

1. P. H. Kasai, J. Chem. Phys., **43**, 3322 (1965)
2. P. P. Edwards, P. A. Anderson, A. R. Armstrong, M. Slaski and L. J. Woodall, Chem. Soc. Rev., **22**, 305 (1993)
3. P. A. Anderson and P. P. Edwards, J. Am. Chem. Soc., **114**, 10608 (1992)
4. V. I. Srdanov, K. Haug, H. Metiu and G. D. Stucky, J. Phys. Chem., **96**, 9039 (1992)
5. A. R. Armstrong, P. A. Anderson and P. P. Edwards, J. Solid State Chem., **111**, 178 (1994)
6. P. A. Anderson, R. J. Singer and P. P. Edwards, J. Chem. Soc. Chem. Commun., **14**, 915 (1991)
7. J. A. Rabo, C. L. Angell, P. H. Kasai and V. Schomaker, Discuss Faraday Soc., **41**, 328 (1966)
8. B. Vu and L. Kevan., J. Chem. Soc. Faraday Trans, **87 (17)**, 2843 (1991)
9. A. R. Armstrong, P. A. Anderson, L. J. Woodall and P. P. Edwards, manuscript submitted for publication.
10. C. P. Ursenbach and P. A. Madden, manuscript submitted for publication.
11. H. Nakayama, D. D. Klug, C. I. Ratcliffe and J. A. Pipmeester, J. Am. Chem. Soc., **116**, 9777 (1994)
12. M. J. Wagner and J. L. Dye, Annun. Rev. Mater. Sci, **23**, 223 (1993)
13. P. P. Edwards, A. Ellaboudy, D. M. Holton and N. C. Pyper, Annun. Reports on NMR Spect., **20**, 35 (1988)
14. A. Monnier, V. Srdanov, G. Stucky and H. Metiu, J. Chem. Phys., **100 (9)**, 6944 (1994)

FERROMAGNETIC BEHAVIOUR IN NANOSCALE COBALT PARTICLES DISPERSED BY ZEOLITE Na-X.

I. HUSSAIN, I. GAMESON, P.A. ANDERSON AND P. P. EDWARDS.
The School of Chemistry, The University of Birmingham, Edgbaston, Birmingham, B15 2TT, UK.

ABSTRACT

This investigation has looked at the preparation of nanoscale cobalt particles by a simple solid state reaction involving cobalt (II) nitrate and zeolite Na-X under vacuum conditions followed by reduction in an hydrogen atmosphere. Samples were characterised by powder x-ray diffraction and scanning/transmission electron microscopy (TEM). Magnetic measurements were performed on the samples below 300 K using a SQUID magnetometer.

INTRODUCTION

Zeolites are a class of crystalline aluminosilicates whose structures are formed by corner linking of AlO_4 and SiO_4 tetrahedra. A very important property of zeolites is that the internal surface consists of interconnecting channels and cavities which can be sufficiently large to allow the entry of small atomic, molecular and metallic clusters of metal. Zeolite Na-X is the synthetic structural analogue of the mineral *faujasite*. In this zeolite the sodalite cages are joined together by hexagonal prisms or double six rings, resulting in the formation of a supercage which has a central diameter of ~12.5 Å. These supercages are accessible through four 12 membered rings which have a free diameter of ~7.4 Å.

Occlusion In Zeolites

The process in which the host zeolite is penetrated by atoms or salt molecules of another substance is generally known as *occlusion*. Interestingly, the pore space within a dehydrated zeolitic framework may be filled from the vapour phase, via salt melts or by heating the zeolite with salt powders[1-4]. The ultimate aim of this research is to successfully insert cobalt atoms, clusters and particles into the supercages of various zeolites and examine the magnetic properties of these new materials. A wide variety of synthetic conditions were used and the resulting materials were investigated using a combination of structural and magnetic techniques.

Experimental

In order for non-framework species to be introduced into the zeolite, its water of crystallisation must first be removed. This is best achieved by heating under vacuum and for this reason the occlusion of cobalt into zeolites must take place under vacuum conditions. A high vacuum system has been designed and constructed for this purpose. Stoichiometric amounts of $Co(NO_3)_2.6H_2O$ (pink powder) and zeolite Na-X (white powder), supplied by Laporte Inorganics, composition $Na_{86}(AlO_2)_{86}(SiO_2)_{106}.264H_2O$ were intimately ground together using a mortar and pestle. The mixture was heated to a temperature of 550°C for 12 hours under vacuum conditions, followed by reduction with 35%H_2/Ar gas mixture for 4 hours at 550°C, to yield the occluded and reduced zeolites. On heating, most anhydrous metal nitrates decompose to the metal oxide, NO_2 and O_2. However, two possible oxides can be formed when cobalt nitrate decomposes, these are CoO and Co_3O_4. When heated under vacuum, the latter of the two oxides is formed. A sample of 10 cobalt atoms per unit cell of NaX was synthesised and its structure and magnetic behaviour examined after occlusion and then reduction.

Results and Discussion

An x-ray powder diffraction (XRD) pattern of the pure zeolite is shown in figure 1, after occlusion and reduction. This was extremely useful in probing the degree of crystallinity of the host material. Basically, the XRD data verified that the zeolite remained intact after it was occluded with cobalt oxide and then reduced.

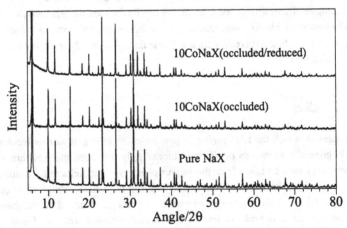

Fig. 1 Powder x-ray diffraction pattern of 10CoNaX after occlusion and then reduction.

No cobalt peaks were observed in the diffraction pattern of the occluded and reduced samples. However a reduction in the intensity of the high 2θ reflections is observed suggesting a loss in crystallinity of the sample.

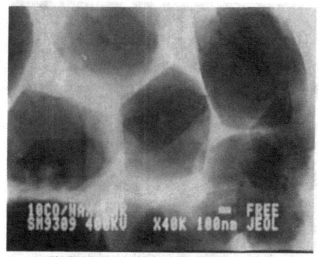

Fig. 2(a) Scanning image of 10CoNaX(occluded).

Fig. 2(b) Transmission image of 10CoNaX(occluded).

The TEM data also proved to be extremely effective as a complementary technique to XRD in revealing the morphology of the zeolite as well as the size and shape of the cobalt clusters formed (figures 2 & 3). It is apparent from our observations that the average size of the particles are in the range 200-250Å. Clearly, the size of the cobalt particles in the zeolitic host will be constrained by the dimensions of the zeolite cage. This implies that particles growing beyond the natural dimensions of the zeolite cavity will eventually break out and appear on the surface.

Fig. 3(a) Scanning image of 10CoNaX(occluded/reduced).

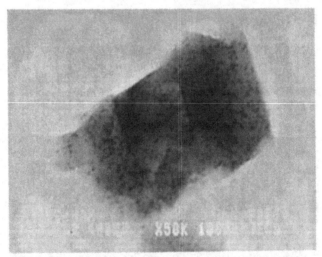

Fig. 3(b) Transmission image of 10CoNaX(occluded/reduced).

SQUID Measurements

The magnetic behaviour of the samples were analysed using a SQUID magnetometer after occlusion and then reduction. Hysteresis loops were taken at 15K with the field being scanned from 0.1 T to -0.1 T and back up to 0.1 T. Susceptibility measurements were taken from 4 K to 300 K. Figure 4(a) shows that there is very little or no hysteresis in the sample. Furthermore the susceptibility plot (figure 4b) shows that the occluded sample behaves as a paramagnet.

Occlusion

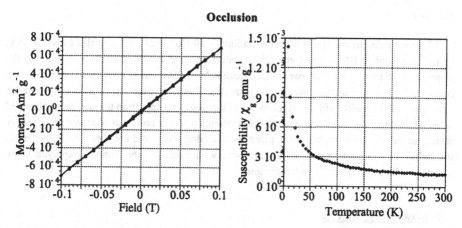

Fig.4(a) Plot of the hysteresis loop for 10CoNaX(occluded) and (b) the susceptibility curve.

Occlusion and reduction

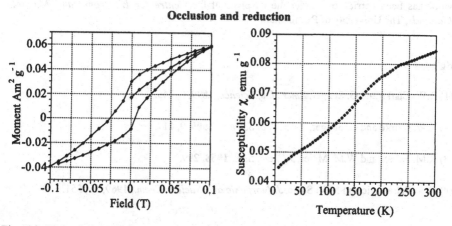

Fig.5(a) Plot of the hysteresis loop for 10CoNaX(occluded and reduced) and (b) is the susceptibility curve.

From the TEM micrographs we observed small particles of cobalt in the size range 200-250Å which were widely dispersed throughout the sample. This evidence suggests that these particles give rise to a hysteresis curve which is indicative of ferromagnetic behaviour. Figure 5b shows that the susceptibility of the reduced sample rises as a function of temperature. This complex behaviour is not normally found for a bulk ferromagnet. The magnetic data indicate the presence of ferromagnetism in the sample which was occluded *and* reduced. This sample was also attracted to a permanent magnet at room temperature. It is apparent that the reduction is a necessary step in the processing to produce this effect.

Summary

The conclusion to emerge is that at a loading of 10 Co atoms per unit cell of zeolite NaX, occlusion by cobalt (II) nitrate followed by reduction appears to be an effective route to the preparation of ultrafine cobalt particles in zeolite NaX. The initial size of these particles is governed by the size of the cavities of the zeolite. Particles which have grown beyond the dimensions of the zeolite will have broken out of the cages and appear on the surface as observed in the TEM. Such a process will reduce the crystallinity of the zeolite as observed in our x-ray data. Also before reduction had taken place the occluded zeolite did not possess any hysteresis and behaved as a paramagnet. After treatment with hydrogen the sample displayed some hysteresis which was typical of a ferromagnet.

ACKNOWLEDGEMENTS

We would like to thank Prof. I.R. Harris and Dr. M. Slaski for support and encouragement. Financial support from Alcan Chemicals Ltd and the EPSRC is gratefully acknowledged. This work has been carried out under the auspices of the *Centre for Electronic and Magnetic Materials*, The University of Birmingham.

References

(1) R.W. Clarke and G. Steiger, *Amer. J. of Science*, **1899**, 8, 245.

(2) R.W. Clarke and G. Steiger, *Amer. J. of Science*, **1900**, 9, 117.

(3) R.M. Barrer and W.M. Meier, *J.Chem. Soc.*, **1958**, 299.

(4) N.A. Petranovic and M.V. Susic, *J. Inorganic and Nuclear Chem.*, **1969**, 31, 551.

MAGNETIC PROPERTIES OF EMBEDDED
Rh CLUSTERS IN Ni MATRIX

Zhi-Qiang LI, Yuichi Hashi*, Jing-Zhi YU, Kaoru OHNO, and Yoshiyuki KAWAZOE
Institute for Materials Research, Tohoku University, Sendai 980-77, Japan
* Hitachi Tohoku Software Ltd., Research and Development Center, Sendai 980, Japan

ABSTRACT

The electronic structure and magnetic properties of rhodium clusters with sizes of 1 - 43 atoms embedded in the nickel host are studied by the first-principles spin-polarized calculations within the local density functional formalism. Single Rh atom in Ni matrix is found to have magnetic moment of $0.45\mu_B$. Rh_{13} and Rh_{19} clusters in Ni matrix have lower magnetic moments compared with the free ones. The most interesting finding is that Rh_{43} cluster, which is bulk-like nonmagnetic in vacuum, becomes ferromagnetic when embedded in the nickel host.

INTRODUCTION

Atomic clusters have opened new prospects in the development of materials science. Taking advantage of the characteristic behavior of small particles, one expects to be able to tailor new materials for specific technological purposes. Consequently, much effort has been invested in the research on the properties of microclusters[1]. Rhodium has specially interesting magnetic properties. It is nonmagnetic in the bulk state. However, the rhodium monolayer on an iron substrate has a measured magnetic moment of 0.82 μ_B per atom[6]. Since Reddy, Khanna and Dunlap[2] predicted that small rhodium clusters show the ferromagnetic properties, many experimental and theoretical investigations have been conducted to explore the unusual magnetic properties of rhodium clusters[3, 4, 5].

Using the local-spin-density (LSD) functional theory, Reddy, Khanna and Dunlap[2] recently calculated the magnetic moments for ruthernium, rhodium, and palladium 13-atom clusters with icosahedral and cubo-octahedral symmetry. They predicted moments of 1.62 μ_B per atom for icosahedral Rh_{13}, 1.02 μ_B per atom for icosahedral Ru_{13}, and $0.12\mu_B$ for icosahedral Pd_{13}. Indeed, Cox et al.[3] observed experimentally giant magnetic moments in small Rh_n clusters with n=12-34. However, their observed value of the average magnetic moment per atom for Rh_{13} is $0.48\mu_B$, only about one-third of the theoretical prediction of Reddy et al.. They also found that the average moment per atom of the Rh clusters depends significantly on the cluster size. There are several sizes, Rh_{15}, Rh_{16}, and Rh_{19} which have magnetic moments per atom that are significantly larger than those of adjacent cluster sizes. The average moment of the rhodium cluster decreases to the bulk value of zero as the cluster size increases. Yang et al.[4] have also performed first principles studies on Rh_n (n=2-19) clusters, and they did not observe the magnetic transition from magnetic state to nonmagnetic state as the cluster size increases, due to small number of atoms in their studies.

We have studied the electronic structure and magnetic properties of Rh_n (n=6, 9, 13, 19, 43) clusters and obtained better results compared with the experiments[5]. Moreover, we found that the magnetic and electronic structure of Rh$_{43}$ cluster have almost the same features as that of the rhodium bulk.

For most technological applications, the properties of embedded clusters (e.g. clusters in a matrix) are more relevant than the free clusters because it is related to the granular or island geometrical arrangements observed in the overlayers, sandwiches and multilayers. Therefore, it is of considerable importance to extend our knowledge on free clusters to the cases where these clusters are embedded in an environment. Comparison between the behavior of the free and embedded clusters would also contribute significantly to the understanding of the specific properties of these materials.

The formation of a magnetic moment on isolated transition metal impurities dissolved into metallic hosts continues to be a topic of experimental as well as theoretical interests. Extensive data are now available for the magnetism of $3d$ ions in various metals[7]. In comparison, much less information is available on the magnetic behavior of isolated $4d$ impurities in transition metals hosts. Using a cluster model, we can study the magnetic impurities in a more flexible way.

As an extension to our previous paper on the free rhodium clusters, we report here a first-principles study on Rh$_n$ clusters embedded in the nickel matrix. We chose nickel as host because it has the same fcc structure as rhodium and it is ferromagnetic. Ni$_{43}$ cluster is used as a model of the nickel matrix, as we will discuss later, which represents well the electronic and magnetic structures of nickel crystal.

METHOD

The electronic structures of the clusters are calculated with the first principles discrete variational method (DVM)[10]. The same method has already been employed in several other studies on metal clusters[11, 12], and described in detail elsewhere[12]. In short, the numerical atomic orbitals are used in the construction of molecular orbitals. In the present work, atomic orbital configurations composed of $4d$, $5s$ and $5p$ for Rh and $3d$, $4s$ and $4p$ for Ni are employed to generate the valence orbitals. The secular equation (H-ES)C=0 is then solved self-consistently using the matrix elements obtained via three-dimensional numerical integrations on a grid of random points by the diophantine method. About 900 sampling points around each site are employed. These points were found to be sufficient for convergence of the electronic spectrum within 0.01eV[11]. Self-consistent-charge (SCC) scheme[13] and von Barth-Hedin[9] exchange-correlation function are used in the calculations.

RESULTS

We first study the Ni$_{43}$ cluster in fcc structure, see Figure 1, which is used as a model of nickel bulk. There are four different sites in this cluster labelled A-D. The calculated results are summaried in Table 1. The central atom in the model (A site) has the lowest magnetic moment of 0.66 μ_B which is in good agreement with the value of nickel bulk. It is noted that while the $3d$ moment is positive, the $4s$ and $4p$ moments are negative.

The local density of states (DOS) of the central atom in this cluster are shown in Figure 2, which are obtained by a Lorentz expansion of the discrete energy levels and a summation over them. In comparison with the DOS of nickel bulk calculated by LDA band-structure method[14], we notice that the main features, namely three large peaks and about 4eV

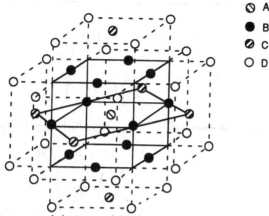

	A	B	C	D

Figure 1: Structure of cluster model, where different sites are labelled by A-D.

Table 1: Magnetic moments of each site of Ni_{43} and $RhNi_{42}$ clusters.

	A	B	C	D
Ni_{43}				
d	0.73	0.74	0.97	1.10
s	-0.03	-0.02	-0.03	0.03
p	-0.04	-0.01	-0.05	-0.01
total	0.66	0.71	0.89	1.12
$RhNi_{42}$				
d	0.55	0.64	0.98	1.08
s	-0.03	-0.02	-0.03	0.02
p	-0.07	0.01	-0.06	-0.02
total	0.45	0.63	0.89	1.08

valence band width, are well reproduced by the Ni_{43} clusters. Therefore, this cluster can be used to represent the nickel bulk for dealing with the local problems.

Next, we discuss the local moment formation of single Rh impurity in Ni host. Zeller[15] has examined the electronic structure of $4d$ impurities in Ni, using a LSD approach based on the Korringa-Kohn-Rostocker (KKR) Green's function method. Here the impurity is described by a single-site perturbated muffin-tin potential in an otherwise perfect periodic lattice. The calculated magnetic moment of Rh atom is $0.57\mu_B$ which is much smaller than the experimental result of $2\mu_B$ observed by the neutron scattering[16]. However, this experimental result is contrary to the expectation that in the $4d$ series the moments are always smaller than the $3d$ series and therefore the Rh moment should be smaller than the Co moment of $1.8\mu_B$. The discrepancy between the theory and experiment is so large that it deserves more studies by different approaches. Here, we use a cluster model to represent one Rh impurity in the Ni host. Similiar approach has been used to calculate the magnetism of single Fe in Al by a 43-atom cluster[17]. The calculated results are listed in

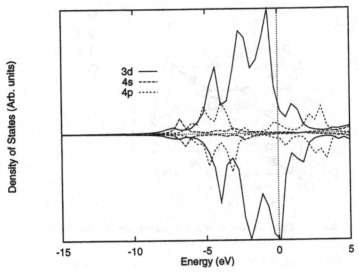

Figure 2: Local density of states of the central atom in the Ni_{43} cluster.

Table 1. The total magnetic moment of Rh atom is $0.45\mu_B$ which is mainly arised from the polarization of d states whereas s and p polarization add only a small negative contribution. It is noted from Table 1 that the effect of the Rh impurity on the Ni host is quite local, only a little change for the next-nearest neighboring Ni atoms. Our result is in good agreement with the KKR calculations but contrary to the neutron scattering experiments. A possible reason for the discrepancy may be the analysis of the experiments. The experimentist uses a two-moment version of the local environment model and the application is incomplete as it has been argued by Hicks[18].

Figure 3 shows the LDOS of the Rh atom in nickel host. It is clear that small exchange splitting between the majority and minority spin results in the magnetic formation of Rh atom. Compared with the LDOS of free Rh_{43} cluster, it is noticed there are extra states for both majority and minority spin at about -7eV below Fermi level, which are due to the hybridization of Ni $3d$ and Rh $4d$ states.

In the previous paper[5], we have shown that small rhodium clusters have finite magnetic moments, and Rh_{43} cluster is non-magnetic exhibiting the bulk-like electronic and magnetic properties. To characterize the effect of matrix on the magnetic moments of Rh clusters, we study the electronic properties of Rh_nNi_{43-n} and $Rh_{43}Ni_{12}$ clusters. Table 2 gives the calculated magnetic moments for each cluster as well as the results of free clusters for comparison. The embedded Rh_{13} and Rh_{19} clusters are also ferromagnetic with reduced magnetic moments compared with the free ones. It is interesting to note that Rh_{19} cluster surrounded by Rh_{24} is bulk-like nonmagnetic, whereas it is ferromagnetic when surrounded by Ni_{24}. To further clarify this point, we study a larger cluster $Rh_{43}Ni_{12}$, which has 12 nickel atoms surrounding 43 Rh cluster in fcc structure. We find that Rh_{43} cluster becomes ferromagnetic in this case. This is very similar with the experimental observation that rhodium monolayer over iron has magnetic moment which is induced by the hybridization of $4d$-$3d$ electronic states[6]. Therefore, one can obtain larger ferromagnetic Rh clusters in

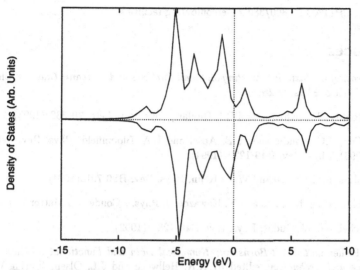

Figure 3: Local density of states of the Rh atom in the $RhNi_{42}$ cluster.

this way and we hope that experiments may check our expectation.

Table 2: Magnetic moments of Rh_n and Rh_nNi_{43-n} clusters at different site A-E.

cluster	A	B	C	D	E
Rh_{13}	1.59	0.63			
Rh_{19}	0.06	0.59	0.16		
Rh_{43}	0.00	0.13	0.04	0.01	
Ni_{43}	0.67	0.71	0.89	1.12	
$RhNi_{42}$	0.45	0.63	0.98	1.08	
$Rh_{13}Ni_{30}$	0.49	0.52	0.987	1.11	
$Rh_{19}Ni_{24}$	0.25	0.37	0.208	1.24	
$Rh_{43}Ni_{12}$	0.02	0.23	0.35	0.53	1.45

To summarize, by using the first-principles self-consistent LSD calculations, we find that single Rh atom impurity in the Ni host has magnetic moment of $0.45\mu_B$ and the embedded Rh_{13} and Rh_{19} clusters are also ferromagnetic though the moments are smaller than the free clusters. While the free-standing Rh_{43} cluster shows the nonmagnetic property, the embedded one has the induced magnetic moments. Those properties might be found in the granular materials and thin films.

Acknowledgments

The authors would like to express their sincere thanks to the Materials Information Science Group of the Institute for Materials Research, Tohoku University, for their continuous support of the HITAC S-3800/380 supercomputing facilities.

References

[1] See *Proc. of the Sixth Int. Meeting on Small Particles and Inorganic Clusters*, Chicrgo, 1992; 1993 Z. Phys. D **26**.

[2] B.V. Reddy, S.N. Khanna and B.I. Dunlap, Phys. Rev. Lett. **70** 3323(1993).

[3] A.J. Cox, J.G. Louderback, S.E. Apsel and L.A. Bloomfield, Phys. Rev. Lett. **71** 923(1993); Phys. Rev. **B49** 12295(1994).

[4] Yang Jinlong, F. Toigo,and Wang Kelin, Phys. Rev. **B50** 7915(1994)

[5] Z.Q. Li, J.Z. Yu, K. Ohno and Y. Kawazoe, J. Phys. : Condensed Matter **7**, 47(1995)

[6] T. Kachel and W. Gudat, Phys. Rev. **B46** 12888(1992).

[7] K.H. Fisher in *Landolt-Bornstein: Numerical Data and Functional Relationship in Science and Technology*, edited by K.H. Hellwege and J.L. Olsen, (Spring-Verlag, Berlin,1982).

[8] W. Kohn and L.J. Sham, Phys. Rev. **140** A1133(1965).

[9] U. von Barth and L. Hedin, J. Phys. C5 1629(1972).

[10] B. Delley, D.E. Ellis, A.J. Freeman, and D. Post, Phys. Rev. **B27** 2132(1983).

[11] M.R. Press, F. Liu, S.N. Khanna and P. Jena, Phys. Rev. **B40** 399(1989).

[12] Z.Q. Li and B.L. Gu, Phys. Rev. **B47** 13611(1993).

[13] D.E. Ellis and G.P. Painter, Phys. Rev. **B2** 2887(1970).

[14] V.L. Moruzzi, J.F. Janak and A.R. Williams, *Calculated Electronic Properties of Metals* (Pergamon, New York,1978)

[15] R. Zeller, J. Phys. F **17** 2123(1987).

[16] J.W. Cable, Phys. Rev. **B15** 3477(1977).

[17] D. Guenzburger and D.E. Ellis, Phys. Rev. Lett. **67** 3832(1991); Phys. Rev. **B45** 285(1992).

[18] T.J. Hicks, J. Phys. F **10** 879(1980).

UNIVERSAL mB/T SCALING OF THE GIANT MAGNETORESISTANCE IN Cu-Co GRANULAR RIBBONS PRODUCED BY CONTROLLED MELT-SPINNING

VICENTE MADURGA*, R.J.ORTEGA*, J.VERGARA*, R.ELVIRA*, V.KORENIVSKI** AND K.V.RAO**.
*Department of Physics. Universidad Pública de Navarra. Campus Arrosadía. E-31006 Pamplona. Spain.
**Department of Condensed Matter Physics. The Royal Institute of Technology. S-10044 Stockholm. Sweden.

ABSTRACT

Studies of the evolution of electrical resistance in an external applied magnetic field, B, as well as with temperature, T, on $Cu_{86}Co_{14}$ and $Cu_{92}Co_8$ as quenched and annealed melt-spun ribbons, reveal that magnetoresistance, MR, $\Delta R(B,T)=R(B)-R(0)$ scales with B/T. Furthermore, it found that annealing up to 600 °C scales the magnetic moment of the Co-rich superparamagnetic nanoparticles such that, the data for the field dependence of the MR obtained at various temperatures collapses onto the same unique and universal curve f(mB/T) with the Langevin variable mB/T governing the overall behaviour.

INTRODUCTION

Ever since the discovery of giant magnetoresistance, GMR, in granular solids containing transition metal elements in a non magnetic metallic matrix[1], there has been extensive efforts both from an experimental as well as theoretical point of view.

The above phenomenon has also been observed in melt-spun ribbons made of similar constituents[2]. GMR in granular solids appears to be concomitant with a superparamagnetic behaviour[3-5]. An attempt for such a correlation has been given by Parker et al.[6] in terms of Langevin function. It has also been shown[7,8] that ΔR scales with B/T. GMR has also been correlated with the square of the macroscopic magnetization[1,5]. Such correlation has been explained in terms of a spin dependent scattering model[9]. GMR in granular solids has also been explained by considering interfacial spin dependent scattering of conduction electrons, using a log-normal distribution function for the cluster size[10]. Also, in some studies GMR has been shown to depend linearly, and not as the square of the magnetization, which is considered to arise from scattering of pairs of blocked and superparamagnetic particles[11].

In the present work, we show a correlation between the magnetic nature of the samples and their GMR, taking into account their dependence on magnetic field and temperature, as well as the modifications which takes place with annealing treatments. The magnetization behaviour is described by a Langevin function. We analyse our data by considering a spin dependent scattering model, and the dependence of GMR to the square of the magnetization. From such an analyses we find that GMR scales with mB/T and thereby the role of number and size of magnetic clusters in describing the temperature dependence of GMR.

EXPERIMENTAL

Controlled inert atmosphere melt-spinning technique has been used to produce $Cu_{92}Co_8$ and $Cu_{86}Co_{14}$ ribbons 30 μm thick and 2.5 mm wide. We have annealed in inert atmosphere samples in a commercial halogen lamps furnace as follow: 500 °C for 60 min., 600 °C for 30 min. and 700 °C for 30 min. X-ray diffraction using Cu-Kα radiation (λ=1.5423 Å) as well as 200 kV transmission electron microscopy (TEM) investigation have been carried out in commercial equipment's. Details of these analyses have been reported before[8].

541

MR measurements in $Cu_{92}Co_8$ have been made using the four probe technique with 1 mA d.c. flowing along the ribbon. The field was applied parallel to the current. The voltage drop was measured over a length of 6.5 mm, from which the resistance can be obtained. The SQUID was used to provide the field and the low temperature capability. $Cu_{86}Co_{14}$ samples have been annealed in a conventional resistance furnace at 425 °C, 450 °C, 475°C, 500 °C and 600 °C for 1h. in inert atmosphere. MR measurements have been made by means of an Automatic Resistance Bridge from A.S.L. model F26. 1 mA current was flowing along the ribbon 12 mm. long, 1 mm width, and the magnetic field was applied perpendicular to the current and perpendicular to the ribbon plane. An electromagnet was used to provide the magnetic field up to 2 Tesla. Due to air gap limitations, measurements at 77 K were made in a maximum field of 1 Tesla. Measurement at temperatures lower than 77 K were not made on these samples.

RESULTS AND DISCUSSION.

Magnetoresistance

In Fig. 1 and Fig. 2 we show the field dependence of MR values corresponding to the $Cu_{92}Co_8$ samples. The maximum value of GMR corresponds to the 600 °C annealed sample with a negative relative increment of 17% for a magnetic field B=1 Tesla at 5K. An increment in GMR effect is produced by annealing; however, for the sample annealed at 700 °C the MR has a very low value and the GMR effect is lost. We have recently shown[4] that the magnetic nature of these samples is superparamagnetic, which is lost simultaneously with GMR above 700 °C anneal. The GMR, for each sample, increases monotonically with decreasing temperatures, as shown in Fig.2. In Fig. 3 and Fig. 4 we show the GMR data obtained for the $Cu_{86}Co_{14}$ samples annealed at different temperatures. The maximum negative relative increment in MR, 12%, corresponds to the 475 °C annealed sample, in a magnetic field B= 1.1 Tesla at 77 K.

To correlate the GMR with the magnetization data we have considered that the origin of GMR is the spin dependent scattering of conduction electrons by the magnetic moment of clusters. This approach has been used in the past to explain the MR in dilute magnetic alloys[12]. In this model the change in resistivity produced by a magnetic field is proportional to the square of the change in the macroscopic magnetization

$$\Delta R = -\beta M^2. \qquad (1)$$

ΔR is measured between the initial demagnetised state, B=0, M=0, and the state of

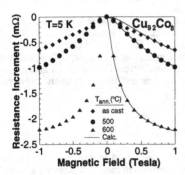

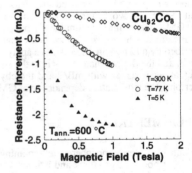

Fig. 1 MR for differently annealed samples. The solid lines correspond to calculated values using the expression (3).

Fig. 2 MR at different temperatures. The maximum GMR value corresponds to a relative change of 17% at 5 K.

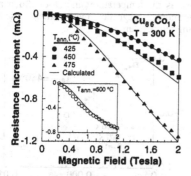

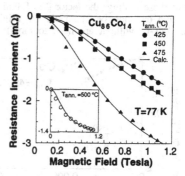

Fig. 3 MR at 300 K. The solid lines correspond to calculated values using the Langevin function. The inset is for clarity in the figure and corresponds to 500 °C annealed sample.

Fig. 4 MR at 77 K. The relative change of GMR is 12%, the maximum for the sample annealed at 475 °C.

magnetization $M(B,T)$ reached at temperature T. The macroscopic magnetization of these superparamagnetic samples is described by the Langevin function

$$M(B,T)=M_s L(\alpha) \qquad (2)$$

where $L(\alpha)$ is the Langevin function and $\alpha=mB/kT$. m is the magnetic moment of each cluster and k the Boltzmann constant.

Hence the increment in electrical resistance produced on application of a magnetic field can be expressed as

$$\Delta R=-\beta m^2 N^2 \{ L \ [mB/k(T-\theta)]\}^2 \qquad (3)$$

where we have considered that N magnetic clusters participating in the phenomenon of MR have magnetic moment m. We use the same ordering temperature $\theta=-13$ K which was obtained on scaling the superparamagnetic behaviour of these samples[4].

The continuous lines shown in Figs. 1, 3 and 4 are the fits of the MR data to expression (3). The values of m and $\beta m^2 N^2$ obtained from these fittings are used in the various scaling analyses discussed below.

ΔR scales with B/T

It can be shown that for a given sample the field dependence of the changes in the resistance measured at various temperatures shown in Figs.2, 3 and 4 can be collapsed in to a single curve with the appropriate scaling. Fig. 5 and 6 show the scaling of magnetoresistance when plotted vs B/T. It can be observed for such scaling it is necessary to introduce a temperature dependence of the factor $\beta(T)$.Thus it is possible to scale ΔR using the same ordering temperature $\theta=-13$K for all of the samples independently of the annealing treatments. This value of θ was obtained from the magnetic data of the same samples.

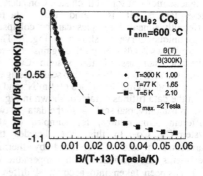

Fig. 5 Scaling of ΔR vs.B/T. The values of β relative to T=300 K, are shown for each temperature.

For these scalings the factor β is found to decrease as the temperature increases. Such an evolution can not be magnetic in origin. Furthermore, magnetic measurements scale with $B/(T-\theta)$ without any need to modify the ordering temperature θ.

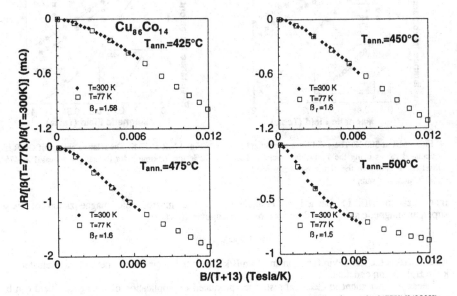

Fig. 6 ΔR vs.B/T scaling for differently annealed samples. Note that the relative factor $\beta_r=\beta(77K)/\beta(300K)$ must be included to scale. Note, also, that this factor is independent of the annealing treatments.

It is interesting to point out that the temperature dependence of β is qualitatively similar to the temperature dependence of the mean free path of conduction electrons[13].

ΔR scales with mB

With annealing, the size and number of the magnetic clusters must change and hence the magnetic moment for each cluster. Considering these factors it is found that data for differently annealed samples can be collapsed in to unique curve as shown in Fig. 7. This figure shows $\Delta R/\beta m^2 N^2$ plotted against mB for different samples of the system $Cu_{92}Co_8$, measured at T=5 K. In fact this scaling is again rather good, taking the values of $\beta m^2 N^2$ and m from the data fittings obtained, shown in Fig. 1. Fig. 8 is a universal scaling of the data , at two different temperatures, for samples of the system $Cu_{86}Co_{14}$. In this case, besides the variation of m and N produced by thermal treatments, the variation with temperature of β (T) has been taken into account. 4 different samples (corresponding to 4 different thermal treatments), measured at two different

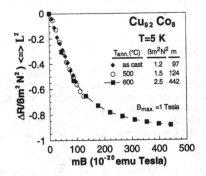

Fig. 7 ΔR vs.mB scaling for differently annealed samples. The values of β, m and N used here, were obtained from fitting the magnetoresistance behaviour shown in Fig. 1. The units of m are 10^{-20} emu and for $\beta m^2 N^2$, mΩ.

temperatures are being represented in Fig. 8. This shows the very good overall scaling of ΔR with mB/T. The influence of changes of m and N produced by annealing have been considered and also the changes in β due to temperature.

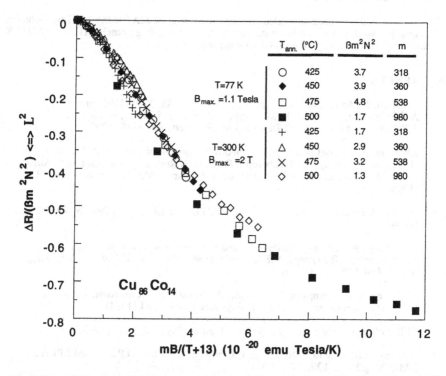

Fig. 8 ΔR vs.mB/T scaling for differently annealed samples and two temperatures. The values of β, m and N used for this scaling were obtained from the fit shown in Fig. 3 and 4. Note the changes in these parameters, m and N for differently annealed samples and the change in β for different temperatures. The units of m are 10^{-20} emu and the one of the $\beta m^2 N^2$ are $m\Omega$.

A model considering a distribution of the size of the magnetic clusters, instead of an unique value, would enable us to better understand quantitatively the different parameters involved in the MR phenomenon.

CONCLUSIONS

The study of evolution of MR with magnetic field and temperature as well as with thermal treatments are shown to be uniquely governed by the Langevin variable mB/T, as long as the samples are in the superparamagnetic state.

The MR is an even function of the macroscopic magnetization.

From the factor of the Langevin function we have deduced: the influence on the MR of the number and size of magnetic clusters participating in the phenomenon, and how their variations produced by thermal treatments change the GMR.

There are two main and distinguishable temperature dependencies of the GMR: 1) through the magnetization process, by different values of magnetization reached at different temperatures.

2) Through the factor β, which is of non-magnetic origin and may be related to electronic conduction processes involved.

ACKNOWLEDGEMENTS

The work in Spain has been supported by CICYT under grant MAT94-0964 and by Gobierno de Navarra. The work in Sweden has been supported by the funding agencies TFR and NUTEK We thank Professor Dahlberg for useful discussion and comments.

REFERENCES

1 J.Q.Xiao, J.S.Jiang and C.L.Chien. Phys. Rev. Lett.**68**, 3749 (1992) and A.E.Berkowitz, J.R.Mitchell, M.J.Carey, A.P.Young, S.Zhang, F.E.Spada, F.T.Parker, A.Hutten and G.Thomas. Phys.Rev.Lett., **68**, 3745 (1992).

2 J.Wecker, R. von Helmolt, L.Schultz and K.Samwer Appl. Phys. Lett. **62**, 1985 (1993). And B.Dieny, A.Chamberod, J.B.Genin, B.Rodmacq, S.R.Teixeira, S.Auffret, P.Gerard, O.Redon, J.Pierre, R.Ferrer and B.Barbara. J.Mag.Mag.Mater. **126**, 433 (1993).

3 H.Wan, A.Tsoukatos, G.C.Hadjipanayis, Z.G.Li and J.Liu. Phys.Rev. B, **49**, 1524 (1994)

4 V.Madurga, R.J.Ortega, V.Korenivski and K.V.Rao.1994 March Meeting of the A P S. and V.Madurga, R.J.Ortega, V.Korenivski, H.Medelius and K.V.Rao. J.Mag.Mag. Mater. **140-144**, 465 (1995)

5 J.F.Gregg, S.M.Thompson, S.J.Dawson, K.Ounadjela, C.R.Staddon, J.Hamman, C.Fermon, G.Saux and K.O'Grady. Phys.Rev.B. **49**, 1064 (1994).

6 M.R.Parker, J.A.Barnard and J.Waknis, J. Appl. Phys. **73**, 5512 (1993)

7 B.Dieny, R.S.Teixeira, B.Rodmacq, C.Cowache, S.Auffret, O.Redon and J.Pierre. J.Mag.Mag.Mater.**130**, 197 (1994)

8 R.J.Ortega, J.Vergara, R.Elvira and V.Madurga. 14 GCCMD of the European Physical Society. (Madrid, March 1994) and V.Madurga, R.J.Ortega, V.Korenivski and K.V.Rao. IVInternational Workshop on Non-Crystalline Solids.(Madrid, September, 1994) to appear in Nanocrystallized and Non-Crystalline Materials. edited by M.Vázquez and A.Hernando (World Scientific Publishing Co. Pte. Ltd., London, 1995)

9 J.Q.Wang and G.Xiao. Phys. Rev. B **49**, 3982 (1994)

10 R.von Helmolt, J.Wecker and K.Samwer. Phys. Stat. Sol. (b) **182** K25 (1994)

11 B.J.Hickey, M.A.Howson, S.O.Musa and N.Wiser. Phys. Rev. B, **51**,667 (1995)

12 M.T. Beal-Monod and R.A. Weiner. Phys.Rev., **170**, 552 (1968)

13 F.Seitz. The Modern Theory of Solids. Dover Publications Inc. (New York, 1987). pag. 533

AUTHOR INDEX

SUBJECT INDEX

Printed in the United States
By Bookmasters